Exosomes

Exosomes

A Clinical Compendium

Edited by

Lawrence Edelstein
John Smythies
Peter Quesenberry
Denis Noble

ELSEVIER

ACADEMIC PRESS
An imprint of Elsevier

Academic Press is an imprint of Elsevier
125 London Wall, London EC2Y 5AS, United Kingdom
525 B Street, Suite 1650, San Diego, CA 92101, United States
50 Hampshire Street, 5th Floor, Cambridge, MA 02139, United States
The Boulevard, Langford Lane, Kidlington, Oxford OX5 1GB, United Kingdom

Library of Congress Cataloging-in-Publication Data
A catalog record for this book is available from the Library of Congress

British Library Cataloguing-in-Publication Data
A catalogue record for this book is available from the British Library

ISBN 978-0-12-816053-4

For information on all Academic Press publications
visit our website at https://www.elsevier.com/books-and-journals

Publisher: Andre Gerhard Wolff
Acquisition Editor: Andre Gerhard Wolff
Editorial Project Manager: Tracy Tufaga
Production Project Manager: Mohana Natarajan
Cover designer: Miles Hitchen

Typeset by SPi Global, India

Contents

Contributors

Numbers in parentheses indicate the pages on which the authors' contributions begin.

Divya Aickara (239), Department of Dermatology and Cutaneous Surgery, University of Miami, Miami, FL, United States

Riccardo Alessandro (1,81), Department of Biomedicine, Neuroscience and Advanced Diagnostics, University of Palermo, Palermo, Italy

Shamila D. Alipoor (383), Molecular Medicine Department, Institute of Medical Biotechnology, National Institute of Genetic Engineering and Biotechnology (NIGEB), Tehran, Iran

Ramaroson Andriantsitohaina (343), SOPAM, U1063, INSERM, UNIV ANGERS, SFR ICAT, Bat IRIS-IBS; Angers University Hospital, Angers, France

Jun Araya (307), Division of Respiratory Diseases, Department of Internal Medicine, The Jikei University School of Medicine, Tokyo, Japan

Koji Asano (101), Department of Urology, Jikei University School of Medicine, Tokyo, Japan

Evangelos Badiavas (239), Department of Dermatology and Cutaneous Surgery, University of Miami, Miami, FL, United States

Matthew A. Bailey (257), University/BHF Centre for Cardiovascular Science, Queen's Medical Research Institute, University of Edinburgh, Edinburgh, United Kingdom

Scott Bonner (285), University of Oxford, Department of Paediatrics, Oxford, United Kingdom

Han Chen (23), Microscopy Imaging Facility, Penn State Hershey Medical Center, Hershey, PA, United States

Yong Cheng (123), Department of Biological Sciences, Eck Institute for Global Health, Center for Rare and Neglected Diseases, University of Notre Dame, Notre Dame, IN, United States

Raul Coimbra (325), Division of Trauma, Surgical Critical Care, Burns and Acute Care Surgery, Department of Surgery, University of California San Diego, San Diego, CA, United States

Alice Conigliaro (1), Department of Biomedicine, Neuroscience and Advanced Diagnostics, University of Palermo, Palermo, Italy

Denis Corbeil (39), Biotechnology Center and Center for Molecular and Cellular Bioengineering, Technische Universität Dresden, Dresden, Germany

Chiara Corrado (1), Department of Biomedicine, Neuroscience and Advanced Diagnostics, University of Palermo, Palermo, Italy

Todd W. Costantini (325), Division of Trauma, Surgical Critical Care, Burns and Acute Care Surgery, Department of Surgery, University of California San Diego, San Diego, CA, United States

Dragos Cretoiu (199), Department of Cell and Molecular Biology and Histology, Carol Davila University of Medicine and Pharmacy; Alessandrescu-Rusescu National Institute of Mother and Child Health, Fetal Medicine Excellence Research Center, Bucharest, Romania

Sanda Maria Cretoiu (199), Department of Cell and Molecular Biology and Histology, Carol Davila University of Medicine and Pharmacy, Bucharest, Romania

André Cronemberger-Andrade (179), Laboratory of Cellular Immunology and Biochemistry of Fungi and Protozoa, Department of Pharmaceutical Sciences, Federal University of São Paulo (UNIFESP), Diadema, SP, Brazil

James W. Dear (257), University/BHF Centre for Cardiovascular Science, Queen's Medical Research Institute, University of Edinburgh, Edinburgh, United Kingdom

Alexandru Florian Deftu (199), Department of Anatomy, Animal Physiology and Biophysics, Faculty of Biology, University of Bucharest; Life, Environmental and Earth Sciences Division, Research Institute of the University of Bucharest (ICUB), Bucharest, Romania

Antonia Teona Deftu (199), Department of Anatomy, Animal Physiology and Biophysics, Faculty of Biology, University of Bucharest; Life, Environmental and Earth Sciences Division, Research Institute of the University of Bucharest (ICUB), Bucharest, Romania

Shin Egawa (101), Department of Urology, Jikei University School of Medicine, Tokyo, Japan

Brian P. Eliceiri (325), Division of Trauma, Surgical Critical Care, Burns and Acute Care Surgery, Department of Surgery, University of California San Diego, San Diego, CA, United States

Simona Fontana (1), Department of Biomedicine, Neuroscience and Advanced Diagnostics, University of Palermo, Palermo, Italy

Yu Fujita (307), Division of Molecular and Cellular Medicine, National Cancer Center Research Institute; Division of Respiratory Diseases, Department of Internal Medicine, The Jikei University School of Medicine, Tokyo, Japan

Andrew F. Hill (285), La Trobe University, La Trobe Institute for Molecular Science, Melbourne, VIC, Australia

Robert W. Hunter (257), University/BHF Centre for Cardiovascular Science, Queen's Medical Research Institute, University of Edinburgh, Edinburgh, United Kingdom

Tsukasa Kadota (307), Division of Molecular and Cellular Medicine, National Cancer Center Research Institute; Division of Respiratory Diseases, Department of Internal Medicine, The Jikei University School of Medicine, Tokyo, Japan

Ju-Seop Kang (467), Department of Pharmacology & Clinical Pharmacology Lab, College of Medicine, Hanyang University, Seoul, South Korea

Nobuyoshi Kosaka (101,307,433), Division of Molecular and Cellular Medicine, National Cancer Center Research Institute; Department of Molecular and Cellular Medicine, Institute of Medical Science; Department of Translational Research for Extracellular Vesicles, Institute of Medical Science, Tokyo Medical University, Tokyo, Japan

Kazuyoshi Kuwano (307), Division of Respiratory Diseases, Department of Internal Medicine, The Jikei University School of Medicine, Tokyo, Japan

Soazig Le Lay (343), SOPAM, U1063, INSERM, UNIV ANGERS, SFR ICAT, Bat IRIS-IBS, Angers, France

Aurelio Lorico (39), College of Medicine, Touro University Nevada, Henderson, NV, United States; Mediterranean Institute of Oncology, Viagrande, Italy

Imre Mäger (285), University of Oxford, Department of Paediatrics, Oxford, United Kingdom

M. Carmen Martinez (343), SOPAM, U1063, INSERM, UNIV ANGERS, SFR ICAT, Bat IRIS-IBS; Angers University Hospital, Angers, France

Jeffrey D. McBride (239), Department of Dermatology and Cutaneous Surgery, University of Miami, Miami, FL, United States

Esmaeil Mortaz (383), Clinical Tuberculosis and Epidemiology Research Center, National Research Institute of Tuberculosis and Lung Diseases (NRITLD); Department of Immunology, Faculty of Medicine, Shahid Beheshti University of Medical Sciences, Tehran, Iran

Soumyalekshmi Nair (357), Exosome Biology Laboratory, Centre for Clinical Diagnostics, UQ centre for Clinical Research, Royal Brisbane and Women's Hospital, The University of Queensland, St Lucia, QLD, Australia

Denis Noble (487), Department of Physiology, Anatomy & Genetics, University of Oxford, Oxford, United Kingdom

Takahiro Ochiya (101,307,433), Division of Molecular and Cellular Medicine, National Cancer Center Research Institute; Department of Molecular and Cellular Medicine, Institute of Medical Science, Tokyo Medical University, Tokyo, Japan

Siew-Wai Pang (147), Department of Medical Sciences, School of Healthcare and Medical Sciences, Sunway University, Petaling Jaya, Malaysia

Beatrice Mihaela Radu (199), Department of Anatomy, Animal Physiology and Biophysics, Faculty of Biology, University of Bucharest; Life, Environmental and Earth Sciences Division, Research Institute of the University of Bucharest (ICUB), Bucharest, Romania

Stefania Raimondo (81), Department of Biomedicine, Neuroscience and Advanced Diagnostics, University of Palermo, Palermo, Italy

Laura Saieva (81), Department of Biomedicine, Neuroscience and Advanced Diagnostics, University of Palermo, Palermo, Italy

Carlos Salomon (357), Exosome Biology Laboratory, Centre for Clinical Diagnostics, UQ centre for Clinical Research, Royal Brisbane and Women's Hospital, The University of Queensland, St Lucia, QLD, Australia; Department of Clinical Biochemistry and Immunology, University of Concepci6n, Concepci6n, Chile; Department of Obstetrics and Gynecology, Ochsner Baptist Hospital, New Orleans, LA, United States

Jeffery S. Schorey (123), Department of Biological Sciences, Eck Institute for Global Health, Center for Rare and Neglected Diseases, University of Notre Dame, Notre Dame, IN, United States

Jeffrey M. Sundstrom (23,415), Department of Ophthalmology, Penn State College of Medicine; Department of Ophthalmology, Penn State Hershey Medical Center, Hershey, PA, United States

Sin-Yeang Teow (147), Department of Medical Sciences, School of Healthcare and Medical Sciences, Sunway University, Petaling Jaya, Malaysia

Ana Claudia Torrecilhas (179), Laboratory of Cellular Immunology and Biochemistry of Fungi and Protozoa, Department of Pharmaceutical Sciences, Federal University of São Paulo (UNIFESP), Diadema, SP, Brazil

Fumihiko Urabe (101), Division of Molecular and Cellular Medicine, National Cancer Center Research Institute; Department of Urology, Jikei University School of Medicine, Tokyo, Japan

Sarah R. Weber (23,415), Department of Ophthalmology, Penn State College of Medicine; Department of Ophthalmology, Penn State Hershey Medical Center, Hershey, PA, United States

Eduard Willms (285), University of Oxford, Department of Paediatrics, Oxford, United Kingdom

Matthew J.A. Wood (285), University of Oxford, Department of Paediatrics, Oxford, United Kingdom

Patricia Xander (179), Laboratory of Cellular Immunology and Biochemistry of Fungi and Protozoa, Department of Pharmaceutical Sciences, Federal University of São Paulo (UNIFESP), Diadema, SP, Brazil

Junjie Xiao (199), Department of Cardiology, The First Affiliated Hospital of Nanjing Medical University, Nanjing; Cardiac Regeneration and Ageing Lab, Experimental Center of Life Sciences, School of Life Science, Shanghai University, Shanghai, People's Republic of China

Zhongdang Xiao (433), State Key Laboratory of Bioelectronics, School of Biological Science and Medical Engineering, Southeast University, Nanjing, China

Yuanjun Zhao (23,415), Department of Ophthalmology, Penn State College of Medicine; Department of Ophthalmology, Penn State Hershey Medical Center, Hershey, PA, United States

Mi Zhou (23,415), Department of Ophthalmology, Penn State College of Medicine; Department of Ophthalmology, Penn State Hershey Medical Center, Hershey, PA, United States

Yueyuan Zhou (433), Division of Molecular and Cellular Medicine, National Cancer Center Research Institute, Tokyo, Japan; State Key Laboratory of Bioelectronics, School of Biological Science and Medical Engineering, Southeast University, Nanjing, China

Editor biography

Lawrence R. Edelstein

President and Founder, Medimark Corporation, Del Mar, CA, United States

Lawrence R. Edelstein, Ph.D. is a neuroscientist and pharmaceutical industry consultant with research interests in multisensory convergence/integration (claustrum) and intercellular communication (exosomes, telocytes). His interest in exosomes was fueled by a theme issue he guest-edited with John Smythies and Denis Noble entitled "Epigenetic information-processing mechanisms in the brain" (2014, https: //royalsocietypublishing.org/toc/rstb/369/1652). That undertaking proved to be the impetus for a series of peer-reviewed articles in which he and his colleagues theorized as to the whys and wherefores of the multifunctional roles played by seemingly omnipresent and phyla-agnostic exosomes. In addition, Dr. Edelstein is co-editor (along with John Smythies and Vilayanur S. Ramachandran, M.B.B.S., Ph.D., Hon. F.R.C.P.) of the book Claustrum - Structural, Functional and Clinical Neuroscience (2014, www.elsevier.com/books/the-claustrum/smythies/978-0-12-404566-8).

John Raymond Smythies

John Raymond Smythies, M.B. B.Chir., M.D., F.R.C.P., F.R.C. Psych. (1922–2019) was the Director of the Integrative Neuroscience Program in the Department of Psychology at the University of California San Diego. As a pre-eminent neuropsychiatrist and neuroscientist he made significant contributions to both these disciplines. Together with Humphry Osmond he developed the first biochemical theory of schizophrenia - the transmethylation hypothesis. This has recently come back into focus following the finding that DNA methylation is abnormal in schizophrenia. He made extensive contributions to knowledge in a number of fields including the neuropharmacology of psychedelic drugs; the functional neuroanatomy of synapses with particular regard to the role of synaptic plasticity, endocytosis and redox factors; the role in the brain of orthoquinone metabolites of catecholamines; and, in particular, theories of brain-consciousness relations. More recently he developed foundational hypotheses and theories specific to the function of exosomes, telocytes and the claustrum, and on epigenetic processes in information processing in the brain. Professor Smythies served as President of the International Society of Psychoneuroendocrinology from 1970–1974, Consultant to the World Health Organization from 1963–1968, and Editor of the International Review of Neurobiology from 1958–1991. He was elected a member of the Athenaeum in 1968. He held the positions of Professor Emeritus and the Charles Byron Ireland Professor of Psychiatric Research at the University of Alabama Medical Center at Birmingham, Visiting Scholar at the Center for Brain and Cognition, University of California San Diego, and Senior Research Fellow at the Institute of Neurology, University College London. He published over 240 scientific papers and sixteen books. https://en.wikipedia.org/wiki/John_Raymond_Smythies

Peter J. Quesenberry
Paul Calabresi Professor of Oncology, Professor of Medicine, The Warren Alpert Medical School of Brown University, Providence, RI, USA
Peter J. Quesenberry, M.D., is the Paul Calabresi Professor of Oncology at The Warren Alpert Medical School of Brown University. He received his medical degree from the University of Virginia, completed residency at University Hospital and Boston City Hospital in Boston, MA, and completed a Hematology/Oncology Fellowship at St. Elizabeth's Hospital.

Professor Quesenberry is a leading investigator in stem cell biology and extracellular vesicle research. He was President of the International Society of Hematology, editor of the journal Experimental Hematology from 1990–1998 and the leukocyte editor for the Year Book of Hematology from 1987–1998. More recently he is a co-editor-in-chief for the Journal of Extracellular Vesicles.

Denis Noble
Emeritus Professor of Cardiovascular Physiology, Department of Physiology, Anatomy, and Genetics, University of Oxford, Oxford, UK
Denis Noble, C.B.E., Ph.D., F.R.S. is a British biologist who held the Burdon Sanderson Chair of Cardiovascular Physiology at the University of Oxford from 1984 to 2004 and was appointed Professor Emeritus and co-Director of Computational Physiology. He is one of the pioneers of systems biology.

Professor Noble developed the first viable mathematical model of the working heart in 1960 using his discovery, with his supervisor Otto Hutter, of two of the main cardiac potassium ion channels. These discoveries were published in Nature (1960) and The Journal of Physiology (1962). The work was later developed with Dick Tsien, Dario DiFrancesco, Don Hilgemann and others to become the canonical models on which more than 100 cardiac cell models are based today.

He was elected President of the International Union of Physiological Sciences (IUPS) at its Congress in Kyoto in 2009 and was re-elected for a second term at the 2013 Congress in Birmingham, UK.

He is the author of the first popular book on systems biology, The Music of Life, and his most recent lectures concern the implications for evolutionary biology.

Professor Noble has published more than 500 papers and 11 books. https://en.wikipedia.org/wiki/Denis_Noble

Preface

Nothing speaks more clearly to the envisaged merits of a major clinical and scientific discovery than the rapidly increasing number of journal articles and their citation in a brief period of time, followed apace by the launch of biopharma start-ups founded on such, each competing to fill their respective pipelines with internally developed and in-licensed compounds. Case in point, exosomes, a seemingly omnipresent and phyla-agnostic extracellular vesicle of endosomal origin which has quickly come to the fore in the context of intercellular communication, with payloads of miRNA, mRNA, lncRNA, and transcription factors at the ready. Arguably, the most compelling aspect of exosomes, and conceivably their raison d'être, is their role in transgenerational epigenetics, the topic of our closing chapter.

My interest in exosomes was sparked by a theme issue I guest-edited with my fellow editors John Smythies and Denis Noble entitled "Epigenetic information-processing mechanisms in the brain" (https://royalsocietypublishing.org/toc/rstb/369/1652). This proved to be the impetus for a series of peer-reviewed articles in which my colleagues and I theorized as to the whys and wherefores of the multifunctional role of exosomes in most if not all living organisms. Over 2 years in the making, we are so very pleased to see this undertaking come to fruition: *Exosomes—A Clinical Compendium.*

Armed with a modicum of publishing experience as a neuroscientist, author, editor and founding editor-in-chief of a scientific journal, I was quick to note that there was a multidisciplinary component missing from the extant narrative. The time was ripe for an exosome book with a rather daunting objective—to bottle lightning by inviting contributions from a global cohort of peer-acknowledged expert clinicians and researchers across a wide range of medical disciplines, affording each an equal voice at the table. *Exosomes—A Clinical Compendium* serves to provide readers with a broad and timely overview of exosomes in health and disease. Within its 21 chapters, our authors have summarized the most recent laboratory and clinical findings, thereby illuminating the path forward for prospective investigative efforts.

To the exosome novitiate I say, welcome to what I consider to be the leading edge of diagnostic, therapeutic and theragnostic research. Have at it! To those for whom exosomes have played and continue to play an integral role in their clinical and research endeavors and, in turn, helping to advance the field, I say thank you for your unwavering efforts as without them this book would have withered on the vine.

Lawrence Edelstein, Ph.D.

Acknowledgments

Early in the evolution of this book it became apparent that the study of extracellular vesicles in general—and of exosomes in particular—has been accelerating at a remarkable pace. Nowhere has this been more evident than in the context of their clinical implications across the spectrum of medical disciplines. First and foremost we are deeply indebted to our authors, each a peer-acknowledged expert in their respective fields of endeavor. We are also grateful for the steadfast guidance and assistance provided by our colleagues at Elsevier, most notably Mica Haley, Tracy Tufaga, Swapna Praveen, Mohana Natarajan, Jyotsna Gopichandran, André Wolff, and Jaclyn Truesdell. Lastly, were it not for a series of theory/hypothesis articles by our late co-editor John Smythies on the raison d'être of exosomes and telocytes in reference to neural coding, repair and cognate mechanisms, it is fair to say that this book would not have seen the light of day.

Special acknowledgment

In memory of a gentleman scientist and *rara avis* of the highest order—John Raymond Smythies, M.B. B. Chir., M.D., F.R.C. Psych., F.R.C.P. (https://en.wikipedia.org/wiki/John_Raymond_Smythies).

Apropos of an early British maritime phrase oft-used by John, who was a Surgeon-Lieutenant with the Royal Naval Volunteer Reserve from 1946–49 aboard H.M.S. Porlock Bay, "All is shipshape and Bristol fashion!" Mission accomplished.

Chapter 1

Exosome basic mechanisms

Alice Conigliaro, Chiara Corrado, Simona Fontana, Riccardo Alessandro
Department of Biomedicine, Neuroscience and Advanced Diagnostics, University of Palermo, Palermo, Italy

1 Exosomes biogenesis and release

Exosomes biogenesis is correlated to intraluminal vesicles (ILVs) formation and starts with a first invagination of the plasmatic membrane, leading to the formation of early endosomes, low-density vesicles very close to the inner membrane surface.

Different materials can be internalized by endocytosis and by receptor mediated-endocytosis and early endosomes are responsible of their sorting; after receptor-ligand interaction on the cell surface, ligand is separated from its receptor, located inside the early endosome and transferred by endosomal transport vesicles to the late endosome. The receptor, instead, can be recycled on the membrane surface or partially degraded via late endosomes-lysosomes pathway, in order to turn off the signal and reduce the receptor concentration on the cell surface.

During their path, mediated by microtubules, inside the cell, early endosomes go through different modifications; they can fuse each other or with vesicles containing acid hydrolases, thus forming intermediate structure, called multivesicular bodies (MVB) and containing ILVs. The formation of ILVs is the second event of membrane invagination; in this case, it involves endosomal membrane.

MVBs originate by two different mechanisms: first, the Endosomal Sorting Complex Request for Transport (ESCRT) is responsible of the MVB formation; alternatively, the MVBs can originate from endosomes that contain specific domains on their membrane, known as lipid rafts (Fig. 1).

At this point, MVB, containing ILVs, can follow two different fates. MVB can fuse with other MVBs or with late endosomes and receive vesicles originating from transGolgi and containing lysosomal enzymes. In this case, during the maturation process, the pH becomes more acid thus activating the hydrolases that digest molecules inside MVB and allow the transformation in lysosomes, in eukaryotic cells. This pathway allows a rapid and consistent turnover

Exosomes. https://doi.org/10.1016/B978-0-12-816053-4.00001-8

of transmembrane proteins and lipids. Alternatively, MVBs migrate toward the membrane, fuse with the plasma membrane and release ILVs outside of the cell, as exosomes. Exosomes are, therefore, the only secreted vesicles of *endo*-lysosomal origin, generated from internal membranes.

2 Endosomal sorting complex request for transport (ESCRT) and its role in exosome biogenesis

The role of ESCRT in exosome biogenesis emerges by the evidence that numerous ESCRT proteins have been discovered in exosomes preparations from different cell types or biological fluids, for this reason, many components are acknowledged in the scientific community as exosomal markers.

ESCRT pathway is, by now, a well-described mechanism to explain ILVs and MVB formation. ESCRT is composed of five multimeric cytosolic complexes: ESCRT 0, I, II, III, and Vps4 [1] (Fig. 1).

Specific ubiquitinated proteins, contained in early endosomes, are recognized, in their ubiquitination site, by ESCRT-0 that places them in specific endosomes' areas enriched in phosphatidyl inositol monophosphate (PI3P). ESCRT-0 is composed of two subunits, called HRS (hepatocyte growth factor regulated tyrosine kinase substrate) and STAM1/2 (signal transducing adaptor molecule 1/2), which bind one to each other and are able to recognize ubiquitin and PI3P enriched domains. This allows recruiting ESCRT-I, a heteromeric complex composed of Tsg101 (tumor susceptibility gene 101), Vps28, Vps37 and Mvb12 (multi-vesicular body 12).

ESCRT-I takes the place of ESCRT-0 and recruits ESCRTII, composed of four subunits: Vps22-EAP30, Vps36-EAP45 and two subunits of Vps25-EAP20. ESCRTI-II can start the endosome's membrane invagination thus allowing internalization of different molecules/cargo, such as nucleic acids or proteins.

ESCRT-III is another heterotetrameric complex (Vps20-CHMP6; Snf7-CHMP4; Vps24-CHMP3; Vps2-CHMP2); activation and its subsequent recruitment to the endosome is mediated by the binding of the ESCRT- II subunit Vps25 with Vps20. ESCRT-III has the role to detach the ubiquitin residues from the protein and to allow the complete invagination of the membrane (membrane budding), thus generating the ILVs. Cargo deubiquitination is mediated also by accessory subunits recruited by ESCRT-III, such as Bro1/Alix (BCK1-like resistance to osmotic shock protein-1/apoptosis linked gene 2 interacting protein X) that, after binding to Snf7, recruits the enzyme Doa4 (degradation of alpha 4), to complete the cargo deubiquitination [2]. Finally, others adaptor proteins assist Vps4-ESCRT III interaction and allow the Vps4 ATPase activity, required for the final membrane budding and scission, for ESCRT subunits removal and for recycling and cargo delivery.

In summary, the ESCRT complex controls the entire process from ILV budding to the selection of cargo, the membrane remodeling and the incorporation of ILVs into MVB. An interesting study of RNAi, on 23 components of ESCRT

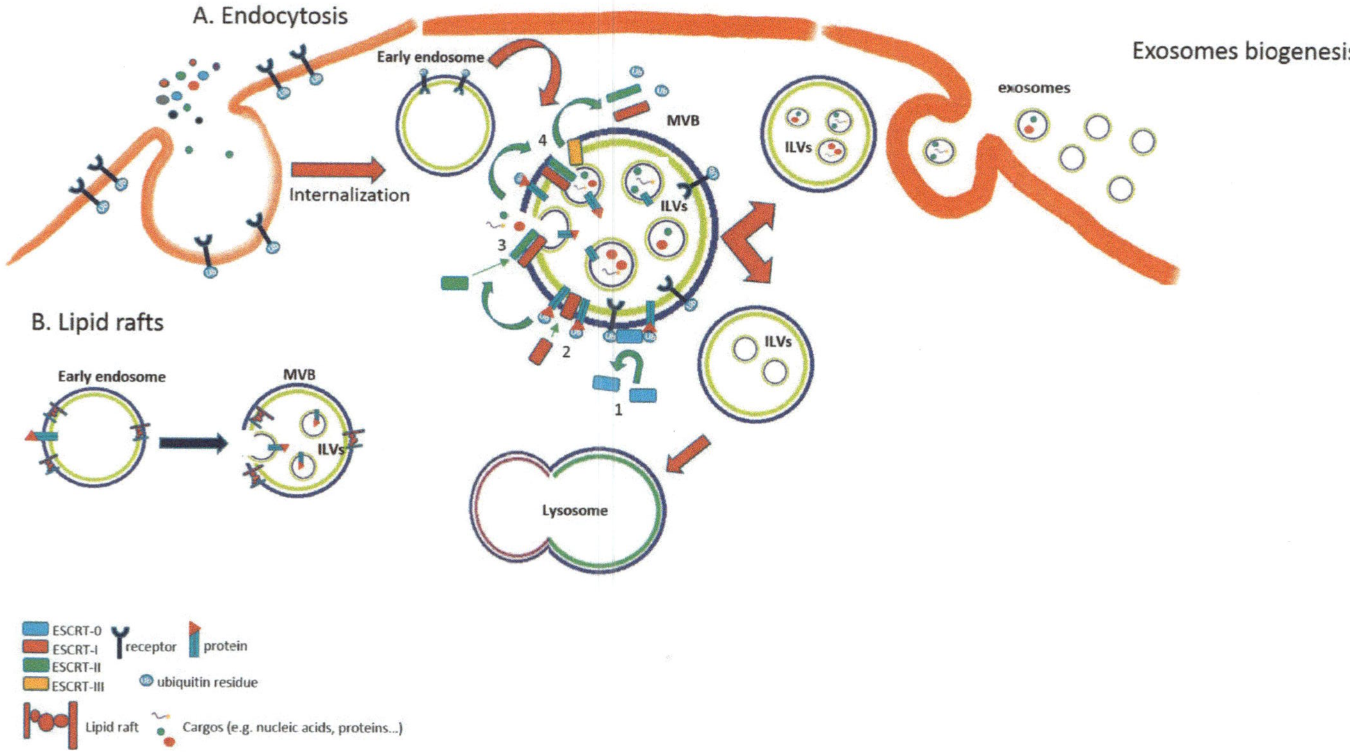

FIG. 1 Exosomes biogenesis. Exosomes biogenesis can occur according different mechanisms. (A) Endocytosis: after receptor-ligand interaction on the cell surface, ligand is separated from its receptor and located inside the early endosome. The receptor, instead, can be recycled on the membrane surface or partially degraded via late endosomes-lysosomes pathway. Early endosome subsequently forms multivesicular bodies (MVB) containing intraluminal vesicles (ILVs). The endosomal sorting complex request for transport (ESCRT) is responsible of the MVB formation. (1) ESCRT0 binds ubiquitin residues thus allowing the recruitment of ESCRTI, which finally undermines ESCRT0 (2). (3) ESCRTII starts endosome's membrane invagination and binds ESCRTIII (4), that detaches the ubiquitin residues from the protein and allows the complete invagination of the membrane (membrane budding), thus generating the ILV. MVBs migrate toward the membrane, fuse with the plasma membrane and ILVs are released as exosomes. Alternatively, MVB can fuse with other MVBs or with late endosomes, receive from transGolgi a set of vesicles containing lysosomal enzymes that allow the transformation in lysosomes. (B) Lipid rafts. MVBs can also originate from endosomes that contain specific domain on their membrane, known as lipid rafts.

complex, allowed clarifying that only few of them are essential in exosome biogenesis. Among these subunits, in fact, Hrs, Tsg101 and STAM1 simultaneous silencing (ESCRT0/I complex) is able to decrease exosome secretion; on the contrary, inhibition of CHMP4C, VPS4B, VTA1 and Alix (ESCRT III complex) increases the same process of exosomes secretion [3]. Recent investigations, demonstrated, in opposition to the role of VPS4 just stated, that its inhibition decreases exosomes release [4]. In addition, Alix, through its interaction with several ESCRT proteins, such as Tsg101 or CHMP4, is involved in protein composition/cargo loading, ILVs' budding and its incorporation into MVBs.

Numerous proteins collaborate with ESCRT complexes in all the steps of exosome biogenesis, from endosomal budding to ILV's formation.

Syntenins are soluble proteins that act as intracellular adaptors, through their PDZ domains, recruiting syndecans, membrane proteins carrying heparan sulfate chains (HS). Syndecans, via HS, bind numerous ligands such as adhesion molecules and growth factors thus allowing them to interact with their receptors and assist the endocytosis process. Recent evidences demonstrated that Alix binds the N-terminal of syntenins and connects syndecans with the ESCRT machinery. Therefore, this heterotrimeric complex is involved not only in exosomal sorting and cargo, as discussed later, but also in endosomal budding and exosomes biogenesis [5, 6].

Related to late endosome trafficking and syntenin-exosomes production, the role of a GTP binding protein is emerged: ADP-ribosylation factor 6 (ARF6). ARF6 is able to activate phosphatidyl-inositol (4)-phosphate 5-kinase (PIPK), involved in PIP2 synthesis. The PIP2 synthesis supports the recruitment of syntenin-syndecan from perinuclear compartment to the plasma membrane [7].

Another kinase, upstream of ARF6, regulates syntenin/syndecan activity: the oncoprotein Src. Src acts on endosomal trafficking by phosphorylation of syntenin/syndecan tyrosine residues thus stimulating endosomal budding and the biogenesis of specific syntenin-dependent exosomes [8].

Heat shock proteins (Hsps) are chaperones generally involved in the control of protein aggregation and folding. These proteins were initially studied for their role in this intracellular pathway, subsequent observations indicated their presence outside the cell. Now it is well known that Hsps are proteins secreted by a non-classical pathway and, most of them, by exosomes. Interestingly, Hsps are also involved in exosomes biogenesis by collaborating and interacting with ESCRTs proteins [9].

3 ESCRT-independent mechanism of exosomes biogenesis

Simultaneous depletion of critical ESCRT proteins demonstrated that MVB biogenesis and ILV formation is also ensured through ESCRT independent mechanisms, thus suggesting that the regulation of this process is more complex than expected. The lack of ESCRT complex, indeed, induces a deep morphological modification of the MVBs that appear enlarged, even if early and late endosome differentiation is maintained.

The existence of ILVs in this circumstance, although different in number and size, suggests that their formation could occur via an ESCRT-independent endocytic sorting mechanisms and allow to hypothesize that both mechanisms of exosomes biogenesis could co- exist and function together, in higher eukaryotes. Cell type and cellular homeostasis could be the elements that drive different subpopulation of exosomes through one of the two mechanisms of biogenesis.

Cells mutated for ESCRTs are still able to form ILVs due also to the lipid composition of their membranes. It is well known that the biophysical characteristics of single membrane lipid, such as the size of head group, length and saturation of acyl chains, are fundamental for membrane curvature.

Endosomes with particular areas enriched in cholesterol and sphingolipids, the lipid rafts, curve inward and may determine the MVBs formation, through the help of pH gradient across the membrane. In this case and without the assistance of ESCRTs, endosomes membrane invagination is due to the synthesis, mediated by phospholipases, of ceramides from sphingolipids. Ceramides alone, due to their cone shape, or in association with cholesterol generate particular domains that favor membrane deformation and ILV budding.

Moreover, ceramide induces exosomal biogenesis through its conversion into sphingosine 1-phosphate (S1P), which binds its receptor on the membrane of MVBs [10].

Neutral Sphingomyelinases (SMases), enzymes involved in sphingomyelin conversion in ceramide, are preferentially located in Golgi-ER but also in plasma membrane thus participating into exosome biogenesis. Inhibition of SMases, in specific cell types, reduces exosomal release of certain proteins thus demonstrating that ceramide is essential to generate microdomains that favor membrane budding [11].

Others lipid modifying enzymes, phospholipase D2 (PLD2) and diacylglycerol kinase α (DGKα), are involved in exosomes biogenesis through the production of phosphatidic acid (PA) that favors, as well as ceramide, membrane invagination [7, 12].

On the other hand, sphingomyelin has a high affinity for cholesterol in the membrane; its hydrolysis increases cholesterol migration from the PM to intracellular membrane and consequently membrane fluidity. It is well known, in fact, that cholesterol molecules cluster in lipid domains thus affecting plasma membrane lipid order and finally vesicle shedding.

Numerous proteins mediate the ESCRT independent mechanism of exosomes biogenesis. Tetraspanins, such as CD9, CD63 and CD81, are transmembrane proteins originally identified in B lymphocytes and generally involved in cell fusion, migration and cell adhesion [13]. Furthermore, tetraspanins are abundant in exosomes and, for this reason, commonly recognized as exosomes markers. In particular, tetraspanins have four transmembrane domains through which they interact with many others proteins, cholesterol and gangliosides, thus generating the TEM domain (tetraspanin-enriched domain). TEM domain can affect, finally, membrane bending and actin polymerization. In addition, tetraspanins mediate cargo sorting and ILV formation, for example CD9 participates in plasma membrane fusion, while CD63, similarly to syndecans, interacts with the PDZ domain of syntenin [14].

4 MVBs transport to the membrane and exosomes release

The mechanism, through which MVBs decide, instead of fuse with lysosomes, to move up to the plasma membrane for exosome release, is not well understood.

However, it is clear that exosomes release outside the cell in the extracellular microenvironment is due to specific protein-protein and protein-lipid interactions during the MVBs fusion with the plasma membrane.

Proteins involved in this event of membrane fusion are, certainly, the SNAREs proteins, tethering factors as well as many small GTPases.

SNARE proteins are generically interested in vesicles fusion to the target membrane that occurs through the formation of a complex of three-four subunits, one R-SNARE in the vesicle and two-three Q-SNAREs in the target membrane. Overexpression of the R-SNARE VAMP7 (vesicle associated membrane protein 7) causes the formation of enlarged MVBs in the cell periphery thus impairing exosomes release [15]. In line with this, knockdown of another R-SNARE, YKT6, reduces the level of Tsg101 secreted in exosomes [16].

Microtubules and cytoskeleton proteins, such as actin and its binding proteins, mediate MVBs transport to the membrane. The role of the actin binding protein cortactin in MVBs trafficking is demonstrated by knockdown or overexpression experiments where cortactin decrease/increase exosomes release [17].

RAB family of small GTPases participates in endosomal traffic and it has been recently involved also in membrane trafficking, vesicles transport along cytoskeleton, MVBs docking to plasma membrane and exosomes release. The main role of RAB27 and RAB35 in docking of MVBs at the plasma membrane has been demonstrated in numerous studies [18–20]. Interestingly, down regulation of RAB7, involved in late endosome traffic, as well as overexpression of RAB5, responsible of large endosomes formation, blocks the release of SDC/syntenin contacting exosomes [5, 21]. Other groups of small GTPases are involved in exosomes release, such as Rho/Rac/cdc42 family [22].

Tetraspanins, previously cited for their role in exosomes biogenesis, are also involved in ESCRT independent exosomes release. Tetraspanins can be found in TEM domain at the plasma membrane but they are able also to interact with the cytoskeleton through other proteins thus affecting their release via exosomes.

We mentioned above that membrane lipids are important for membrane curvature thus modulating MVBs formation; moreover, lipids play a key role in MVBs fusion to the plasma membrane thus increasing exosomes secretion.

Addition of an ether lipid precursor that increases cellular lipids, is able to increase exosomal release [23, 24]; in line with this experimental observation, addition of cholesterol increases typical exosomal proteins, such as Alix or CD63 [23, 24]. On the contrary, metabolic inhibition of cholesterol synthesis, in another cell model, increased secretion of several exosomal proteins [25].

Exosomes release is affected also by other components as calcium, or by different mechanisms. Calcium is able to interfere with the activity of enzymes that regulate plasma membrane symmetry such as translocases or lipid scramblases. Translocases, that allow the inversion of phosphatidylserine and phosphatidylethanolamine from the outer to the inner layer of plasma membrane or lipid scramblases, that promote lipids movement across the membrane, affect exosomes release.

Treatment with a calcium ionophore, that increases intracellular level of calcium, augments exosomes secretion [26]. Moreover, the calcium sensor proteins synaptotagmins, generically implicated in vesicular transport, are able to regulate exosomes secretion [27].

Recent evidences identify ISGylation as a novel ubiquitin-like modification that is able to control exosomes release through promoting protein aggregation and enhancing MVB degradation by the autophagosome-lysosome compartment. In particular, ISGylation of TSG101 and, consequently, its degradation is sufficient to affect exosomes secretion [28].

Cellular stresses such as irradiation, chemotherapy oxidative stress or hypoxia are signals to increase exosomes release. Interestingly, recent evidence from stress condition allowed to identify new actors as putative "balance needle" in the MVB fate, sorting them to exosomes or endolysosomal pathway. An example could be the extracellular small heat shock protein αB-Crystallin (αBC). Exosomes released under oxidative stress condition are enriched of αBC; interestingly Gangalum and collaborators demonstrated that αBC inhibition results in an increased expression of the lysosome marker LAMP1 and of the late endosome marker RAB7 indicating the activation of *endo*-lysosomal pathway [29]. These data allow us to hypothesize a pivotal role of αBC in exosomes release.

5 Basal composition and cargo

The numerous -omics studies performed in the last decade have clearly demonstrated that exosomes contain and transport multiple types of biological macromolecules that maintain their whole activity when delivered to target cells. This bioactive cargo, including nucleic acids (both DNA and all types of RNAs) lipids, and soluble or membrane-bound proteins, is strictly related to the type and functional state of the producing cells although it is not an identical subset of their contents. Growing evidence of last years clearly indicates that the internalization of macromolecules into exosomes is not a random process but the biological and molecular mechanisms driving this process are still far from being fully understood [30]. Since the functional properties of exosomes as mediators of cell-to-cell communication and their capability to modify the behavior of target cells is specifically related to their cargo, the deep characterization of their molecular components as well as the understanding of the pathways leading to internalization process currently represent crucial aims in the field of exosome research [31]. Moreover, since exosome's cargo and biomolecule internalization machinery are specifically related to disease states, exosomes are widely considered as promising potential source for the discovery of novel biomarkers.

In the following sections, a detailed insight of exosome molecular composition will be given and the known mechanisms related to their incorporation into vesicles are described.

Proteins: Exosomes contain a complex set of proteins (cytosolic, nuclear, mitochondrial, ribosomal and membrane-bound proteins) derived from the parent cell. Data from multiple proteomics studies have clearly demonstrated that among these exosome proteins some are irrespective of their cell origin and can therefore be considered "exosomal markers", while others define a unique exosome signature specifically related to the producing cell, determining the exosome properties and activities.

Specifically, among the vesicle-specific proteins often used as markers there are cytosolic proteins such as 14-3-3 proteins and specific heat stress proteins (HSPs) as well as several proteins related to the biogenesis process of EVs such as tetraspanins (CD9, CD63, CD81), lectins, GTPases, major histocompatibility complex (MHC) molecules and proteins of ESCRT complexes (Alix and TSG101) [32, 33].

Beyond to contain these well-maintained proteins, exosomes contain a discrete subset of proteins specifically related to the phenotype of originating cell, through which exosomes are able to differentially reprogram the properties of proximal and distal recipient cells. Accumulating evidence derived from studies on tumor-derived exosomes (TDEs) indicates that depending on their protein content these nanovesicles have a peculiar role in regulating cell survival, tumor progression, metastasis and chemoresistance [31]. A well-known study form Lyden's group showed that changes in integrin composition differentially drive TDEs to a tissue-specific colonization inducing an organ-specific pre-metastatic niche formation [34].

SWATH-based quantitative proteomic analysis highlighted that exosomes released by metastatic colon cancer cells are significant enriched in several cytoskeletal-associated proteins as well as in proteins related to RhoA/ROCK signaling, such as RacGAP1 and thrombin, when compared to those released by less aggressive tumor cells. It has been demonstrated that metastatic tumor cell derived exosomes are able to spread the malignant properties in tumor microenvironment, affecting both the tumor cell plasticity and endothelial cell behavior, and this ability is specifically related to their protein signature. RacGAP1 and thrombin have indeed identified as key mediators of the effects induced in target cells by metastatic exosomes [35].

Several studies have also evidenced that treatment of cancer cells with antitumor compounds can alter the basal protein composition of TDEs reverting their pro-tumor actions. Taverna et al. reported that exosomes released by Chronic Myeloid Leukemia cells (curcumin/CML-exos) were significantly modified in their protein cargo after treatment with curcumin, a plant-derived compound well known for its anticancer effects. In particular, curcumin/CML-exos were depleted in pro-angiogenic proteins and enriched in proteins with anti-angiogenic activity in comparison to exosomes released by no-treated CML cells. These changes determined the loss of CML-exosome's ability to promote the angiogenic phenotype and to alter the endothelial barrier organization [36].

Although accumulating studies have provided many details on protein composition of exosomes clearly indicating that is cell type dependent and that can also be influenced by different cellular conditions or treatments, the mechanisms of protein loading are not yet fully understood. The most described and characterized system involved in exosomal protein sorting is that mediated by ESCRT family members, but growing evidence suggests that several ESCRT-independent pathways can be also involved [37]. Moreover, the enrichment of selected sets of proteins in exosomes suggests that their sorting can be driven by specific mechanisms. Many proteins detected in exosomes have post-translational modifications (PTMs), such as glycosylation, phosphorylation, ubiquitination or SUMOylation. This observation suggests that PTMs can confer specific properties to proteins playing a critical role in regulation of protein sorting into exosomes [38]. Moreover, among the ESCRT-independent mechanisms of proteins loading into exosomes, those mediated by lipid raft and ceramide have been also described. Interestingly it has been reported that some proteins lose their exosome localization following the disruption of lipid rafts (chemokins, αB-crystallin, stem cell surface markers) or when metabolic pathway of ceramide is blocked (CD63) [37].

Nucleic acids: Exosomes contain different types of nucleic acids, such as single-stranded (ssDNA) and double-stranded DNA (dsDNA), mitochondrial DNA (mtDNA), mRNA, micro RNA (miRNA) and long non-coding RNA (lncRNA) [31].

DNA: The presence of DNA molecules in exosomes (ExoDNA) has been widely reported [39]. DNA, both mitochondrial and genomic, was found in exosomes isolated from cell culture supernatants as well as in human and mouse biological fluids and DNAse treatment has revealed that ds-DNA (unlike ss-DNA and RNAs) is mostly found inside exosomes rather than outside [40–42]. The ability of exosomes to mediate the horizontal transfer of DNA has been supported by the detection in normal recipient human neutrophils of ExoDNA containing the BCR/ABL hybrid gene derived from K562 cells [43]. Even if the physiological significance of DNA exosome-mediated transfer is not yet fully clear, there is evidence showing that in recipient cells exoDNA can localize to nucleus where is transcripted [44]. Moreover, it has been recently proposed that secretion of genomic DNA fragments via exosome can play an important role in maintaining cellular homeostasis by avoid the cytoplasmic accumulation of DNA that can elicit cellular senescence or apoptosis [45]. The presence of genomic DNA in exosomes reflecting the mutational status of parental tumor cells for genes as P53, KRAS and EGFR strongly supports the diagnostic value of exoDNA and its potential role in clinic [40, 42, 46]. Although the presence of DNA in exosomes is well documented and its characteristics have been widely described, numerous doubts remain about the mechanisms leading to DNA loading into exosomes. Since it has reported that based on their originating cell exosomes can contain distinct types of DNA (e.g. mtDNA was found in exosomes derived from astrocytes and glioblastoma but not in others) [42, 47], it has been hypothesized that DNA packaging into exosomes could be dynamically regulated by cell type-specific mechanisms. On the other side, the presence inside exosomes of DNA fragments equally distributed over the whole genome without bias for specific regions could indicate that DNA sorting in exosomes is a random process [40, 42].

RNAs: All cell types release exosomes enriched both in coding mRNAs and in non-coding RNAs such as micro RNAs (miRNAs), long noncoding RNAs (lncRNAs), ribosomal RNA (rRNA) and circular RNAs (circRNAs) [37, 48]. Over the last few years several interesting data is emerging about the selective package of RNA molecules within exosomes [48–50]. Due to the growing interest in miRNAs as key regulators of gene expression able to drive cell phenotype, particular attention is now focused on the sorting mechanisms of these small RNAs. The evidence that miRNA profiles of secreted exosomes are distinct from those of the originating cells has strongly indicated that their sorting into exosomes cannot occur randomly [51, 52].

To date several pathways and molecules are reported to be involved in miRNAs sorting in exosomes but many aspects of this complex system remain to be further explored.

Data from current research indicates that miRNA incorporation into exosomes can be guided by specific sequences present in certain miRNAs and by their interaction with some enzymes or other proteins [31, 53]. RNA-binding proteins are among the proteins mainly involved in regulating exosomal miRNA content. Y-box protein I (YBX1), appears to be required for the sorting of specific miRNA in exosomes as well as the ubiquitous heterogeneous nuclear ribonucleoprotein A2B1 (hnRNPA2B1) binding the miRNAs EXO-motif (GGAG) or protein SYNCRIP that is a crucial component of exosomal miRNA sorting machinery in hepatocytes [31, 37, 54, 55].

Beside the RNA-binding proteins, other proteins have been reported to have a critical role in regulating miRNA packaging in exosomes. Knock out experiments for proteins as Argonaute 2 (Ago2), Alix and Neutral Sphingomyelinase 2 (nSMase 2) showed their direct involvement in regulation of exosomal miRNA levels [29]. Post-transcriptional modifications have been also described to drive the sorting of miRNA into EVs. Koppers-Lalic et al. demonstrated that non-templated nucleotide additions are associated with enrichment of miRNAs in EVs (3′-urydilation) or retention within the cell of origin (3′-adenylation) [56]. Moreover, it has been reported that miRNA is retained in cytoplasm when expression levels of its target transcript are high, showing that mRNA-miRNA interaction can modulate incorporation into EVs [57]. The role of raft-like regions of MVB as a target for miRNAs has been also suggested [58]. Finally, Melo et al. have reported that breast cancer associated exosomes contain pre-miRNAs associated with the RISC-Loading Complex thus showing the cell-independent capacity to process precursor miRNAs into mature miRNAs [59].

Packaging of miRNAs into exosomes physically protects them from enzymatic degradation, assuring their effective horizontal transfer to other cells where they can induce the activation of different physiological and pathological processes. Many studies carried out in the last decade have focused on the role of exosomal miRNA in modulating tumor microenvironment. Several exosomal miRNAs as miR-9, miR-105 or miR-21 have been widely described for their significant function in regulating tumor proliferation, vascularization, immune system activity, metastasis and other biological characteristics supporting cancer progression [31, 53].

In addition to transport miRNAs, exosomes carry a broad range of lncRNAs known to modulate gene expression by translational inhibition or by acting as competitive endogenous RNA [31].

As the other biomolecules, lncRNAs are probably selectively sorted in exosomes, since it has been observed that some of them are enriched in exosomes, while others are less present. Also in this case, the mechanisms determining the exosomal lncRNA content are not well known. Specific proteins seem work as lncRNAs carriers driving their internalization into exosomes, but not clear data is still available [53]. Exosome-derived long noncoding RNAs have shown to be involved in the regulation of several steps of tumor progression [53].

Finally, an interesting aspect to consider is that exosomes, for their extreme stability in blood, urine, and other body fluids of patients all body, provide a consistent source for detecting miRNA and lncRNAs as disease biomarkers. Numerous studies have reported interesting data concerning different amount and composition of exosomal miRNAs and lncRNAs between patients with disease and healthy individuals suggesting that circulating exosomal miRNA and lncRNAs may be used as liquid biopsies and noninvasive biomarkers for the early detection, diagnosis, and clinical management of patients [48, 53–55].

Lipids: A complete dataset of exosomal lipids has not yet available, but is clear that the lipid bilayer of exosomes has an exclusive lipid composition with respect to the parental cells. It is characterized by selective enrichment in sphingomyelins, cholesterol, phosphatidylserine, phosphatidylcholine, phosphatidylethanolamine and ganglioside GM3 [31]. The exosomal lipids are asymmetrically distributed on the two side of the bilayer and they are organized to form lipid raft-like domains that seem have a specific role in exosome structure and formation [31, 60].

It has been also observed that exosomes from hepatocarcinoma cells (Huh7) and Mesenchymal Stem Cells are enriched in cardiolipins, while glioblastoma cells (U87) exosomes are enriched in sphingomyelins, indicating that exosomal lipid composition can depend on the cell type of origin [61]. Exosomes are also enriched in ceramide, involved in the budding of exosomes into the lumen of multivesicular bodies and it has reported that the inhibition of the synthesis of neutral sphingomyelin, a ceramide precursor, induces a significant decrease of exosome release. Vesicular lipids are not essential for exosome biogenesis, release and interaction with target cells, but they have to be also considered as bioactive components able to regulate pathophysiological pathways. Already in 2002 Kim and colleagues showed that extracellular membrane vesicles from tumor cells promote angiogenesis via sphingomyelin they transport [62]. Exosomes have been also described as intercellular signalosomes carrying GTP-activatable phospholipases and prostaglandins from activated to resting cells [63].

Recently, lipidomic studies performed on urinary exosomes collected from prostate cancer patients and healthy volunteers have provided preliminary but interesting results supporting the use of exosomal lipids as biomarkers [60].

Further characterization of exosome composition and cargo sorting will allow to better understand the biological relevance of these natural nanocarriers and will provide new knowledge for developing innovative diagnostic and therapeutic strategies.

6 Mechanisms of action

6.1 Exosome mechanism of action

Firstly, identified as garbage disposal, the high number of scientific publications about exosomes revealed their pivotal role in regulating physiological and pathological states by transferring information in a horizontal way. Macromolecules and bioactive compounds, packaged inside these phospholipidic spheres, are transferred from exosome producing cells to compliant receiving cells, being these near or far along the body. Exosomes are listed among the strategies adopted by cells to communicate each other showing a remarkable role in homeostasis maintenance so that their study has been defined as "new endocrinology" [63]. Moreover, the identification of these vesicles in different body fluids including semen, blood, urine, cerebrospinal fluid and milk, corroborates this definition.

A growing number of research groups are currently engaged into a deeper investigation of exosomes' world in order to understand their communication strategies and take advantages from this delivery system.

Unfortunately, the final effects of exosomes following the interaction with receiving cells are not easy predictable, since many factors affect cells/exosomes interaction.

First, exosomes are selective couriers that bind preferentially with specific cell types which compliance is required. An interesting study, aiming to investigate the correlation between exosomes and cancer metastases, demonstrated that among tissue-resident stromal cells only few of them uptake tumor exosomes and that this ability depends on exosomes integrins [33]. A proteomic approach was adopted to characterize exosomes released from tumors that, although different for histological origin, shared the preferential site of metastases. Thanks to this study, Hoshino et al. demonstrated that TDEs, driven by their integrins (ITGs), chose metastatic site targeting specific organs and selecting stromal cells inside the tissues. Exosomes exposing ITGα6β4 and ITGα6β1, in fact, were found internalized by S100A4-positive cells, while the expression of ITG$\alpha_v\beta_5$ allowed the exosome uptake by F4/80$^+$ macrophages [33].

Exosomes, once reached the correct site and recognized the acceptor cytotypes, can adopt different strategies to transform receiving cells, depending on the molecular interactors that physically mediate cell-exosome contacts. *Activation of cell receptors and downstream signaling pathways, membranes fusion, exosomes internalization and nuclear translocation are among the molecular mechanisms by which exosomes deliver their messages.*

Proceeding from the outside, the first step of an exosome-receiving cell interaction could be the ligand/receptor binding.

Ligands exposed on exosome's surface, binding specific membrane receptors on receiving cells, can activate them turning on signal transduction pathways.

The best-studied example is the exosome-mediated cell death, this is a strategy used by several cancer cells to promote immune tolerance. Tumor derived exosomes, through the surface expression of death signals, as the PD-L1 (programmed death-ligand 1) or Fas Ligand, promote a systemically suppression of the immune system by inducing apoptosis in receiving T cells and NK cells [64, 65]. Also in physiological condition exosomes cooperate with APC (antigen presenting cell) as mediators of immune regulatory signals; an interesting paper demonstrated, for example, that dendritic cells of immunized mice release in the plasma MHCII+/FasL+ exosomes that are able to suppress immune response in Ag-specific manner through Fas activation [66]. In line with collected data, engineered exosomes exposing TRAIL are now proposed as strategy for the delivery of pro-apoptotic signals to tumor cells [67].

Decorated with cell membrane ligands and receptors, exosomes take part to processes commonly mediated by cell-cell contacts such as development, organogenesis and tissue homeostasis [68]. Endothelial cells, for example, produce exosomes exposing the Notch ligand Dll4, that, once on the surface of receiving cells, promote Notch-cleavage and activation, finally resulting in cells vessel branching and increase of vessel density in vitro. Moreover, as the vesicles remain linked to the activated receptor, these are transported inside the cell together with receptor during the internalization phase [69]. In addition, also secreted proteins of both Hedgehog and WNT families, were found able to promote morphogenetic signals, directly loaded in the exosome bilayer or carried as exosome-associated proteins, while exosomes with a membrane-associated TGF-β1 have been found able to activate TGF-β receptor in receiving cells [70, 71].

Others example of exosomes that activate signal transduction can be found among tumor derived exosomes such as amphiregulin (AREG) expressing exosomes which, isolated from several tumor cells, activate EGFR in receiving cells thus affecting bone marrow microenvironment [72] or promoting bone metastases [73].

Although exosomes are able to turn on intracellular pathways from the outside, *in most cases the interaction with cell membrane proteins, drives vesicles internalization in receiving cells.* Clear evidences of the internalization processes come from the identification of murine proteins inside human mast cells pre-treated with mouse exosomes. Equally important was the experiment designed by Montecalvo et al. [74], in which treatment with luciferin-loaded exosomes was found able to induce luminescence in luciferase expressing cells.

To date, several mechanisms have been described as involved in driving exosome internalization [75, 76] and probably others will be identified (Fig. 2).

The use of proteolytic treatment or specific antibodies that block exosome internalization allowed to identify some protein-protein interactions involved in the activation of an endocytic pathway.

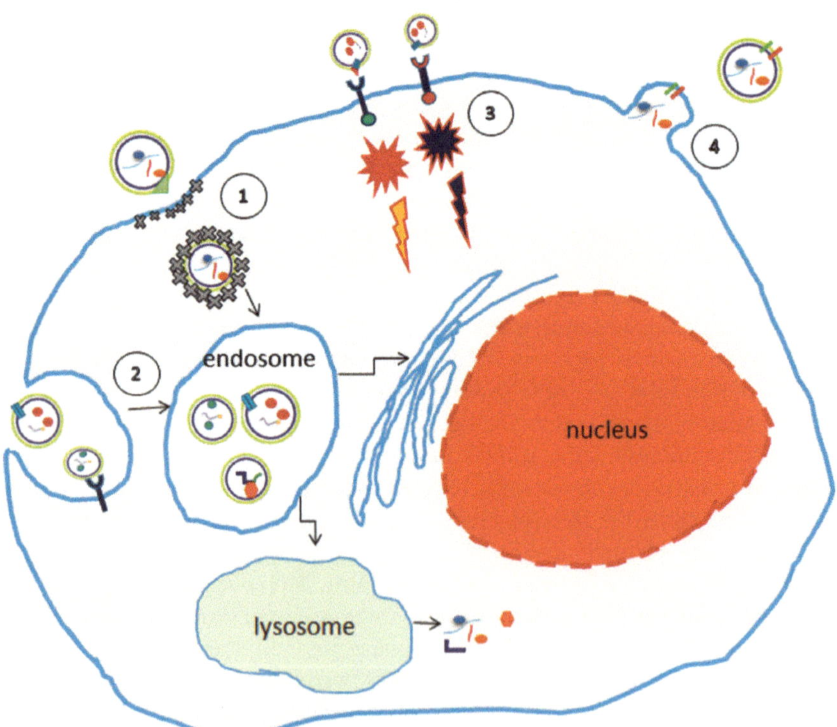

FIG. 2 Mechanisms of action. *Internalization (1–2)*: the chlatrin dependent—(1) or independent- endocytic vesicles (2), converge to endosome and, from here, to lysosomes or to the nucleus. *Intracellular pathway activation* (3) ligand/receptor interaction activates signal transduction. *Fusion* (4): release of bioactive molecules after membrane's fusion.

The *receptor mediated endocytosis* is one of the mechanisms involved in the internalization of external material thus generating early endosomes, first step of exosomes biogenesis, as previously discussed. On the other side, receptor mediated endocytosis requires the binding between a ligand and its specific receptor, respectively exposed on exosome's surface and on the cellular plasma membrane. This binding is sufficient to promote the assembling of coat proteins on the inner surface of cell membrane thus forming clathrin-coated vesicles that, detached from the cell membrane thanks to a large GTPase dynamin helical collar, will proceed through clathrin-mediated endocytosis pathway [77]. Formal proof of exosome uptake through clathrin-dependent endocytosis was obtained by Tian and collaborators, which demonstrated that the loss of clathrin, induced by treatment of cells with a cationic amphipathic drug, strongly inhibits exosomes internalization [78].

In addition, exosomes can enter cells through *non-classical endocytosis pathways* as that mediated by lipid rafts, cholesterol and sphingolipid-rich micro domains in the plasma membrane. Lipid rafts, as discussed above, move long the membrane and organize smooth invaginations of this, by either caveolar or noncaveolar pathways; interestingly, experiments performed with membrane cholesterol depletion or by inhibiting cholesterol biosynthesis, demonstrated a dose dependent reduction in exosomes uptake, up to 60%, in HUVEC [79].

Phagocytosis of exosomes occurs in professional phagocytes and other immune cells (i.e. macrophages, dendritic cells and γδ T cells) and it is dependent on the actin cytoskeleton, phosphatidylinositol 3-kinase (PI3K), and dynamin 2. Moreover, a feature common to extracellular vesicles is the exposition of phosphatidyl serine in outer membrane; this is recognized by numerous plasma membrane receptors that, involved in apoptotic bodies' engulfment, mediate exosomes internalization.

At this point is interesting to note that, physiologically, both endocytosis and phagocytosis drive the internalized cargo to endosomal-lysosomal degradative pathway, however a massive number of collected evidences indicate that exosome cargo is bioactive in receiving cells.

How internalized exosome escapes lysosomal degradation is up to date under investigation. Recently a possible strategy was suggested by Lorico's group that identified in mesenchymal stromal cells and in breast cancer cells an intermediate compartment for the delivery of endocytosed EVs, or part of these (e.g. CD9/CD133 protein complexes), directly to the nucleus of receiving cell. In particular they identified a tripartite complex named VOR, constituted by the ER protein VAP-A and the late endosome proteins ORP1L and Rab7 that drive late endosome close to nuclear pore and, from there, to the nucleoplasm [80].

However, different mechanisms of internalization could have higher delivery cargo efficiency.

The characterization of exosome's proteins on the surface of receiving cells identified *membrane fusion* among the mechanisms adopted by exosomes for cargo delivery. Tetraspanins, for example, interacting themselves and with a variety of molecules including proteins, lipids and carbohydrates, organize TEM domains that, as previously described, are involved in several biological functions including cell membrane fusion. Then, a role of TEMs in exosome internalization was largely proposed even if, to date, only few evidences support this hypothesis [31]. In addition, also integrins take part to membrane fusion and the use of specific antibodies masking for example CD61, CD51 and CD54, demonstrated their pivotal role in exosomes uptake by dendritic cells.

As described above, exosomes are containers that collect and delivery bioactive compounds. Exosomes are able to transform receiving cells and to regulate tissue-specific and/or whole-body metabolism through the delivery of different cargo molecules to adjacent and remote tissues; the phospholipidic structures protect exosomes' cargo by degradative attacks and preserve its biological activity.

The multiple mechanisms by which exosomes affect receiving cells are then strictly associated to exosomes cargo. For example, as previously described, exosomes may activate an intracellular signaling through activation of cell receptor on outer membrane but also by the exosome-mediated delivery, inside cells, of an activated receptor. Song and collaborators demonstrated that Cancer Cell Derived Exosomes are enriched in phosphorylated receptor tyrosine kinases, such as phosphorylated epidermal growth factor receptor (EGFR) and human epidermal growth factor receptor 2 (HER-2) that, once in tumor associated monocytes, activate MAPK pathway increasing survival rate [81]. Moreover, exosomes can deliver functional multiprotein complexes thus inducing a complete rearrangement of transcriptome. It was demonstrated that cancer cell derived exosomes contain microRNAs (miRNAs) associated with the RISC-Loading Complex (RLC) including Dicer, TRBP, and AGO2, which process pre-miRNAs into mature miRNAs and induce Dicer dependent mRNA degradation in target cells [58].

Exosomes are largely studied by endocrinologists for their ability to amplify hormone signaling. Endometrial cells, for example, load in exosomes proteins that internalized by trophoblast cells affect adhesion, migration, invasion, and extracellular matrix remodeling thus promoting successful implantation and favoring pregnancy establishment [82]. The recent data of Crewe et al. [83] confirmed the role of exosomes in metabolic signals. The authors, taking advance from a novel mouse model, which allowed them to track exosomes' movement in vivo, demonstrated that endothelial cells transmit to the adipose tissue a “real time update” of the systemic energy balance. In particular, they showed that endothelial cells from adipose tissue captured signal molecules from blood, that is, soluble or albumin-bound nutrients and hormones, and packaged them inside exosomes that sent, preferentially, to neighbor adipocytes.

In addition, exosomes participate into maintenance of tissue homeostasis; in particular mesenchymal stem cells (MSCs) derived exosomes. These vesicles carry enzymes able to restore the glycolytic deficit and ATP production in the reperfused myocardium, growth factors (i.e. Platelet Derived Growth Factor,

Epidermal Growth Factor, Fibroblast Growth Factor) strongly involved in tissue repair and bioactive molecules with anti-inflammatory role. Interestingly, the collected data revealed that MSC-derived exosomes are as potent as parental stem cells in the regeneration of various organs in injury tissue models, so to suggest MSC-secreted exosomes as a novel cell-free therapy [84].

Although the great number of bioactive macromolecules that, delivered by exosomes, are able to modulate receiving cells, most of the experimental evidences refer to the effects induced by exosome carried RNA and, in particular, non-coding RNA. The best-studied mechanism, adopted by exosomes to modify receiving cell behavior, is definitely the miRNA delivery. Today we know that different pathways and molecules are involved in a selective miRNAs sorting inside exosome with a cell and tissue specificity [30]. Thanks to exosomes loading, protected from RNase degradation, mature miRNA can be transported along the body and, as firstly demonstrated by Pegtel et al. [85], released in receiving cell cytoplasm, and recognize their specific targets thus inducing host's mRNA degradation. Interestingly, a deeper analysis of nucleic acids loaded in exosomes, recognized a great quantity of RNA fragments with size distribution between 25 and 700 nucleotides and, among these, miRNAs represent the minor part, while both tRNA fragments and Y-RNAs seem to be the most abundant, however their role in receiving cells is to define [86]. In addition, a growing number of evidences attributed a role to the exosome-delivered long non coding RNA (lncRNA) in the modulation of phenotype of target cell. These are single strand RNA fragments longer than 200 nucleotides, which although numerically less abundant inside exosomes than small RNAs, retain pleiotropic roles into controlling gene expression. For instance, exosomes released by hypoxic bladder cancer cells are enriched in the oncogenic lncRNA-UCA1 able to induce a remodeling of tumor microenvironment to facilitate tumor growth [60]. Similarly, glioma cells and liver cancer stem cells promote angiogenesis through the release of exosomes containing respectively the lncRNA POU3F3 and lncRNA H19 [61, 62]. LncRNAs participate in epigenetic regulation by recruiting and driving multiprotein complex to specific loci, controlling alternative splicing and protein translation and, finally, through complementary regions, by sponging several miRNAs thus resetting their cytoplasmic pattern. All the data, collected up to date, let us to hypothesize that our knowledge about exosomes' effects are only the tip of the iceberg and that from "cell garbage" they are becoming a "cell treasure".

References

[1] Henne WM, Buchkovich NJ, Emr SD. The ESCRT pathway. Dev Cell 2011;21(1):77–91.

[2] Luhtala N, Odorizzi G. Bro1 coordinates deubiquitination in the multivesicular body pathway by recruiting Doa4 to endosomes. J Cell Biol 2004;166(5):717–29.

[3] Colombo M, et al. Analysis of ESCRT functions in exosome biogenesis, composition and secretion highlights the heterogeneity of extracellular vesicles. J Cell Sci 2013;126(Pt 24): 5553–65.

[4] Jackson CE, et al. Effects of inhibiting VPS4 support a general role for ESCRTs in extracellular vesicle biogenesis. Biophys J 2017;113(6):1342–52.

[5] Baietti MF, et al. Syndecan-syntenin-ALIX regulates the biogenesis of exosomes. Nat Cell Biol 2012;14(7):677–85.

[6] Friand V, David G, Zimmermann P. Syntenin and syndecan in the biogenesis of exosomes. Biol Cell 2015;107(10):331–41.

[7] Ghossoub R, et al. Syntenin-ALIX exosome biogenesis and budding into multivesicular bodies are controlled by ARF6 and PLD2. Nat Commun 2014;5:3477.

[8] Imjeti NS, et al. Syntenin mediates SRC function in exosomal cell-to-cell communication. Proc Natl Acad Sci U S A 2017;114(47):12495–500.

[9] Reddy VS, et al. Extracellular small heat shock proteins: exosomal biogenesis and function. Cell Stress Chaperones 2018;23(3):441–54.

[10] Kajimoto T, et al. Ongoing activation of sphingosine 1-phosphate receptors mediates maturation of exosomal multivesicular endosomes. Nat Commun 2013;4:2712.

[11] Trajkovic K, et al. Ceramide triggers budding of exosome vesicles into multivesicular endosomes. Science 2008;319(5867):1244–7.

[12] Alonso R, et al. Diacylglycerol kinase alpha regulates the secretion of lethal exosomes bearing Fas ligand during activation-induced cell death of T lymphocytes. J Biol Chem 2005;280(31):28439–50.

[13] Charrin S, et al. Tetraspanins at a glance. J Cell Sci 2014;127(Pt 17):3641–8.

[14] Latysheva N, et al. Syntenin-1 is a new component of tetraspanin-enriched microdomains: mechanisms and consequences of the interaction of syntenin-1 with CD63. Mol Cell Biol 2006;26(20):7707–18.

[15] Fader CM, et al. TI-VAMP/VAMP7 and VAMP3/cellubrevin: two v-SNARE proteins involved in specific steps of the autophagy/multivesicular body pathways. Biochim Biophys Acta 2009;1793(12):1901–16.

[16] Ruiz-Martinez M, et al. YKT6 expression, exosome release, and survival in non-small cell lung cancer. Oncotarget 2016;7(32):51515–24.

[17] Sinha S, et al. Cortactin promotes exosome secretion by controlling branched actin dynamics. J Cell Biol 2016;214(2):197–213.

[18] Ostrowski M, et al. Rab27a and Rab27b control different steps of the exosome secretion pathway. Nat Cell Biol 2010;12(1):19–30: sup. pp 1–13.

[19] Bobrie A, et al. Rab27a supports exosome-dependent and -independent mechanisms that modify the tumor microenvironment and can promote tumor progression. Cancer Res 2012;72(19):4920–30.

[20] Hsu C, et al. Regulation of exosome secretion by Rab35 and its GTPase-activating proteins TBC1D10A-C. J Cell Biol 2010;189(2):223–32.

[21] Wegner CS, et al. Ultrastructural characterization of giant endosomes induced by GTPase-deficient Rab5. Histochem Cell Biol 2010;133(1):41–55.

[22] Loomis RJ, et al. Citron kinase, a RhoA effector, enhances HIV-1 virion production by modulating exocytosis. Traffic 2006;7(12):1643–53.

[23] Phuyal S, et al. The ether lipid precursor hexadecylglycerol stimulates the release and changes the composition of exosomes derived from PC-3 cells. J Biol Chem 2015;290(7):4225–37.

[24] Strauss K, et al. Exosome secretion ameliorates lysosomal storage of cholesterol in Niemann-Pick type C disease. J Biol Chem 2010;285(34):26279–88.

[25] Llorente A, van Deurs B, Sandvig K. Cholesterol regulates prostasome release from secretory lysosomes in PC-3 human prostate cancer cells. Eur J Cell Biol 2007;86(7):405–15.

[26] Savina A, et al. Exosome release is regulated by a calcium-dependent mechanism in K562 cells. J Biol Chem 2003;278(22):20083–90.
[27] Hoshino D, et al. Exosome secretion is enhanced by invadopodia and drives invasive behavior. Cell Rep 2013;5(5):1159–68.
[28] Villarroya-Beltri C, et al. ISGylation controls exosome secretion by promoting lysosomal degradation of MVB proteins. Nat Commun 2016;7:13588.
[29] Gangalum RK, et al. Inhibition of the expression of the small heat shock protein alphaB-crystallin inhibits exosome secretion in human retinal pigment epithelial cells in culture. J Biol Chem 2016;291(25):12930–42.
[30] Conigliaro A, et al. Exosomes: nanocarriers of biological messages. Adv Exp Med Biol 2017;998:23–43.
[31] Andreu Z, Yanez-Mo M. Tetraspanins in extracellular vesicle formation and function. Front Immunol 2014;5:442.
[32] Hurley JH, Odorizzi G. Get on the exosome bus with ALIX. Nat Cell Biol 2012;14(7):654–5.
[33] Hoshino A, et al. Tumour exosome integrins determine organotropic metastasis. Nature 2015;527(7578):329–35.
[34] Schillaci O, et al. Exosomes from metastatic cancer cells transfer amoeboid phenotype to non-metastatic cells and increase endothelial permeability: their emerging role in tumor heterogeneity. Sci Rep 2017;7(1):4711.
[35] Taverna S, et al. Curcumin modulates chronic myelogenous leukemia exosomes composition and affects angiogenic phenotype via exosomal miR-21. Oncotarget 2016;7(21):30420–39.
[36] Li SP, et al. Exosomal cargo-loading and synthetic exosome-mimics as potential therapeutic tools. Acta Pharmacol Sin 2018;39(4):542–51.
[37] Moreno-Gonzalo O, Fernandez-Delgado I, Sanchez-Madrid F. Post-translational add-ons mark the path in exosomal protein sorting. Cell Mol Life Sci 2018;75(1):1–19.
[38] Kalluri R, LeBleu VS. Discovery of double-stranded genomic DNA in circulating exosomes. Cold Spring Harb Symp Quant Biol 2016;81:275–80.
[39] Kahlert C, et al. Identification of double-stranded genomic DNA spanning all chromosomes with mutated KRAS and p53 DNA in the serum exosomes of patients with pancreatic cancer. J Biol Chem 2014;289(7):3869–75.
[40] Lazaro-Ibanez E, et al. Different gDNA content in the subpopulations of prostate cancer extracellular vesicles: apoptotic bodies, microvesicles, and exosomes. Prostate 2014;74(14): 1379–90.
[41] Thakur BK, et al. Double-stranded DNA in exosomes: a novel biomarker in cancer detection. Cell Res 2014;24(6):766–9.
[42] Cai J, et al. Functional transferred DNA within extracellular vesicles. Exp Cell Res 2016;349(1):179–83.
[43] Cai J, et al. Extracellular vesicle-mediated transfer of donor genomic DNA to recipient cells is a novel mechanism for genetic influence between cells. J Mol Cell Biol 2013;5(4):227–38.
[44] Takahashi A, et al. Exosomes maintain cellular homeostasis by excreting harmful DNA from cells. Nat Commun 2017;8:15287.
[45] Wan JCM, et al. Liquid biopsies come of age: towards implementation of circulating tumour DNA. Nat Rev Cancer 2017;17(4):223–38.
[46] Guescini M, et al. Astrocytes and glioblastoma cells release exosomes carrying mtDNA. J Neural Transm (Vienna) 2010;117(1):1–4.
[47] Sun Z, et al. Emerging role of exosome-derived long non-coding RNAs in tumor microenvironment. Mol Cancer 2018;17(1):82.

[48] Bolukbasi MF, et al. miR-1289 and "zipcode"-like sequence enrich mRNAs in microvesicles. Mol Ther Nucleic Acids 2012;1:e10.
[49] Gao T, Shu J, Cui J. A systematic approach to RNA-associated motif discovery. BMC Genomics 2018;19(1):146.
[50] Baglio SR, et al. Human bone marrow- and adipose-mesenchymal stem cells secrete exosomes enriched in distinctive miRNA and tRNA species. Stem Cell Res Ther 2015;6:127.
[51] Hessvik NP, et al. Profiling of microRNAs in exosomes released from PC-3 prostate cancer cells. Biochim Biophys Acta 2012;1819(11–12):1154–63.
[52] Sun Z, et al. Effect of exosomal miRNA on cancer biology and clinical applications. Mol Cancer 2018;17(1):147.
[53] Momen-Heravi F, Getting SJ, Moschos SA. Extracellular vesicles and their nucleic acids for biomarker discovery. Pharmacol Ther 2018;192:170–87.
[54] Garcia-Romero N, et al. Extracellular vesicles compartment in liquid biopsies: clinical application. Mol Aspects Med 2018;60:27–37.
[55] Koppers-Lalic D, et al. Nontemplated nucleotide additions distinguish the small RNA composition in cells from exosomes. Cell Rep 2014;8(6):1649–58.
[56] Squadrito ML, et al. Endogenous RNAs modulate microRNA sorting to exosomes and transfer to acceptor cells. Cell Rep 2014;8(5):1432–46.
[57] Janas T, Janas MM, Sapon K. Mechanisms of RNA loading into exosomes. FEBS Lett 2015;589(13):1391–8.
[58] Melo SA, et al. Cancer exosomes perform cell-independent microRNA biogenesis and promote tumorigenesis. Cancer Cell 2014;26(5):707–21.
[59] Skotland T, Sandvig K, Llorente A. Lipids in exosomes: current knowledge and the way forward. Prog Lipid Res 2017;66:30–41.
[60] Haraszti RA, et al. High-resolution proteomic and lipidomic analysis of exosomes and microvesicles from different cell sources. J Extracell Vesicles 2016;5:32570.
[61] Yi X, et al. The feasibility of using mutation detection in ctDNA to assess tumor dynamics. Int J Cancer 2017;140(12):2642–7.
[62] Subra C, et al. Exosomes account for vesicle-mediated transcellular transport of activatable phospholipases and prostaglandins. J Lipid Res 2010;51(8):2105–20.
[63] Guay C, Regazzi R. Exosomes as new players in metabolic organ cross-talk. Diabetes Obes Metab 2017;19(Suppl 1):137–46.
[64] Chen G, et al. Exosomal PD-L1 contributes to immunosuppression and is associated with anti-PD-1 response. Nature 2018;560(7718):382–6.
[65] Lugini L, et al. Immune surveillance properties of human NK cell-derived exosomes. J Immunol 2012;189(6):2833–42.
[66] Kim SH, et al. MHC class II+ exosomes in plasma suppress inflammation in an antigen-specific and Fas ligand/Fas-dependent manner. J Immunol 2007;179(4):2235–41.
[67] Rivoltini L, et al. TNF-related apoptosis-inducing ligand (TRAIL)-armed exosomes deliver proapoptotic signals to tumor site. Clin Cancer Res 2016;22(14):3499–512.
[68] McGough IJ, Vincent JP. Exosomes in developmental signalling. Development 2016;143(14):2482–93.
[69] Sheldon H, et al. New mechanism for Notch signaling to endothelium at a distance by Delta-like 4 incorporation into exosomes. Blood 2010;116(13):2385–94.
[70] Raimondo S, et al. Chronic myeloid leukemia-derived exosomes promote tumor growth through an autocrine mechanism. Cell Commun Signal 2015;13:8.
[71] Yu L, et al. Exosomes with membrane-associated TGF-beta1 from gene-modified dendritic cells inhibit murine EAE independently of MHC restriction. Eur J Immunol 2013;43(9): 2461–72.

[72] Corrado C, et al. Chronic myelogenous leukaemia exosomes modulate bone marrow microenvironment through activation of epidermal growth factor receptor. J Cell Mol Med 2016;20(10):1829–39.

[73] Taverna S, et al. Amphiregulin contained in NSCLC-exosomes induces osteoclast differentiation through the activation of EGFR pathway. Sci Rep 2017;7(1):3170.

[74] Montecalvo A, et al. Mechanism of transfer of functional microRNAs between mouse dendritic cells via exosomes. Blood 2012;119(3):756–66.

[75] Mulcahy LA, Pink RC, Carter DR. Routes and mechanisms of extracellular vesicle uptake. J Extracell Vesicles 2014;3:24641, 1–14.

[76] McKelvey KJ, et al. Exosomes: mechanisms of uptake. J Circ Biomark 2015;4:7.

[77] Kaksonen M, Roux A. Mechanisms of clathrin-mediated endocytosis. Nat Rev Mol Cell Biol 2018;19(5):313–26.

[78] Tian T, et al. Exosome uptake through clathrin-mediated endocytosis and macropinocytosis and mediating miR-21 delivery. J Biol Chem 2014;289(32):22258–67.

[79] Svensson KJ, et al. Exosome uptake depends on ERK1/2-heat shock protein 27 signaling and lipid raft-mediated endocytosis negatively regulated by caveolin-1. J Biol Chem 2013;288(24):17713–24.

[80] Santos MF, et al. VAMP-associated protein-A and oxysterol-binding protein-related protein 3 promote the entry of late endosomes into the nucleoplasmic reticulum. J Biol Chem 2018;293(36):13834–48.

[81] Song X, et al. Cancer cell-derived exosomes induce mitogen-activated protein kinase-dependent monocyte survival by transport of functional receptor tyrosine kinases. J Biol Chem 2016;291(16):8453–64.

[82] Greening DW, et al. Human endometrial exosomes contain hormone-specific cargo modulating trophoblast adhesive capacity: insights into endometrial-embryo interactions. Biol Reprod 2016;94(2):38.

[83] Crewe C, et al. An endothelial-to-adipocyte extracellular vesicle axis governed by metabolic state. Cell 2018;175(3):695–708. e13.

[84] Lou G, et al. Mesenchymal stem cell-derived exosomes as a new therapeutic strategy for liver diseases. Exp Mol Med 2017;49(6):e346.

[85] Pegtel DM, et al. Functional delivery of viral miRNAs via exosomes. Proc Natl Acad Sci U S A 2010;107(14):6328–33.

[86] Tosar JP, et al. Assessment of small RNA sorting into different extracellular fractions revealed by high-throughput sequencing of breast cell lines. Nucleic Acids Res 2015;43(11):5601–16.

Further reading

[87] Yanez-Mo M, et al. Biological properties of extracellular vesicles and their physiological functions. J Extracell Vesicles 2015;4:27066.

Chapter 2

Methods for exosome isolation and characterization

Mi Zhou[a], Sarah R. Weber[a], Yuanjun Zhao[a], Han Chen[b], Jeffrey M. Sundstrom[a]
[a]*Department of Ophthalmology, Penn State Hershey Medical Center, Hershey, PA, United States,*
[b]*Microscopy Imaging Facility, Penn State Hershey Medical Center, Hershey, PA, United States*

1 Introduction

Extracellular vesicles are released from nearly all cell types under both physiological and pathological conditions [1]. Exosomes are a class of extracellular vesicles that are 30–150 nm in diameter and originate as intraluminal vesicles within multivesicular bodies (MVBs). Exosomes are found in numerous types of body fluid, including blood, urine, cerebrospinal fluid (CSF), vitreous, ascites, and breast milk [2–7]. The mechanisms of exosome formation are debated. A common theory is that exosomes form through an endosomal sorting complex required for transport (ESCRT)-dependent pathway [8]. Another theory of MVB formation and exosome biogenesis is the ceramide-dependent pathway [9]. Ceramide, a lipid, is thought to facilitate invagination of the MVB membrane. Rab27, a member of the Rab family of small GTPases, has also been shown to control the exosome secretion pathway, specifically functioning in MVB docking at the plasma membrane [10, 11]. Due to their heterogeneity, it is critical to select an appropriate method for isolation and detection of these small vesicles.

A large body of evidence has shown that exosomes play a critical role in intercellular communication, as these vesicles encapsulate 'messages' from their parental cells in the form of biomolecules, including proteins, lipids, and nucleic acids. These messages manifest in diverse physiological and pathological contexts. For example, exosomes derived from dendritic cells and T cells have been shown to carry cytokines to their recipient cells [12]. Exosomes derived from cancer cells are involved in multiple steps in tumorigenesis and tumor metastasis, including tumor cell proliferation, tumor cell migration, immune escape, and angiogenesis [13]. In addition, several studies have demonstrated that exosomes are involved in the development of neurodegenerative diseases, such as Alzheimer's disease and Parkinson's disease, by spreading neural toxicity [14, 15].

Exosomes. https://doi.org/10.1016/B978-0-12-816053-4.00002-X

In addition to their role in cell-cell communication, exosomes may have useful clinical applications as well; they are increasingly being investigated as biomarkers sources for diagnosis and as drug-delivery vectors for therapeutic applications [16, 17]. Exosome Diagnostics, a biotechnology company, has developed several breakthrough exosome-based diagnostic tools for lung and prostate cancer. Kamerkar et al. found that exosomes facilitate therapeutic targeting of oncogenic KRAS in pancreatic cancer. Compared to liposomes, engineered exosomes demonstrated enhanced efficacy of cancer suppression via delivery of RNA interference (RNAi) to specific targets [18].

To advance exosome research in both basic science and clinical contexts, proper preparation and characterization techniques are critical. Here we review current techniques for exosome isolation and characterization and address their experimental limitations. Our discussion of isolation techniques includes differential centrifugation, ultrafiltration, density gradient centrifugation, precipitation, and immunoaffinity capture-based isolation. For exosome characterization, we discuss common methods for exosome visualization with a focus on transmission electron microscopy techniques (TEM). Our discussion also includes several techniques for exosome quantification, including nanoparticle tracking analysis (NTA), asymmetric flow field-flow fractionation (AF4), and resistance pulse sensing (RPS). We address flow cytometry and the ExoView (NanoView Biosciences, Boston, MA) platform as tools for exosome surface marker detection and subset characterization. Lastly, we discuss methods for tracking exosomes and defining exosome cargo, including proteomic and microRNA sequencing techniques.

2 Methods for exosome isolation

2.1 Differential centrifugation

Differential centrifugation was initially developed to isolate exosomes from reticulocytes and is currently the most commonly used method for exosome isolation [19]. The basis of this method is the removal of cells, cell debris, and large vesicles by successive centrifugation steps. Briefly, experimental samples are centrifuged at 300, 2000, and 10,000×*g* to remove cells, dead cells, and cell debris, respectively. Exosomes are obtained by centrifugation at 100,000×*g*. This method is suitable for experiments involving a large volume of the initial sample, as quite a substantial number of vesicles are lost during the process. According to a study from Kowal et al., 70% of the exosome fraction obtained by this method consists of vesicles 50–150 nm in diameter, with the remaining vesicles having diameters of >150 nm (20%) or <50 nm (10%) [20].

2.2 Filtration and ultracentrifugation

Filtration is another common method for exosome isolation. The principle behind this method is that the use of membranous filters enables separation of exosomes from other sample components based on their molecular weights and sizes.

This method has two main steps: (1) separation of exosomes from larger particles, such as cells, cell debris, and microparticles, using membrane filters with pore diameters of 0.1–0.22 μm and (2) separation of exosomes from smaller particles, such as soluble proteins and protein aggregates, using filters with a molecular weight cut-off of 3–100 kDa. Several commercial membranous filters have been developed, such as the Corning Disposable Bottle-Top Filter with a 500 mL volume and a pore size of 0.22 μm and the Amicon Centrifugal Filter Ultra for proteins ranging from 3 to 100 kDa. If additional volume reduction is needed, samples containing exosomes can be centrifuged at 100,000–200,000×*g* to pellet exosomes.

2.3 Density gradient centrifugation

Density gradient centrifugation takes advantage of differences in vesicle sizes and mass densities to isolate exosomes via creation of a discontinuous density gradient. During the centrifugation step, the particles travel through the gradient until they reach the point at which their density matches that of the surrounding solution. Sucrose and iodixanol are two common media used to create the gradient. Iodixanol is more stable and less viscous relative to sucrose [21]. There are two ways to load the samples: top-loading and bottom-loading. Bottom-loading has an advantage over top loading as soluble proteins will remain at the bottom during ultracentrifugation, whereas soluble proteins will sediment through the gradient if the sample is top-loaded.

Greening et al. reported a density-based separation method using iodixanol from OptiPrep (MilliporeSigma, US) that uses sample top-loading [22]. In this method, the experimental sample is first ultracentrifuged at 100,000×*g* for 2 h. Crude exosome pellets are then resuspended in phosphate-buffered saline (PBS) and top-loaded into an iodixanol gradient buffer. After ultracentrifugation at 100,000×*g* for 18 h, 12 fractions remain in the tube, with exosomes in the seventh fraction from the top. The 12 fractions are then collected and subjected to downstream analyses (Fig. 1A).

In contrast, Choi et al. reported a method for exosome isolation and downstream proteomic profiling using sample bottom-loading instead [23]. In this method, experimental samples are added to the 0.8 and 2 M sucrose cushions, and ultracentrifugation is conducted. After repeating this process twice, exosomes can be found at the interface between two sucrose cushion buffer layers. The interface solution is collected and placed at the bottom of an iodixanol gradient buffer. An additional ultracentrifugation yields 10 remaining fractions, with exosomes in the third fraction from the top. The 10 fractions are then collected and can be subjected to downstream analyses (Fig. 1B).

2.4 Precipitation

Exosome precipitation is a faster and more efficient method for exosome isolation relative to the those described above. The objective of this method is to capture exosomes by incubating them with polymers, which allows exosomes

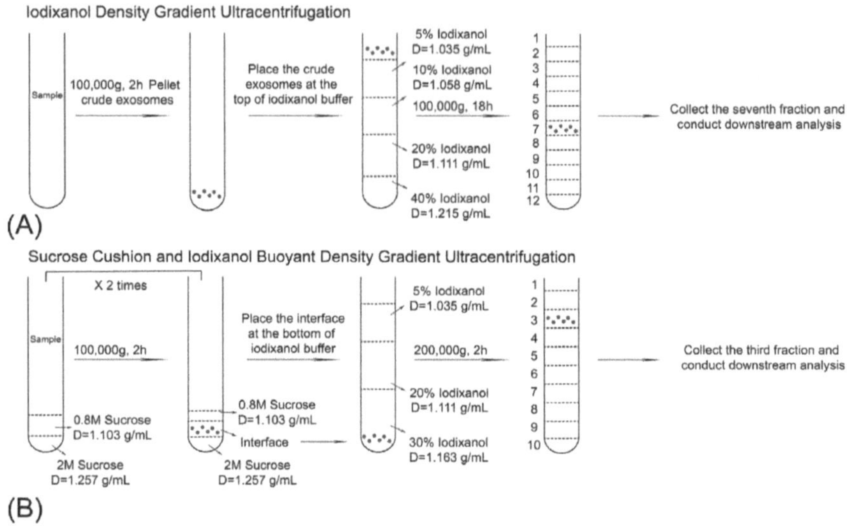

FIG. 1 Schematic illustration of the process of (A) iodixanol density gradient ultracentrifugation and (B) sucrose cushion and iodixanol buoyant gradient ultracentrifugation.

to be obtained at low speeds of centrifugation in combination with polymers. The most commonly used polymer for this purpose is polyethylene glycol (PEG). ExoQuick Exosome Precipitation Solution (System Biosciences, Palo Alto, CA) and Invitrogen Total Exosome Isolation Reagent (Thermo Fisher Scientific, Waltham, MA) are two popular commercial products for exosome precipitation. According to the ExoQuick user manual, experimental samples are pre-cleared to remove cells and cellular debris. The cleared solution is then incubated with an appropriate volume of ExoQuick for 0.5–12 h, depending on sample type. Lastly, exosomes are collected by centrifugation at 1500 ×*g* for 30 min.

2.5 Immunoaffinity capture-based isolation

Neither ultracentrifugation nor precipitation allows enrichment for exosome subtypes. To date, immunoaffinity capture remains the only method for isolation of exosome subgroups. By coating magnetic beads with specific antibodies, exosomes with those surface antigens are able to be captured. CD9, CD63, and CD81 are the most common markers and are expressed on nearly all exosomes [24]. Additional markers may be used to isolate exosomes derived from a specific cell type. For example, chondroitin sulfate peptidoglycan 4 antibody-coated beads have been used to capture melanoma cell-derived exosomes [25], while magnetic beads coated with CD56 or CD171 antibodies have been used to capture neuronal cell-derived exosomes [26, 27].

2.6 Summary of methods for exosome isolation

Selection of the proper exosome isolation method is usually dependent on the volume and origin of samples, as well as the goals of downstream analyses. The most relevant characteristics to consider in selecting the best exosome isolation method for a particular study are: (1) the exosome recovery rate or yield of the method, (2) the purity of the obtained exosome sample, and (3) the efficiency of the method in terms of time and labor. In the following section, we focus on these characteristics in the context of the methods described above.

2.6.1 Recovery rate

Several studies have shown that PEG-based precipitation has the highest recovery rate compared to other methods, at around 80–90% [28–30]. Differential centrifugation and ultrafiltration have a relatively low-to-medium yield. Multiple studies have shown that differential centrifugation has a recovery rate of only 20–40% due to loss of a large fraction of exosomes during the process [29, 31, 32]. Ultrafiltration has a higher recovery rate relative to differential centrifugation, at around 60% [31]. Both differential centrifugation and ultrafiltration are commonly used in experiments, and selection of these two methods tends to be based on individual laboratory habits. Density gradient ultracentrifugation has the lowest exosome recovery rate at around only 10% [28]. As PEG-based precipitation has the highest exosome recovery rate, this method is commonly used in experiments with a small starting sample volume, such as those involving clinical biofluid samples (e.g. plasma, urine, amniotic fluid, vitreous, and CSF). Differential centrifugation and ultrafiltration are usually used for isolation of exosomes from cell culture media, as these studies tend to involve a large starting sample volume.

2.6.2 Purity

Density gradient ultracentrifugation is currently considered the gold-standard method for achieving the highest-purity exosome samples, as they remove nonspecifically bound proteins from vesicles (Fig. 2) [28]. As such, density gradient ultracentrifugation is commonly used to isolate exosomes for exosomal proteomics and RNA sequencing studies [23]. Although differential centrifugation and ultrafiltration can obtain relatively high sample purities, exosome aggregates are a major artifact in these methods. As such, vigorous re-suspending of exosome pellets is recommended. PEG-based precipitation tends to yield the lowest-purity exosome samples. Fig. 3 shows exosomes isolated from human vitreous using ExoQuick under TEM. It is difficult to separate exosomes from associated polymers. As protein contaminants may skew experimental results, data interpretation in downstream studies should be performed with caution.

2.6.3 Time and labor cost

Of these methods, density gradient centrifugation requires the most time and labor, involving 2–3 days for completion of the process in its entirety. Although

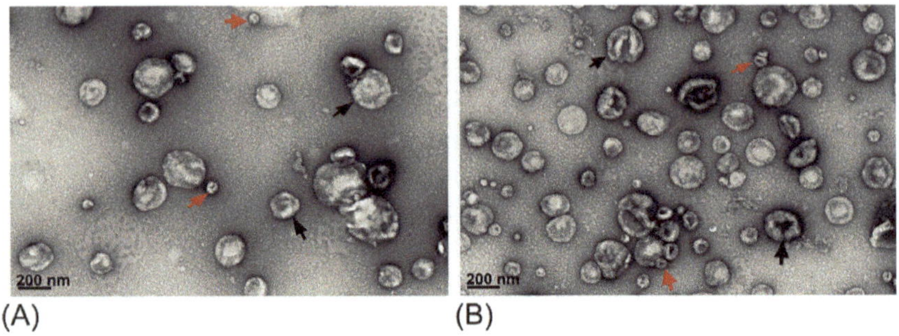

FIG. 2 TEM showed two subpopulations of exosomes, small vesicles (red arrows) and large vesicles (black arrows), isolated from (A) retinal pigment epithelial cells expressing wild-type fibulin-3 and (B) retinal pigment epithelial cells expressing R345W-fibulin-3, using sucrose cushion and iodixanol buoyant density gradient ultracentrifugation.

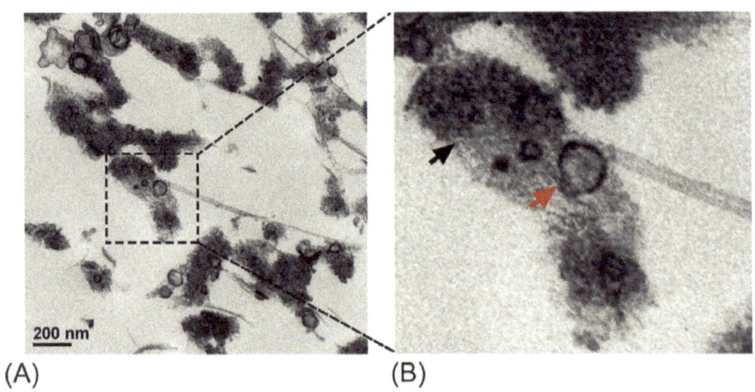

FIG. 3 (A) TEM image of exosomes isolated from human vitreous using ExoQuick. (B) Magnification of the indicated area shows that vesicles (red arrow) attach to the polymer (black arrow).

differential centrifugation requires more time than ultrafiltration, membrane filters and other ultrafiltration equipment cost more than the supplies required for differential centrifugation, provided basic laboratory equipment has already been obtained. PEG-based precipitation is the least costly of these methods in terms of both labor and time.

3 Methods for exosome characterization

3.1 Ultrastructural analysis

3.1.1 TEM

Due to the small sizes of exosomes and the ease of sample preparation, TEM is the gold-standard method to study exosome morphology. The resolution for TEM is about 1 nm, and the procedure of negative staining is simple and rapid, lasting only 2–3 h. Briefly, exosomes are fixed in 2% paraformaldehyde, deposited on Formvar carbon-coated TEM grids, and incubated for 20 min. The grids are then washed with PBS, incubated with glutaraldehyde, and washed with water. The vesicles are then stained with uranyl acetate solution and air dried [33]. Regular TEM can be used to: (1) validate the existence of exosomes in the solution, (2) assess the quality of exosomes, and (3) study the morphology of exosomes. TEM is rarely used for exosome quantification due to limited reproducibility and inefficiency. Immuno-EM has been used to investigate exosome surface markers and involves incubating exosomes with primary antibodies and secondary antibodies attached to 5–40 nm gold particles.

3.1.2 Scanning electron microscopy (SEM)

Rather than using a broad beam as in TEM, SEM uses a fine point beam that scans samples line by line. As such, rather than the two-dimensional image generated by TEM, SEM focuses on the surface of samples to provide a three-dimensional image of exosomes. Briefly, exosomes are fixed with glutaraldehyde and dehydrated with an ascending sequence of ethanol. After samples are air dried at room temperature, exosomes are ready for SEM analysis [34]. Sharma et al. reported that, unlike the cup-shaped morphology observed under TEM, SEM showed round, bulging exosomes without a central depression [35].

3.1.3 Cryogenic electron microscopy (Cryo-EM)

Cryo-EM is a type of TEM. In contrast to air drying samples as is done for TEM and SEM, cryo-EM allows samples to remain in their native aqueous environments. For cryo-EM analysis, suspended exosomes are placed on a grid, which is then rapidly immersed in liquid ethane to allow vitrification of the sample. After samples are vitrified, they can then be analyzed under cryo-EM or transferred in liquid nitrogen for storage [36]. Yuana et al. showed that, under cryo-EM, exosomes have a clear bilayer structure and are sometimes surrounded by smaller vesicles [37].

3.1.4 Atomic force microscopy (AFM)

AFM has a high resolution of approximately 1 nm and is suitable for exosome topography studies. Sharma et al. identified multiple CD63 receptor sites on exosomes using AFM immunogold imaging [38]. Briefly, the exosome suspension is placed on a mica substrate and allowed to air dry at room temperature. Samples are then washed with ultrapure water and dried with nitrogen gas. Samples are viewed using AFM with silicon probes and analyzed with AFM software (Fig. 4) [5].

3.2 NTA

NTA is a sophisticated method for the measurement of exosome concentration and size distribution. Compared to flow cytometry and TEM, NTA

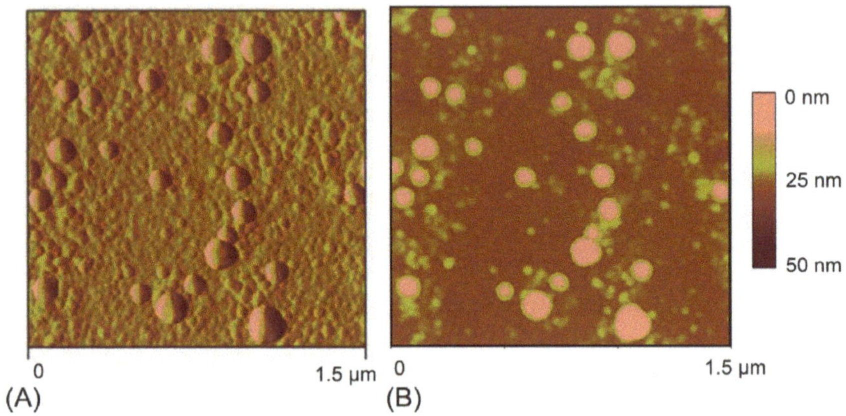

FIG. 4 (A) AFM image and (B) AFM height image of exosomes in human vitreous.

has better reproducibility. NTA has high resolution and is able to detect vesicles with diameters of 30–1000 nm. This technique takes advantage of dynamic light scattering and the Stokes-Einstein equation to quantify particle size and concentration. According to the NanoSight NS300 (Malvern Panalytical, Malvern, UK) user manual, under the scattered setting, each sample is loaded by syringe pump into the machine and five 60-s videos are generated. The entire process of sample measurement takes about 15 min. Fig. 5 shows NTA readouts of exosomes derived from ARPE-19 cells, a human retinal pigment epithelial cell line, isolated using filtration and ultracentrifugation.

3.3 AF4

AF4 is one of the few methods that is able to separate distinct subsets of extracellular vesicles. In this technique, exosomes are separated based on their density and hydrodynamic properties. Exosomes flow through a forward laminar channel and, based on their Brownian motion, are sorted into different populations. Smaller particles have higher diffusion rates and tend to move faster; in contrast, larger particles have lower diffusion rates and tend to move more slowly. Currently, there are two major companies that produce AF4 systems: Wyatt Technology and Postnova. Several studies have compared AF4 and NTA for exosome quantification. In contrast to NTA, which in prior studies resolved only a single, broad peak from 50 to 150 nm, AF4 was able to distinguish two separate exosome subpopulations, large exosome vesicles with sizes between 90 and 150 nm and small exosome vesicles with sizes around 30 nm [39–41]. Therefore, AF4 is better able to address the size heterogeneity of exosome samples and can be considered an advanced analytical technique for exosome subset characterization.

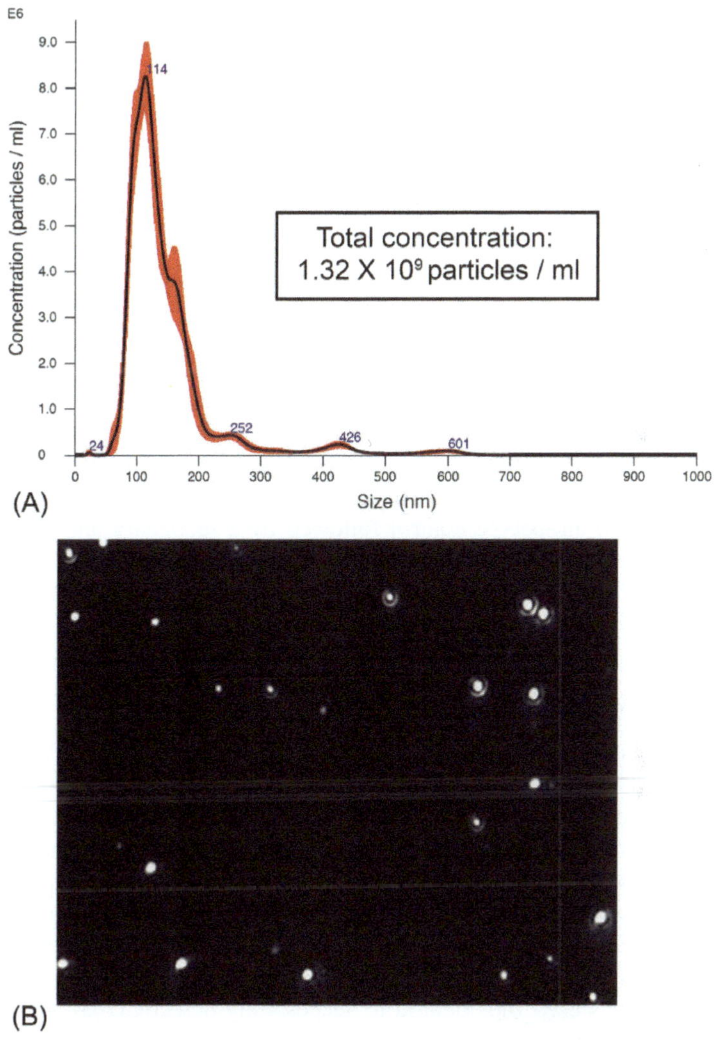

FIG. 5 Representative NTA distribution profiles of exosomes derived from ARPE-19 cells, including a size distribution plot (A) and a screen shot of the corresponding video (B).

3.4 RPS

RPS measures the size of vesicles based on their electrical resistance when they pass through a small orifice. RPS is able to detect vesicles with diameters of 50–1000 nm. Spectradyne, LLC is a leading company in the microfluidic measurement of nanoparticles. Compared to dynamic light scattering and NTA, RPS has both higher size resolution and better accuracy in measuring particle size distribution. Relative to NTA, exosome concentrations obtained by RPS are closer to the count obtained from TEM. Grabarek et al. reported that exosomes measured by NTA showed 5- to 10-fold higher concentration than RPS, as it failed to distinguish exosomes from protein aggregates, liposomes, and bacteria [42].

3.5 Flow cytometry

As exosomes fall below the resolution limit of standard flow cytometers, which is between 300 and 500 nm, exosomes cannot be detected directly by this method. Several studies have reported methods to semi-quantitatively measure exosome subgroups by detecting specific membrane markers on their surface [43, 44]. Briefly, exosomes are attached to aldehyde/sulfate-latex beads via a 15-min incubation with continuous rotation. The reaction is stopped by adding glycine and bovine serum albumin (BSA) solution. Exosome-bound beads are washed with PBS and blocked with BSA. Primary antibodies and fluorescence-labeled secondary antibodies are added in sequence for specific membrane marker detection. A negative control is obtained by incubating the exosome-bound beads with an isotype control followed by a secondary antibody, or in the absence of primary antibody. Using this method, Melo et al. found that exosomes derived from pancreatic cancer cells carry more glypican-1 compared to exosomes derived from non-tumorigenic cells [44].

3.6 ExoView characterization platform

Recently, antibody-based exosome arrays have been developed by NanoView Biosciences. This method enables fractionation of exosome subpopulations with a very small sample volume. Briefly, antibodies against exosome surface markers are arrayed on silicon chips. Exosome suspensions or biofluids containing exosomes are incubated with the chips overnight. After incubation, chips are washed with PBS on a shaker and air dried. Captured exosomes are detected using Single Particle Interferometric Reflectance Imaging Sensor technology. This technique allows enhanced contrast in the signal from particles. Daaboul et al. has used this method to successfully detect CD63-, CD8-, and CD9-positive exosomes derived from a human embryonic kidney cell line, and CD171-positive exosomes from human cerebrospinal fluid [45]. Compared to flow cytometry, this method requires a smaller sample volume and is less time-consuming.

3.7 In vivo exosome tracking

It is believed that exosomes mediate intercellular communication in part by transporting their microRNA cargo from cell to cell. However, the details of how exosomes carry out this function remain largely unknown. Two methods currently exist for analyzing the fate of exosomes after leaving their parental cells. These exosome labeling and tracking methods include: (1) staining exosomes with lipophilic carbocyanine dyes, including PKH67 (green) and PKH26 (red) (Millipore Sigma, USA); and (2) expressing exosomal markers, for example CD63, with green fluorescent protein (GFP) or mCherry in cells [46]. To study the mechanisms by which exosomes are endocytosed, confocal microscopy allows observation of the interaction between fluorescence-labeled exosomes and the plasma membrane, as well as the co-localization of exosomes and the recipient cell's organelles. To track the fate of circulating exosomes in vivo, fluorescence-labeled exosomes have been injected into mice via the tail vein and detected using the IVIS Spectrum system [47]. GFP-CD63 genetic rats have been generated by transfecting GFP-CD63 fusion protein into rat embryonic stem cells [48]. In these transgenic rats, GFP-CD63-labeled extracellular vesicles have been found in blood serum, breast milk, and amniotic fluid.

4 Exosome cargo

4.1 Exosomal proteins

4.1.1 Proteomics

Due to the heterogeneity of exosomes, exosomal proteins have been analyzed by mass spectrometry (MS) in several types of cells and biofluids in an effort to gain a better understanding of exosomes and to investigate whether they can be used as potential biomarkers for diagnosis [2, 3, 49]. According to ExoCarta, a platform that provides information about exosomal proteins, 9769 proteins have been identified in association with exosomes. Advanced gene ontology analysis has shown that exosomal proteins are involved in several biological functions, including subcellular localization, protein binding, molecular transport, and others.

There are two common methods that are been employed for sample preparation in MS analysis: in-gel digestion and in-solution digestion. Compared to in-solution digestion, in-gel digestion has the advantage that contaminants can be removed from samples during electrophoresis. For the in-gel digestion method, exosomal proteins can be separated by sodium dodecyl sulfate-polyacrylamide gel electrophoresis. Following electrophoresis, the gel lanes are excised into multiple, equal-sized segments and processed by in-gel digestion with trypsin. The digested peptides can then be analyzed by MS coupled to a high-performance liquid chromatography system. Bioinformatic approaches such as pathway analysis and gene ontology can then be applied for further analysis of the exosomal protein profile [50].

4.1.2 Exosomal proteins: apical versus basolateral

Proper function of epithelial cells is highly contingent upon their polarity. Multiple studies have compared the cargo of apically versus basolaterally secreted epithelial cell-derived exosomes. Klingeborn et al. found 299 unique proteins in exosomes released apically from retinal epithelial cells, and 94 unique proteins in basolaterally secreted exosomes [51]. In intestinal epithelial cells, apical exosomes were found to carry proteins involved in endosomal trafficking, whereas basolateral exosomes contained proteins that function in adhesion and stimulation [52]. These studies suggest that epithelial polarity impacts exosome cargo and that exosomes secreted from different directions conduct polarity-specific functions. Additional studies will be needed to investigate the underlying mechanisms and clarify the functional roles of these directionally secreted exosome subpopulations.

4.1.3 Exosome as biomarkers

Exosomal contents are frequently interrogated for identification of potential disease biomarkers. Prostate cancer antigen 3 and transmembrane protease serine 2 are two exosomal proteins used for prostate cancer diagnosis [53], while exosomal epidermal growth factor receptor VIII was shown to be expressed a higher level in patients with glioblastoma [54]. Exosomal glypican-1 was shown to be elevated in pancreatic cancer patients [44], and exosomal CD26 and CD10 were suggested to be potential markers of liver injury [55]. Proteomic analysis combined with validation via western blot or enzyme-linked immunosorbent assay is a widely used method for the discovery of exosomal biomarkers.

4.1.4 Exosomal proteins: Surface-bound versus encapsulated

As with transmembrane and cytosolic proteins, surface-bound and encapsulated exosomal proteins should have distinct biogenesis mechanisms and carry out different functions. Fitzgerald et al. developed a method to systematically analyze the positions of cytokines in exosomes [56]. Triton X-100 and sonication are able to lyse the exosome membrane. By comparing the protein concentration before and after Triton X-100 or sonication treatment, Fitzgerald et al. revealed the exact association between 33 cytokines and exosomes. Additionally, Skliar et al. found that exosome sizes become smaller after treatment with proteinase K or trypsin due to digestion of surface proteins [57]. In our lab, we found that low-concentration trypsin (0.1 g trypsin/1 g exosomal protein) only digests proteins located on the external surface of exosomes without perturbing membrane-spanning or encapsulated proteins. In contrast, high-concentration trypsin (0.3–1 g trypsin/1 g exosomal protein) can destabilize the exosome membrane, enabling access to the vesicle interior and digestion of proteins associated with both its external and internal aspects (unpublished data). This method provides an alternative approach for determination of the manner in which a given protein is associated with exosomes.

4.2 Exosomal microRNA (miRNA; miR)

Like exosomal proteins, exosomal miRNAs have been investigated in several pathological contexts, including cancer, inflammation, and age-related degenerations. ExRNA is an open access platform that provides novel information, protocols, recent publications in the field of exosomal RNAs. miR-21 has been found to be increased in tumor-derived exosomes in several cancers, including ovarian, lung, breast, and esophageal squamous cell cancers [58]. Exosomal miR-155 was shown to coordinate inflammatory responses [59]. Exosomal miR-9, miR-107, and miR-124 have been shown to play a role in neuronal differentiation, and their downregulation is associated with Alzheimer's disease pathogenesis [60].

Tang et al. compared different isolation methods for exosomal RNA and found that the Invitrogen Total Exosome RNA and Protein Isolation kit (Thermo Fisher Scientific) results in high extraction efficiency and purity compared to the SeraMir Exosome RNA Column Purification Kit (System Biosciences) and the Invitrogen TRIzol LS Reagent (Thermo Fisher Scientific) [29]. Next-generation sequencing techniques can generate global sequencing data that map all exosomal microRNA positions. miRNA arrays use hybridization to detect hundreds to thousands of miRNAs simultaneously. Real-time polymerase chain reaction and commercial miRNA and anti-miRNA antibody products are commonly used to validate results and mechanistic studies.

5 Summary

In summary, emerging evidence supports the significant contributions of exosomes and exosomal cargo to cell-cell communication. This review has summarized the principles, procedures, advantages, and disadvantages of commonly used methods for exosome isolation and characterization. Exosome research is a new and rapidly progressing field. New methodologies are continuously being developed and will facilitate the translation of exosome research to clinical applications.

References

[1] Thery C, Zitvogel L, Amigorena S. Exosomes: composition, biogenesis and function. Nat Rev Immunol 2002;2:569–79.

[2] Kalra H, et al. Comparative proteomics evaluation of plasma exosome isolation techniques and assessment of the stability of exosomes in normal human blood plasma. Proteomics 2013;13:3354–64.

[3] Gonzales PA, et al. Large-scale proteomics and phosphoproteomics of urinary exosomes. J Am Soc Nephrol 2009;20:363–79.

[4] Stuendl A, et al. Induction of alpha-synuclein aggregate formation by CSF exosomes from patients with Parkinson's disease and dementia with Lewy bodies. Brain 2016;139:481–94.

[5] Zhao Y, et al. Liquid biopsy of vitreous reveals an abundant vesicle population consistent with the size and morphology of exosomes. Transl Vis Sci Technol 2018;7:6.

[6] Runz S, et al. Malignant ascites-derived exosomes of ovarian carcinoma patients contain CD24 and EpCAM. Gynecol Oncol 2007;107:563–71.

[7] Torregrosa Paredes P, et al. Differences in exosome populations in human breast milk in relation to allergic sensitization and lifestyle. Allergy 2014;69:463–71.

[8] Colombo M, et al. Analysis of ESCRT functions in exosome biogenesis, composition and secretion highlights the heterogeneity of extracellular vesicles. J Cell Sci 2013;126:5553–65.

[9] Katarina Trajkovic CH, Chiantia S, Rajendran L, Wenzel D, Felix Wieland PS, Brügger B, Simons M. Ceramide triggers budding of exosome vesicles into multivesicular endosomes. Science 2008;319:1244–7.

[10] Ostrowski M, et al. Rab27a and Rab27b control different steps of the exosome secretion pathway. Nat Cell Biol 2010;12(Suppl. 11–13):19–30.

[11] Bobrie A, Colombo M, Krumeich S, Raposo G, Thery C. Diverse subpopulations of vesicles secreted by different intracellular mechanisms are present in exosome preparations obtained by differential ultracentrifugation. J Extracell Vesicles 2012;1.

[12] Mittelbrunn M, et al. Unidirectional transfer of microRNA-loaded exosomes from T cells to antigen-presenting cells. Nat Commun 2011;2:282.

[13] Azmi AS, Bao B, Sarkar FH. Exosomes in cancer development, metastasis, and drug resistance: a comprehensive review. Cancer Metastasis Rev 2013;32:623–42.

[14] Rajendran L, et al. Alzheimer's disease beta-amyloid peptides are released in association with exosomes. Proc Natl Acad Sci U S A 2006;103:11172–7.

[15] Alvarez-Erviti L, et al. Lysosomal dysfunction increases exosome-mediated alpha-synuclein release and transmission. Neurobiol Dis 2011;42:360–7.

[16] Lin J, et al. Exosomes: novel biomarkers for clinical diagnosis. Scientific World J 2015;2015:657086.

[17] Phinney DG, Pittenger MF. Concise review: MSC-derived exosomes for cell-free therapy. Stem Cells 2017;35:851–8.

[18] Kamerkar S, et al. Exosomes facilitate therapeutic targeting of oncogenic KRAS in pancreatic cancer. Nature 2017;546:498–503.

[19] Johnstone RM, Adam M, Hammond JR, Orr L, Turbide C. Vesicle formation during reticulocyte maturation. Association of plasma membrane activities with released vesicles (exosomes). J Biol Chem 1987;262:9412–20.

[20] Kowal J, et al. Proteomic comparison defines novel markers to characterize heterogeneous populations of extracellular vesicle subtypes. Proc Natl Acad Sci U S A 2016;113:E968–77.

[21] Van Veldhoven PP, Baumgart E, Mannaerts GP. Iodixanol (Optiprep), an improved density gradient medium for the iso-osmotic isolation of rat liver peroxisomes. Anal Biochem 1996;237:17–23.

[22] Greening DW, Xu R, Ji H, Tauro BJ, Simpson RJ. A protocol for exosome isolation and characterization: evaluation of ultracentrifugation, density-gradient separation, and immunoaffinity capture methods. Methods Mol Biol 2015;1295:179–209.

[23] Choi D, Gho YS. Isolation of extracellular vesicles for proteomic profiling. Proteomic profiling: methods and protocols. Methods Mol Biol 2016;1295:167–77.

[24] Clayton A, et al. Analysis of antigen presenting cell derived exosomes, based on immunomagnetic isolation and flow cytometry. J Immunol Methods 2001;247:163–74.

[25] Sharma P, et al. Immunoaffinity-based isolation of melanoma cell-derived exosomes from plasma of patients with melanoma. J Extracell Vesicles 2018;7:1435138.

[26] Lugini L, et al. Immune surveillance properties of human NK cell-derived exosomes. J Immunol 2012;189:2833–42.

[27] Mustapic M, et al. Plasma extracellular vesicles enriched for neuronal origin: a potential window into brain pathologic processes. Front Neurosci 2017;11(278).

[28] Van Deun J, et al. The impact of disparate isolation methods for extracellular vesicles on downstream RNA profiling. J Extracell Vesicles 2014;3.

[29] Tang YT, et al. Comparison of isolation methods of exosomes and exosomal RNA from cell culture medium and serum. Int J Mol Med 2017;40:834–44.

[30] Kim J, Shin H, Kim J, Kim J, Park J. Isolation of high-purity extracellular vesicles by extracting proteins using aqueous two-phase system. PLoS ONE 2015;10:e0129760.

[31] Lobb RJ, et al. Optimized exosome isolation protocol for cell culture supernatant and human plasma. J Extracell Vesicles 2015;4:27031.

[32] Yuana Y, et al. Handling and storage of human body fluids for analysis of extracellular vesicles. J Extracell Vesicles 2015;4:29260.

[33] Jung MK, Mun JY. Sample preparation and imaging of exosomes by transmission electron microscopy. J Vis Exp 2018.

[34] Sokolova V, et al. Characterisation of exosomes derived from human cells by nanoparticle tracking analysis and scanning electron microscopy. Colloids Surf B Biointerfaces 2011;87:146–50.

[35] Sharma S, et al. Structural-mechanical characterization of nanoparticle exosomes in human saliva, using correlative AFM, FESEM, and force spectroscopy. ACS Nano 2010;4:1921–6.

[36] Tatischeff I, Larquet E, Falcon-Perez JM, Turpin PY, Kruglik SG. Fast characterisation of cell-derived extracellular vesicles by nanoparticles tracking analysis, cryo-electron microscopy, and Raman tweezers microspectroscopy. J Extracell Vesicles 2012;1:.

[37] Yuana Y, et al. Cryo-electron microscopy of extracellular vesicles in fresh plasma. J Extracell Vesicles 2013;2.

[38] Sharma S, Rasool HI, Palanisamy V, Mathisen C, Schmidt M, Wong DT, Gimzewski JK. Structural-mechanical characterization of nanoparticle exosomes in human saliva, using correlative AFM, FESEM, and force spectroscopy. ACSNANO 2010;4:1921–6.

[39] Zhang H, Freitas D, Kim HS. Identification of distinct nanoparticles and subsets of extracellular vesicles by asymmetric flow field-flow fractionation. Nat Cell Biol 2018;20:332–43.

[40] Petersen KE, et al. A review of exosome separation techniques and characterization of B16-F10 mouse melanoma exosomes with AF4-UV-MALS-DLS-TEM. Anal Bioanal Chem 2014;406:7855–66.

[41] Sitar S, et al. Size characterization and quantification of exosomes by asymmetrical-flow field-flow fractionation. Anal Chem 2015;87:9225–33.

[42] Grabarek AD, Weinbuch D, Jiskoot W, Hawe A. Critical evaluation of microfluidic resistive pulse sensing for quantification and sizing of nanometer- and micrometer-sized particles in biopharmaceutical products. J Pharm Sci 2019;108:563–73.

[43] Suarez H, et al. A bead-assisted flow cytometry method for the semi-quantitative analysis of extracellular vesicles. Sci Rep 2017;7:11271.

[44] Melo SA, et al. Glypican-1 identifies cancer exosomes and detects early pancreatic cancer. Nature 2015;523:177–82.

[45] Daaboul GG, et al. Digital detection of exosomes by interferometric imaging. Sci Rep 2016;6:37246.

[46] Takahashi Y, et al. Visualization and in vivo tracking of the exosomes of murine melanoma B16-BL6 cells in mice after intravenous injection. J Biotechnol 2013;165:77–84.

[47] Suetsugu A, et al. Imaging exosome transfer from breast cancer cells to stroma at metastatic sites in orthotopic nude-mouse models. Adv Drug Deliv Rev 2013;65:383–90.

[48] Yoshimura A, et al. Generation of a novel transgenic rat model for tracing extracellular vesicles in body fluids. Sci Rep 2016;6:31172.

[49] Epple LM, et al. Medulloblastoma exosome proteomics yield functional roles for extracellular vesicles. PLoS ONE 2012;7:e42064.

[50] Schey KL, Luther JM, Rose KL. Proteomics characterization of exosome cargo. Methods 2015;87:75–82.

[51] Klingeborn M, et al. Directional exosome proteomes reflect polarity-specific functions in retinal pigmented epithelium monolayers. Sci Rep 2017;7:4901.
[52] van Niel G, Heyman M. The epithelial cell cytoskeleton and intracellular trafficking. II. Intestinal epithelial cell exosomes: perspectives on their structure and function. Am J Physiol Gastrointest Liver Physiol 2002;283:G251–5.
[53] Nilsson J, et al. Prostate cancer-derived urine exosomes: a novel approach to biomarkers for prostate cancer. Br J Cancer 2009;100:1603–7.
[54] Skog J, et al. Glioblastoma microvesicles transport RNA and proteins that promote tumour growth and provide diagnostic biomarkers. Nat Cell Biol 2008;10:1470–6.
[55] Momen-Heravi GSF. Extracellular vesicles in liver disease and potential as biomarkers and therapeutic targets. Nat Rev Gastroenterol Hepatol 2017;14:455–66.
[56] Fitzgerald W, et al. A system of cytokines encapsulated in extracellular vesicles. Sci Rep 2018;8:8973.
[57] Skliar M, et al. Membrane proteins significantly restrict exosome mobility. Biochem Biophys Res Commun 2018;501:1055–9.
[58] Taylor DD, Gercel-Taylor C. MicroRNA signatures of tumor-derived exosomes as diagnostic biomarkers of ovarian cancer. Gynecol Oncol 2008;110:13–21.
[59] Bala S, et al. Circulating microRNAs in exosomes indicate hepatocyte injury and inflammation in alcoholic, drug-induced, and inflammatory liver diseases. Hepatology 2012;56:1946–57.
[60] Van Giau V, An SS. Emergence of exosomal miRNAs as a diagnostic biomarker for Alzheimer's disease. J Neurol Sci 2016;360:141–52.

Chapter 3

Exosomes, microvesicles, and their friends in solid tumors

Denis Corbeil[a], Aurelio Lorico[b,c]
[a]*Biotechnology Center and Center for Molecular and Cellular Bioengineering, Technische Universität Dresden, Dresden, Germany,* [b]*College of Medicine, Touro University Nevada, Henderson, NV, United States,* [c]*Mediterranean Institute of Oncology, Viagrande, Italy*

1 Introduction

Membrane-enclosed extracellular vesicles (EVs) are released into the extracellular milieu by virtually all cell types. Initial assessment of their function was to expunge "cellular dust" outside the cells, allowing their homeostasis. This scavenger function did not draw much attention and consequently, has never been conclusively demonstrated. By contrast, the Geuze's group elegantly demonstrated in 1996 that EVs discharged by lymphoblastoid cells induced antigen-specific, Major Histocompatibility Complex (MHC) class II-restricted T cell responses, suggesting for the first time the involvement of EVs in intercellular communication [1]. A decade later, EVs were shown to envelope different types of functionally active ribonucleic acid (RNA) molecules, acting as vehicles in cell-cell communication. Two independent groups found messenger RNA (mRNA) and microRNA (miRNA) in EVs and excitingly demonstrated their functionality upon uptake by target or recipient cells [2, 3]. Given the accumulating evidence that EV-associated molecules mediate phenotypic alterations after internalization, EVs are now considered as mediators of intercellular communication in multicellular organisms acting at a limited range or long distance during their development and across the lifespan under physiological and pathological conditions [3–6], as described in other Chapters of this book. Furthermore, the presence of EV-derived cargo proteins and nucleic acids in the nucleus of host cell suggest that a fraction of them are shuttled to nuclear compartment and regulate gene expression, hence transforming the biochemistry of target cells notably in cancer [7–11].

EVs are generally classified based on their biogenesis as exosomes when derived from multivesicular bodies (MVB); ectosomes or microvesicles when directly budding from plasma membrane; apoptotic bodies when released

Exosomes. https://doi.org/10.1016/B978-0-12-816053-4.00003-1

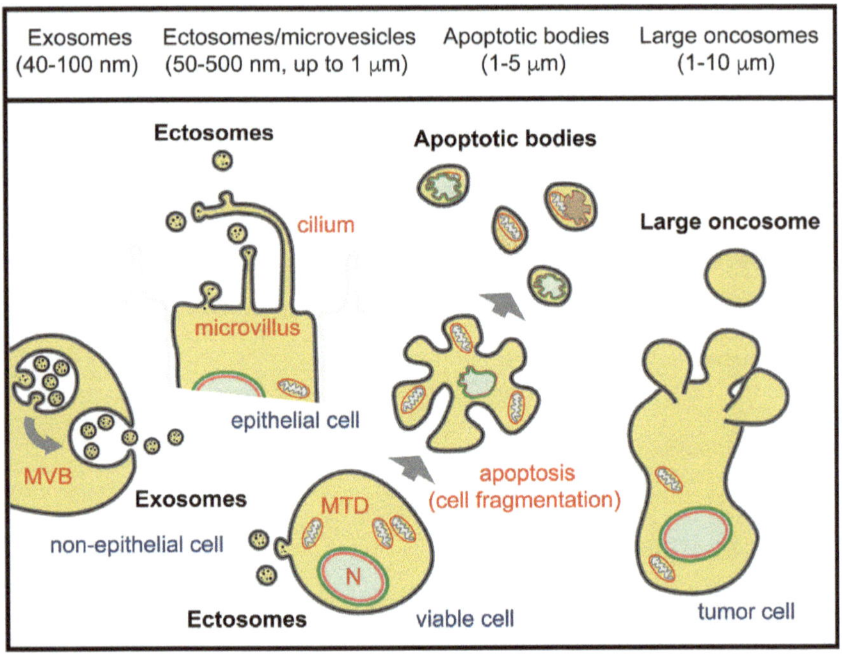

FIG. 1 Biogenesis and heterogeneity of EVs. EVs represent a heterogeneous population of soluble particles that are categorized according to their size (top panel) and cellular origin (bottom panel). The relative diameter of a given type of EVs is indicated in brackets. Various types of EVs can be found together in an extracellular milieu, as a single cell can simultaneously release them as exosomes upon the fusion of multivesicular bodies (MVB) with plasma membrane and ectosomes/microvesicles that bud directly from plasma membrane and its protrusions such as microvilli and cilium, as observed in epithelial and non-epithelial cells [16, 17]. Within a given EV type, various entities with different profiles can also co-exist. The complexity of EV population is even increased in vivo, where numerous cell types can contribute to their presence in a given biological fluid. Although EVs carry several types of cellular information, in contrast to cells (or even viruses) these entities cannot self-renew. The release of EVs can be stimulated by cellular differentiation and transformation. Voluminous EVs known as apoptotic bodies and large oncosomes, which are released by apoptotic cells and tumor cells, respectively, complete a large spectrum of EVs found in the extracellular milieu. N, nucleus; MTD, mitochondria.

upon cell fragmentation during apoptotic cell death; and large oncosomes when shed from non-apoptotic membrane blebs of tumor cells with an amoeboid phenotype [12–15]. The heterogeneity of EVs is reflected by their size that varies considerably, which somehow summarizes their mode of formation (Fig. 3.1). Within a given category, individual EVs might differ from each other, and their composition mirrors the donor cell and their physiological status. Particular constituents (soluble or membranous) can be enriched in EVs [12]. Given the lack of specific markers for different EV categories [14], we will refer to them collectively as EVs unless their biogenesis has been indicated by the authors.

In general, cancer cells release a higher number of EVs than normal proliferating cells, and EV levels are usually elevated in plasma from patients with cancer [18, 19]. The role of EVs as a vehicle-like structure in cell-to-cell communication appears as an important multifaceted regulator of tumor progression. For instance, cancer cell-derived EVs can modulate the immune responses at different levels, transfer oncogenic proteins and nucleic acids, reprogram stromal cells, promote neo-angiogenesis, and transfer the drug-resistance phenotype. Despite abundant reports highlighting these EV-mediated events in cancer, there is little mechanistic insight into the intracellular path(s) of EVs upon their internalization, the mechanism(s) of cargo release, and their molecular target(s) located in cytoplasm and nuclear compartment of host cells [20].

In the present chapter, we will discuss the function(s) and mechanism(s) of action of cancer cell-derived EVs, their interplay with normal cells in the tumor microenvironment and EVs derived thereof, and how the growing body of knowledge on cancer-related EVs can be exploited to improve cancer diagnosis, monitor disease progression and therapeutic response, and importantly to develop successful anticancer strategies.

2 Cellular uptake of (cancer) cell-derived EVs

Numerous mechanisms were proposed to explain the transfer of bioactive molecules carried by EVs to the target cells, and those enveloped by cancer cell-derived EVs can follow similar pathways to physiological EVs released by healthy cells [21]. Although the transfer mechanisms are not mutually exclusive, some could predominate over others depending on cell type and physiological conditions of cells. At first, EVs can fuse directly with the plasma membrane of target cells and release their soluble signaling and/or regulatory molecules (proteins, lipids and nucleic acids) into the cytoplasm, which in turn could trigger a cascade of cellular responses (Fig. 3.2A, i). The EV-cell fusion can be promoted by an acidic pH of extracellular milieu as found in the tumor microenvironment [23]. Under these hard conditions, variations in the membrane fluidity of both EVs and host plasma membrane, as well as unmasking fusogenic proteins could explain the EV-cell fusion. The initial interaction of EVs with target cells can be done in a specific manner via receptor and/or adhesion proteins (for an exhaustive list see Ref. [21]), although the molecular bases for this selective cellular targeting remain poorly described [24]. Likewise, the detailed information about the EV-cell fusion machinery requires more investigation. In contrast to the sophisticated biogenesis mechanism and cargo selection of EVs [25] (see also other chapters), this mode of delivery results in a random cytoplasmic distribution of bioactive molecules, notably at the cellular periphery, which might not be sufficient to engage a desired biological effect. This is particularly true for EV-derived miRNAs that need to find specific RNA transcript targets. Alternatively, EVs can bind to host cells and promote cellular responses without membrane fusion—a mechanism similar to a ligand-receptor interaction [26].

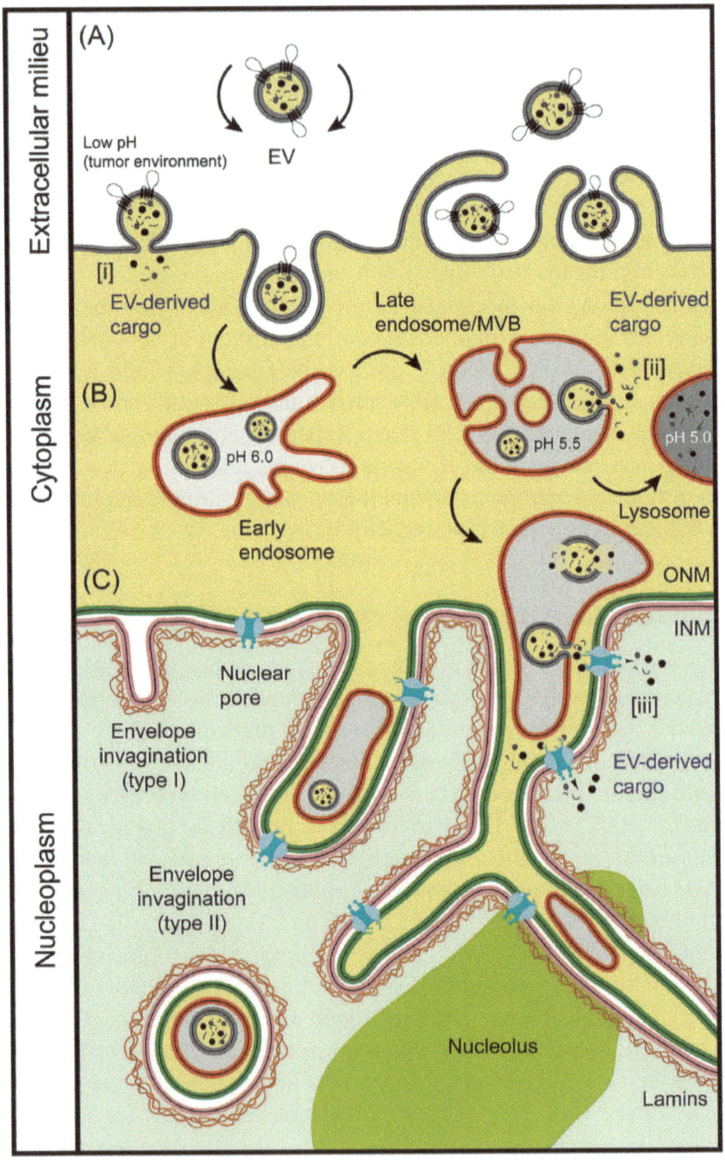

FIG. 2 See the legend in facing page

Soluble ligands directly derived from EV membrane proteins upon a proteolytic cleavage could bind to host cell receptors as well, and consequently promote downstream signaling responses [27].

The transfer of bioactive molecules could occur by internalization of EVs themselves, and various mechanisms of endocytosis were described [28]. For instance, the endocytosis can occur by a clathrin-dependent or cholesterol-rich membrane microdomain (lipid raft)-dependent mechanisms [29] (Fig. 3.2A). Chemical drugs or treatments interfering with biogenesis and dynamics of clathrin-coated pits, for example, potassium depletion and cytosol acidification, can block the EV entry [30]. Similarly, compounds that perturb membrane fluidity and/or membrane microdomain organization, such as cholesterol-sequestering agent methyl-β-cyclodextrin [31], can interfere with the uptake of EVs as we demonstrated for melanoma-derived EVs [11]. The EV endocytosis might also proceed via caveolae in which cholesterol-binding caveolin protein plays a major role [32]. Caveolae are rich in cholesterol and sphingolipids and membrane cholesterol is essential for their formation [33]. The use of distinct machineries regulating the endocytosis of EVs can be cell type dependent and possibly co-exist in a given cell [34]. Lastly, macropinocytosis and phagocytosis are two other mechanisms that could lead to the internalization of EVs [35, 36] (Fig. 3.2A).

FIG. 2 Entry, intracellular routes of EVs, and delivery of their cargo content. (A) Various mechanisms were described to explain the delivery of EV cargo molecules (proteins, nucleic acids) in target cells. The cellular entry of EVs can occur by various processes, notably cell-EV fusion, which can be stimulated by the low pH observed in the extracellular milieu of tumor microenvironment, endocytosis, micropinocytosis, and phagocytosis. These mechanisms are not mutually exclusive. The direct fusion of an EV with plasma membrane will allow the immediate delivery of its cargo into the cytoplasm of targeted cell (i) which might provide a signal transduction cascade. The initial EV-cell binding can be mediated by specific receptor and/or adhesion proteins, while fusogenic proteins may account for the fusion ability. These processes might be similar to the entry of certain viruses. The endocytosis can be mediated by clathrin-dependent, cholesterol-rich lipid rafts mechanisms and/or caveolae (not shown). The use of these distinct machineries can be cell type dependent and possibly co-exist. (B) Once internalized, EVs (and their content) are delivered to the endocytic pathway, notably early endosomes and late endosomes "en route" to lysosomes where they will be subjected to degradation. The low pH in late endosomes (i.e., about 5.5) would favor the fusion of EV with endosomal membrane, leading to the release of its content into the cytoplasm (ii). (C) The subcellular position of late endosomes might pre-dispose the targeted delivery of EV cargo. A fraction of the late endosome population can also translocate into the nuclear compartment by means of nuclear envelope invaginations of type II, which are created by invaginations of both outer (ONM) and inner (INM) nuclear membranes. These structures can be superficial or reach deep regions of the nucleus. They often come in close contact with nucleoli, organelles responsible for ribosome biogenesis, cell cycle control, and cell signaling [11, 22]. The type I nuclear envelope invaginations are generated only by INM and hence do not allow the cellular organelles to penetrate therein. After the fusion of EV with the membrane of late endosomes associated with nuclear invaginations (iii), the EV content will be transported from cytoplasm to the nucleoplasm. The nuclear pores are involved. It remains to be determined how EV-associated membrane proteins (e.g., multispan protein CD9, CD133) can be extracted from the endosomal membrane and translocated to the nucleoplasm, as the luminal content of EVs does.

Upon internalization, EVs and their content enter the endocytic pathway that leads to their intracellular journey in host cells. Therein, they will be delivered to early endosomes, the first endocytic compartment to receive incoming EV cargo from plasma membrane [37] (Fig. 3.2B). These dynamic structures that often appear with thin tubular elongations act as a sorting platform from which cargo proteins can traffic back to plasma membrane or Golgi apparatus via various routes or to lysosomes for degradation via late endosomes (reviewed in Ref. [38]). The latter have a lower pH by comparison to the early endosomes and appear more rounded with numerous internal vesicles, hence explaining why they are also referred to as multivesicular endosomes or MVB [39, 40] (Fig. 3.2B). Proton-pumping vacuolar-ATPase, and perhaps the fission of tubular endosomes, might explain the gradual intraluminal acidification of the endosomal compartment [41, 42]. Early and late endosomes differ to each other not only by their morphology and pH, but also in protein and lipid composition [43]. The internalized signaling molecules and their receptors at the plasma membrane are sequentially sorted to early and late endosomes prior to their degradation into lysosomes, which contain all necessary enzymes to complete such function under acidic condition [44]. Similar fate might occur for endocytosed EVs and their cargo (Fig. 3.2B). As a role in intercellular communication, the lysosomal degradation of EV cargo would not provide any cellular response. Alternatively, internalized EVs can fuse with late endosomal membrane and release their content into the cytoplasm (Fig. 3.2B, ii). EV-late endosome fusion could be promoted by the acidic microenvironment in late endosomes compared to early endosomes. In contrast to the fusion of EVs with plasma membrane, the subcellular localization of late endosomes might pre-dispose the delivery of EV content to their molecular targets. This is particularly true for late endosomes located at the perinuclear region [45, 46] (Fig. 3.2C). The interaction of late endosomes with different organelles, such as endoplasmic reticulum (ER) and mitochondria may also target their cargo to the site of action [47–49]. The recent publication showing that late endosomes by their association with translational machinery being the site of local translation of RNA transcripts in axonal outgrowth may bring a new spatiotemporal dimension in EV life and their fate after internalization [50]. Whether mRNA cargo carried by EVs can be translated immediately upon EV fusion with late endosomal membrane remains to be explored.

2.1 Positioning role of early and late endosomes in nuclear transfer of extracellular information

By following the intracellular pathway of internalized EVs, we recently observed that a fraction of EV-loaded late endosomes (as defined by the presence of small GTPase Rab7) reached deep regions of the nuclear compartment by invading nuclear envelope invaginations [11] (Fig. 3.2C). Two types of nuclear invagination were described: the type I are created by invagination of inner nuclear membrane (INM) into the nucleoplasm, whereas the type II contains both outer nuclear membrane (ONM) and INM (Fig. 3.2C). The latter can host various cytoplasmic organelles [51]. The presence of late endosomes in type II nuclear envelope invagination provides a new route to target EV cargo to the nuclear compartment. They have been observed in vitro with numerous cell lines, including melanoma cells and stromal cells and in vivo with breast cancer samples [11]. After the fusion of endocytosed EV with the membrane of late endosomes, the EV content is in an appropriate position within nuclear invaginations to be transported to the nucleoplasm through the nuclear pores where they can access the genome and possibly modulate the gene expression [11] (Fig. 3.2C, iii). The vicinity of nuclear envelope invaginations with the nucleolus [11, 51], a subnuclear structure, suggests also that the EV cargo might interfere with its activities, such as ribosome biogenesis and nucleolar protein sequestration [52]. It might be more than a coincidence that nucleolus is often hijacked by tumors to promote the growth of transformed cells [53] and hence to be the target for cancer intervention [54]. This novel path could explain, at least in part, the nuclear localization of multispan membrane proteins associated with EVs, such as CD9 and CD133 (see below), and possibly epidermal growth factor receptor (EGFR), notably the oncogenic form of the latter [10, 55–60]. Although it is not demonstrated yet, we cannot exclude that this pathway is involved in the shuttling of plasma membrane proteins, including receptors, to nuclear compartment [61]. In both cases, it remains to be determined how membrane proteins can be extracted from endosomal membrane and pass-through the nuclear pores. Besides the late endosomal compartment, the docking and fusion of early endosomes (as defined by the presence of the early endosome antigen 1) with peripheral nuclear envelope might also participate in the delivery of extracellular biomaterials to nucleus of host cells [62]. In the latter, nuclear envelope proteins SUN1 and SUN2, as well as the Sec61 translocon complex, were described to participate in the nuclear transfer of cell surface proteins. Altogether, nuclear envelope-associated (early) endosomes (NAE) [62] and nuclear envelope invagination-associated late endosomes (N-ALE) [11] are in strategic positions to rely extracellular information to host cell nuclei. Because the appearance of N-ALE has some similarity upon double Rab7-SUN2 immunofluorescence labeling with a sword inside its scabbard we named this structure "spathasome" from the Greek/Latin words "spathi/spatha" for sword [11].

These two newly described pathways of intracellular trafficking deserve further attention as they can provide potential targets to interfere with the EV-mediated intercellular communication between cancer cells and healthy cells notably those associated with cancer stem cell niche.

2.2 VOR protein complex

The interaction between membrane-bound organelles has emerged as a new theme in molecular cell biology of eukaryotic cells, and their crosstalk through membrane contacts plays an underestimated role in membrane dynamics and cell signaling (reviewed in Refs [63, 64]). The interaction of late endosomes with ER is well documented [48], and its knowledge was useful to determine part of the mechanism that regulates the entry of late endosomes into nucleoplasmic reticulum and their docking to ONM [65]. The ER-localized vesicle-associated membrane protein (VAMP)-associated protein A (VAP-A) was found determinant to mediate the late endosome–nuclear membrane interaction. Together with R-Ras-interacting oxysterol-binding protein (OSBP)-related protein 3 (ORP3), which regulates cell adhesion and is overexpressed in numerous cancers, VAP-A sets the interaction with late endosome-associated Rab7 [65]. The VAP-A-ORP3 complex was previously shown to mediate the membrane contact of ER to plasma membrane [66, 67]. The tripartite complex formed by the interactions of VAP-A/ORP3/Rab7 (named VOR complex) is essential to transfer EV cargo to nucleoplasm. Silencing VAP-A or ORP3 impeded the entry of late endosomes in nucleoplasmic reticulum [65] and consequently the presence of nuclear EV cargo [65]. Given that nuclear translocation of late endosomes is dependent of microtubules unidentified motor proteins should be involved in this process [65]. Altogether, the VOR complex plays an important role in relying information from the EVs to nuclear compartment of target cells and represents a potential target to inhibit the intercellular communication in cancer. The implication of VOR complex during the development of a multicellular organism and its homeostasis where EV-mediated intercellular communication is involved remains to be investigated.

3 EVs and cancer microenvironment

The pleiotropic effects of EVs is well illustrated in carcinogenesis where they promoted significant alterations in normal intercellular communication between healthy cells, leading to the tumor growth and metastases. This is particularly true in a given tissue or organ which contains numerous cell types that release EVs and uptake those originating from cancer cells. EVs are now considered as a major player of the tumor progression, notably in the establishment of metastasis microenvironment. Thus, they merit a special consideration not only from a cell biological point of view, but also clinically as therapeutic targets.

3.1 Role of EVs in metastasis

Metastasis is the leading cause of cancer-related death. It is considered the result of a complex cellular process initiated by detachment of cancer cells, followed by their dissemination via lymph and blood circulation and engraftment at sites distant from the primary tumor [68–70]. Tumor cells exchange bioactive molecules, such as growth factors, integrins, and coding and non-coding RNAs, with nearby as well as distant mesenchymal cells, endothelial cells, and immune cells, establishing favorable conditions for cancerous growth and the formation of metastases [71]. These molecular exchanges modulate signaling pathways, dampen the immune response to the unregulated cell growth, recruit naïve neighboring cells, and favor the local and distant spreading of cancer cells [72]. Although the main actors and their respective roles in the process leading to the establishment of distant tumor metastases are still debated, the implication for EVs as modulators of the microenvironment that orchestrate events critical to cancer growth and formation of distant metastasis, including the regulation of angiogenesis and immune response, has been substantiated by several studies in recent years. By mediating an indirect interaction between tumor, stroma, and immune cells, EVs appear to have a dominant role in preparing the tumor niche at metastatic sites. Such transformation would have a negative impact on native cellular constituents found therein (Fig. 3.3).

How can cancer cell-derived EVs colonize a specific organ site and predispose a favorable cancer microenvironment at future metastatic sites? In line with data showing that malignant transformation can be stimulated by adhesion and extracellular matrix proteins such as integrin and tenascin molecules [73, 74], integrins carried by EVs can play a significant role in this context [75]. Indeed, EV-associated integrins can promote cancer progression by determining the

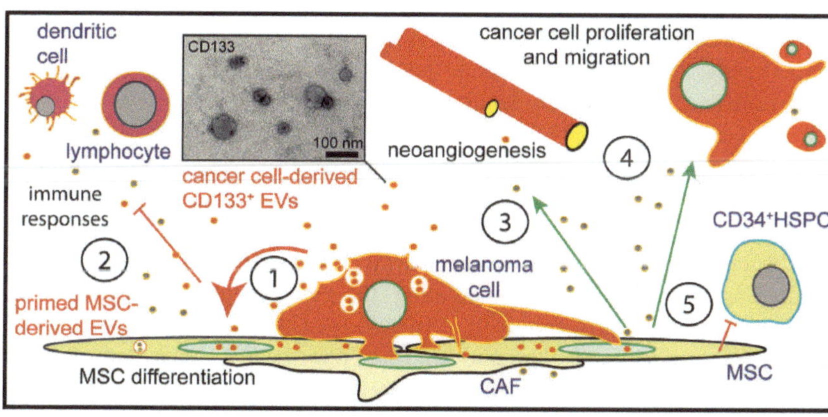

FIG. 3 Impact of EVs in intercellular communication under cancer conditions. In melanoma as in other types of cancer, the malignant cells might use EVs, notably those expressing CD133 to transform the bone marrow hematopoietic stem cell niche into a cancerous one. Upon endocytosis by MSCs and other stromal cell types such fibroblasts, cancer cell-derived EVs can modify their biochemistry, leading to their transformation into cancer-associated fibroblasts (CAF) (#1). The melanoma-derived EVs, EV-primed stromal cells and/or the EVs derived therefrom can also negatively modulate the immune responses by acting on dendritic cells and lymphocytes among others (#2), stimulate the formation of blood vessels (#3), and boost cancer cell population by stimulating cell migration and proliferation (#4). All these EV-mediated modifications would reduce the supportive capacity of stromal cells for resident CD34^{+} hematopoietic stem and progenitor cells (HSPC), while increasing that for cancerous cells (#5). An example of CD133 immunogold-labeling electron microscopy of human melanoma-derived EVs is displayed. *(Micrograph is taken from our previous publication: Rappa G, et al. Wnt interaction and extracellular release of prominin-1/CD133 in human malignant melanoma cells. Exp Cell Res 2013;319:810–9).*

target tissues to form new tumor niches during metastatic spread and the direct transfer of bioactive molecules to target cells [75]. A repertoire of integrins has been described to guide the EVs to specific organs. For instance, Hoshino and colleagues have demonstrated that $\alpha_6\beta_1$ and $\alpha_6\beta_4$ integrins mediated the selective adhesion of EVs to the extracellular matrix at their predicted lung metastatic destination, followed by their uptake by resident fibroblasts and epithelial cells [76]. Likewise, a specific binding of EVs containing integrin αvβ5 to Kupffer cells was reported to mediate liver metastasis [77]. Languino's group described the transfer of αvβ6 and αvβ3 integrins among prostate cancer cells via EVs [78, 79]. The αvβ6, an integrin absent in normal prostate, is required for monocyte M2 polarization in prostate cancer, while αvβ3 integrin from tumorigenic to non-tumorigenic prostate cells promoted their migratory ability [80, 81]. Altogether, particular adhesion proteins carried by circulating EVs can regulate their specific spreading and predispose a given tissue and/or organ to metastases. Clinically, a prospective analysis of the EV-associated integrin signature can thus make it possible to predict their metastatic propensity.

Upon their arrival at the predetermined distant site, specific EVs and their cargo promote the formation of the pre-metastatic niche. Therein, they can accomplish a process of "education" of normal cells toward a pro-metastatic phenotype, as illustrated in bone marrow stem cell niche where hematopoietic stem and progenitor cells (HSPCs) are hosted (Fig. 3.3). As major targets, the resident cells such as fibroblasts and mesenchymal stromal cells (MSCs) are susceptible to transformation, leading to cancer-associated fibroblasts (CAFs) [82, 83]. Innate and adaptive immune inflammatory cells and endothelial cells are also contributing to tumor microenvironment. Moreover, cancer cells themselves can exchange EVs, contributing to cancer cell heterogeneity.

Melanoma cell-derived EVs are good example of particles leading to the transformation of bone marrow niche toward a pro-metastatic phenotype, inducing vascular leakiness by mediating the transfer of oncogene c-MET and reprogramming progenitors toward a pro-vasculogenic phenotype [84]. Although a replication of that study did not find the pro-metastatic effect of c-MET-expressing EVs to be statistically significant [85], the former investigation should nonetheless be kept in mind (see below). Subsequently, it was found that injection of EVs from cells expressing high levels of c-MET increased the metastatic burden of low c-MET-expressing cells [86]. In two independent reports, cancer cell-derived EVs were found to promote metastases to the liver by regulating the hepatic microenvironment and initiating the formation of a pre-metastatic niche [77, 87]. Several other investigations have established that tumor-derived EVs administered to mice prior to tumor cell injection increased the metastatic burden by inducing pre-metastatic niche formation [88–91].

The implication of the stemness marker, pentaspan membrane glycoprotein CD133 (also known as prominin-1) in the cancer stem cell niche has been extensively documented, which includes promoting the invasive properties of cancer cells and inducing vasculogenic mimicry formation [92–96]. As a biomarker of EVs such as ectosomes and exosomes [16, 97] (reviewed in Ref. [98]), two studies reported the release of CD133$^+$ EVs from cancer cells, that is, colon cancer and melanoma [16, 99] (Fig. 3.3, inset). A pro-metastatic role of CD133$^+$ EVs in FEMX-I melanoma was proposed [99], consistent with an earlier study showing that short interfering (sh)RNA-mediated downregulation of CD133 resulted in loss of metastatic potential of melanoma [94]. The exposure of MSCs to CD133$^+$ melanoma-derived EVs increased their invasiveness suggesting that EV cargo modified the biological properties of target cells and hence their phenotype [100]. As an organizer of plasma membrane protrusions (i.e., microvilli, filopodia, primary cilium), CD133 itself can contribute to these morphological transformations [60, 101]. In line with these observations, Alessandro's group elegantly demonstrated that EVs released by colon SW620 metastatic cells modulated the surface activity, motility, and invasiveness of non-metastatic SW480 cells by means of Rac GTPase-activating protein 1 (RacGAP1)-mediated pathway [102]. RacGAP1 is a Rho GTPase-activating protein, which is involved in controlling cellular phenomena such as cytokinesis, transformation, cell migration and metastasis. It has the ability to suppress Rac1 activity and activate RhoA at the cell front edge promoting the invasive migration [103]. The SW480 and SW620 cell lines represent two distinct stages of tumor development and can be considered representative of diverse sub-clones that constitute within a tumor of the same patient. CD133 is strongly expressed in SW620 but not in SW480 cells [104, 105]. Similarly, it was shown using intravital imaging that less aggressive breast tumor cells exhibit increased migratory abilities upon internalization of EVs originating from highly aggressive cancer cells, suggesting that cancer cell-derived EVs can act as messengers of malignancy [4] (Fig. 3.3). The transfer of aggressive phenotype can be related to oncogenic forms of EGFR and Met 72 tumor antigen [106, 107].

Through the release of cytokines and extracellular matrix (ECM) components, CAFs are important to establish pre-metastatic niches [108, 109] (Fig. 3.3). For an exhaustive list of molecular and cellular factors involved in pre-metastatic niche formation we invite the readers to access an excellent review [110]. As mentioned above, they derived from tissue resident fibroblasts and MSCs often in response to platelet-derived growth factor (PDGF) and transforming growth factor beta (TGF-β) signaling [111, 112]. The advantage of CAFs vs. normal resident cells in promoting tumor growth and metastasis was demonstrated by several groups [113, 114]. Interestingly, cancer cell-derived EVs from distinct tumor types have been reported to promote differentiation of normal fibroblasts into CAFs by transferring TGF-β, miR-125b, or miR-155 [115–117]. In particular, EV-mediated transfer of miR-125b from breast cancer cells to normal fibroblasts within the tumor microenvironment contributed to their differentiation, acting at least in part through its target such as the tumor suppressor p53 protein (p53; transformation-related protein 53). Similarly, amphiregulin-containing EVs derived from non-small-cell lung cancer activate osteoclastogenesis in MSCs through EGFR-mediated pathway in vitro and in vivo [118]. Melanoma cell-derived EVs has shown to modify the gene expression profile of MSCs, notably genes encoding for players involved in inflammation processes. The EV-mediated transformation of MSCs was demonstrated to be dependent on spathasome pathway described above [11].

The role of EV-associated non-coding RNAs in the context of metastases and stem cell niche transformation was also explored. While under non-neoplastic conditions where endogenous RNAs are shielded by RNA-binding proteins [119], cancer cells reportedly induced CAFs within the tumor microenvironment to produce EVs containing unshielded endogenous RNAs (e.g., RN7SL1), which in turn facilitated cancer growth, metastasis, and therapy resistance [120]. Several effects of cancer cell-derived EVs and their associated miRNAs on the cancer stem cell niche and tumor progression were reported by Wong's laboratory: (1) miR-105-induced metabolic plasticity in CAFs to convert cancer-produced metabolic waste (e.g., lactate and ammonia) into energy-rich metabolites to re-enter cancer bioenergetics [121]; (2) suppressed glucose utilization by lung fibroblasts and astrocytes in favor of glucose uptake by metastasized cancer cells mediated by cancer-secreted miR-122 [88]; and (3) vascular leakiness and enhanced metastasis caused by cancer-secreted miR-105, which downregulates endothelial tight junctions [91].

Recent experimental evidences suggest that in some cases, the metastatic process might not be determined by the migration of cancer cells, but by transfer of malignant factors contained in cancer cell-derived EVs to target non-cancerous cells. Thus, cancer cell-derived EVs taken up by oncosuppressor-mutated cells (OMC) of epithelial and mesenchymal origin, that is, cells with mutated BRCA1 or with a phosphatase and tensin homolog (PTEN) deletion caused malignant transformation, resulted in a phenotype compatible with that of the cancer cells from which the EVs had originated [122–124]. When exposed to the serum of cancer patients, cancer cell-conditioned media or cancer EVs, OMCs displayed characteristics of transformed cells both in vitro (anchorage-independent growth assay) and in vivo (tumor formation in immunodeficient mice) [122, 124]. It appears that a single oncosuppressor mutation is a pre-requisite and represents the first hit to allow malignant transformation following exposure to cancer factors contained in EVs. These findings, as well as the description of an in vivo model of cancer progression mediated by horizontal gene transfer [125] and the observation of oncogenic transformation mediated by cell-free nucleic acids circulating in the plasma of patients with colorectal carcinoma [126], support the existence of a non-classical pathway based on the exchange of cancer traits between malignant and non-malignant cells that may have implications in cancer progression and metastasis.

Although all cancer cells release small EVs, only certain types release atypically large EVs known as large oncosomes. They were first identified in vitro in cancer cells transitioning from a mesenchymal mode of migration to an amoeboid type as a consequence of Akt activation and Diaphanous related formin 3 (DIAPH3) loss [127]. The latter protein is involved in actin remodeling and regulates adhesion and cell movement. Large oncosomes were successively identified in tumor tissues and in plasma of patients with metastatic prostate cancer [128]. Technically, large oncosomes could be separated from other small EVs by differential ultracentrifugation followed by a density gradient. This approach enabled the identification of a set of distinct proteins [128]. The internalization of large oncosomes into fibroblasts resulted in functional reprogramming mediated by activation of transcription factor MYC that promoted growth of cancer cells, reactive stroma and tubular morphogenesis [129]. Large oncosomes can also mediate intercellular transfer of functional miRNA and enhance migration of CAFs to secondary sites for pre-metastatic niche [130]. Further studies are needed to dissect all facets of large oncosomes in transforming cancer stem cell niche.

3.2 Cancer cell-derived EVs and inhibition of the anti-cancer immune response

As for most biological systems, the interplay between EVs and the immune system is complex. EVs produced by non-immune and immune cells can mediate both immune suppression and stimulation [131]. In terms of immune suppression, EVs can inhibit (i) the differentiation of dendritic cells from myeloid precursors [132]—dendritic cells are antigen presenting cells with the capacity to induce primary and secondary immune responses; (ii) the cytotoxic activity of natural killer (NK) cells [133]; and (iii) proliferation and cytotoxic activity of T-cells [132]. In particular, EVs expressing MHC class I and II along with FasL released from intestinal epithelial cells acquire antigens after feeding and prevent commonly encountered foreign antigens from causing chronic inflammation and autoimmunity [134]. In terms of immune stimulation, EVs can carry (i) antigens to dendritic cells for processing and subsequent presentation to T-lymphocytes; (ii) MHC class II molecules to stimulate T-clones; (iii) signals that promote activation of the acceptor cell into antigen-presenting cell; and (iv) regulate the formation of the immunological synapse between T-lymphocytes and antigen-presenting cells [132].

It is now evident that tumor-derived EVs are key players in the crosstalk between malignant and transformed cells and the immune system. They are also part of the process employed by cancer cells to inhibit immune surveillance, that is, specific identification and elimination of neoplastic cells based on the expression of tumor-specific antigens, in order to grow and metastasize [135] (Fig. 3.3). In the tumor microenvironment however, immune cells from both the adaptive and innate systems are present, with even more complex effects on tumor growth and metastasis. In fact, chronic inflammatory states can favor cancer initiation, promotion, neo-angiogenesis, and metastasis [136], while at the same time, the immune system is also responsible for cancer immune surveillance [137, 138]. Thus, EVs can mediate the release of inflammatory mediators through a tumor necrosis factor (TNF)-alpha-mediated pathway, potentially promoting a chronic inflammatory state [139]. In most cases, cancer cells employ their self-made EVs to hijack anti-tumor immune response, for example, inducing defects in antigen presentation, increasing the immune-suppression activity of myeloid cells, enhancing the function of regulatory T cells, suppressing NK cells, and inhibiting tissue macrophages [140]. Several studies have found that melanoma, ovarian, and colon cancer cell-derived EVs caused apoptosis in Fas-ligand^{+} lymphocytes by the expression of Fas ligand on their own surface [141–143]. Employing Cre-lox recombination-based tracing of EV-mediated RNA transfer from tumor to host cells in vivo, Ridder and colleagues showed changes in immunosuppressive phenotype and miRNA profiles after EV uptake and demonstrated in tumor models of carcinoma and glioma that myeloid-derived suppressor cells (MDSC) represent a major cell population targeted by tumor released EVs [144]. These findings suggest that tumor EVs could be responsible for the expansion and activation of MDSC immunosuppressive functions in cancer patients.

3.3 EVs, angiogenesis, endothelial cells, and platelets

To grow and metastasize, tumors often generate their own vasculature. Through delivery of proteins and RNAs to endothelial cells, their progenitors and supporting cells, cancer-derived EVs contribute to tumor neo-angiogenesis, that is, the generation of new vessels inside and around the tumor mass (Fig. 3.3). Cancer-derived EVs stimulate also the pro-inflammatory and the pro-coagulant activities of endothelial cells and platelets, which favor different steps of the metastatic process, including tumor extravasation, the formation of microthrombi of tumor cells, extracellular matrix remodeling as well as platelet aggregation [145]. Besides cancer cell-derived EVs, those released by non-cancerous cells in the tumor microenvironment contribute also to the angiogenetic process and ultimately to the metastatic process. Most circulating EVs in plasma originate from platelets and endothelial cells [146], and increased platelet-derived EVs are associated with the severity of gastric cancer [147]. To gain more insight about platelets and EVs in cancer, please consult an excellent review [148]. Endothelial cell-derived EVs can be pro- or anti-angiogenic, depending on the stimuli causing their production.

Several studies have established that EV secretion is not constitutive, but modulated by alteration in the cellular environment. For instance, changes in the microenvironmental acidic pH and exposure of cells to hypoxia can alter the EV secretion and their uptake [23, 149–151]. This is particularly relevant for cancer, where the tumor microenvironment and metastatic niche harbor these harsh conditions. In particular, under hypoxia, lung cancer cells induced an increased level of EV-associated miR-23a, which in turn promoted tumor angiogenesis and increased vascular leakiness through suppression of its targets prolyl hydroxylase 1 and 2 and inhibition of tight junction protein ZO-1 of endothelial cells [152]. Among the pro-angiogenetic proteins delivered by cancer cell-derived EVs, the EV-associated Tspan8 tetraspanin protein was reported to trigger endothelial branching and elevate levels of factors driving endothelial cell proliferation, migration, sprouting and progenitor maturation, including von Willebrand factor, vascular endothelial growth factor (VEGF), and VEGF receptor 2 (VEGF-R2) [153, 154]. Mutated EGFR delivered to endothelial cells via lung cancer cell-derived EVs can drive the secretion of autocrine VEGF and elevated VEGF-R2 through mitogen-activated protein kinase (MAPK) and Akt, and enhance endothelial cell responses to VEGF [106]. The pro-angiogenetic effects of cancer cell-derived EV-associated coding and non-coding RNAs were also demonstrated in several studies. The enrichment of transcripts related to angiogenesis was shown in glioblastoma EVs as well as their subsequent translation in target cells [155]. Similarly, it was found that cell cycle-related mRNAs associated with colon carcinoma EVs enhanced the proliferation of endothelial cells as well as tubule formation in tri-dimensional (3D)-culture system [156]. In regard to miRNA, the transmission of miR-17-92 cluster from leukemic cell derived EVs to endothelial cells resulted in downregulation of integrin alpha5 and consequently promote endothelial cell migration and tube formation [157]. The EV-associated miR-135b secreted by hypoxic multiple myeloma cells can enhance angiogenesis by targeting hypoxia-inducible factor (HIF)-1, a dimeric protein complex that plays a role in the body's response to low oxygen concentration [158]. In renal carcinoma, cancer cell-derived EVs carried both mRNAs and miRNAs with pro-angiogenic activity that stimulated the formation of new vessel and increased metastatic potential [159]. Secretion of specific EV-associated miRNAs regulated by neutral sphingomyelinase 2 promoted angiogenesis and subsequently the formation of metastasis [160]. Finally, long non-coding RNAs that stimulate a pro-angiogenetic activity was reported [161]. For instance, EVs released by $CD90^+$ stem cell-like hepatocarcinoma cells contain H19 long non-coding RNA molecules, which induced a pro-angiogenetic phenotype upon their uptake in endothelial cells [161]. For more details and information about cancer cell-derived EVs and their components in angiogenesis, as well as the underlying molecular mechanisms please see a recent article review by Song and colleagues [162].

3.4 EVs and cancer drug resistance

The resistance of malignant tumors to currently available chemotherapeutic agents, biologicals, and radiotherapy limits the efficacy of cancer therapy. Besides promoting tumor growth and the formation of metastases, EVs play an important role in fostering tumors resistant to different types of therapy. They do so by transferring from resistant to sensitive cells drug exporters and mRNAs or non-coding RNAs, including miRNAs, which increase their expression. Alternatively, they can sequester anti-cancer monoclonal antibodies in the peripheral circulation [163], or they can expulse small molecules via their shedding from cancer cells as a novel drug efflux mechanism [164].

EV-mediated delivery of P-glycoprotein contributed to docetaxel resistance of MCF-7 breast cancer cells [165], while the transfer of ABCA3 transporters caused resistance to humoral immunotherapy in lymphoma [166, 167]. In another study, EVs released from doxorubicin-resistant MCF-7 cells were shown to transfer a Ca(2+)-permeable protein channel TrpC5 to microvessel-associated human endothelial cells, resulting in the induction of P-glycoprotein expression through the activation of the transcription factor NFATc3 (nuclear factor of activated T cells isoform c3) [168].

EVs can also mediate resistance to immunotherapy. Human breast cancer cells overexpressing epidermal growth factor receptor 2 (HER2) released EVs containing it, which mediated the resistance to the anti-cancer activity of the HER2 inhibitor trastuzumab and contributed to cancer malignancy [169]. The release of these EVs was modulated by two HER2 receptor-activating ligands (i.e., EGF and heregulin) that are normally present in the surrounding cancer cell microenvironment. Trastuzumab is a monoclonal antibody currently used to treat breast cancer. Thus, EVs derived from the $HER2^+$ SKBR3 and BT474 cell lines sequestered the anti-HER2 monoclonal antibody, reducing its bioavailability [169]. Of note, a significantly higher number of advanced-stage vs. early-stage breast cancer patients had $HER2^+$ EVs that bind to trastuzumab, which might explain the mechanism of resistance to anti-HER2 therapy in metastatic setting [170].

EVs were also found to carry out and eliminate cisplatin (also called diamminedichloroplatinum(II)) from melanoma and lung cancer cells exposed to the chemotherapeutic drug [171, 172]. Cisplatin is a metallic coordination compound with a square planar geometry, and its action is linked to the ability to crosslink with the purine bases on the deoxyribonucleic acid (DNA) and hence interferes with DNA repair mechanisms, causing DNA damage and finally inducing apoptosis in cancer cells [173]. Interestingly, the acidic tumor microenvironment resulted in enhanced export of EV-associated cisplatin [171], consistent with the observation that microenvironmental pH is a key factor for exosome trafficking in tumor cells [23]. Acquisition of resistance to cisplatin was also induced both in vitro and in vivo by EV-carried miR-155 through telomeric repeat-binding factor (TERF)1 targeting and subsequent increased in telomerase activity [174]. In human osteosarcoma, miR-221 induced resistance to cisplatin through phosphoinositide 3-kinase/Akt signaling pathway [175]. The resistance to other drugs, such as tamoxifen, doxorubicin and docetaxel was reportedly transferred to sensitive MCF-7 breast cancer cells by EV-associated miR-221/222 [176, 177]. Yet, the mechanism behind this miRNA-induced resistance was not elucidated.

Exosome exchanges from stromal cells to cancer cells can also regulate therapy resistance pathways. Transfer of non-coding RNAs and transposable elements mediated by stroma-derived EVs to breast cancer stem cells was shown to stimulate retinoic acid-inducible gene 1 protein (RIG1)-like receptor, which in turn activates transcription factor STAT1-dependent antiviral signaling, and together with NOTCH3 pathway, induce resistance to both chemotherapeutic drugs and radiation therapy [178]. Thus, stromal cells orchestrate via EVs a crosstalk with cancer cells, leading the latter to resist to therapy and promote their growth. As well, EVs released by irradiated head and neck cancer cell lines (i.e., BHY and FaDu cells) induced radiation resistance and, at the same time, increased the proliferation of non-irradiated target cells [179]. For more recent information about EVs and drug resistance, we invite the readers to consult an excellent and comprehensive review [180].

4 Environmental clues and carcinogenicity of anticancer drugs—Impact on EVs

A large body of evidence indicates that the release of EVs can be included among the genotoxic or mutagenic changes induced by environmental toxins and carcinogens. The exposure to environmental *noxae* triggers the discharge of EVs containing DNA, coding and non-coding RNA molecules, integrins, cytokines, or chemokines that modulate the cellular microenvironment and potentiate carcinogenesis, as demonstrated in certain cancers such as liver, lung, ovarian and tracheal carcinomas [181].

4.1 Air pollutants

Air pollutants such as inorganic arsenic, radon, and asbestos are well known carcinogens just like tobacco smoke [182]. In smokers, bronchoalveolar macrophages released EVs that induced the production of proinflammatory cytokines by lung epithelial cells [183]. Due to the established link between chronic inflammation and cancer [184], this observation suggests that EVs may contribute to the carcinogenicity of tobacco smoke. EV-containing medium from arsenite-transformed human L-02 hepatic epithelial cells transferred miR-155 to normal liver cells, which in turn downregulated the tumor suppressor gene QKI [185]. QK1 is an RNA-binding protein, which belongs in the signaling transduction and activation of RNA (STAR) family of proteins. This resulted in activation of NF-κB and increased production of interleukin (IL)-6 and IL-8, thus supporting oncogenesis [185]. The latter effect was caused by EVs, given their depletion from conditioned medium was devoid of oncogenic properties. Another report has shown that arsenic can disrupt stem cell dynamics during the carcinogenic process [186]. Thus, arsenite-transformed prostate epithelial cells recruit prostate stem cells into rapidly acquiring a cancer stem cell phenotype via EV-mediated signaling. Astonishingly, arsenic-transformed prostate epithelial cells secreted 700% more EVs than parental cells, and they are enriched with oncogenic factors, including inflammation-related transcripts, and oncogenesis-associated miRNAs [186].

4.2 Carcinogenic potential of anticancer drugs and treatments

Although cytotoxic chemotherapy has an anti-neoplastic and anti-metastatic activity in several types of cancer, many agents have carcinogenetic potential and in some cases, may have a pro-metastatic effect [187–189]. De Palma's laboratory reported that treatment of breast cancer cells with two classes of cytotoxic drugs largely used in pre-operative breast cancer therapy, taxanes and anthracyclines, elicited the production and release of pro-metastatic EVs both in vitro and in vivo with two different murine models [190]. Mechanistically, chemotherapy-elicited EVs are enriched in annexin A6, a Ca^{2+}-dependent protein that promotes NF-κB-dependent endothelial cell activation, C-C motif chemokine ligand 2 (CCL2) induction and $Ly6C^{+}CCR2^{+}$ monocyte expansion in the pulmonary pre-metastatic niche to facilitate the establishment of lung metastasis [190]. Lead sulfide quantum dots, which are fluorescent bio-imagers used for theranostic applications [191], can generate reactive oxygen species leading to oxidative stress and cell death. They were found to induce the release of EVs containing markers of DNA damage as well as factors promoting inflammation, including p53, IL-8 and C-X-C motif (CXC) chemokine 5, which are potent activators of neutrophil chemotaxis, and thus inflammation and carcinogenesis at the site of quantum dot accumulation [192].

Given it is now documented that the secretion and/or composition of EVs may be affected by the exposure to air pollutants and clinically used carcinogens, EVs can allow to measure the carcinogenic effects at the individual level and on a large scale among population. For instance, drawing blood from a cohort of people and measuring changes in EVs (and the associated cargo markers) can allow to track the development of cancer in relation to the exposure from a potentially carcinogenic substance or condition [181].

5 Clinical applications of EVs

The clinical interest of circulating EVs is growing exponentially, and appropriate and effective methods of isolating these particles, and eventually their quantification, are becoming the "Holy Grail" in various fields, particularly in oncology [193, 194]. As diagnostic and predictive biomarkers, it is essential to identify robust EV markers; proteins, and nucleic acids [195], to document the current status of patients in a non-invasive diagnostic context. Of course, the sensitivity of detection methods and the tools available to detect them are at the heart of these investigations. Proteomic profiling of EVs by mass spectrometry and/or their transcriptome by deep sequencing may become the norm in the near future [196]. In addition to the utility of EVs and their content as biomarkers, these biological nano-entities can be designed to interfere with the actions of EVs derived from cancer cells and/or as vectors for drug delivery, among others. The limit of their use seems to be restricted to our knowledge of them and their involvement in the different cellular systems. In the following subsections, these issues will be addressed.

5.1 Liquid biopsy: Cancer cell-derived EV-associated proteins and nucleic acids as biomarkers

The discovery of EVs has revolutionized the relevance of liquid biopsy for cancer patients, providing specific and highly stable biomarkers easily recoverable from a given biological fluid. This is particularly true given that EV cargo reflects the pathological conditions of donor cell source. The main advantage of liquid vs. tissue biopsy is that the former, being minimally invasive, is universally feasible and allows repeated sampling—a characteristic particularly important for early diagnosis and therapeutic monitoring. Identification of easily accessible materials like proteins and nucleic acids in circulating cancer EVs, coupled with their capacity to protect their cargo, produce unprecedented advantages for early detection, diagnosis, prognosis, disease monitoring and treatment of patients with different types of cancer [19, 197]. Circulating EVs are more than 10^{10} per mL of blood in non-neoplastic patients while this amount can double and even more in cancer patients [19, 198, 199]. This may be due to the fact that conditions known to stimulate EV production and release, such as genotoxic, hypoxic, and metabolic stress, are generally present during neoplastic growth [200, 201]. Although plasma/serum EVs include not only cancer-derived EVs, but also those released by inflammatory cells and other normal cells in the tumor microenvironment, like blood cells, endothelial cells, and stromal cells, whose behavior changes because of the presence of cancer growth, the contribution of cancer cells to the total EV population is highly significant [164]. A recent investigation has demonstrated that human glioblastoma cells orthotopically implanted in immune-deficient mice contributed to 35–50% of all circulating EVs [19].

5.1.1 Proteins

In 2008, the constitutively active form of the EGFR (EGFRvIII), which contains an in-frame deletion of 267 residues in the extracellular ligand-binding domain, was found in EVs released from glioblastoma cells [106], indicating that EVs may carry oncogenic proteins extracellularly that may be of diagnostic and/or prognostic utility in the context of liquid biopsy.

Several cancer-associated proteins with a potential value as (cancer) biomarkers have been identified in circulating EVs from cancer patients. For instance, an increase in $CD147^{+}$ EVs in colorectal cancer patients compared to healthy controls was observed [202], as well as the findings reported by three independent studies of the glypican-1 or of a panel of 29 EV-associated proteins differentially regulated in patients with pancreatic cancer vs. healthy volunteers [203–205]. In the latter study, 18 exosomal proteins (e.g., tetraspanin protein CD9, a disintegrin and metalloproteinase domain-containing protein 10 (ADAM10, CD156c), integrin beta (CD29)) were upregulated while 11 others were downregulated in patient sera. Monitoring their expression might find application in the evaluation of cancer progression. Kimura and colleagues found that the cytoskeleton-associated protein 4, a receptor for Dickkopf-related protein 1, in EVs from serum of patients with pancreatic cancer represents a biomarker as well as a potential therapeutic target [206]. Active signaling molecules relevant to prostate carcinoma, such as proto-oncogene c-Src, insulin-like growth factor I receptor, G-protein-coupled receptor kinases, and focal adhesion kinase are also enriched in EVs, suggesting that signaling network may provide useful biomarkers detectable by liquid biopsy [207]. The combination of EV-associated alpha-2-HS-glycoprotein (fetuin-A), extracellular matrix protein 1, and carcinoembryonic antigen were described to have a high diagnostic potential for non-small-cell lung cancer [208].

As mentioned above, CD133 might deserve a particular attention as a biomarker of EVs found in extracellular fluids. Indeed, ectosomes and exosomes harboring CD133 were reported in human urine, tear, seminal fluid, cerebrospinal fluid [16, 209]. Recently, $CD133^+$ EVs were also found in pancreatic cancer ascites [210, 211]. Given the expression of CD133 by cancer cells, including cancer stem cells, the upregulation of $CD133^+$ EVs in a given body fluid might be related to cancer growth, and potentially cancer cell differentiation. Indeed, $CD133^+$ EV release was correlated to stem cell differentiation as demonstrated with stem cells from neural and hematopoietic systems and colon cancer cells [16, 97]. In pilot experimetal screenings, the authors could observe that $CD133^+$ EVs were upregulated in cerebrospinal fluids of patients with gliomablastoma, normal pressure hydrocephalus, parkinsonism, relapsing-remitting and secondary-progressive multiple sclerois, or other neuronal diseases such as partial epilepsy [209, 212, 213]. In glioblastoma patients, the variation of $CD133^+$ EVs level was linked to disease progression, suggesting its utility for studying human diseases [209]. Larger cohort of patients with defined clinical and neuroradiological parameters are needed to validate CD133 as a clinically recognized neuronal biomarker of EVs. An exhaustive characterization (omic approaches) of the $CD133^+$ EVs isolated from each patient group might also provide additional markers (proteins, lipids, nucleic acids) and be instructive about disease origin and progression [100]. Again, we should keep in mind that $CD133^+$ EVs could represent a mixture of distinct EVs such as ectosomes budding from plasma membrane or exosomes released after the fusion of MVB with the plasmalemma.

Finally, numerous other proteins associated with EVs and/or specific post-translational modifications, such as phosphorylation and glycosylation were reported to be upregulated/modified in physiological fluids, or conditioned media, contacting cancer cells [214]. For instance, a differential expression pattern of ascites-derived exosomal CD133 was proposed as potential prognostic biomarker in patient with advanced pancreatic cancer [210]. The large volume of information concerning cancer cell-derived EV biomarkers (>2500 entries in PubMed (15 June 2019) using exosome/protein/cancer as keywords) and the mercantile excitation in developing kits that allow the purification of EVs are explained by the clinical relevance in the discovery of EV-associated protein biomarkers from liquid biopsies.

5.1.2 Nucleic acids

Due to the susceptibility of RNAs to degradation by ubiquitous RNAses and their protection conferred by the lipid bilayer membrane of EVs, several investigators focused their attention on EV-associated RNAs as cancer biomarkers. EV-associated RNAs are generally free of endogenous RNA contaminants such as ribosomal RNA [215]. Technically, they remain stable for years when EVs are properly conserved at minus 20 °C [216], which facilitates their storage when a large clinical screen is initiated. As biomarkers, variation of levels of specific miRNAs compared to the producer cells were reported in different types of tumors. For instance, a specific accumulation of 20 miRNAs with cancer-related function was found in melanoma-derived EVs by comparison to the miRNAome of the producer cells [100]. In melanoma patients, a low level of plasma EV-associated miR-125b was found to correlate with disease progression [217]. For ovarian cancer, eight miRNA signatures (i.e., miR-21, miR-141, miR-200a, miR-200b, miR-200c, miR-203, miR-205 and miR-214) were described and proposed for cancer screening in the general population [218]. Analogously, potential circulating EV-associated miRNA signatures were reported for patients with pancreatic ductal adenocarcinoma, non-small-cell lung cancer, colon carcinoma, glioblastoma, esophageal cancer, prostate cancer, meningioma, hepatocellular carcinoma and other types of cancer [219–230]. The report that EVs released from breast cancer cells contained the complete machinery for miRNA biogenesis, including Dicer, TRBP and AGO2 indicates that certain types of EVs may be able to transform pre-miRNAs into mature miRNAs [231]. Interestingly, EVs derived from HIV-1-infected cells were also found to contain Dicer and other components of the host miRNA machinery [232].

Although presently most studies on EV-associated RNA molecules as cancer biomarkers have focused on miRNAs, most RNAs contained in EVs belong to other species of non-coding RNA, such as signal recognition particles (SRP) RNA, P-element-induced wimpy testis (PIWI)-interacting RNA (piRNA), small nuclear RNA (snRNA), small nucleolar RNA (snoRNA), and others [120, 233, 234]. A systematic study of other species of non-coding RNAs in cancer cell-derived EVs may provide additional markers and signatures that would be clinically useful.

Lastly, EV-associated RNA transcripts as tumor biomarkers may also have potential value for diagnosis and therapy management of patients, including those with metastatic breast cancer [235]. For instance, in plasma-associated EVs of glioblastoma patients, the mRNA levels of O(6)-methylguanine DNA methyltransferase and alkylpurine-DNA-N-glycosylase, two enzymes that induce resistance to temozolomide, correlated with levels found in producer cells. Upon treatment with temozolomide, a drug employed in glioblastoma therapy, their transcript levels changed considerably and could be associated to treatment response [236].

5.2 EV-based anti-cancer strategies

The field of EV is young and much remains to be learned about their heterogeneity, function, specificity, intracellular pathway and regulation. Although the rationale use of EVs in cancer therapy may need to wait for an increase of our knowledge about them and their effective characterization, several studies suggest that various therapeutic approaches, although apparently contradictory, may indeed work in the clinics [237]. Due to their heterogeneity, it is conceivable that anti-cancer/anti-metastatic effects might be obtained by administration of certain EV subclasses, EVs derived from specific cells or by ablation of specific cargo molecules. In other instances, selective inhibition of EV production and/or of their uptake by target cells may be beneficial to patients with cancer. Interference with the intracellular route of endocytosed EVs in target cells within the tumor microenvironment or in the pre-metastatic niche may have therapeutic potential. With this regard, the recent discovery of novel nuclear transport pathways of extracellular cargo involving the early or late endosome-nuclear membrane interaction (see above), as well as the interference with endocytosis of EVs may open new possibilities for targeted intervention [11, 65, 238, 239].

In the past decade, EVs have been intensely studied as natural delivery agents for treatment of different medical conditions. Due to their ability to carry and transfer bioactive molecules to the intracellular compartment of target cells while protecting them from proteases/nucleases degradation and shielding from immune detection as foreign antigens, the lipid bilayer membrane-encapsulated EVs represent optimal vehicles to carry and selectively deliver, upon sophisticated genetic or chemical engineering, drugs and other therapeutic materials that target specific cancer cells. By exploiting the endocytic-exocytic pathways, EVs and their modified cargo can bypass the immune system with minimal cytotoxic effects and hence provide new therapeutic opportunities to control primary tumors and metastatic disease [240, 241]. Their natural origin and their intrinsic function to deliver molecules to target cells contrast the artificiality of all vehicles investigated so far in anti-cancer therapy, including liposomes [242]. Moreover, proper engineering of exosomes or ectosomes loaded with therapeutic molecules may target them to specific tissues in the body. Given that therapeutics administered in vivo go mainly to the liver, which is the primary site of detoxification, EVs would increase the effectiveness of therapeutics, while reducing potential toxicity and other complications. So far, diverse sources, methods of EV isolation, engineering strategies, and loading techniques have been tested with generally promising results, although the ultimate goal of consistently transforming EV biological messages into therapeutic ones has not yet been reached [243].

Overall, many scenarios, which are not mutually exclusive, can be considered to interfere with EV actions in cancer progression. Targeting the biogenesis and release of EVs and their assimilation by a given cell type as well as inhibiting the delivery of their cargo to target molecules are potential pathways that require further investigation. Their use as Trojan horses to target and administer selective anti-cancer drugs is in addition to clinical intervention options. The following six themes could provide new insights into the utilization of EVs to fight cancer.

5.2.1 Inhibition of EV production and release

Formation and release of exosomes are derived from the concerted action of several proteins belonging to the endosomal sorting complexes required for transport (ESCRT)-0, I, II, and III, or associated proteins. The inhibition of ESCRT or ESCRT-associated proteins, such as syndecan, syntenin, and alix has resulted in decrease in exosome production and inhibition of cancer progression [244]. Similar effects have been observed with inhibition of neutral sphingomyelinases, which are enzymes that catalyze the breakdown of sphingomyelin to ceramide and phosphorylcholine [245]. In fact, the same authors discovered that exosomal cargo is segregated into distinct subdomains on the endosomal membrane and that the transfer of exosome-associated domains into the lumen of endosomes requires ceramide. The inhibition of the small GTPase Rab27a, a protein regulating the docking of multivesicular endosomes onto the plasma membrane, and hence exosome secretion, resulted in decreased primary tumor growth and lung dissemination of a metastatic mammary carcinoma [246].

In addition to impeding the biogenesis of EVs, stimulating their release could also provide a new way to limit their intervention as observed in $CD133^+$ cancer cells. The histone deacetylase 6 (HDAC6) inhibitor tubacin promoted the extracellular release of $CD133^+$ EVs from human metastatic melanoma and colorectal carcinoma cells, with a concomitant downregulation of intracellular CD133. Interestingly, this effect was specific for tubacin, as inhibition of HDAC6 deacetylase activity by another selective HDAC6 inhibitor ACY-1215 or the pan-HDAC inhibitor trichostatin A, or downregulation of HDAC6 did not enhance the release of $CD133^+$ EVs. The tubacin-induced EV release was associated with changes in cellular lipid composition, loss of clonogenic capacity, and decrease in the ability to form multicellular aggregates, indicating a novel potential anti-tumor mechanism for tubacin involving EVs in CD133-expressing malignancies [247]. The CD133-HDAC6 interaction and its involvement in primary cilium renewal and cell cycle may merit more attention in the future [60, 248].

Collectively, these few examples show the importance to decipher deeper our knowledge about the biogenesis mechanisms of EVs, as all molecular players involved in them are potential targets to interfere with the roles of EVs in pathological conditions. Similar thought can be applied to the mechanisms of internalization and the delivery of EV cargo to intracellular target. The discovery of the spathasome pathway described in the previous section may provide alternative targets to inhibit the EV function, and further investigation should reveal novel drugs that interfere with intracellular trafficking of EVs upon their uptake.

5.2.2 Inhibition of EV uptake

Several mechanisms were described to explain the cellular uptake of EVs (see above). Their contribution to EV internalization is variable and both EV and target cell-dependent. Therefore, it is conceivable that possible intervention to selectively inhibit internalization has an anti-tumor/anti-metastatic effect. For instance, chlorpromazine, an inhibitor of clathrin-mediated endocytosis, was found to inhibit EV uptake by endocytosis or macropinocytosis, and consequently impede cancer malignancy in vitro [30]. The specific glycosylation pattern of surface proteins of tumor-derived EVs was found to be involved in the regulation of EV uptake by target cells, suggesting that alteration in the glycosylation of EV-associated proteins can hinder cancer progression [249]. The identification of specific EV-associated surface molecules as well as their receptor/adhesion partners at the plasma membrane of target cells could thus be critical to develop such therapeutic strategy. We have recently proposed a novel immunotherapy approach based on a monovalent F(ab) fragment (fragment-antigen binding; referred hereafter as CD9 Fab) derived from mouse monoclonal antibody 5H9 directed against the tetraspanin CD9 [250]. CD9 (alias tetraspanin-29, motility-related protein-1) is a transmembrane protein involved in cell fusion, adhesion, and motility [251, 252]. Depending on the circumstances, CD9 functions have a metastasis promoter or suppressor activity (reviewed in Ref. [252]). Accordingly, it has been broadly proposed as a potential therapeutic target. In the context of EV-cell interaction, CD9 could promote the initial binding, and its dimerization (cis or trans) or oligomerization as well as its interaction with other partners such as adhesion proteins might stimulate it and promote the endocytosis of EVs (Fig. 3.4A). The spatial organization of CD9 molecules could be facilitated by its association with the tetraspanin web, also called tetraspanin-enriched microdomains [253]. In agreement with such a role, the addition of an anti-CD9 antibody stimulated the uptake of melanoma $CD9^+$ EVs by target cells, while its silencing in EVs, host cells or both impeded the endocytosis of EVs [11]. Divalent antibody might promote the cross-link of CD9 proteins associated with host cells and EVs, and consequently stimulate the endocytosis of EVs (Fig. 3.4B). In contrast to the full-length antibody, CD9 Fab was shown to impede with the endocytosis of $CD9^+$ EVs at doses achievable in vivo [250]. Under these conditions, CD9 Fab could saturate the CD9 molecules located on the surface of EVs and cells, and consequently interfere negatively with its role (Fig. 3.4C). This hypothesis is in line with a study demonstrating that CD9 Fab can inhibit the transfer of biomaterials between $CD9^+$ EVs, called epididymosomes, and maturing epididymal spermatozoa [254]. Besides the exact mechanism regulating EV adhesion to target cells and their internalization, it will be important to evaluate whether other anti-CD9 antibodies interfere with EV uptake, as reported for CD9 Fab derived from 5H9 antibody [250]. The proper localization of their epitope might be essential to support these effects. Other tetraspanin proteins enriched in EVs, such as CD63 and CD81, should also be evaluated in this respect. Synergic effects might be

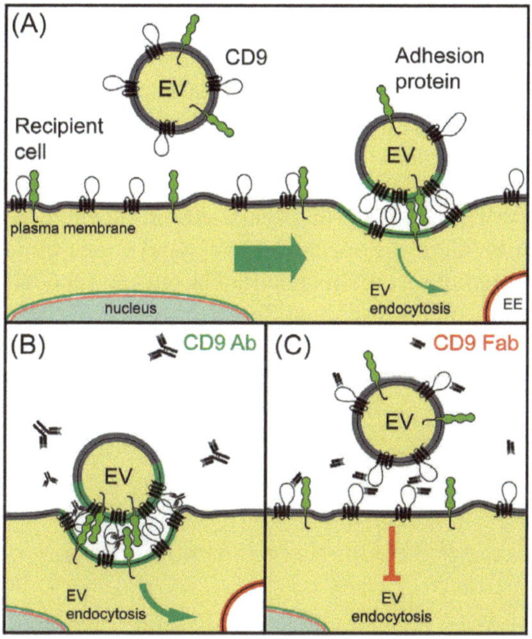

FIG. 4 The endocytosis of EVs and the nuclear transfer of their cargo proteins are impeded by an anti-CD9 Fab. (A) Schematic representation of an EV and a target cell, both containing the tetraspanin CD9 protein on their membrane, and their interaction leading to the endocytosis of the EV (A, green arrow). In this context, cis- and/or trans-dimerization of CD9, its oligomerization, and/or its interaction with other protein partners might organize the protein and lipid components of plasma membrane of host cell and/or those in EV membrane into a specific tetraspanin web (green segment), whose constituents, notably potential adhesion proteins, would regulate or favor the cell-EV interaction and promote the endocytosis of EVs. Upon endocytosis, EVs and its content will be transferred to the endosomal compartment, notably early endosomes (EE) and late endosomes (not depicted), and a fraction of them will end in the nucleus (see also Fig. 3.2). The role of CD9 in this cellular process was substantialized by its silencing in target cells and/or EVs, which interferes with the endocytosis of EVs [11]. (B, C) Representation showing the positive and negative impact of the intact CD9 Ab (clone 5H9) (B) and Fab derived therefrom (C), respectively, on the endocytosis of $CD9^+$ EVs. The divalent CD9 Ab would favor the cross-link of CD9 proteins associated with target cells and EVs and consequently stimulate the EV endocytosis (B, green arrow) [11]. In contrast, CD9 Fab will saturate CD9 proteins present at the surface of target cell and EVs and hence interfere with the CD9 function, thus blocking EV endocytosis (C, red bar) [250]. The outer (green) and inner (pink) nuclear membranes are illustrated.

generated as previously observed for two anti-Fab antibodies against CD9 and CD26 that blocked EV-spermatozoa interaction [254]. By impeding the intercellular communication in the tumor microenvironment, Fab-mediated inhibition of EV uptake, combined with direct targeting of cancerous cells, could lead to the development of novel anti-cancer therapeutic strategies. Finally, it is worth mentioning that the extracorporeal hemofiltration of EVs from the blood circulation by affinity plasmapheresis may constitute an additional therapeutic strategy to interfere in cancer progression [255].

5.2.3 Blocking the pro-angiogenetic effect of cancer cell-derived EVs

Activated EGFR plays an oncogenic role in several human malignant tumors, and oncogenic EGFR released from cancer cells in association with EVs can interact with other cells notably endothelial cells in which they elicit a cellular response, notably the activation of MAPK and Akt signaling pathways. This cascade of events can be blocked using annexin V homodimer that binds to the surface of EVs, and hence impede their interaction with endothelial cells. This resulted in interference with downstream signaling pathways and subsequent secretion of autocrine VEGF and elevated VEGF-R2, causing a reduction in microvascular density in vivo and a reduction of tumor growth [256]. These observations provide early evidence that targeting the pro-angiogenetic effect induced by EVs has therapeutic potential in cancer, particularly when tumor growth is dependent on neo-angiogenesis.

5.2.4 Harnessing the EV-mediated immune response for cancer therapeutic purposes

Dendritic cells release antigen-presenting exosomes that express functional MHC class I and II and T-cell co-stimulating molecules, which can suppress cancer progression in a T cell-dependent manner [257]. This superb study published in Nature Medicine has established that it is possible to harness the immune role of EVs for anti-cancer/metastatic purposes. Tumor peptide-loaded dendritic-derived exosomes induce cytotoxic T lymphocytes priming and suppress growth of several tumors in murine models. Since then, several reports have confirmed that EVs derived from antigen-presenting cells harbor MHC molecules and are able to induce specific anti-cancer immune responses through the activation of T lymphocytes [258, 259]. Moreover, EVs expressing heat shock protein (Hsp) 70 on their surface were reported to induce an anti-cancer response through stimulation of both NK cells and macrophages, as well as apoptosis-inducing protease granzyme B [260, 261].

EVs are being investigated as natural tools to develop cell free-based cancer vaccines, via antigen-presenting cell technology, to induce recognition and killing of cancer cells by the immune system [262]. In numerous studies, dendritic cells have been chosen as EV donor cells because of their low immunogenicity. They capture, process, and present tumor antigens to cognate T cells to generate anti-cancer T cell response [263]. Their properties can be harnessed to generate immune responses against cancer cells. The main advantage of EV- vs. dendritic cell-based immunotherapy is the greater stability of EVs and, importantly, the lack of risks associated with potential in vivo replication of injected materials [264].

Immune checkpoint protein inhibitors, such as antibodies against programmed death ligand 1 (PD-L1) and PD-1 receptor are currently employed in the clinics, having shown effectiveness against melanoma, non-small-cell lung cancer, renal cancer, among others [265]. PD-L1 on the surface of tumor cells binds its receptor PD-1 on effector T cells, thereby suppressing their activity. Antibody blockade of PD-L1 can activate an anti-tumor immune response, leading to durable remissions in a subset of cancer patients. Cancer cell-derived EVs were found to transfer noncoding RNAs to monocytes that modulate the PD-L1 expression and subsequent immune escape [266]. The EV-associated PD-L1 and FasL were found to promote the apoptosis of immune cells allowing tumors to grow undetected [267]. Likewise, they suppress T cell activation in the draining lymph node. Interestingly, depletion of exosomes containing PD-L1 inhibits tumor growth, while their systemic administration rescues growth of tumors unable to secrete their own, suggesting that exosomal PD-L1 represents an unexplored therapeutic target that could overcome resistance to current antibody approaches [268]. Of note, autologous dendritic cell-derived EVs containing the melanoma-associated antigen (MAGE)-3 peptide have shown promising results in malignant melanoma, with minimal toxicity [241]. Antigens encoded by *MAGE* genes are of interest for cancer immunotherapy, given their strict tumoral specificity, and expression by numerous cancers [269].

5.2.5 Delivery of anti-cancer drugs by engineered EVs

An important advantage compared with anticancer drugs alone would be to engineer EVs appropriately for tissue-specific targeting to achieve the maximum concentration of the carried drug in the target tissue, while at the same time minimizing the general toxicity. Another potential advantage of EVs as carriers of therapeutic agents including anti-cancer drugs lies in the observation that cells incubated with chemotherapeutic compounds successfully package molecules into EVs, which can be subsequently collected and used for therapeutic delivery [270–272]. Interestingly, small anti-cancer molecules, such as paclitaxel and doxorubicin, "in vitro" encapsulated into EVs derived from different glioblastoma cell lines, pass through the blood-brain barrier and are successfully delivered to the brain of zebrafish embryos, with greater effect for the drug alone, and with a low immune response [273]. Drug-loaded EVs were prepared by mixing drugs and EVs followed by a two-hour incubation at 37 °C [273, 274]. This and other experiments, indicate that drug encapsulation into EVs may overcome problems of poor solubility and toxicity. Another exiting study has demonstrated the successfully delivery of doxorubicin in a mouse tumor model [275]. Interestingly, a specific accumulation of doxorubicin-containing EVs at the targeted organ and major suppression of tumor growth were observed. In another report, curcumin, a natural polyphenol with anti-neoplastic and anti-inflammatory properties found in the rhizomes of turmeric, was complexed with EVs to enhance its effectiveness both in the lab and in cancer clinical trials [276, 277]. Both efficacy and safety were reported, although bioavailability was suboptimal because of curcumin's hydrophobicity [277, 278].

5.2.6 Delivery of anti-cancer proteins and RNAs by bio-engineered EVs

Qian Lu's laboratory showed that the arrestin-domain containing protein 1 [ARRDC1]-mediated microvesicles (ARMMs) can selectively recruit, package, and deliver into target cells a myriad of biologically active macromolecules, including the tumor suppressor p53 protein, NOTCH2 receptor, RNAs, and the genome-editing CRISPR-Cas9/guide RNA complex [279–281]. ARMMs are EVs originating from plasma membrane, and their budding requires ARRDC1 and ECSCRT I complex TSG101 protein. Although the function of ARMMs is currently unknown, their strategic position at the plasmalemma made them ideal biological tools to engineer EVs with therapeutic purposes.

For example, p53 delivered via ARMMs induces DNA damage-dependent apoptosis in multiple tissues in mice, while ectosomal NOTCH2 initiates in target cells a very specific receptor signaling, which normally requires cell-to-cell contact, facilitating NOTCH receptor signaling at a long distance. By swapping molecules in the ARMMs with therapeutic cargos, such exciting findings have established ARMMs as a flexible platform for packaging and intracellular delivery of therapeutic macromolecules [280]. EVs derived from normal fibroblast-like mesenchymal cells were also bio-engineered to carry shRNA specific to oncogenic $Kras^{G12D}$, a common mutation in pancreatic cancer, and found to suppress cancer in multiple pancreatic cancer mouse models and consequently increase overall survival significantly [282]. Human group O erythrocytes can produce EVs for RNA therapies. Besides being accessible in large quantities through blood banks, the fact that they do not contain DNA avoids the risk of horizontal gene transfer. Upon isolation and electroporation with antisense oligonucleotides directed to miR-125b-2, or Cas9 mRNA and guide RNA (gRNA) targeting the miR125b-2 human locus, erythrocyte-derived EVs were able to inhibit the growth of leukemia and breast cancer cells [283]. Altogether, these few examples point out to the emergence of new therapeutic interventions in which bio-engineered EVs would play a central role as drug delivery vehicles.

6 EV-based clinical cancer therapy studies

The growing knowledge of EVs and their role in intercellular communication has led to a rapid evolution of translational applications of EVs for cancer treatment. Nevertheless, we are still at the beginning of a long adventure towards cell-free therapy. The first clinical trials evaluating the safety and efficacy of EV-based cancer vaccines have revealed promising data that will encourage the development of better therapeutic models [284] (reviewed in Refs [237, 285]). In particular, safety and efficacy of EVs derived from dendritic cells containing MHC/peptide complexes have been evaluated in Phase I and II clinical trials in advanced malignant tumors, demonstrating the safety of such an approach and the propensity of these natural particles to act as intermediates in T- and

NK-cell-based immune responses [241, 286, 287]. Despite the encouraging results however, some limitations were encountered, such as low or variable responses in patients. Further research to determine the in vivo trafficking of injected EVs and their fate will greatly improve the effectiveness of EV-mediated treatment. It is also important to note that the appropriate source of donor cells and potential EV modifications or priming should be considered to improve clinical outcomes. A combination treatment regimen can be of additional help [288]. Finally, quality control in the preparation of EVs is particularly important in this context, as well as all stages of their manufacture in a GMP (Good Manufacturing Practices) cell therapy laboratory [289].

7 Conclusion and perspective

An impressive, recent, and rapidly growing body of evidence attests to the involvement of EVs in solid tumors. Their roles in tumor growth and dissemination are established as well as the possibility to harness their natural characteristics of immune tolerance and targeting of specific cells/tissues/organs to engineer them as vehicles of anti-cancer molecules. Their analysis in the context of liquid biopsy may significantly contribute to tumor diagnosis, classification, and prognostic assessment. However, much remains to be discovered, especially in regard to the distinction of EV sub-classes based on their content and in regard to the understanding of their distribution and selective content release in target cells. Future studies of interference with their production, uptake, and/or targeting could lead to novel and effective therapeutic strategies for metastatic diseases.

Acknowledgments

Given the large volume of publications in the field, we apologize to the individuals whose studies were not specifically cited in this chapter due to reference limitations. We are grateful to all members of our groups for valuable discussion and suggestion and to Mark F. Santos for assistance editing the manuscript.

References

[1] Raposo G, et al. B lymphocytes secrete antigen-presenting vesicles. J Exp Med 1996;183:1161–72.

[2] Ratajczak J, et al. Embryonic stem cell-derived microvesicles reprogram hematopoietic progenitors: evidence for horizontal transfer of mRNA and protein delivery. Leukemia 2006;20:847–56.

[3] Valadi H, et al. Exosome-mediated transfer of mRNAs and microRNAs is a novel mechanism of genetic exchange between cells. Nat Cell Biol 2007;9:654–9.

[4] Zomer A, et al. In Vivo imaging reveals extracellular vesicle-mediated phenocopying of metastatic behavior. Cell 2015;161:1046–57.

[5] van Niel G, D'Angelo G, Raposo G. Shedding light on the cell biology of extracellular vesicles. Nat Rev Mol Cell Biol 2018;19:213–28.

[6] Crewe, C. *et al.* An endothelial-to-adipocyte extracellular vesicle axis governed by metabolic state. Cell 2018 175, 695-708 e613.
[7] Waldenstrom A, Genneback N, Hellman U, Ronquist G. Cardiomyocyte microvesicles contain DNA/RNA and convey biological messages to target cells. PLoS ONE 2012;7, e34653.
[8] Cai J, et al. Extracellular vesicle-mediated transfer of donor genomic DNA to recipient cells is a novel mechanism for genetic influence between cells. J Mol Cell Biol 2013;5:227–38.
[9] Dovrat S, et al. 14-3-3 and beta-catenin are secreted on extracellular vesicles to activate the oncogenic Wnt pathway. Mol Oncol 2014;8:894–911.
[10] Read J, et al. Nuclear transportation of exogenous epidermal growth factor receptor and androgen receptor via extracellular vesicles. Eur J Cancer 2017;70:62–74.
[11] Rappa G, et al. Nuclear transport of cancer extracellular vesicle-derived biomaterials through nuclear envelope invagination-associated late endosomes. Oncotarget 2017;8:14443–61.
[12] Raposo G, Stoorvogel W. Extracellular vesicles: exosomes, microvesicles, and friends. J Cell Biol 2013;200:373–83.
[13] Meehan B, Rak J, Di Vizio D. Oncosomes—large and small: what are they, where they came from? J Extracell Vesicles 2016;5:33109.
[14] Théry C, et al. Minimal information for studies of extracellular vesicles 2018 (MISEV2018): a position statement of the International Society for Extracellular Vesicles and update of the MISEV2014 guidelines. J Extracell Vesicles 2018;7:1535750.
[15] Gyorgy B, et al. Membrane vesicles, current state-of-the-art: emerging role of extracellular vesicles. Cell Mol Life Sci 2011;68:2667–88.
[16] Marzesco AM, et al. Release of extracellular membrane particles carrying the stem cell marker prominin-1 (CD133) from neural progenitors and other epithelial cells. J Cell Sci 2005;118:2849–58.
[17] Dubreuil V, Marzesco AM, Corbeil D, Huttner WB, Wilsch-Bräuninger M. Midbody and primary cilium of neural progenitors release extracellular membrane particles enriched in the stem cell marker prominin-1. J Cell Biol 2007;176:483–95.
[18] Szczepanski MJ, Szajnik M, Welsh A, Whiteside TL, Boyiadzis M. Blast-derived microvesicles in sera from patients with acute myeloid leukemia suppress natural killer cell function via membrane-associated transforming growth factor-beta1. Haematologica 2011;96:1302–9.
[19] Osti D, et al. Clinical significance of extracellular vesicles in plasma from glioblastoma patients. Clin Cancer Res 2019;25:266–76.
[20] Schneider A, Simons M. Catching filopodia: exosomes surf on fast highways to enter cells. J Cell Biol 2016;213:143–5.
[21] Gonda A, Kabagwira J, Senthil GN, Wall NR. Internalization of exosomes through receptor-mediated endocytosis. Mol Cancer Res 2019;17:337–47.
[22] Leung AKL. The Whereabouts of microRNA actions: cytoplasm and beyond. Trends Cell Biol 2015;25:601–10.
[23] Parolini I, et al. Microenvironmental pH is a key factor for exosome traffic in tumor cells. J Biol Chem 2009;284:34211–22.
[24] Prada I, Meldolesi J. Binding and fusion of extracellular vesicles to the plasma membrane of their cell targets. Int J Mol Sci 2016;17, pii: E1296.
[25] Colombo M, Raposo G, Théry C. Biogenesis, secretion, and intercellular interactions of exosomes and other extracellular vesicles. Annu Rev Cell Dev Biol 2014;30:255–89.
[26] Mulcahy LA, Pink RC, Carter DR. Routes and mechanisms of extracellular vesicle uptake. J Extracell Vesicles 2014;3:24641.

[27] Hakulinen J, Junnikkala S, Sorsa T, Meri S. Complement inhibitor membrane cofactor protein (MCP; CD46) is constitutively shed from cancer cell membranes in vesicles and converted by a metalloproteinase to a functionally active soluble form. Eur J Immunol 2004;34:2620–9.
[28] McKelvey KJ, Powell KL, Ashton AW, Morris JM, McCracken SA. Exosomes: mechanisms of Uptake. J Circ Biomark 2015;4:7.
[29] Lamaze C, et al. Interleukin 2 receptors and detergent-resistant membrane domains define a clathrin-independent endocytic pathway. Mol Cell 2001;7:661–71.
[30] Tian T, et al. Exosome uptake through clathrin-mediated endocytosis and macropinocytosis and mediating miR-21 delivery. J Biol Chem 2014;289:22258–67.
[31] Klein U, Gimpl G, Fahrenholz F. Alteration of the myometrial plasma membrane cholesterol content with beta-cyclodextrin modulates the binding affinity of the oxytocin receptor. Biochemistry 1995;34:13784–93.
[32] Pelkmans L, Burli T, Zerial M, Helenius A. Caveolin-stabilized membrane domains as multifunctional transport and sorting devices in endocytic membrane traffic. Cell 2004;118:767–80.
[33] Rothberg KG, et al. Caveolin, a protein component of caveolae membrane coats. Cell 1992;68:673–82.
[34] Subtil A, Hemar A, Dautry-Varsat A. Rapid endocytosis of interleukin 2 receptors when clathrin-coated pit endocytosis is inhibited. J Cell Sci 1994;107:3461–8.
[35] Fitzner D, et al. Selective transfer of exosomes from oligodendrocytes to microglia by macropinocytosis. J Cell Sci 2011;124:447–58.
[36] Feng D, et al. Cellular internalization of exosomes occurs through phagocytosis. Traffic 2010;11:675–87.
[37] Gruenberg J, Griffiths G, Howell KE. Characterization of the early endosome and putative endocytic carrier vesicles in vivo and with an assay of vesicle fusion in vitro. J Cell Biol 1989;108:1301–16.
[38] Jovic M, Sharma M, Rahajeng J, Caplan S. The early endosome: a busy sorting station for proteins at the crossroads. Histol Histopathol 2010;25:99–112.
[39] Piper RC, Luzio JP. Late endosomes: sorting and partitioning in multivesicular bodies. Traffic 2001;2:612–21.
[40] Falguieres T, Luyet PP, Gruenberg J. Molecular assemblies and membrane domains in multivesicular endosome dynamics. Exp Cell Res 2009;315:1567–73.
[41] Lafourcade C, Sobo K, Kieffer-Jaquinod S, Garin J, van der Goot FG. Regulation of the V-ATPase along the endocytic pathway occurs through reversible subunit association and membrane localization. PLoS ONE 2008;3, e2758.
[42] Mesaki K, Tanabe K, Obayashi M, Oe N, Takei K. Fission of tubular endosomes triggers endosomal acidification and movement. PLoS ONE 2011;6, e19764.
[43] Wallroth A, Haucke V. Phosphoinositide conversion in endocytosis and the endolysosomal system. J Biol Chem 2018;293:1526–35.
[44] DiCiccio JE, Steinberg BE. Lysosomal pH and analysis of the counter ion pathways that support acidification. J Gen Physiol 2011;137:385–90.
[45] Rocha N, et al. Cholesterol sensor ORP1L contacts the ER protein VAP to control Rab7-RILP-p150 Glued and late endosome positioning. J Cell Biol 2009;185:1209–25.
[46] Menard L, Parker PJ, Kermorgant S. Receptor tyrosine kinase c-Met controls the cytoskeleton from different endosomes via different pathways. Nat Commun 2014;5:3907.
[47] Friedman JR, Dibenedetto JR, West M, Rowland AA, Voeltz GK. Endoplasmic reticulum-endosome contact increases as endosomes traffic and mature. Mol Biol Cell 2013;24:1030–40.

[48] Raiborg C, et al. Repeated ER-endosome contacts promote endosome translocation and neurite outgrowth. Nature 2015;520:234–8.
[49] Das A, Nag S, Mason AB, Barroso MM. Endosome-mitochondria interactions are modulated by iron release from transferrin. J Cell Biol 2016;214:831–45.
[50] Cioni JM, et al. Late endosomes act as mRNA translation platforms and sustain mitochondria in axons. Cell 2019;176:56–72, e15.
[51] Malhas A, Goulbourne C, Vaux DJ. The nucleoplasmic reticulum: form and function. Trends Cell Biol 2011;21:362–73.
[52] Audas TE, Jacob MD, Lee S. Immobilization of proteins in the nucleolus by ribosomal intergenic spacer noncoding RNA. Mol Cell 2012;45:147–57.
[53] Hetman M. Role of the nucleolus in human diseases [Preface]. Biochim Biophys Acta 2014;1842:757.
[54] Quin JE, et al. Targeting the nucleolus for cancer intervention. Biochim Biophys Acta 2014;1842:802–16.
[55] Zenali MJ, Tan D, Li W, Dhingra S, Brown RE. Stemness characteristics of fibrolamellar hepatocellular carcinoma: immunohistochemical analysis with comparisons to conventional hepatocellular carcinoma. Ann Clin Lab Sci 2010;40:126–34.
[56] Rappa G, Green TM, Lorico A. The nuclear pool of tetraspanin CD9 contributes to mitotic processes in human breast carcinoma. Mol Cancer Res 2014;12:1840–50.
[57] Cantile M, et al. Nuclear localization of cancer stem cell marker CD133 in triple-negative breast cancer: a case report. Tumori 2013;99:e245–50.
[58] Nunukova A, et al. Atypical nuclear localization of CD133 plasma membrane glycoprotein in rhabdomyosarcoma cell lines. Int J Mol Med 2015;36:65–72.
[59] Huang M, Zhu H, Feng J, Ni S, Huang J. High CD133 expression in the nucleus and cytoplasm predicts poor prognosis in non-small cell lung cancer. Dis Markers 2015;2015:986095.
[60] Singer D, et al. Prominin-1 controls stem cell activation by orchestrating ciliary dynamics. EMBO J 2019;38:e99845.
[61] De Angelis Campos AC, et al. Epidermal growth factor receptors destined for the nucleus are internalized via a clathrin-dependent pathway. Biochem Biophys Res Commun 2011;412:341–6.
[62] Chaumet A, et al. Nuclear envelope-associated endosomes deliver surface proteins to the nucleus. Nat Commun 2015;6:8218.
[63] Schrader M, Godinho LF, Costello JL, Islinger M. The different facets of organelle interplay—an overview of organelle interactions. Front Cell Dev Biol 2015;3:56.
[64] Cohen S, Valm AM, Lippincott-Schwartz J. Interacting organelles. Curr Opin Cell Biol 2018;53:84–91.
[65] Santos MF, et al. VAMP-associated protein-A and oxysterol-binding protein-related protein 3 promote the entry of late endosomes into the nucleoplasmic reticulum. J Biol Chem 2018;293:13834–48.
[66] Lehto M, et al. Targeting of OSBP-related protein 3 (ORP3) to endoplasmic reticulum and plasma membrane is controlled by multiple determinants. Exp Cell Res 2005;310:445–62.
[67] Weber-Boyvat M, et al. OSBP-related protein 3 (ORP3) coupling with VAMP-associated protein A regulates R-Ras activity. Exp Cell Res 2015;331:278–91.
[68] Jemal A, et al. Global cancer statistics. CA Cancer J Clin 2011;61:69–90.
[69] Kim MY, et al. Tumor self-seeding by circulating cancer cells. Cell 2009;139:1315–26.
[70] Nguyen DX, Bos PD, Massague J. Metastasis: from dissemination to organ-specific colonization. Nat Rev Cancer 2009;9:274–84.

[71] Figueroa J, et al. Exosomes from glioma-associated mesenchymal stem cells increase the tumorigenicity of glioma stem-like cells via transfer of miR-1587. Cancer Res 2017;77:5808–19.
[72] Chiang AC, Massagué J. Molecular basis of metastasis. N Engl J Med 2008;359:2814–23.
[73] Weaver VM, et al. Reversion of the malignant phenotype of human breast cells in three-dimensional culture and in vivo by integrin blocking antibodies. J Cell Biol 1997;137:231–45.
[74] Oskarsson T, et al. Breast cancer cells produce tenascin C as a metastatic niche component to colonize the lungs. Nat Med 2011;17:867–74.
[75] Paolillo M, Schinelli S. Integrins and exosomes, a dangerous liaison in cancer progression. Cancers (Basel) 2017;9:95.
[76] Hoshino A, et al. Tumour exosome integrins determine organotropic metastasis. Nature 2015;527:329–35.
[77] Costa-Silva B, et al. Pancreatic cancer exosomes initiate pre-metastatic niche formation in the liver. Nat Cell Biol 2015;17:816–26.
[78] Fedele C, Singh A, Zerlanko BJ, Iozzo RV, Languino LR. The alphavbeta6 integrin is transferred intercellularly via exosomes. J Biol Chem 2015;290:4545–51.
[79] Krishn SR, et al. Prostate cancer sheds the alphavbeta3 integrin in vivo through exosomes. Matrix Biol 2019;77:41–57.
[80] Lu H, et al. Exosomal alphavbeta6 integrin is required for monocyte M2 polarization in prostate cancer. Matrix Biol 2018;70:20–35.
[81] Singh A, et al. Exosome-mediated transfer of alphavbeta3 integrin from tumorigenic to non-tumorigenic cells promotes a migratory phenotype. Mol Cancer Res 2016;14:1136–46.
[82] Ridge SM, Sullivan FJ, Glynn SA. Mesenchymal stem cells: key players in cancer progression. Mol Cancer 2017;16:31.
[83] Borriello L, et al. Cancer-associated fibroblasts share characteristics and protumorigenic activity with mesenchymal stromal cells. Cancer Res 2017;77:5142–57.
[84] Peinado H, et al. Melanoma exosomes educate bone marrow progenitor cells toward a pro-metastatic phenotype through MET. Nat Med 2012;18:883–91.
[85] Kim J, et al. Replication study: melanoma exosomes educate bone marrow progenitor cells toward a pro-metastatic phenotype through MET. elife 2018;7, e39944.
[86] Adachi E, et al. Different growth and metastatic phenotypes associated with a cell-intrinsic change of Met in metastatic melanoma. Oncotarget 2016;7:70779–93.
[87] Zhang H, et al. Exosome-delivered EGFR regulates liver microenvironment to promote gastric cancer liver metastasis. Nat Commun 2017;8:15016.
[88] Fong MY, et al. Breast-cancer-secreted miR-122 reprograms glucose metabolism in premetastatic niche to promote metastasis. Nat Cell Biol 2015;17:183–94.
[89] Liu Y, et al. Tumor exosomal RNAs promote lung pre-metastatic niche formation by activating alveolar epithelial TLR3 to recruit neutrophils. Cancer Cell 2016;30:243–56.
[90] Plebanek MP, et al. Pre-metastatic cancer exosomes induce immune surveillance by patrolling monocytes at the metastatic niche. Nat Commun 2017;8:1319.
[91] Zhou W, et al. Cancer-secreted miR-105 destroys vascular endothelial barriers to promote metastasis. Cancer Cell 2014;25:501–15.
[92] Wei Y, et al. Activation of PI3K/Akt pathway by CD133-p85 interaction promotes tumorigenic capacity of glioma stem cells. Proc Natl Acad Sci U S A 2013;110:6829–34.
[93] Grosse-Gehling P, et al. CD133 as a biomarker for putative cancer stem cells in solid tumours: limitations, problems and challenges. J Pathol 2013;229:355–78.
[94] Rappa G, Fodstad O, Lorico A. The stem cell-associated antigen CD133 (Prominin-1) is a molecular therapeutic target for metastatic melanoma. Stem Cells 2008;26:3008–17.

[95] Wang SS, et al. CD133+ cancer stem-like cells promote migration and invasion of salivary adenoid cystic carcinoma by inducing vasculogenic mimicry formation. Oncotarget 2016;7:29051–62.

[96] Liu C, et al. The Interaction between cancer stem cell marker CD133 and Src protein promotes focal adhesion kinase (FAK) phosphorylation and cell migration. J Biol Chem 2016;291:15540–50.

[97] Bauer N, et al. Haematopoietic stem cell differentiation promotes the release of prominin-1/CD133-containing membrane vesicles—a role of the endocytic-exocytic pathway. EMBO Mol Med 2011;3:398–409.

[98] Marzesco AM. Prominin-1-containing membrane vesicles: origins, formation, and utility. Adv Exp Med Biol 2013;777:41–54.

[99] Rappa G, et al. Wnt interaction and extracellular release of prominin-1/CD133 in human malignant melanoma cells. Exp Cell Res 2013;319:810–9.

[100] Rappa G, Mercapide J, Anzanello F, Pope RM, Lorico A. Biochemical and biological characterization of exosomes containing prominin-1/CD133. Mol Cancer 2013;12:62.

[101] Thamm K, et al. Prominin-1 (CD133) modulates the architecture and dynamics of microvilli. Traffic 2019;20:39–60.

[102] Schillaci O, et al. Exosomes from metastatic cancer cells transfer amoeboid phenotype to non-metastatic cells and increase endothelial permeability: their emerging role in tumor heterogeneity. Sci Rep 2017;7:4711.

[103] Fackler OT, Grosse R. Cell motility through plasma membrane blebbing. J Cell Biol 2008;181:879–84.

[104] Muraro MG, et al. CD133+, CD166+CD44+, and CD24+CD44+ phenotypes fail to reliably identify cell populations with cancer stem cell functional features in established human colorectal cancer cell lines. Stem Cells Transl Med 2012;1:592–603.

[105] Wang C, et al. Evaluation of CD44 and CD133 as cancer stem cell markers for colorectal cancer. Oncol Rep 2012;28:1301–8.

[106] Al-Nedawi K, et al. Intercellular transfer of the oncogenic receptor EGFRvIII by microvesicles derived from tumour cells. Nat Cell Biol 2008;10:619–24.

[107] Hao S, et al. Epigenetic transfer of metastatic activity by uptake of highly metastatic B16 melanoma cell-released exosomes. Exp Oncol 2006;28:126–31.

[108] Quail DF, Joyce JA. Microenvironmental regulation of tumor progression and metastasis. Nat Med 2013;19:1423–37.

[109] Karagiannis GS, et al. Cancer-associated fibroblasts drive the progression of metastasis through both paracrine and mechanical pressure on cancer tissue. Mol Cancer Res 2012;10:1403–18.

[110] Liu Y, Cao X. Characteristics and significance of the pre-metastatic niche. Cancer Cell 2016;30:668–81.

[111] Cirri P, Chiarugi P. Cancer-associated-fibroblasts and tumour cells: a diabolic liaison driving cancer progression. Cancer Metastasis Rev 2012;31:195–208.

[112] Erez N, Truitt M, Olson P, Arron ST, Hanahan D. Cancer-associated fibroblasts are activated in incipient neoplasia to orchestrate tumor-promoting inflammation in an NF-kappaB-dependent manner. Cancer Cell 2010;17:135–47.

[113] Olumi AF, et al. Carcinoma-associated fibroblasts direct tumor progression of initiated human prostatic epithelium. Cancer Res 1999;59:5002–11.

[114] LeBleu VS, Kalluri R. A peek into cancer-associated fibroblasts: origins, functions and translational impact. Dis Model Mech 2018;11, pii: dmm029447.

[115] Pang W, et al. Pancreatic cancer-secreted miR-155 implicates in the conversion from normal fibroblasts to cancer-associated fibroblasts. Cancer Sci 2015;106:1362–9.
[116] Webber J, Jenkins RH, Meran S, Phillips A, Steadman R. Modulation of TGFbeta1-dependent myofibroblast differentiation by hyaluronan. Am J Pathol 2009;175:148–60.
[117] Vu LT, et al. Tumor-secreted extracellular vesicles promote the activation of cancer-associated fibroblasts via the transfer of microRNA-125b. J Extracell Vesicles 2019;8, 1599680.
[118] Taverna S, et al. Amphiregulin contained in NSCLC-exosomes induces osteoclast differentiation through the activation of EGFR pathway. Sci Rep 2017;7:3170.
[119] Devarkar SC, et al. Structural basis for m7G recognition and 2′-O-methyl discrimination in capped RNAs by the innate immune receptor RIG-I. Proc Natl Acad Sci U S A 2016;113:596–601.
[120] Nabet BY, et al. Exosome RNA unshielding couples stromal activation to pattern recognition receptor signaling in cancer. Cell 2017;170:352–66, e313.
[121] Yan W, et al. Cancer-cell-secreted exosomal miR-105 promotes tumour growth through the MYC-dependent metabolic reprogramming of stromal cells. Nat Cell Biol 2018;20:597–609.
[122] Hamam D, et al. Transfer of malignant trait to BRCA1 deficient human fibroblasts following exposure to serum of cancer patients. J Exp Clin Cancer Res 2016;35:80.
[123] Abdouh M, et al. Exosomes isolated from cancer patients' sera transfer malignant traits and confer the same phenotype of primary tumors to oncosuppressor-mutated cells. J Exp Clin Cancer Res 2017;36:113.
[124] Abdouh M, et al. Oncosuppressor-mutated cells as a liquid biopsy test for cancer-screening. Sci Rep 2019;9:2384.
[125] Trejo-Becerril C, et al. Cancer progression mediated by horizontal gene transfer in an in vivo model. PLoS ONE 2012;7, e52754.
[126] Garcia-Olmo DC, et al. Cell-free nucleic acids circulating in the plasma of colorectal cancer patients induce the oncogenic transformation of susceptible cultured cells. Cancer Res 2010;70:560–7.
[127] Di Vizio D, et al. Oncosome formation in prostate cancer: association with a region of frequent chromosomal deletion in metastatic disease. Cancer Res 2009;69:5601–9.
[128] Di Vizio D, et al. Large oncosomes in human prostate cancer tissues and in the circulation of mice with metastatic disease. Am J Pathol 2012;181:1573–84.
[129] Minciacchi VR, et al. MYC mediates large oncosome-induced fibroblast reprogramming in prostate cancer. Cancer Res 2017;77:2306–17.
[130] Morello M, et al. Large oncosomes mediate intercellular transfer of functional microRNA. Cell Cycle 2013;12:3526–36.
[131] Robbins PD, Morelli AE. Regulation of immune responses by extracellular vesicles. Nat Rev Immunol 2014;14:195–208.
[132] Bobrie A, Colombo M, Raposo G, Théry C. Exosome secretion: molecular mechanisms and roles in immune responses. Traffic 2011;12:1659–68.
[133] Viaud S, et al. Dendritic cell-derived exosomes promote natural killer cell activation and proliferation: a role for NKG2D ligands and IL-15Ralpha. PLoS ONE 2009;4, e4942.
[134] Ostman S, Taube M, Telemo E. Tolerosome-induced oral tolerance is MHC dependent. Immunology 2005;116:464–76.
[135] Zhang HG, Grizzle WE. Exosomes and cancer: a newly described pathway of immune suppression. Clin Cancer Res 2011;17:959–64.
[136] Grivennikov SI, Greten FR, Karin M. Immunity, inflammation, and cancer. Cell 2010;140:883–99.

[137] de Visser KE, Eichten A, Coussens LM. Paradoxical roles of the immune system during cancer development. Nat Rev Cancer 2006;6:24–37.
[138] Vesely MD, Kershaw MH, Schreiber RD, Smyth MJ. Natural innate and adaptive immunity to cancer. Annu Rev Immunol 2011;29:235–71.
[139] Obregon C, Rothen-Rutishauser B, Gerber P, Gehr P, Nicod LP. Active uptake of dendritic cell-derived exovesicles by epithelial cells induces the release of inflammatory mediators through a TNF-alpha-mediated pathway. Am J Pathol 2009;175:696–705.
[140] Andaloussi ELS, Mager I, Breakefield XO, Wood MJ. Extracellular vesicles: biology and emerging therapeutic opportunities. Nat Rev Drug Discov 2013;12:347–57.
[141] Andreola G, et al. Induction of lymphocyte apoptosis by tumor cell secretion of FasL-bearing microvesicles. J Exp Med 2002;195:1303–16.
[142] Taylor DD, Gercel-Taylor C, Lyons KS, Stanson J, Whiteside TL. T-cell apoptosis and suppression of T-cell receptor/CD3-zeta by Fas ligand-containing membrane vesicles shed from ovarian tumors. Clin Cancer Res 2003;9:5113–9.
[143] Huber V, et al. Human colorectal cancer cells induce T-cell death through release of proapoptotic microvesicles: role in immune escape. Gastroenterology 2005;128:1796–804.
[144] Ridder K, et al. Extracellular vesicle-mediated transfer of functional RNA in the tumor microenvironment. Oncoimmunology 2015;4, e1008371.
[145] Mitrugno A, Tormoen GW, Kuhn P, McCarty OJ. The prothrombotic activity of cancer cells in the circulation. Blood Rev 2016;30:11–9.
[146] Berckmans RJ, et al. Cell-derived microparticles circulate in healthy humans and support low grade thrombin generation. Thromb Haemost 2001;85:639–46.
[147] Kim HK, et al. Elevated levels of circulating platelet microparticles, VEGF, IL-6 and RANTES in patients with gastric cancer: possible role of a metastasis predictor. Eur J Cancer 2003;39:184–91.
[148] Dovizio M, Bruno A, Contursi A, Grande R, Patrignani P. Platelets and extracellular vesicles in cancer: diagnostic and therapeutic implications. Cancer Metastasis Rev 2018;37:455–67.
[149] Taraboletti G, et al. Bioavailability of VEGF in tumor-shed vesicles depends on vesicle burst induced by acidic pH. Neoplasia 2006;8:96–103.
[150] Ban JJ, Lee M, Im W, Kim M. Low pH increases the yield of exosome isolation. Biochem Biophys Res Commun 2015;461:76–9.
[151] Fan GC. Hypoxic exosomes promote angiogenesis. Blood 2014;124:3669–70.
[152] Hsu YL, et al. Hypoxic lung cancer-secreted exosomal miR-23a increased angiogenesis and vascular permeability by targeting prolyl hydroxylase and tight junction protein ZO-1. Oncogene 2017;36:4929–42.
[153] Gesierich S, Berezovskiy I, Ryschich E, Zoller M. Systemic induction of the angiogenesis switch by the tetraspanin D6.1A/CO-029. Cancer Res 2006;66:7083–94.
[154] Nazarenko I, et al. Cell surface tetraspanin Tspan8 contributes to molecular pathways of exosome-induced endothelial cell activation. Cancer Res 2010;70:1668–78.
[155] Skog J, et al. Glioblastoma microvesicles transport RNA and proteins that promote tumour growth and provide diagnostic biomarkers. Nat Cell Biol 2008;10:1470–6.
[156] Hong BS, et al. Colorectal cancer cell-derived microvesicles are enriched in cell cycle-related mRNAs that promote proliferation of endothelial cells. BMC Genomics 2009;10:556.
[157] Umezu T, Ohyashiki K, Kuroda M, Ohyashiki JH. Leukemia cell to endothelial cell communication via exosomal miRNAs. Oncogene 2013;32:2747–55.
[158] Umezu T, et al. Exosomal miR-135b shed from hypoxic multiple myeloma cells enhances angiogenesis by targeting factor-inhibiting HIF-1. Blood 2014;124:3748–57.

[159] Grange C, et al. Microvesicles released from human renal cancer stem cells stimulate angiogenesis and formation of lung premetastatic niche. Cancer Res 2011;71:5346–56.

[160] Kosaka N, et al. Neutral sphingomyelinase 2 (nSMase2)-dependent exosomal transfer of angiogenic microRNAs regulate cancer cell metastasis. J Biol Chem 2013;288:10849–59.

[161] Conigliaro A, et al. CD90+ liver cancer cells modulate endothelial cell phenotype through the release of exosomes containing H19 lncRNA. Mol Cancer 2015;14:155.

[162] Song W, et al. Tumor-derived extracellular vesicles in angiogenesis. Biomed Pharmacother 2018;102:1203–8.

[163] Soekmadji C, Nelson CC. The emerging role of extracellular vesicle-mediated drug resistance in cancers: implications in advanced prostate cancer. Biomed Res Int 2015;2015:454837.

[164] Shedden K, Xie XT, Chandaroy P, Chang YT, Rosania GR. Expulsion of small molecules in vesicles shed by cancer cells: association with gene expression and chemosensitivity profiles. Cancer Res 2003;63:4331–7.

[165] Lv MM, et al. Exosomes mediate drug resistance transfer in MCF-7 breast cancer cells and a probable mechanism is delivery of P-glycoprotein. Tumour Biol 2014;35:10773–9.

[166] Aung T, et al. Exosomal evasion of humoral immunotherapy in aggressive B-cell lymphoma modulated by ATP-binding cassette transporter A3. Proc Natl Acad Sci U S A 2011;108:15336–41.

[167] Bebawy M, et al. Membrane microparticles mediate transfer of P-glycoprotein to drug sensitive cancer cells. Leukemia 2009;23:1643–9.

[168] Dong Y, et al. Tumor endothelial expression of P-glycoprotein upon microvesicular transfer of TrpC5 derived from adriamycin-resistant breast cancer cells. Biochem Biophys Res Commun 2014;446:85–90.

[169] Ciravolo V, et al. Potential role of HER2-overexpressing exosomes in countering trastuzumab-based therapy. J Cell Physiol 2012;227:658–67.

[170] Tagliabue E, Balsari A, Campiglio M, Pupa SM. HER2 as a target for breast cancer therapy. Expert Opin Biol Ther 2010;10:711–24.

[171] Federici C, et al. Exosome release and low pH belong to a framework of resistance of human melanoma cells to cisplatin. PLoS ONE 2014;9, e88193.

[172] Xiao X, et al. Exosomes: decreased sensitivity of lung cancer A549 cells to cisplatin. PLoS ONE 2014;9:e89534.

[173] Dasari S, Tchounwou PB. Cisplatin in cancer therapy: molecular mechanisms of action. Eur J Pharmacol 2014;740:364–78.

[174] Challagundla KB, et al. Exosome-mediated transfer of microRNAs within the tumor microenvironment and neuroblastoma resistance to chemotherapy. J Natl Cancer Inst 2015;107:.

[175] Zhao G, et al. MicroRNA-221 induces cell survival and cisplatin resistance through PI3K/Akt pathway in human osteosarcoma. PLoS ONE 2013;8, e53906.

[176] Wei Y, et al. Exosomal miR-221/222 enhances tamoxifen resistance in recipient ER-positive breast cancer cells. Breast Cancer Res Treat 2014;147:423–31.

[177] Chen WX, et al. Exosomes from drug-resistant breast cancer cells transmit chemoresistance by a horizontal transfer of microRNAs. PLoS ONE 2014;9, e95240.

[178] Boelens MC, et al. Exosome transfer from stromal to breast cancer cells regulates therapy resistance pathways. Cell 2014;159:499–513.

[179] Mutschelknaus L, et al. Exosomes derived from squamous head and neck cancer promote cell survival after ionizing radiation. PLoS ONE 2016;11, e0152213.

[180] Namee NM, O'Driscoll L. Extracellular vesicles and anti-cancer drug resistance. Biochim Biophys Acta Rev Cancer 2018;1870:123–36.

[181] Harischandra DS, Ghaisas S, Rokad D, Kanthasamy AG. Exosomes in toxicology: relevance to chemical exposure and pathogenesis of environmentally linked diseases. Toxicol Sci 2017;158:3–13.

[182] Chung JY, Yu SD, Hong YS. Environmental source of arsenic exposure. J Prev Med Public Health 2014;47:253–7.

[183] Cordazzo C, et al. Rapid shedding of proinflammatory microparticles by human mononuclear cells exposed to cigarette smoke is dependent on Ca^{2+} mobilization. Inflamm Res 2014;63:539–47.

[184] Chai EZ, Siveen KS, Shanmugam MK, Arfuso F, Sethi G. Analysis of the intricate relationship between chronic inflammation and cancer. Biochem J 2015;468:1–15.

[185] Chen C, et al. NF-kB-regulated exosomal miR-155 promotes the inflammation associated with arsenite carcinogenesis. Cancer Lett 2017;388:21–33.

[186] Ngalame NNO, Luz AL, Makia N, Tokar EJ. Arsenic alters exosome quantity and cargo to mediate stem cell recruitment into a cancer stem cell-like phenotype. Toxicol Sci 2018;165:40–9.

[187] Shaked Y. Balancing efficacy of and host immune responses to cancer therapy: the yin and yang effects. Nat Rev Clin Oncol 2016;13:611–26.

[188] Volk-Draper L, et al. Paclitaxel therapy promotes breast cancer metastasis in a TLR4-dependent manner. Cancer Res 2014;74:5421–34.

[189] Daenen LG, et al. Chemotherapy enhances metastasis formation via VEGFR-1-expressing endothelial cells. Cancer Res 2011;71:6976–85.

[190] Keklikoglou I, et al. Chemotherapy elicits pro-metastatic extracellular vesicles in breast cancer models. Nat Cell Biol 2019;21:190–202.

[191] Fang M, Peng CW, Pang DW, Li Y. Quantum dots for cancer research: current status, remaining issues, and future perspectives. Cancer Biol Med 2012;9:151–63.

[192] Kim JH, et al. Carcinogenic activity of PbS quantum dots screened using exosomal biomarkers secreted from HEK293 cells. Int J Nanomedicine 2015;10:5513–27.

[193] Witwer KW, et al. Standardization of sample collection, isolation and analysis methods in extracellular vesicle research. J Extracell Vesicles 2013;2:20360.

[194] Guo SC, Tao SC, Dawn H. Microfluidics-based on-a-chip systems for isolating and analysing extracellular vesicles. J Extracell Vesicles 2018;7:1508271.

[195] Quinn JF, et al. Extracellular RNAs: development as biomarkers of human disease. J Extracell Vesicles 2015;4:27495.

[196] Turchinovich A, Drapkina O, Tonevitsky A. Transcriptome of extracellular vesicles: state-of-the-art. Front Immunol 2019;10:202.

[197] Mitchell PS, et al. Circulating microRNAs as stable blood-based markers for cancer detection. Proc Natl Acad Sci U S A 2008;105:10513–8.

[198] Kalluri R. The biology and function of exosomes in cancer. J Clin Invest 2016;126:1208–15.

[199] Caradec J, et al. Reproducibility and efficiency of serum-derived exosome extraction methods. Clin Biochem 2014;47:1286–92.

[200] Webber J, Yeung V, Clayton A. Extracellular vesicles as modulators of the cancer microenvironment. Semin Cell Dev Biol 2015;40:27–34.

[201] Park JE, et al. Hypoxic tumor cell modulates its microenvironment to enhance angiogenic and metastatic potential by secretion of proteins and exosomes. Mol Cell Proteomics 2010;9:1085–99.

[202] Tian Y, et al. Protein profiling and sizing of extracellular vesicles from colorectal cancer patients via flow cytometry. ACS Nano 2018;12:671–80.

[203] Jiao YJ, et al. Characterization and proteomic profiling of pancreatic cancer-derived serum exosomes. J Cell Biochem 2019;120:988–99.

[204] Melo SA, et al. Glypican-1 identifies cancer exosomes and detects early pancreatic cancer. Nature 2015;523:177–82.

[205] Frampton AE, et al. Glypican-1 is enriched in circulating-exosomes in pancreatic cancer and correlates with tumor burden. Oncotarget 2018;9:19006–13.

[206] Kimura H, et al. CKAP4, a DKK1 receptor, is a biomarker in exosomes derived from pancreatic cancer and a molecular target for therapy. Clin Cancer Res 2019;25:1936–47.

[207] DeRita RM, et al. c-Src, insulin-like growth factor I receptor, G-protein-coupled receptor kinases and focal adhesion kinase are enriched into prostate cancer cell exosomes. J Cell Biochem 2017;118:66–73.

[208] Niu L, et al. Tumor-derived exosomal proteins as diagnostic biomarkers in non-small cell lung cancer. Cancer Sci 2019;110:433–42.

[209] Huttner HB, et al. The stem cell marker prominin-1/CD133 on membrane particles in human cerebrospinal fluid offers novel approaches for studying central nervous system disease. Stem Cells 2008;26:698–705.

[210] Sakaue T, et al. Glycosylation of ascites-derived exosomal CD133: a potential prognostic biomarker in patients with advanced pancreatic cancer. Med Mol Morphol 2019; https://doi.org/10.1007/s00795-019-00218-5.

[211] Fargeas CA, Karbanová J, Corbeil D. Assessment of CD133-positive extracellular membrane vesicles in pancreatic cancer ascites and beyond. Med Mol Morphol 2019; https://doi.org/10.1007/s00795-019-00221-w.

[212] Huttner HB, et al. Increased membrane shedding—indicated by an elevation of CD133-enriched membrane particles—into the CSF in partial epilepsy. Epilepsy Res 2012;99:101–6.

[213] Bobinger T, et al. CD133-positive membrane particles in cerebrospinal fluid of patients with inflammatory and degenerative neurological diseases. Front Cell Neurosci 2017;11(77).

[214] Weeraphan C, et al. Phosphoproteome profiling of isogenic cancer cell-derived exosome reveals HSP90 as a potential marker for human cholangiocarcinoma. Proteomics 2019;e1800159.

[215] Cheng L, Sharples RA, Scicluna BJ, Hill AF. Exosomes provide a protective and enriched source of miRNA for biomarker profiling compared to intracellular and cell-free blood. J Extracell Vesicles 2014;3, 23743.

[216] Sanz-Rubio D, et al. Stability of circulating exosomal miRNAs in healthy subjects. Sci Rep 2018;8, 10306.

[217] Alegre E, et al. Study of circulating microRNA-125b levels in serum exosomes in advanced melanoma. Arch Pathol Lab Med 2014;138:828–32.

[218] Taylor DD, Gercel-Taylor C. MicroRNA signatures of tumor-derived exosomes as diagnostic biomarkers of ovarian cancer. Gynecol Oncol 2008;110:13–21.

[219] Madhavan B, et al. Combined evaluation of a panel of protein and miRNA serum-exosome biomarkers for pancreatic cancer diagnosis increases sensitivity and specificity. Int J Cancer 2015;136:2616–27.

[220] Dejima H, Iinuma H, Kanaoka R, Matsutani N, Kawamura M. Exosomal microRNA in plasma as a non-invasive biomarker for the recurrence of non-small cell lung cancer. Oncol Lett 2017;13:1256–63.

[221] Wang J, et al. Circulating exosomal miR-125a-3p as a novel biomarker for early-stage colon cancer. Sci Rep 2017;7, 4150.

[222] Manterola L, et al. A small noncoding RNA signature found in exosomes of GBM patient serum as a diagnostic tool. Neuro-Oncology 2014;16:520–7.

[223] Alhasan AH, et al. Circulating microRNA signature for the diagnosis of very high-risk prostate cancer. Proc Natl Acad Sci U S A 2016;113:10655–60.

[224] Warnecke-Eberz U, Chon SH, Holscher AH, Drebber U, Bollschweiler E. Exosomal onco-miRs from serum of patients with adenocarcinoma of the esophagus: comparison of miRNA profiles of exosomes and matching tumor. Tumour Biol 2015;36:4643–53.
[225] Nilsson J, et al. Prostate cancer-derived urine exosomes: a novel approach to biomarkers for prostate cancer. Br J Cancer 2009;100:1603–7.
[226] Thind A, Wilson C. Exosomal miRNAs as cancer biomarkers and therapeutic targets. J Extracell Vesicles 2016;5, 31292.
[227] Sohn W, et al. Serum exosomal microRNAs as novel biomarkers for hepatocellular carcinoma. Exp Mol Med 2015;47, e184.
[228] Duijvesz D, Luider T, Bangma CH, Jenster G. Exosomes as biomarker treasure chests for prostate cancer. Eur Urol 2011;59:823–31.
[229] Huang X, et al. Exosomal miR-1290 and miR-375 as prognostic markers in castration-resistant prostate cancer. Eur Urol 2015;67:33–41.
[230] Lai X, et al. A microRNA signature in circulating exosomes is superior to exosomal glypican-1 levels for diagnosing pancreatic cancer. Cancer Lett 2017;393:86–93.
[231] Melo SA, et al. Cancer exosomes perform cell-independent microRNA biogenesis and promote tumorigenesis. Cancer Cell 2014;26:707–21.
[232] Narayanan A, et al. Exosomes derived from HIV-1-infected cells contain trans-activation response element RNA. J Biol Chem 2013;288:20014–33.
[233] Rappa G, Conigliaro A, Santos MF, Alessandro R, Lorico A. Cancer relevance of signal recognition particle and other non-coding RNAs in extracellular vesicles. Transl Cancer Res 2017;6:S1257–60.
[234] Nolte-'t Hoen EN, et al. Deep sequencing of RNA from immune cell-derived vesicles uncovers the selective incorporation of small non-coding RNA biotypes with potential regulatory functions. Nucleic Acids Res 2012;40:9272–85.
[235] Keup C, et al. RNA profiles of circulating tumor cells and extracellular vesicles for therapy stratification of metastatic breast cancer patients. Clin Chem 2018;64:1054–62.
[236] Shao H, et al. Chip-based analysis of exosomal mRNA mediating drug resistance in glioblastoma. Nat Commun 2015;6, 6999.
[237] Tai YL, Chen KC, Hsieh JT, Shen TL. Exosomes in cancer development and clinical applications. Cancer Sci 2018;109:2364–74.
[238] Mathieu M, Martin-Jaular L, Lavieu G, Théry C. Specificities of secretion and uptake of exosomes and other extracellular vesicles for cell-to-cell communication. Nat Cell Biol 2019;21:9–17.
[239] Heusermann W, et al. Exosomes surf on filopodia to enter cells at endocytic hot spots, traffic within endosomes, and are targeted to the ER. J Cell Biol 2016;213:173–84.
[240] Lakhal S, Wood MJ. Exosome nanotechnology: an emerging paradigm shift in drug delivery: exploitation of exosome nanovesicles for systemic in vivo delivery of RNAi heralds new horizons for drug delivery across biological barriers. Bioessays 2011;33:737–41.
[241] Escudier B, et al. Vaccination of metastatic melanoma patients with autologous dendritic cell (DC) derived-exosomes: results of the first phase I clinical trial. J Transl Med 2005;3:10.
[242] Malam Y, Loizidou M, Seifalian AM. Liposomes and nanoparticles: nanosized vehicles for drug delivery in cancer. Trends Pharmacol Sci 2009;30:592–9.
[243] Liu C, Su C. Design strategies and application progress of therapeutic exosomes. Theranostics 2019;9:1015–28.
[244] Baietti MF, et al. Syndecan-syntenin-ALIX regulates the biogenesis of exosomes. Nat Cell Biol 2012;14:677–85.
[245] Trajkovic K, et al. Ceramide triggers budding of exosome vesicles into multivesicular endosomes. Science 2008;319:1244–7.

[246] Bobrie A, et al. Rab27a supports exosome-dependent and -independent mechanisms that modify the tumor microenvironment and can promote tumor progression. Cancer Res 2012;72:4920–30.

[247] Chao OS, et al. The HDAC6 inhibitor tubacin induces release of CD133(+) extracellular vesicles from cancer cells. J Cell Biochem 2017;118:4414–24.

[248] Mak AB, et al. Regulation of CD133 by HDAC6 promotes beta-catenin signaling to suppress cancer cell differentiation. Cell Rep 2012;2:951–63.

[249] Escrevente C, Keller S, Altevogt P, Costa J. Interaction and uptake of exosomes by ovarian cancer cells. BMC Cancer 2011;11(108).

[250] Santos MF, et al. Anti-human CD9 antibody Fab fragment impairs the internalization of extracellular vesicles and the nuclear transfer of their cargo proteins. J Cell Mol Med 2019;23:4408–21.

[251] Schmid E, et al. Antibodies to CD9, a tetraspan transmembrane protein, inhibit canine distemper virus-induced cell-cell fusion but not virus-cell fusion. J Virol 2000;74:7554–61.

[252] Zöller M. Tetraspanins: push and pull in suppressing and promoting metastasis. Nat Rev Cancer 2009;9:40–55.

[253] Zuidscherwoude M, et al. The tetraspanin web revisited by super-resolution microscopy. Sci Rep 2015;5, 12201.

[254] Caballero JN, Frenette G, Belleannee C, Sullivan R. CD9-positive microvesicles mediate the transfer of molecules to Bovine Spermatozoa during epididymal maturation. PLoS ONE 2013;8, e65364.

[255] Marleau AM, Chen CS, Joyce JA, Tullis RH. Exosome removal as a therapeutic adjuvant in cancer. J Transl Med 2012;10(134).

[256] Al-Nedawi K, Meehan B, Kerbel RS, Allison AC, Rak J. Endothelial expression of autocrine VEGF upon the uptake of tumor-derived microvesicles containing oncogenic EGFR. Proc Natl Acad Sci U S A 2009;106:3794–9.

[257] Zitvogel L, et al. Eradication of established murine tumors using a novel cell-free vaccine: dendritic cell-derived exosomes. Nat Med 1998;4:594–600.

[258] Théry C, et al. Indirect activation of naive CD4+ T cells by dendritic cell-derived exosomes. Nat Immunol 2002;3:1156–62.

[259] Morelli AE, et al. Endocytosis, intracellular sorting, and processing of exosomes by dendritic cells. Blood 2004;104:3257–66.

[260] Gastpar R, et al. Heat shock protein 70 surface-positive tumor exosomes stimulate migratory and cytolytic activity of natural killer cells. Cancer Res 2005;65:5238–47.

[261] Vega VL, et al. Hsp70 translocates into the plasma membrane after stress and is released into the extracellular environment in a membrane-associated form that activates macrophages. J Immunol 2008;180:4299–307.

[262] Tan A, De La Pena H, Seifalian AM. The application of exosomes as a nanoscale cancer vaccine. Int J Nanomedicine 2010;5:889–900.

[263] Pitt JM, et al. Dendritic cell-derived exosomes for cancer therapy. J Clin Invest 2016;126:1224–32.

[264] Zhang B, Yin Y, Lai RC, Lim SK. Immunotherapeutic potential of extracellular vesicles. Front Immunol 2014;5(518).

[265] Salmaninejad A, et al. PD-1/PD-L1 pathway: basic biology and role in cancer immunotherapy. J Cell Physiol 2019;234:16824–37.

[266] Haderk F, et al. Tumor-derived exosomes modulate PD-L1 expression in monocytes. Sci Immunol 2017;2.

[267] Peng P, Yan Y, Keng S. Exosomes in the ascites of ovarian cancer patients: origin and effects on anti-tumor immunity. Oncol Rep 2011;25:749–62.

[268] Poggio M, et al. Suppression of exosomal PD-L1 induces systemic anti-tumor immunity and memory. Cell 2019;177:414–27.
[269] Schultz ES, et al. A MAGE-3 peptide recognized on HLA-B35 and HLA-A1 by cytolytic T lymphocytes. Tissue Antigens 2001;57:103–9.
[270] Ha D, Yang N, Nadithe V. Exosomes as therapeutic drug carriers and delivery vehicles across biological membranes: current perspectives and future challenges. Acta Pharm Sin B 2016;6:287–96.
[271] Alvarez-Erviti L, et al. Delivery of siRNA to the mouse brain by systemic injection of targeted exosomes. Nat Biotechnol 2011;29:341–5.
[272] Sun D, et al. A novel nanoparticle drug delivery system: the anti-inflammatory activity of curcumin is enhanced when encapsulated in exosomes. Mol Ther 2010;18:1606–14.
[273] Yang T, et al. Exosome delivered anticancer drugs across the blood-brain barrier for brain cancer therapy in Danio rerio. Pharm Res 2015;32:2003–14.
[274] Jang SC, et al. Bioinspired exosome-mimetic nanovesicles for targeted delivery of chemotherapeutics to malignant tumors. ACS Nano 2013;7:7698–710.
[275] Tian Y, et al. A doxorubicin delivery platform using engineered natural membrane vesicle exosomes for targeted tumor therapy. Biomaterials 2014;35:2383–90.
[276] Anand P, Sundaram C, Jhurani S, Kunnumakkara AB, Aggarwal BB. Curcumin and cancer: an "old-age" disease with an "age-old" solution. Cancer Lett 2008;267:133–64.
[277] Dhillon N, et al. Phase II trial of curcumin in patients with advanced pancreatic cancer. Clin Cancer Res 2008;14:4491–9.
[278] Anand P, Kunnumakkara AB, Newman RA, Aggarwal BB. Bioavailability of curcumin: problems and promises. Mol Pharm 2007;4:807–18.
[279] Nabhan JF, Hu R, Oh RS, Cohen SN, Lu Q. Formation and release of arrestin domain-containing protein 1-mediated microvesicles (ARMMs) at plasma membrane by recruitment of TSG101 protein. Proc Natl Acad Sci U S A 2012;109:4146–51.
[280] Wang Q, et al. ARMMs as a versatile platform for intracellular delivery of macromolecules. Nat Commun 2018;9:960.
[281] Wang Q, Lu Q. Plasma membrane-derived extracellular microvesicles mediate non-canonical intercellular NOTCH signaling. Nat Commun 2017;8:709.
[282] Kamerkar S, et al. Exosomes facilitate therapeutic targeting of oncogenic KRAS in pancreatic cancer. Nature 2017;546:498–503.
[283] Usman WM, et al. Efficient RNA drug delivery using red blood cell extracellular vesicles. Nat Commun 2018;9:2359.
[284] Dai S, et al. Phase I clinical trial of autologous ascites-derived exosomes combined with GM-CSF for colorectal cancer. Mol Ther 2008;16:782–90.
[285] Pitt JM, Kroemer G, Zitvogel L. Extracellular vesicles: masters of intercellular communication and potential clinical interventions. J Clin Invest 2016;126:1139–43.
[286] Morse MA, et al. A phase I study of dexosome immunotherapy in patients with advanced non-small cell lung cancer. J Transl Med 2005;3:9.
[287] Besse B, et al. Dendritic cell-derived exosomes as maintenance immunotherapy after first line chemotherapy in NSCLC. Oncoimmunology 2016;5, e1071008.
[288] Taieb J, et al. Chemoimmunotherapy of tumors: cyclophosphamide synergizes with exosome based vaccines. J Immunol 2006;176:2722–9.
[289] Viaud S, et al. Updated technology to produce highly immunogenic dendritic cell-derived exosomes of clinical grade: a critical role of interferon-gamma. J Immunother 2011;34:65–75.

Chapter 4

Hematologic malignancies: The exosome contribution in tumor progression

Stefania Raimondo[a], Laura Saieva[a], Riccardo Alessandro
Department of Biomedicine, Neuroscience and Advanced Diagnostics, University of Palermo, Palermo, Italy.

1. Introduction: Hematologic malignancies and the tumor microenvironment

Cancer is regulated by intrinsic and extrinsic signals that promote its progression mainly by the following phases: initiation, promotion, progression and metastasis. These processes are driven by genetic alterations of cancer cells together with the deregulated communication between surrounding cells, in order to create what is now defined as "tumor microenvironment" [1].

While past researches were mainly focused on defining cancer cell genetic and molecular profile, in order to identify targeted therapies, today we know that this is not enough. In fact, cancer cells are not isolated but they are part of a complex network in which tumor cells, as well as neighboring normal cells, exchanges information that contributes to tumor progression and survival.

Therefore, studies focused on the understanding of how cells communicate in the tumor microenvironment are now the basis for the development of new therapeutic approaches.

The important role played by the microenvironment in tumorigenesis has now been widely described for both solid [2] and hematological tumors [3]. Interaction of leukemic blasts with endothelial and stromal cells represents a crucial process for tumor cells proliferation and survival as well as for the development of drug resistance [4]. This mechanism is mainly regulated by the CXCL12-CXCR4 axis; in fact, the chemokine CXCL12, produced by stromal cells in the bone marrow microenvironment, triggers growth promoting and anti-apoptotic signals in leukemic blasts by the interaction with the cell surface receptor CXCR4 [5]. In multiple myeloma (MM), the proliferation and survival of plasma cells is regulated by the

a. These authors contributed equally to this work.

Exosomes. https://doi.org/10.1016/B978-0-12-816053-4.00004-3

bone microenvironment, in particular, angiogenesis is believed to play a key role in the pathogenesis and progression of the disease [6].

Among the factors responsible for the crosstalk between hematologic tumor cells and the bone marrow microenvironment, extracellular vesicles, and in particular exosomes, have recently interested the scientific community. These lipoproteic structures are in fact considered as containers of cellular messages and vehicles of these information among the cellular components of the microenvironment [7]. Here we will discuss the current knowledge on the EV-mediated crosstalk between tumor and normal cells, in order to better understand how vesicles can contribute to tumor progression.

2. Physiological regulation and pathological alterations of hematopoiesis by extracellular vesicles

Hematopoiesis is the physiological process of blood cell production. Hematopoietic stem cells (HSCs) and stromal cells, together with the extracellular matrix, where HSC growth and division occurs, represent the main actors. HSCs are multipotent stem cells that give rise to all blood lineages, including myeloid and lymphoid lineages, and are characterized by self-renewal and differentiation ability; a balance of these two processes is required for the blood and immune system functions [8].

Accumulating evidences suggest that microenvironmental stimuli take part to the regulation of the characteristics of stemness. The bone marrow (BM), composed of cells, extracellular matrix, and soluble factors, such as cytokines, chemokines and signaling molecules, provides a favorable microenvironment for HSCs and is the major location where hematopoiesis occurs. Although only partially defined, there are different mechanisms and factors contributing to the maintenance of hematopoietic stem compartment homeostasis. For example, the interaction with other cell populations of the BM, such as mesenchymal stromal cells, is crucial for HSC self-renewal, survival, and behavior [9].

Recent studies, in the attempt to elucidate the mechanisms underlying the regulatory activity of the BM niche, have highlighted the existence of a complex communication system within the BM cell populations. Chemokines, cytokines, adhesion molecules, enzymes, receptors and signal transduction molecules contribute to define this molecular frame.

Recently, extracellular vesicles have emerged as important modulators within the hematopoietic niche both at physiological as well as pathological levels [10]. Studies have shown that extracellular vesicles represent a novel communication system within the hematopoietic niche for cell homeostasis regulation (Fig. 1).

The first evidence was published in 2006 in a study by Ratajczak and colleagues; authors demonstrated that embryonic stem cells delivered exosomal mRNAs and proteins to hematopoietic stem and progenitor cells, increasing their pluripotency [11]. Ekstrom demonstrated that human mast cell-derived exosomes contain RNA that can be shuttled to other human mast cells and to

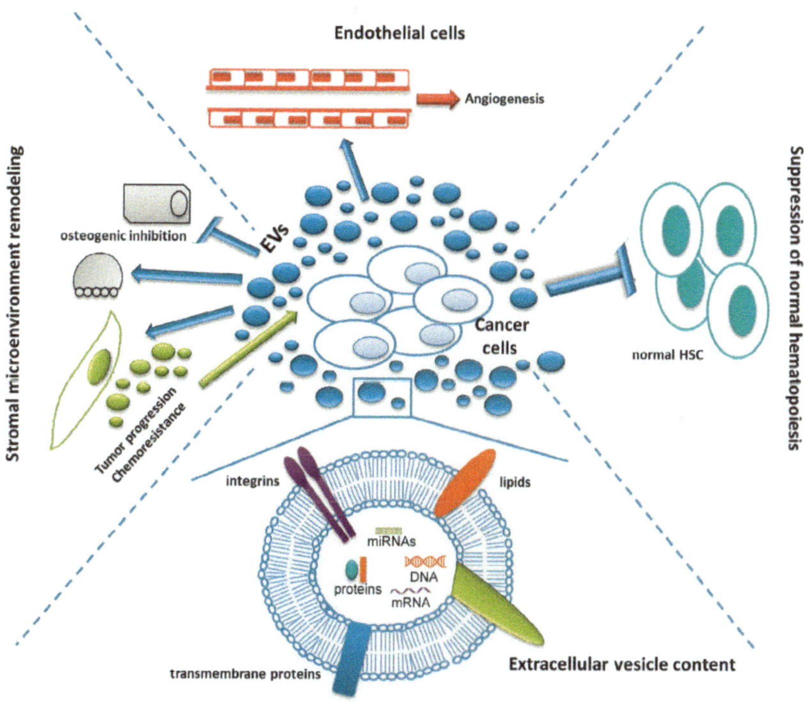

FIG. 1 Schematic representation of cancer cell-derived extracellular vesicles in tumor microenvironment. In hematologic malignancies the crosstalk between cells of the bone marrow and neoplastic cells is deregulated and it contributes to the creation of a permissive microenvironment for tumor growth and progression.

human CD34+ hematopoietic progenitor cells. These findings suggested that the exosomal shuttle RNA can play a role in the communication between immune cells, including mast cells and CD34+ progenitor cells, implying a role in cells maturation process [12].

More recent studies elucidated the role of EVs released by BM stroma cells, including endothelial and MSCs, within the niche. Goloviznina et al. showed that MSCs are able to regulate hematopoiesis by EVs secretion. In particular they observed that hematopoietic progenitors are activated by in vitro exposure to MSC EVs; in fact, exosomes activated myeloid progenitors via TLR4-MyD88-NF-κB pathways [13]. Furthermore, a study by Stik and colleagues demonstrated that stromal cell-derived EVs support the hematopoietic stem and progenitor cells by maintaining cell survival and clonogenic potential and preventing them from apoptosis. Authors concluded that these effects are due to extracellular vesicle miRNA cargoes able to modulate target cell phenotype [14].

Accumulating evidences suggest that EVs are responsible for the deregulation of the balance between proliferation and differentiation of committed progenitors that results in hematologic malignancies. Most of the studies published up to now, highlighting the role of tumor exosomes in inhibiting hematopoiesis, are referred to Acute Myeloid Leukemia (AML). In 2015 Kurre's group reported that AML derived exosomes are able to regulate hematopoietic stem and progenitor cells both directly and indirectly; in particular they demonstrated that exosome treatment downregulates CXCL12 in stromal cells leading to the mobilization of hematopoietic stem cells from the bone marrow. In addition, AML exosomes directly reduce the expression of hematopoietic transcription factors, including c-Myb, Cebp-β and Hoxa-9, in hematopoietic stem and progenitor cells. By a proteomic approach authors identified the candidate pathways involved in the exosome-mediated modulation of stem cell functions [15]. The same group further investigated the molecular mechanism underlying exosomes-mediated hematopoiesis inhibition. They found that AML exosomes contain miRNAs that target master regulators of hematopoiesis, including c-Myb; among these, the exosomal miR-150 and miR-155 were sufficient to suppress hematopoietic stem and progenitor cell clonogenicity [16]. More recently Razmkhah et al. isolated microvesicles from AML patients and they found that the addition for one week of these EVs to normal hematopoietic stem cells leads to the over-expression of the oncomiRs miR-21 and miR-29a in target cells [17]. Finally, Kumar and collaborators observed that AML-derived exosomes reduce the stromal expression of hematopoietic stem cell-supporting factors, including CXCL12, KITL and IGF1, thus suppressing normal hematopoiesis [18].

In addition to AML, it has been demonstrated that EVs affect hematopoiesis also in myelodysplastic syndromes (MDS); in fact, microvesicles from MDS patients increase the viability and clonogenicity of hematopoietic stem cells [19].

3. Endothelial remodeling by extracellular vesicles

Even though increased angiogenesis has long been recognized in solid tumors, today it is known that it also contributes to the progression of hematological diseases [20]. The increase in BM vascular density, as a consequence of stimulated angiogenesis, has been described in several hematologic disorders as leukemia, myelodysplastic syndromes and multiple myeloma.

The induction of angiogenesis results from activating and inhibitory driver forces in the bone microenvironment. Accumulating evidences support the hypothesis that extracellular vesicles, released by tumor cells, are able to stimulate endothelial cells to remodel BM vasculature, and that this effect depends on EV content. In particular, EVs deliver angiogenic proteins and miRNAs thus inducing changes in recipient cells. Studies from Taverna et al. have investigated the role of chronic myeloid leukemia (CML)-derived exosomes in the angiogenic process. In 2012 authors demonstrated for the first time that exosomes from CML cell lines, as well as from patients, promote tumor angiogenesis by increasing endothelial cell motility, pro-angiogenic cytokine secretion and cell adhesion [21]. Further studies showed that this tumor-endothelial crosstalk is mediated by miR-126 transfer from CML exosomes to endothelial cells [22].

Similarly, Mineo and colleagues confirmed the CML exosome-mediated increase of the angiogenic process, attributing this effect to the activation of Src pathway [23]. Since hypoxia plays a pivotal role in tumor progression, Tadokoro and colleagues investigated the ability of exosomes derived from hypoxic tumor cells in modulating the tumor microenvironment; they found that hypoxic CML cells secrete exosomes containing pro-angiogenic miRNAs that target the receptor tyrosine kinase ligand Ephrin-A3, leading to the enhanced tube formation [24]. Likewise, different studies highlighted a role of multiple myeloma-derived exosomes in modulating tumor angiogenesis. Liu's group found that MM cell line RPMI8226 release exosomes with a diameter ranging from 100 to 1000nm that promote the proliferation of endothelial cells as well as cancer cell invasion and migration. Exosomes-endothelial cell interaction resulted in the increased secretion of IL6 and VEGF [25]. In 2014, in order to mimic the in vivo bone marrow microenvironment, Umezu established a model of MM cell line grown continuously under chronic hypoxia (HR-MM cells); they demonstrated that HR-MM cells produced more exosomes that the same cell line under normoxia or acute hypoxia conditions. In addition, authors found that HR-MM cell-derived exosomes enhanced endothelial tube formation and that this effect was due to the exosomal miR-135b [26]. Increased angiogenesis has been extensively observed in Acute Myeloid Leukemia (AML) patients. Kurre's group observed that primary AML vesicle RNA was enriched in the Insulin-like growth factor 1 receptor (IGF-IR) mRNA and that this led to an increase of the angiogenic potential of stromal cells [27].

4. Extracellular vesicles reprogram the bone marrow stromal cells

Bone marrow microenvironment plays a pivotal role in the maintenance of hemopoietic stem cells, in particular in supporting cell renewal and differentiation [28]. In hematologic malignancies the crosstalk between bone marrow stromal cells and neoplastic cells is deregulated and it contributes to the creation of a permissive microenvironment for tumor growth and progression [29].

While it was previously thought that the direct contact between tumor and stromal cells was the only one responsible for the establishment of the tumor microenvironment, today it is known that the release of soluble factors influences the cells of the microenvironment. In this context, extracellular vesicles, released by tumor cells as well as by stromal cells, have been largely recognized as responsible for the changes observed in the BM niche [30, 31]. In particular, BM mesenchymal stromal cells (BMSCs) have been shown to support tumor cell growth by exosomes release; in fact, BMSC-derived exosomes can transfer nucleic acids as well as proteins to recipient cancer cells, thus influencing tumor growth, metastasis and response to therapy [31].

Roccaro recently demonstrated that Multiple Myeloma BMSC-derived exosomes contain and deliver, compared to normal BMSC, high levels of oncogenic proteins, responsible for tumor cell adhesion and migration. These exosomes are internalized by MM cells, leading to cancer cell growth and dissemination [32]. Crompot investigated the role of extracellular vesicles from the BM of patients affected by chronic lymphocytic leukemia (CLL) showing that they enhanced tumor cell migratory capacities and chemoresistance and decreased leukemic cell apoptosis [33]. Barrera-Ramirez et al. performed a miRNA sequencing of exosome RNA content isolated from marrow-derived MSCs from patients with AML and from controls. Authors found that two miRNAs were significantly increased in AML-derived samples and three were significantly decreased. By prediction analysis authors found that these miRNAs modulate the expression of genes involved in leukemogenesis [34].

More evidences have accumulated on the role of cancer cell-derived exosomes in modulating stromal cell phenotype to create a more permissive microenvironment for tumor progression.

We have provided evidences on the direct effect of chronic myeloid leukemia-derived exosomes on bone marrow stromal cells. We found that exosomes stimulate bone marrow stromal cells to produce IL8 that is able to modulate both in vitro and in vivo the leukemia cell malignant phenotype, promoting cancer cell proliferation and adhesion [7]. Further studies demonstrated that these mechanisms are mediated by the activation of the EGFR pathway in stromal cells [35]. Similarly, Ghosh et al. found microvesicles in the plasma of B-cell chronic lymphocytic leukemia (CLL) that promote, in BMSC, the activation of AKT pathway and the production of VEGF [36].

Cheng and colleagues found that MM-derived exosomes promote the proliferation of MSC and activate IL6 release through the delivery of miR-21 and miR146a [37].

Interestingly, recent studies investigated the role of exosomes in reprogramming stromal cell metabolism [38]. Human melanoma-derived exosomes are able to reprogram stromal fibroblasts metabolism, causing extracellular acidification, a condition that favors pre-metastatic niche formation [39]. Johnson reported that acute lymphoblastic leukemia (ALL) cells released EVs that are internalized by mesenchymal stromal cells. This interaction leads to a reduction of mitochondrial respiration in recipient stromal cells and to an increase of the glycolytic rate. Overall authors conclude that EVs are able to induce a metabolic switch from oxidative phosphorylation to aerobic glycolysis in order to support cancer cell requirements [40].

5. Extracellular vesicles alter bone homeostasis

Bone homeostasis is a dynamic equilibrium maintained by the opposite activities of bone-forming cells (osteoblasts, OBs) and bone-resorbing cells (osteoclast, OCs); both cell types are involved in *bone remodeling* that consists of continuous bone destruction and formation. These two different processes are finely regulated by the secretion of paracrine/autocrine factors [41].

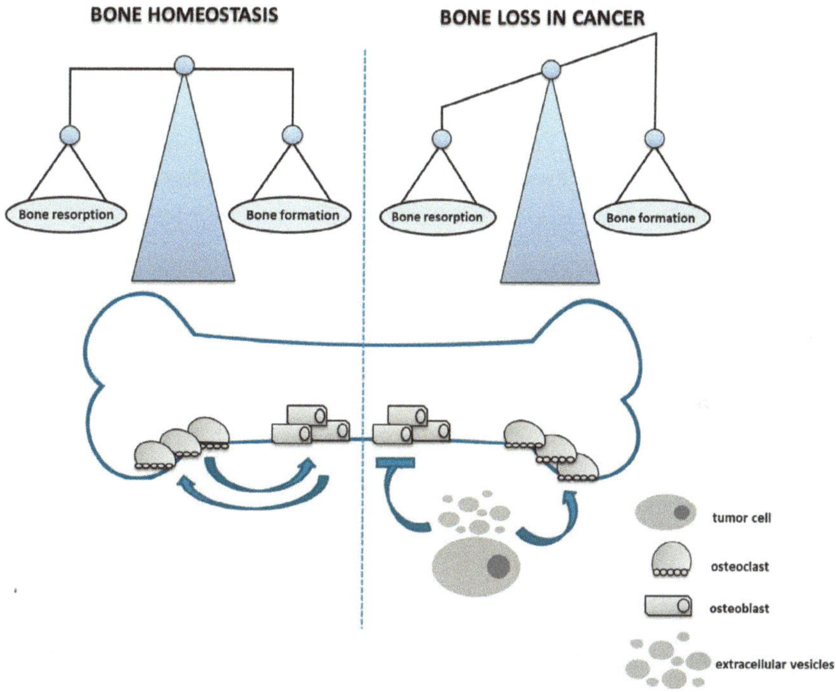

FIG. 2 Schematic representation of bone homeostasis in physiological and pathological conditions. Bone homeostasis is maintained by the balanced action of osteoblastic bone formation and osteoclastic bone resorption. In hematological cancer, extracellular vesicles have been recognized as responsible for the de-regulated bone homeostasis, promoting bone resorption.

The deregulation of bone homeostasis causes the progression of various disorders including hematological cancers [42] in which the resorption prevails over the neoformation leading to the reduction of bone mass, impairment of tissue microarchitecture and an increased risk of fractures (Fig. 2).

Experimental evidences on the role of extracellular vesicles in the modulation of bone homeostasis have recently accumulated. Studies have shown that exosomes released by OCs and OBs regulate bone remodeling by transferring molecules involved in both cell type differentiation. Sun and colleagues reported that osteoclasts secrete exosomes enriched in miR-214; the transfer of this microRNA into osteoblasts led to the inhibition of cell differentiation with the consequent in vivo bone loss [43]. On the other side Cui et al. observed that mineralizing osteoblasts release exosomes that promote bone marrow stromal cell differentiation into osteoblasts by the activation of *Wnt* signaling pathway [44]. Recent experimental evidences have shown that exosomes released by tumor cells are responsible for the altered balance between resorption and new bone formation, particularly in multiple myeloma (MM) [45, 46] and acute myeloid leukemia (AML) [18]. Bone disease and specifically osteolytic lesions is the most common complication of MM patients; in 2015 our group demonstrated that the exposure of pre-osteoclast cells to exosomes from MM cell lines as well as from MM patients increased OC differentiation markers such as Cathepsin K (CTSK), Matrix Metalloproteinases 9 (MMP9) and Tartrate-resistant Acid Phosphatase (TRAP) [45]. In addition, we recently observed that exosomes from MM cell lines as well as from the BM aspirates of patients are enriched in the EGFR ligand amphiregulin that is responsible for the exosome-induced osteoclastogenesis (*manuscript submitted*). We further found that MM-derived exosomes block osteogenic differentiation of MSCs. Our results are in line with data provided by Li et al. that have recently reported that MM cell-derived exosomes decreased the osteogenic potential of MSCs by transferring the lncRNA RUNX2-AS; in addition authors demonstrated that the inhibition of exosome secretion, in in vivo mouse models, prevent bone loss [46]. Similarly, Kumar and collaborators, by evaluating the role of acute myeloid leukemia-derived exosomes in remodeling the bone marrow niche, found that AML exosomes suppress the osteogenic differentiation of mesenchymal stromal progenitors, decreasing osteocalcin gene expression and increasing the negative regulator of OB differentiation, DKK1 [18]. Overall these recent evidences have shed light on the role of tumor exosomes in promoting bone loss; further studies will be needed in order to better characterize the molecular content of exosome responsible for bone impairment.

6. Extracellular vesicles regulate immune cell functions

The immune system is essential in tumor progression modulation and killing of cancer cells. Immune cells, from both the innate and adaptive immune systems, are able to recognize, attack, and eliminate tumor cells. They collaborate and stimulate each other to induce a potent immune response against the tumor cells. However, cancer still occurs. Tumor cells are able to fight back or hide from the immune system surveillance. The interaction between tumor and immune system consists of three phases and is known as cancer immunoediting [47]. During the first phase (the elimination or cancer immunosurveillance phase) the immune system is able to protect the host from a developing tumor. However, some transformed cells may escape the immunological pressure, thereby entering the second phase (the equilibrium phase). During this period of immune-mediated latency, the tumor persists and accumulates new mutations. This may allow the tumor to enter the third phase (tumor escape) during which cancer becomes clinically manifest [48, 49]. Tumor cells have developed numerous strategies to change their phenotype and mislead the immune system. Tumor cells can change their cell surface phenotype to prevent recognition and binding by immune effector cells. Recently, increasing evidence focused on exosomes as major participants in the crosstalk with immune system (Fig. 3). Tumor-derived exosomes, which carry several membrane-bound and soluble factors, may induce immune evasion with different mechanisms. They are able to participate

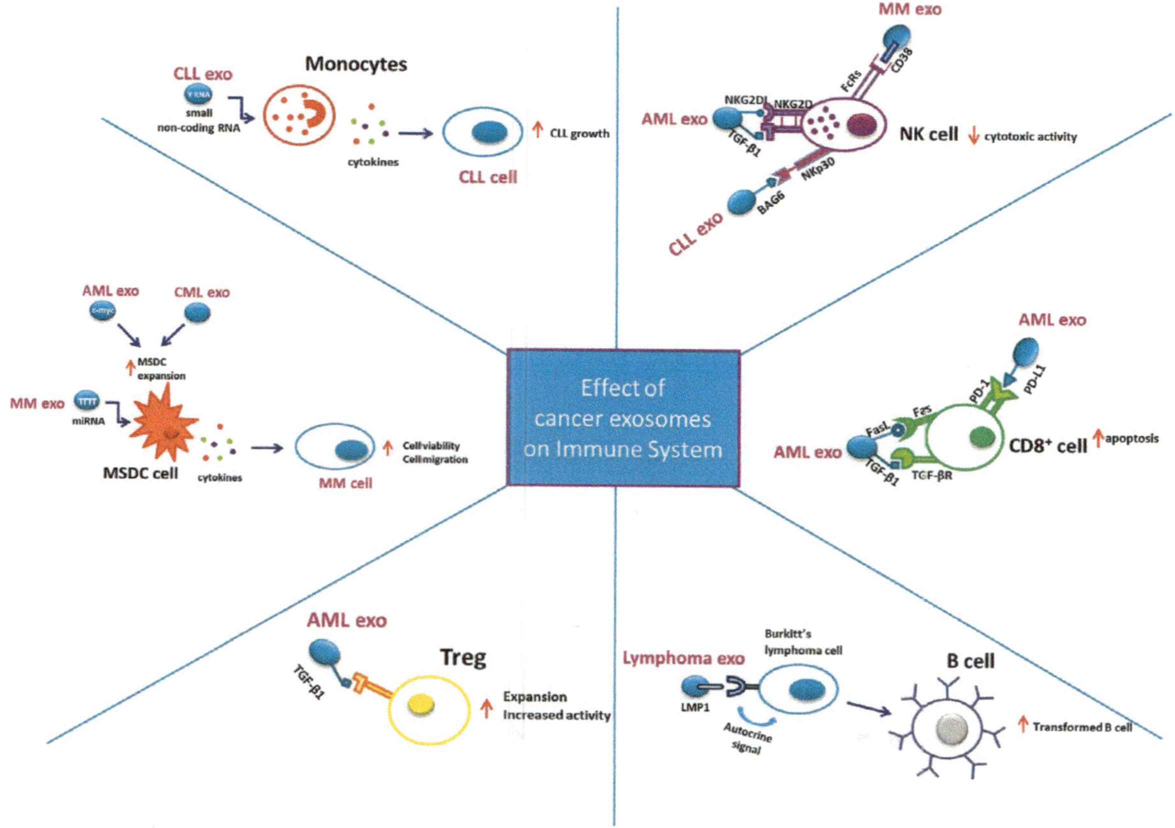

FIG. 3 Schematic representations of cancer cell-derived extracellular vesicles in the regulation of immune cell functions. Tumor extracellular vesicles participate in various stages of the immune response affecting the different immune cell types.

in various stages of immune response and help tumors to escape immune surveillance affecting the different type of immune cells: natural killer (NK) cells, CD8+ T cells, T-reg, monocytes and myeloid-derived suppressor cells. The role of exosomes on immune cells behavior has been extensively studied in solid tumor while in hematological diseases further studies are needed.

6.1 Exosomes and NK cells

Natural killer cells (NK) are cytotoxic lymphocytes, part of the innate immune system that act as a first line of defense against viral-infected cells and tumor cells. The main feature of these cells is their ability to kill tumor cells without any priming or prior activation (in contrast to cytotoxic T cells, which need priming by antigen presenting cells). NK cells secrete cytokines such as IFNγ and TNFα, which act on other immune cells as macrophages and dendritic cells to enhance the immune response. Another key feature is the ability of these cells to discriminate between self and non-self. The first activating signal for NK cells is the loss of MHC class I molecules on the surface of cancer cells [50]. MHC class I molecules act as inhibitory signals for NK cells, thus making MHC class I- deficient tumor cells a target for NK cells. NK cells use their large repertoire of inhibitory and activating receptors to detect the transformed cell. Many tumor cells express activating NK ligands (like MICA and MICB) recognized by NKG2D, on NK surface, and are thus efficiently eliminated by NK cells and γδ T cells [51]. After successful recognition of tumor cells, NK cells utilize a pore-forming protein, perforin, and granzymes that activate caspases, thus leading to apoptosis. The apoptotic program can also be activated after the binding of TRAIL or Fas-ligand, expressed by effector cells, to one of its cognate receptors on cancer cells. These NK cells are able to kill tumor cells via cytokines, such as IFN-γ, TNF-α and TNF-β. Cancer impairs NK cell functions by implementing different immune escape strategies, that in hematological disease causes an inadequate NK cell function such as low NK cell number and impaired cytotoxicity. Accumulating evidence has shown the relevance of the NKG2D/NKG2DL system in patients with leukemia/lymphoma. Cell surface-expressed NKG2D ligands mark tumor cell for attack by NK cells, but tumor cells employ the strategy of exosomal secretion of these ligands to escape the immune system. In fact, Leukemia/lymphoma cells constitutively secrete exosomes, carrying NKG2DL of both the MIC and ULBP families, that enhance the suppression of the NKG2D-dependent NK cell cytotoxicity, as described by Hedlund et al. [52]. Exosomes derived by sera of AML patients, showed membrane-associated TGF-β1, which impairs the competence of NK lymphocytes to kill leukemic cells by reducing NKG2D expression and activating the SMAD pathway [53]. Recently, Huang and colleagues demonstrated that exosomes from TGF-β1-silenced lymphoblastic leukemia cells promoted DC maturation and immune function through CD4+ T cell proliferation, and cytotoxic T lymphocyte response [54].

In chronic lymphocytic leukemia (CLL) NK cells display a poor cytolytic activity, due to the exosomal downregulation of BAG6, a ligand for NKp30 activating receptor on NK cells. In physiological condition, BAG 6 is expressed on exosomes surface, engages NKp30 receptor and triggers NK cell cytotoxic activity. CLL cells fail to release BAG6-positive exosomes, and contribute to immune escape in this disease [55]. In addition, CD38-positive exosomes, released by Multiple Myeloma cells, could represent a strategy, for cancer cell, to escape from immune system. The presence of CD38 on MM exosomes is able to generate an anergic immune system by converting nucleotides to adenosine, which is an immune-suppressor and reduces the cytotoxic activity of NK cells against MM cells. Taken together, this evidence suggests that exosome-mediated NK cell dysfunction compromises the immune surveillance to eliminate cancer cells in various hematologic malignancies.

6.2 Exosomes and CD8+ cytotoxic lymphocytes

CD8+ cytotoxic T cell responses are critical for antitumor immunity. After binding to the target cells, CD8+ cells integrate perforin in the cell membrane and cytoplasmic granules containing granzymes pass through the pores from the T cell into the cytoplasm of target cell where induce apoptosis. CD8+ cells are able to induce apoptosis also through the FAS/FASL system. Tumor cells have developed a number of mechanisms to escape from CD8+ activity, through strategies that affect proliferation, activation and apoptosis of these T cells. One of the most well-known example of immune escape strategy is represented by the presence of apoptosis-inducing molecules on exosome surface. Most of CD8+ T lymphocytes in the circulation of cancer patients express Fas, and many express programmed death 1 molecule (PD-1); they are therefore sensitive to apoptosis by exosomes carrying the membrane form of FasL or programmed death ligand 1 (PD-L1). AML exosomes isolated from patients' plasma are enriched in proteins that are able to induce immune suppression, such as TGF-β1, PD-L1 or FasL [56].

6.3 Exosomes and MDSC

Tumor-derived exosomes promote other immune escape mechanisms, for example by triggering myeloid suppressive cells (MDSC) proliferation. These cells are found in large number in lymphoid organs, blood and tumor tissues in cancer patients and are immature myeloid cells. These cells express myeloid marker stimulatory molecules (CD14 and CD11b) and are devoid of costimulatory molecules (HLA DR, CD80, CD86). They spontaneously secret TGFβ and have suppressive activity on activated T lymphocytes since they are able to inhibit T cells proliferation and cytolytic functions [57]. It was shown that tumor derived exosomes alter the monocyte differentiation and maturation into dendritic cells, that are specialized antigen-presenting cells able to initiate an early anti-tumor T cell responses, leading to the generation of myeloid suppressive cells [57]. Furthermore, it was demonstrated that MDSC-mediated promotion of tumor progression depends on exosomal TGF-β, but also on the lipid mediator prostaglandin E2 (PGE2) transported by tumor exosomes [58]. In particular, activated MDSCs reduce the number of CD4+ and CD8+ lymphocytes, as well as NK cells, and down-modulate their cytotoxic capabilities as a result of the interaction between exosomal tumor HSP72 (heat shock protein 27), myeloid-derived TLR-2 (Toll-like receptor 2) and MyD88 (myeloid differentiation primary response protein 88) [59]. In CML, exosomes promote the expansion of these myeloid suppressive cells that are able to reduce T cell proliferation and then facilitate leukemia progression [60, 61]. AML blasts aberrantly expressed the MUC1 oncogene [62] that supports tumor cell proliferation and resistance to apoptosis [63]. The presence of MUC1 is pivotal for MSDC expansion in AML and it is also responsible for the release of exosomes containing c-myc. Moreover, exosomes carrying c-myc led to an upregulation of cyclin D2 and cyclin E1, c-myc downstream targets, and drive the MSDC proliferation [64]. Recent studies demonstrated that CLL-derived exosomes contain miR-155 that contribute to MSDC expansion, through NF-κB activation [65]. Multiple Myeloma cell-derived exosomes promote MDSC viability and proliferation, and increase their immunosuppressive capacity through the activation of STAT3 pathway. Modulation of this pathway induces high levels of both arginase 1 and iNOS, which can suppress T-cell function thus contributing to immune evasion of MM cells [66]. Finally, miRNA146a in exosomes from MM cells is transferred in MSCs inducing the secretion of elevated levels of cytokines, which improve both MM cell viability and migration [67]. The ability of exosomes to expand MSDC, creating an immune-tolerant environment, results in T cell anergy and facilitates tumor growth.

6.4 Exosomes and Treg

$CD4^+CD25^{high}Foxp3^+$ regulatory T cells (Treg), have an important role in maintaining self-tolerance and modulating the immune response [68]. Tumor cells can efficiently recruit Tregs, and promote their expansion to evade the host immune response. Exosomes shed by cancer cells are involved in this mechanism, especially via TGF-*β. Studies have shown that TGF-β1 is increased in AML exosomes, and TGF-β1 is known to play a role in promoting differentiation and expansion of Tregs* [69].

6.5 Exosomes and monocytes

Exosomes shed by leukemic cells have a prominent role in reprogramming monocytes into tumor-associated macrophages (TAMs) [70]. Monocytes enter into the tumor tissue, where they support tumor initiation, local progression, and distant metastasis [71]. In CLL, malignant B cells and their exosomes accumulate in lymph nodes and blood, resulting in a chronic stimulation and activation of myeloid cells. These exosome are able to transfer Y RNA hY4, a non-coding RNA, to monocytes that induce a release of cytokines thus supporting cancer growth and more importantly causing immune escape through the expression of PDL1 ligand, an immunosuppressive protein [72]. Exosomes shed by lymphoma B cells contain a mutated MYD88 sequence, which helps to reprogram the cell of the microenvironment, such as monocytes-macrophages thus promoting pro-inflammatory signaling pathways [73].

6.6 Exosomes and B cells

B cells are classically known to modulate immune responses and inflammation through antibody production and to promote T-cell activation and proliferation through antigen presentation [74]. Recently, it has been demonstrated a role of B cells in carcinogenesis and tumor progression. The mechanisms involve the secretion of antibodies that promote tumorigenesis, and the release of pro-tumorigenic factors that favor growth and metastasis of cancer cells. Tumor exosomes help B cells in their immunosuppressive function. Exosomes shed by B-cell lymphoma promote cancer progression by carrying components of the Wnt signaling pathway that facilitates the malignant phenotype [75]. Lymphoma- derived exosomes carrying LMP1 oncogene interact with Burkitt's lymphoma cells, triggering an autocrine signal that makes lymphoma cells able to bind B-cells; this interaction induces B-cell proliferation and differentiation toward plasma blast-like phenotype [76]. Furthermore, there are few studies on the ability of cancer exosomes in stimulating regulatory B cells that, in turns, inhibit T cell activity and provide a new immune escape mechanism. These pleiotropic effects lead to hypothesize that interfering with exosome released by tumor cells may perhaps represent a novel strategy for simultaneously recovering multiple immune functions in cancer patients; then, the knowledge on the exosomes-mediated immune evasion could help for the development of a successful, multilevel and multi-target anti-cancer therapy.

7. Extracellular vesicles mediate drug resistance

Despite the advances of conventional therapies in the treatment of solid and hematological tumors, many therapies are not effective in eradicating the disease.

One of the main causes of the failures of anticancer therapies, in addition to the side effects on normal tissues, is the development of drug resistance [77].

Numerous scientific evidences have shown that the bidirectional crosstalk between tumor cells and the cells of the microenvironment plays a fundamental role in the acquisition of pharmacological resistance. In fact, stromal cells, "*educated*" by cancer cells, contribute to the establishment of a protective environment that allows tumor cells to escape the drug treatment.

In this scenario, extracellular vesicles, released by stromal cells, as well as by tumor cells, contribute to pharmacological resistance by transferring nucleic acid and multidrug resistance protein cargo [78, 79]. The first study on the role of extracellular vesicles from hematopoietic malignancies in inducing drug resistance was published in 2009 by Bebawy et al.; in this study authors showed that microparticles from drug-resistant acute lymphoblastic leukemia cell contain the P-glycoprotein, a plasma membrane transporter whose overexpression has been correlated to multidrug resistance. In addition, they found that these microparticles were internalized by drug-sensitive recipient cells and that they were able to confer drug resistance by transferring the protein [80]. Few years later, the same group observed that acute lymphoblastic leukemia cells, that overexpress the Multidrug Resistance-Associated Protein 1 (MRP1), released microparticles enriched in the protein. In addition, they found that exosomal-MRP1 was internalized by drug sensitive cells, leading to the acquisition of MRP1-mediated multidrug resistance in recipient cells [81]. Crompot et al. found that extracellular vesicles released by BM-MSCs of patients with Chronic lymphocytic leukemia (CLL) protect tumor cells from the spontaneous apoptosis, increasing cell viability and migratory capacities, and from the apoptosis induced by the treatment with conventional drugs, such as cladribine and bortezomib [33].

In addition to protein transfer, recent studies have shown that extracellular vesicles can modulate the response to therapies through microRNA transfer. Viola and colleagues found that exosomes derived from the bone marrow aspirates of AML patients are able to protect tumor cells by delivering TGFB1, miR155 and miR375, and hypothesized that they can promote the chemoresistance in AML cells [82]. Min et al. provided evidences on the exosome-mediated drug resistance in chronic myeloid leukemia. Authors found that exosomes released by imatinib-resistant CML cells are internalized by drug-sensitive cells, conferring drug resistance through the delivery of exosomal miR-365; in fact, the transfer of the miRNA leads to the inhibition of pro-apoptotic proteins in recipient cells [83]. Similarly, exosomes from the BM of multiple myeloma patients promote tumor cell survival, by activating c-Jun, p38, p53, and Akt pathways, inducing resistance to bortezomib treatment in MM cells [84].

8. Clinical relevance of exosome for disease diagnosis and drug therapy monitoring

The need to detect cancers at an early stage is generating continuous advances in the area of molecular diagnostics and in our ability to screen tissue or blood samples for tumor-specific genomic, proteomic, and epigenetic signatures. The goal of modern medicine is to identify tumor biomarkers that can allow tumor detection at early stage in an accurate and non-invasive way.

Recent studies suggest that the combination of patient clinical data and extracellular vesicle profile may represent a valid "liquid biopsies" in hematologic malignancies; in fact, EVs, by protecting their cargo from degradation, can provide clinical information about tumor stage and treatment outcome [85]. Caivano and colleagues have performed a correlation between the amount of plasma EVs and hematological cancer; authors isolated serum EVs from patients with different hematological tumors, including chronic lymphocytic leukemia, multiple myeloma, acute myeloid leukemia, and controls. They observed that microvesicle levels were higher in patient samples compared to normal ones, thus highlighting the possibility to use EVs as biomarkers [86].

Hong et al. compared the amount of TGF-β1 in exosomes from newly diagnosed acute myeloid leukemia patients with the quantity of the exosomal proteins among patient cohorts undergoing treatment. Interestingly they found that the level of exosomal-TGF-β1 reflects the response to therapy and that it was higher in newly diagnosed patients compared to healthy controls. Overall this study provide a significant evidence of the use of exosome-associated TGF-β1 as potential diagnostic or prognostic biomarker in acute myeloid leukemia [69]. In addition to exosomal proteins, exosome-associated microRNAs may be considered as biomarkers [87]; Hornick et al. identified a group of miRNAs enriched in AML exosomes from both leukemic blasts and marrow stromal cells, that can be used in clinical diagnosis [88]. In line with these results, more recently, Caivano observed that the EV-associated miR155 was higher in chronic lymphocytic leukemia (CLL) and acute myeloid leukemia (AML) compared to normal samples [89].

9. Conclusions

Advances in the understanding of the role of cell-derived EVs in tumor microenvironment, together with the development of new strategies that allow the rapid and accurate exosome isolation and characterization, will provide the basis for the identification of relevant biomarkers for cancers, including hematological diseases. Results from already present and future studies would make the use of these vesicles accessible in the clinic in the next future, thus improving disease outcome and survival.

Acknowledgment

Riccardo Alessandro received a grant (no. 18783) from the Associazione Italiana per la Ricerca sul Cancro (AIRC). Stefania Raimondo is supported by a "FIRC" (Fondazione Italiana Ricerca sul Cancro) fellowship.

Conflict of interests

The authors declare no conflict of interest.

References

[1] Hanahan D, Weinberg RA. Hallmarks of cancer: the next generation. Cell 2011;144:646–74.

[2] Belli C, Trapani D, Viale G, D'Amico P, Duso BA, Della Vigna P, Orsi F, Curigliano G. Targeting the microenvironment in solid tumors. Cancer Treat Rev 2018;65:22–32.

[3] Sison EAR, Kurre P, Kim YM. Understanding the bone marrow microenvironment in hematologic malignancies: a focus on chemokine, integrin, and extracellular vesicle signaling. Pediatr Hematol Oncol 2017;34:365–78.

[4] Sison EA, Brown P. The bone marrow microenvironment and leukemia: biology and therapeutic targeting. Expert Rev Hematol 2011;4:271–83.

[5] Cho BS, Kim HJ, Konopleva M. Targeting the CXCL12/CXCR4 axis in acute myeloid leukemia: from bench to bedside. Korean J Intern Med 2017;32:248–57.

[6] Hideshima T, Bergsagel PL, Kuehl WM, Anderson KC. Advances in biology of multiple myeloma: clinical applications. Blood 2004;104:607–18.

[7] Corrado C, Raimondo S, Saieva L, Flugy AM, De Leo G, Alessandro R. Exosome-mediated crosstalk between chronic myelogenous leukemia cells and human bone marrow stromal cells triggers an interleukin 8-dependent survival of leukemia cells. Cancer Lett 2014;348:71–6.

[8] Arai F, Hirao A, Suda T. Regulation of hematopoiesis and its interaction with stem cell niches. Int J Hematol 2005;82:371–6.

[9] Mendelson A, Frenette PS. Hematopoietic stem cell niche maintenance during homeostasis and regeneration. Nat Med 2014;20:833–46.

[10] Laurenzana I, Lamorte D, Trino S, De Luca L, Ambrosino C, Zoppoli P, Ruggieri V, Del Vecchio L, Musto P, Caivano A, Falco G. Extracellular vesicles: a new prospective in crosstalk between microenvironment and stem cells in hematological malignancies. Stem Cells Int 2018;2018:9863194.

[11] Ratajczak J, Miekus K, Kucia M, Zhang J, Reca R, Dvorak P, Ratajczak MZ. Embryonic stem cell-derived microvesicles reprogram hematopoietic progenitors: evidence for horizontal transfer of mRNA and protein delivery. Leukemia 2006;20:847–56.

[12] Ekstrom K, Valadi H, Sjostrand M, Malmhall C, Bossios A, Eldh M, Lotvall J. Characterization of mRNA and microRNA in human mast cell-derived exosomes and their transfer to other mast cells and blood CD34 progenitor cells. J Extracell Vesicles 2012;1. https://doi.org/10.3402/jev.v1i0.18389.

[13] Goloviznina NA, Verghese SC, Yoon YM, Taratula O, Marks DL, Kurre P. Mesenchymal stromal cell-derived extracellular vesicles promote myeloid-biased multipotent hematopoietic progenitor expansion via toll-like receptor engagement. J Biol Chem 2016;291:24607–17.

[14] Stik G, Crequit S, Petit L, Durant J, Charbord P, Jaffredo T, Durand C. Extracellular vesicles of stromal origin target and support hematopoietic stem and progenitor cells. J Cell Biol 2017;216:2217–30.

[15] Huan J, Hornick NI, Goloviznina NA, Kamimae-Lanning AN, David LL, Wilmarth PA, Mori T, Chevillet JR, Narla A, Roberts Jr. CT, et al. Coordinate regulation of residual bone marrow function by paracrine trafficking of AML exosomes. Leukemia 2015;29:2285–95.

[16] Hornick NI, Doron B, Abdelhamed S, Huan J, Harrington CA, Shen R, Cambronne XA, Chakkaramakkil Verghese S, Kurre P. AML suppresses hematopoiesis by releasing exosomes that contain microRNAs targeting c-MYB. Sci Signal 2016;9:ra88.

[17] Razmkhah F, Soleimani M, Mehrabani D, Karimi MH, Amini Kafi-Abad S, Ramzi M, Iravani Saadi M, Kakoui J. Leukemia microvesicles affect healthy hematopoietic stem cells. Tumour Biol 2017;39:1010428317692234.

[18] Kumar B, Garcia M, Weng L, Jung X, Murakami JL, Hu X, McDonald T, Lin A, Kumar AR, DiGiusto DL, et al. Acute myeloid leukemia transforms the bone marrow niche into a leukemia-permissive microenvironment through exosome secretion. Leukemia 2018;32:575–87.
[19] Muntion S, Ramos TL, Diez-Campelo M, Roson B, Sanchez-Abarca LI, Misiewicz-Krzeminska I, Preciado S, Sarasquete ME, de Las Rivas J, Gonzalez M, et al. Microvesicles from mesenchymal stromal cells are involved in HPC-microenvironment crosstalk in myelodysplastic patients. PLoS One 2016;11:e0146722.
[20] Moehler TM, Ho AD, Goldschmidt H, Barlogie B. Angiogenesis in hematologic malignancies. Crit Rev Oncol Hematol 2003;45:227–44.
[21] Taverna S, Flugy A, Saieva L, Kohn EC, Santoro A, Meraviglia S, De Leo G, Alessandro R. Role of exosomes released by chronic myelogenous leukemia cells in angiogenesis. Int J Cancer 2012;130:2033–43.
[22] Taverna S, Amodeo V, Saieva L, Russo A, Giallombardo M, De Leo G, Alessandro R. Exosomal shuttling of miR-126 in endothelial cells modulates adhesive and migratory abilities of chronic myelogenous leukemia cells. Mol Cancer 2014;13:169.
[23] Mineo M, Garfield SH, Taverna S, Flugy A, De Leo G, Alessandro R, Kohn EC. Exosomes released by K562 chronic myeloid leukemia cells promote angiogenesis in a Src-dependent fashion. Angiogenesis 2012;15:33–45.
[24] Tadokoro H, Umezu T, Ohyashiki K, Hirano T, Ohyashiki JH. Exosomes derived from hypoxic leukemia cells enhance tube formation in endothelial cells. J Biol Chem 2013;288:34343–51.
[25] Liu Y, Zhu XJ, Zeng C, Wu PH, Wang HX, Chen ZC, Li QB. Microvesicles secreted from human multiple myeloma cells promote angiogenesis. Acta Pharmacol Sin 2014;35:230–8.
[26] Umezu T, Tadokoro H, Azuma K, Yoshizawa S, Ohyashiki K, Ohyashiki JH. Exosomal miR-135b shed from hypoxic multiple myeloma cells enhances angiogenesis by targeting factor-inhibiting HIF-1. Blood 2014;124:3748–57.
[27] Huan J, Hornick NI, Shurtleff MJ, Skinner AM, Goloviznina NA, Roberts Jr. CT, Kurre P. RNA trafficking by acute myelogenous leukemia exosomes. Cancer Res 2013;73:918–29.
[28] Morrison SJ, Scadden DT. The bone marrow niche for haematopoietic stem cells. Nature 2014;505:327–34.
[29] Tripodo C, Sangaletti S, Piccaluga PP, Prakash S, Franco G, Borrello I, Orazi A, Colombo MP, Pileri SA. The bone marrow stroma in hematological neoplasms—a guilty bystander. Nat Rev Clin Oncol 2011;8:456–66.
[30] Ohyashiki JH, Umezu T, Ohyashiki K. Extracellular vesicle-mediated cell-cell communication in haematological neoplasms. Philos Trans R Soc Lond B Biol Sci 2018;373(1737).
[31] Zhou J, Tan X, Tan Y, Li Q, Ma J, Wang G. Mesenchymal stem cell derived exosomes in cancer progression, metastasis and drug delivery: a comprehensive review. J Cancer 2018;9:3129–37.
[32] Roccaro AM, Sacco A, Maiso P, Azab AK, Tai YT, Reagan M, Azab F, Flores LM, Campigotto F, Weller E, et al. BM mesenchymal stromal cell-derived exosomes facilitate multiple myeloma progression. J Clin Invest 2013;123:1542–55.
[33] Crompot E, Van Damme M, Pieters K, Vermeersch M, Perez-Morga D, Mineur P, Maerevoet M, Meuleman N, Bron D, Lagneaux L, Stamatopoulos B. Extracellular vesicles of bone marrow stromal cells rescue chronic lymphocytic leukemia B cells from apoptosis, enhance their migration and induce gene expression modifications. Haematologica 2017;102:1594–604.
[34] Barrera-Ramirez J, Lavoie JR, Maganti HB, Stanford WL, Ito C, Sabloff M, Brand M, Rosu-Myles M, Le Y, Allan DS. Micro-RNA profiling of exosomes from marrow-derived mesenchymal stromal cells in patients with acute myeloid leukemia: implications in leukemogenesis. Stem Cell Rev 2017;13:817–25.

[35] Corrado C, Saieva L, Raimondo S, Santoro A, De Leo G, Alessandro R. Chronic myelogenous leukaemia exosomes modulate bone marrow microenvironment through activation of epidermal growth factor receptor. J Cell Mol Med 2016;20:1829–39.
[36] Ghosh AK, Secreto CR, Knox TR, Ding W, Mukhopadhyay D, Kay NE. Circulating microvesicles in B-cell chronic lymphocytic leukemia can stimulate marrow stromal cells: implications for disease progression. Blood 2010;115:1755–64.
[37] Cheng Q, Li X, Liu J, Ye Q, Chen Y, Tan S. Multiple myeloma-derived exosomes regulate the functions of mesenchymal stem cells partially via modulating miR-21 and miR-146a. Stem Cells Int 2017;2017:9012152.
[38] Chiarugi P, Cirri P. Metabolic exchanges within tumor microenvironment. Cancer Lett 2016;380:272–80.
[39] La Shu S, Yang Y, Allen CL, Maguire O, Minderman H, Sen A, Ciesielski MJ, Collins KA, Bush PJ, Singh P, et al. Metabolic reprogramming of stromal fibroblasts by melanoma exosome microRNA favours a pre-metastatic microenvironment. Sci Rep 2018;8:12905.
[40] Johnson SM, Dempsey C, Chadwick A, Harrison S, Liu J, Di Y, McGinn OJ, Fiorillo M, Sotgia F, Lisanti MP, et al. Metabolic reprogramming of bone marrow stromal cells by leukemic extracellular vesicles in acute lymphoblastic leukemia. Blood 2016;128:453–6.
[41] Rodan GA. Bone homeostasis. Proc Natl Acad Sci U S A 1998;95:13361–2.
[42] Terpos E, Ntanasis-Stathopoulos I, Gavriatopoulou M, Dimopoulos MA. Pathogenesis of bone disease in multiple myeloma: from bench to bedside. Blood Cancer J 2018;8:7.
[43] Sun W, Zhao C, Li Y, Wang L, Nie G, Peng J, Wang A, Zhang P, Tian W, Li Q, et al. Osteoclast-derived microRNA-containing exosomes selectively inhibit osteoblast activity. Cell Discov 2016;2:16015.
[44] Cui Y, Luan J, Li H, Zhou X, Han J. Exosomes derived from mineralizing osteoblasts promote ST2 cell osteogenic differentiation by alteration of microRNA expression. FEBS Lett 2016;590:185–92.
[45] Raimondi L, De Luca A, Amodio N, Manno M, Raccosta S, Taverna S, Bellavia D, Naselli F, Fontana S, Schillaci O, et al. Involvement of multiple myeloma cell-derived exosomes in osteoclast differentiation. Oncotarget 2015;6:13772–89.
[46] Li B, Xu H, Han H, Song S, Zhang X, Ouyang L, Qian C, Hong Y, Qiu Y, Zhou W, et al. Exosome-mediated transfer of lncRUNX2-AS1 from multiple myeloma cells to MSCs contributes to osteogenesis. Oncogene 2018;37:5508–19.
[47] De La Pena H, Madrigal JA, Rusakiewicz S, Bencsik M, Cave GW, Selman A, Rees RC, Travers PJ, Dodi IA. Artificial exosomes as tools for basic and clinical immunology. J Immunol Methods 2009;344:121–32.
[48] Dunn GP, Bruce AT, Ikeda H, Old LJ, Schreiber RD. Cancer immunoediting: from immunosurveillance to tumor escape. Nat Immunol 2002;3:991–8.
[49] Dunn GP, Old LJ, Schreiber RD. The immunobiology of cancer immunosurveillance and immunoediting. Immunity 2004;21:137–48.
[50] Ljunggren HG, Karre K. Host resistance directed selectively against H-2-deficient lymphoma variants. Analysis of the mechanism. J Exp Med 1985;162:1745–59.
[51] Groh V, Rhinehart R, Secrist H, Bauer S, Grabstein KH, Spies T. Broad tumor-associated expression and recognition by tumor-derived gamma delta T cells of MICA and MICB. Proc Natl Acad Sci U S A 1999;96:6879–84.
[52] Hedlund M, Nagaeva O, Kargl D, Baranov V, Mincheva-Nilsson L. Thermal- and oxidative stress causes enhanced release of NKG2D ligand-bearing immunosuppressive exosomes in leukemia/lymphoma T and B cells. PLoS One 2011;6:e16899.

[53] Szczepanski MJ, Szajnik M, Welsh A, Whiteside TL, Boyiadzis M. Blast-derived microvesicles in sera from patients with acute myeloid leukemia suppress natural killer cell function via membrane-associated transforming growth factor-beta1. Haematologica 2011;96:1302–9.

[54] Huang F, Wan J, Hao S, Deng X, Chen L, Ma L. TGF-beta1-silenced leukemia cell-derived exosomes target dendritic cells to induce potent anti-leukemic immunity in a mouse model. Cancer Immunol Immunother 2017;66:1321–31.

[55] Reiners KS, Topolar D, Henke A, Simhadri VR, Kessler J, Sauer M, Bessler M, Hansen HP, Tawadros S, Herling M, et al. Soluble ligands for NK cell receptors promote evasion of chronic lymphocytic leukemia cells from NK cell anti-tumor activity. Blood 2013;121:3658–65.

[56] Hong CS, Sharma P, Yerneni SS, Simms P, Jackson EK, Whiteside TL, Boyiadzis M. Circulating exosomes carrying an immunosuppressive cargo interfere with cellular immunotherapy in acute myeloid leukemia. Sci Rep 2017;7:14684.

[57] Gabrilovich DI, Nagaraj S. Myeloid-derived suppressor cells as regulators of the immune system. Nat Rev Immunol 2009;9:162–74.

[58] Chen W, Jiang J, Xia W, Huang J. Tumor-related exosomes contribute to tumor-promoting microenvironment: an immunological perspective. J Immunol Res 2017;2017:1073947.

[59] Zhang HG, Grizzle WE. Exosomes and cancer: a newly described pathway of immune suppression. Clin Cancer Res 2011;17:959–64.

[60] Whiteside TL. Immune modulation of T-cell and NK (natural killer) cell activities by TEXs (tumour-derived exosomes). Biochem Soc Trans 2013;41:245–51.

[61] Giallongo C, Parrinello NL, La Cava P, Camiolo G, Romano A, Scalia M, Stagno F, Palumbo GA, Avola R, Li Volti G, et al. Monocytic myeloid-derived suppressor cells as prognostic factor in chronic myeloid leukaemia patients treated with dasatinib. J Cell Mol Med 2018;22:1070–80.

[62] Liu S, Yin L, Stroopinsky D, Rajabi H, Puissant A, Stegmaier K, Avigan D, Kharbanda S, Kufe D, Stone R. MUC1-C oncoprotein promotes FLT3 receptor activation in acute myeloid leukemia cells. Blood 2014;123:734–42.

[63] Yin L, Kosugi M, Kufe D. Inhibition of the MUC1-C oncoprotein induces multiple myeloma cell death by down-regulating TIGAR expression and depleting NADPH. Blood 2012;119:810–6.

[64] Pyzer AR, Stroopinsky D, Rajabi H, Washington A, Tagde A, Coll M, Fung J, Bryant MP, Cole L, Palmer K, et al. MUC1-mediated induction of myeloid-derived suppressor cells in patients with acute myeloid leukemia. Blood 2017;129:1791–801.

[65] Bruns H, Bottcher M, Qorraj M, Fabri M, Jitschin S, Dindorf J, Busch L, Jitschin R, Mackensen A, Mougiakakos D. CLL-cell-mediated MDSC induction by exosomal miR-155 transfer is disrupted by vitamin D. Leukemia 2017;31:985–8.

[66] Wang J, De Veirman K, Faict S, Frassanito MA, Ribatti D, Vacca A, Menu E. Multiple myeloma exosomes establish a favourable bone marrow microenvironment with enhanced angiogenesis and immunosuppression. J Pathol 2016;239:162–73.

[67] De Veirman K, Wang J, Xu S, Leleu X, Himpe E, Maes K, De Bruyne E, Van Valckenborgh E, Vanderkerken K, Menu E, Van Riet I. Induction of miR-146a by multiple myeloma cells in mesenchymal stromal cells stimulates their pro-tumoral activity. Cancer Lett 2016;377:17–24.

[68] Sakaguchi S, Yamaguchi T, Nomura T, Ono M. Regulatory T cells and immune tolerance. Cell 2008;133:775–87.

[69] Hong CS, Muller L, Whiteside TL, Boyiadzis M. Plasma exosomes as markers of therapeutic response in patients with acute myeloid leukemia. Front Immunol 2014;5:160.

[70] Ciardiello C, Cavallini L, Spinelli C, Yang J, Reis-Sobreiro M, de Candia P, Minciacchi VR, Di Vizio D. Focus on extracellular vesicles: new frontiers of cell-to-cell communication in cancer. Int J Mol Sci 2016;17:175.

[71] Richards DM, Hettinger J, Feuerer M. Monocytes and macrophages in cancer: development and functions. Cancer Microenviron 2013;6:179–91.

[72] Haderk F, Schulz R, Iskar M, Cid LL, Worst T, Willmund KV, Schulz A, Warnken U, Seiler J, Benner A, et al. Tumor-derived exosomes modulate PD-L1 expression in monocytes. Sci Immunol 2017;2:eaah5509.

[73] Mancek-Keber M, Lainscek D, Bencina M, Chen JG, Romih R, Hunter ZR, Treon SP, Jerala R. Extracellular vesicle-mediated transfer of constitutively active MyD88(L265P) engages MyD88(wt) and activates signaling. Blood 2018;131:1720–9.

[74] LeBien TW, Tedder TF. B lymphocytes: how they develop and function. Blood 2008;112:1570–80.

[75] Koch R, Demant M, Aung T, Diering N, Cicholas A, Chapuy B, Wenzel D, Lahmann M, Guntsch A, Kiecke C, et al. Populational equilibrium through exosome-mediated Wnt signaling in tumor progression of diffuse large B-cell lymphoma. Blood 2014;123:2189–98.

[76] Siravegna G, Marsoni S, Siena S, Bardelli A. Integrating liquid biopsies into the management of cancer. Nat Rev Clin Oncol 2017;14:531–48.

[77] Housman G, Byler S, Heerboth S, Lapinska K, Longacre M, Snyder N, Sarkar S. Drug resistance in cancer: an overview. Cancers (Basel) 2014;6:1769–92.

[78] Jaiswal R, Raymond Grau GE, Bebawy M. Cellular communication via microparticles: role in transfer of multidrug resistance in cancer. Future Oncol 2014;10:655–69.

[79] Gong J, Jaiswal R, Mathys JM, Combes V, Grau GE, Bebawy M. Microparticles and their emerging role in cancer multidrug resistance. Cancer Treat Rev 2012;38:226–34.

[80] Bebawy M, Combes V, Lee E, Jaiswal R, Gong J, Bonhoure A, Grau GE. Membrane microparticles mediate transfer of P-glycoprotein to drug sensitive cancer cells. Leukemia 2009;23:1643–9.

[81] Lu JF, Luk F, Gong J, Jaiswal R, Grau GE, Bebawy M. Microparticles mediate MRP1 intercellular transfer and the re-templating of intrinsic resistance pathways. Pharmacol Res 2013;76:77–83.

[82] Viola S, Traer E, Huan J, Hornick NI, Tyner JW, Agarwal A, Loriaux M, Johnstone B, Kurre P. Alterations in acute myeloid leukaemia bone marrow stromal cell exosome content coincide with gains in tyrosine kinase inhibitor resistance. Br J Haematol 2016;172:983–6.

[83] Min QH, Wang XZ, Zhang J, Chen QG, Li SQ, Liu XQ, Li J, Liu J, Yang WM, Jiang YH, et al. Exosomes derived from imatinib-resistant chronic myeloid leukemia cells mediate a horizontal transfer of drug-resistant trait by delivering miR-365. Exp Cell Res 2018;362:386–93.

[84] Wang J, Hendrix A, Hernot S, Lemaire M, De Bruyne E, Van Valckenborgh E, Lahoutte T, De Wever O, Vanderkerken K, Menu E. Bone marrow stromal cell-derived exosomes as communicators in drug resistance in multiple myeloma cells. Blood 2014;124:555–66.

[85] Marrugo-Ramirez J, Mir M, Samitier J. Blood-based cancer biomarkers in liquid biopsy: a promising non-invasive alternative to tissue biopsy. Int J Mol Sci 2018;19:E2877.

[86] Caivano A, Laurenzana I, De Luca L, La Rocca F, Simeon V, Trino S, D'Auria F, Traficante A, Maietti M, Izzo T, et al. High serum levels of extracellular vesicles expressing malignancy-related markers are released in patients with various types of hematological neoplastic disorders. Tumour Biol 2015;36:9739–52.

[87] Li Q, Liu L, Li W. Identification of circulating microRNAs as biomarkers in diagnosis of hematologic cancers: a meta-analysis. Tumour Biol 2014;35:10467–78.

[88] Hornick NI, Huan J, Doron B, Goloviznina NA, Lapidus J, Chang BH, Kurre P. Serum exosome MicroRNA as a minimally-invasive early biomarker of AML. Sci Rep 2015;5:11295.

[89] Caivano A, La Rocca F, Simeon V, Girasole M, Dinarelli S, Laurenzana I, De Stradis A, De Luca L, Trino S, Traficante A, et al. MicroRNA-155 in serum-derived extracellular vesicles as a potential biomarker for hematologic malignancies—a short report. Cell Oncol (Dordr) 2017;40:97–103.

Chapter 5

Physiological and pathological functions of prostasomes: From basic research to clinical application

Fumihiko Urabe[a,b], Nobuyoshi Kosaka[a,c], Koji Asano[b], Shin Egawa[b], Takahiro Ochiya[a,c]

[a]*Division of Molecular and Cellular Medicine, National Cancer Center Research Institute, Tokyo, Japan,* [b]*Department of Urology, Jikei University School of Medicine, Tokyo, Japan,* [c]*Department of Molecular and Cellular Medicine, Institute of Medical Science, Tokyo Medical University, Tokyo, Japan*

1 Introduction

In recent years, it has become clear that extracellular vesicles (EVs) play important roles in intercellular communication in addition to conventional signaling mechanisms, such as direct cell-to-cell contact or the transfer of soluble signaling molecules [1]. Although EVs were originally described in 1983, interest in these vesicles has increased dramatically in the last ten years after the finding that EVs contain messenger RNA (mRNA) and microRNA (miRNA) and can function within the recipient cells [2–5]. EVs are often categorized as exosomes, microvesicles, and apoptotic bodies based on their secretory origin. Exosomes are referred to as intraluminal vesicles of multivesicular bodies (MVBs), which are released when the MVBs fuse with the cellular membrane [6]. Microvesicles, which are larger than exosomes, are directly shed from the plasma membrane under normal circumstances or in response to stimuli [7]. Apoptotic bodies are formed when the cell undergoes programmed cell death [8]. It is often unclear whether the reported EVs are exosomes, microvesicles or apoptotic bodies. Therefore, it is more adequate to use the more comprehensive term EVs by the recommendation of the International Society for Extracellular Vesicles (ISEV) [9].

In addition to these EVs, prostasomes, which are derived from prostate epithelial cells, are also well-known EVs in reproductive biology and urology. The normal prostate is a walnut-size gland situated between the bladder and

Exosomes. https://doi.org/10.1016/B978-0-12-816053-4.00005-5

pelvic floor in men, and it is the largest male reproductive gland, which plays a key role in the male reproductive system. It secretes prostatic fluid, a milky liquid that constitutes approximately 30% of the semen volume, and the fluid contains prostasomes [10]. One of the first studies reporting the functional interaction for EVs within cells described the promotion of spermatozoa mobility by prostasomes from the prostate epithelial cells [11]. By now, many reports have shown that prostasomes play a multifunctional role in the normal fertilizing process.

On the other hand, prostasomes have been reported not only to function in normal physiological processes but to also be associated with the pathogenesis of prostate cancer [12]. Prostate cancer cells have also been reported to secrete prostasomes, and the secreted prostasomes contribute to the disease progression. In addition, prostasomes reflect and carry the dynamic molecular changes in prostate cancer, allowing for the acquisition of crucial information about the disease status. Therefore, prostasomes hold great potential as useful clinical biomarkers.

In this review, we focus on two types of prostasomes, which are derived from benign epithelial cells and prostate cancer cells, and summarize physiological and pathological contributions of prostasomes in prostates. Furthermore, we discuss the potential of prostasomes for clinical application in prostate cancer treatment.

2 Prostasomes derived from normal prostatic epithelial cells

Prostasomes in seminal plasma and prostatic fluid were first reported in the 1970s [13, 14]. Prostasomes, which have a diameter between 30 and 200 nm, are components of prostatic fluid and are emitted together with the spermatozoa during ejaculation [15, 16]. The acinar epithelial cells mainly generate prostasomes secreted into the prostate glandular lumen and, finally, into the seminal fluid.

Prostasomes correspond in size to vesicles found inside the so-called "storage vesicles" within prostate epithelial cells [15, 17]. The storage vesicles are a source of prostasomes, which was shown by electron microscopic images from both benign and malignant prostate epithelial cells [18]. The storage vesicles are equivalent to MVBs found in other kinds of cells, and the release of prostasomes into the prostate ducts is the result of a fusion process between the storage vesicle and plasma membrane of the prostate acinar cell [15, 17]. Therefore, one subpopulation of prostasomes is considered to be similar to exosomes. In addition, the storage vesicles in prostate epithelial cells are not the only source of prostasomes. Zijlstra et al. proposed that prostasomes also derive by direct shedding from the plasma membrane of prostate epithelial cells and, in this aspect, another subpopulation of prostasomes is similar to that of microvesicles [19].

Prostasomes are reported to contain various kinds of molecules, such as proteins, nucleic acids and lipids. For protein, two reports have shown by comprehensive proteomic analysis that >400 kinds of proteins are contained in prostasomes [20, 21]. Since then, many papers have specified the kinds of proteins contained in prostasomes. Notably, the presence of a ubiquitous EV marker, CD9, and prostate cancer candidate markers, such as prostate-specific antigen (PSA), type 2 transmembrane serine protease (TMPRSS2) and prostate stem cell antigen (PSCA), were repeatedly reported [20–22].

The presence of DNA in prostasomes has also been reported. Ronquist et al. revealed that human prostasomes contain chromosomal DNA, and nuclease treatment demonstrated that prostasome-associated DNA was protected from an enzyme attack, indicating that the DNA was present inside the prostasomes [23]. In addition, they also showed the transfer of DNA from prostasomes to spermatozoa [24]. Regarding other types of nucleic acids, such as miRNAs or long non-coding RNAs, although these are frequently found in EVs derived from cancer, there have been no reports to date on their presence in seminal prostasomes derived from normal prostatic normal epithelial cells.

Prostasomes have a specific lipid composition. While the plasma membrane of conventional mammalian cells contains more abundant quantities of phosphatidylcholine and phosphatidylethanolamine, the main phospholipid in the prostasome membrane is sphingomyelin. In addition, prostasomes have a high cholesterol:phospholipid ratio close to 2:1 [25], whereas the human sperm cell plasma membrane has a ratio of 0.7 [26]. This peculiar lipid composition probably accounts for the very high stability of the prostasomal membrane [27] and has also been reported to contribute to fertilization [28].

Thinking of their size and a variety of molecules in them, it might be a little hard to believe that a single type of prostasome carries all kinds of molecules. As several researchers have focused on the heterogeneity of EVs in cancer [29, 30] the heterogeneity of prostasomes has also been reported. The heterogeneity of prostasomes was first proposed by Poliakov et al. in 2009 [21]. They reported the morphological diversity of seminal prostasomes observed by cryo-electron microscopy [21]. After that, Alberts et al. confirmed a heterologous population of prostasomes with regard to the size by electron microscopy and protein composition by immunoblotting [22]. They reported two distinct populations of prostasomes with characteristic sizes, which are separated in sucrose gradients. CD9, the ubiquitous EV marker, and PSCA, a prostate specific protein, were present in both populations. However, the GLIPR2 protein was found to be specifically enriched in smaller vesicles, while annexin A1 was uniquely associated with the surface of larger vesicles [22]. This report suggested that the structural heterogeneity of prostasomes may translate into functional diversity, with different types of prostasomes reflecting various and specific functions.

3 Physiological contribution of prostasomes

Prostasomes are brought into contact with sperm cells during ejaculation. As we have mentioned, prostasomes contain various kinds of molecules; therefore, the transfer of these molecules via prostasomes provides both stimulatory and inhibitory effects on spermatozoa function. In this section, we summarize the roles of prostasomes in fertility (Fig. 1).

3.1 Sperm motility and prostasomes

Spermatozoa must migrate through the female genital tract, particularly, by moving through cervical mucus and penetrating the zona pellucida [31]. Therefore, the sperm motility is an important factor in evaluating semen quality, and the motility pattern influences the fertilizing capacity of spermatozoa. Several studies have reported that prostasomes may adhere to and fuse with spermatozoa and increase the overall motility [32, 33].

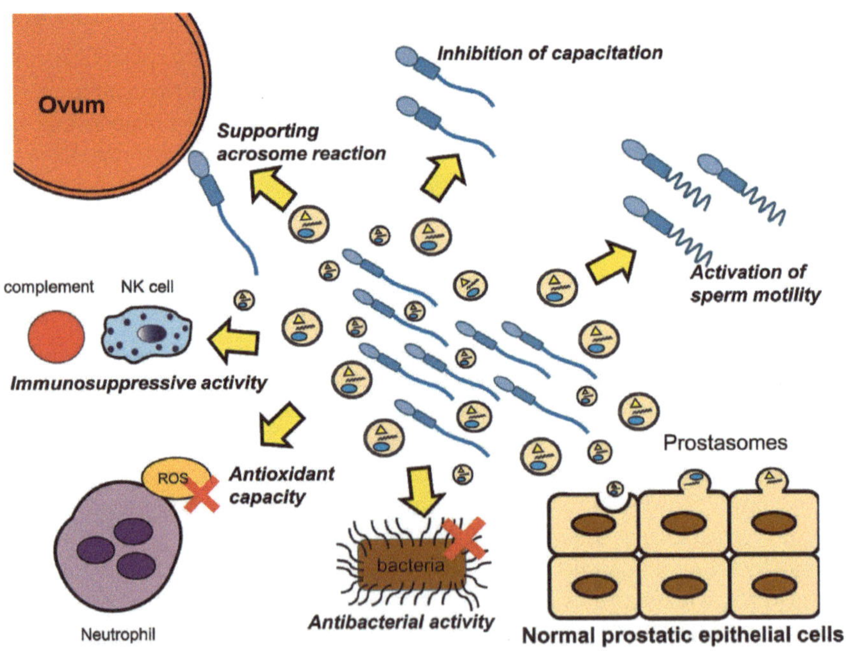

FIG. 1 The role of prostasomes in the process of fertilization. Prostasomes increase sperm motility and inhibit capacitation, which causes it to not occur prior to reaching the ovum. Prostasomes have also been reported to be immunosuppressive, which is important to prevent sperm destruction by the female immune system. In addition, prostasomes themselves have antibacterial activity. Furthermore, prostasomes support the acrosome reaction to promote fertilization. The interaction between prostasomes and spermatozoa seem to be complicated and essential for fertilization at appropriate times and places.

Fabiani et al. reported that prostasomes vigorously promote sperm movement compared to albumin at 37°C [32]. Arienti et al. showed that prostasomes influence the number of motile spermatozoa surviving even the exposure to low pH values [33]. Sperm motility is regulated by Ca^{2+} signaling; therefore, the long-lasting activation of Ca^{2+} signaling is required. However, spermatozoa are tiny cells with few organelles, and they do not produce or contain Ca^{2+} signaling protein. Although the mechanisms had been controversial, Park et al. elucidated the details of sperm motility promotion by prostasomes in 2011. The authors reported that spermatozoa can attain Ca^{2+} signaling through the transfer of a range of calcium ion-signaling tools from prostasomes [34]. This acquisition of Ca^{2+} signaling machinery seems to be required for the continuous motility of spermatozoa and male fertility.

3.2 Immunosuppressive activity of prostasomes

The female genital tract is equipped with a well-balanced immune system. Spermatozoa are considered invading pathogens in the female reproductive tract and, indeed, macrophages, neutrophil granulocytes and natural killer (NK) cells are reported to attack sperm cells [35]. Therefore, spermatozoa need to evade female immune effectors to survive. Several studies reported that prostasomes have a variety of immunomodulatory capacities [36–39]. Skibinsiki et al. reported that prostasomes bind to neutrophils and directly inhibit their phagocytic ability [36]. Tarazona et al. reported that prostasomes expressed a high level of CD48, which is the ligand for activating receptor CD244. The interactions between NK cells and prostasomes resulted in a decrease in the CD244 expression and inhibited the activation of NK [37]. These results suggested that human prostasomes play a role in regulating neutrophils and NK cell activity in the female reproductive tract. In addition, prostasomes have been described to interfere with the complement system [38, 39]. Prostasomes contain CD59 and CD46, which are inhibitors of the membrane attack complex of the complement system, and CD59 and CD46 are transferred from prostasomes to spermatozoa [38, 39].

3.3 Antioxidant capacity of prostasomes

Reactive oxygen species (ROS) are one of the major causes of idiopathic male infertility. Indeed, in 40% of semen samples from infertile men, an abnormally high production of ROS has been shown; however, no ROS production was found in samples from fertile donors [40]. Although the origin of ROS in semen remains controversial, the major source of ROS generation is thought to be leucocyte infiltration to semen [41]. Spermatozoa have little defense against ROS, which damage their membrane and DNA. Saez et al. reported that prostasomes inhibit ROS production [42]. They proposed that prostasomes inhibit the NADPH oxidase activity of polymorphonuclear neutrophils by lipid transfer from prostasomes to the plasma membrane of these cells, resulting in the inhibition of ROS production [42].

3.4 Antibacterial activity of prostasomes

Several reports have shown that prostasomes have antibiotic-like roles. Strub et al. reported that prostasomes contain an antibacterial peptide [43]. The neuroendocrine markers chromogranin A and chromogranin B were reported to be contained in prostasomes, and these chromogranins had an unusual composition, with chromogranin B in excess over chromogranin A [44]. A C-terminal fragment of chromogranin B was found to have a potent antibacterial activity. This peptide forms a three-dimensional structure similar to that in the insect-derived proteins and cecropins and provides prostasomes with antibacterial activity [43]. On the other hand, Carlsson et al. revealed that the antibacterial activity of prostasomes was related to bacterial membrane deformation [45]. These processes are mechanistically different from the antibacterial effect of neutrophil granulocytes, in which the mechanism involves the generation of ROS. As we have referred above, spermatozoa have little defense against ROS; therefore, prostasomes may be more appropriate than neutrophil granulocytes to serve as antibacterial agents in semen.

3.5 Capacitation and acrosome reaction and prostasomes

Spermatozoa that are delivered in the female reproductive tract are mature but not completely primed for fertilization. To acquire the fertilizing potential, spermatozoa need to undergo a series of modifications in a process referred to as capacitation [46, 47]. During this process, spermatozoa acquire the capacity to bind to the zona pellucida of the oocyte and are primed to undergo the acrosome reaction [10]. The intracellular signaling pathway implicated in capacitation has been characterized, and prostasomes affect capacitation. A decrease in the cholesterol content of the sperm membrane is one of the first events leading to capacitation. However, since only a few spermatozoa manage to reach the oocyte, capacitation should not occur prematurely. Seminal plasma has been identified as having an inhibitory influence on capacitation [48]. Specifically, as prostasomes are rich in cholesterol, these vesicles are most-responsible for this inhibition of capacitation [49]. On the other hand, several studies have reported that prostasomes could support the acrosome reaction. Basically, progesterone secreted by cumulus cells is one of the major factors that activate the acrosome reaction [50]. Palmerini et al. reported that prostasome-sperm fusion can stimulate the acrosome reaction making sperm cells more sensitive to the effect of progesterone [51]. In addition, as Park et al. reported, prostasomes also confer progesterone receptors to spermatozoa, and contribute to acrosome reaction [34].

TABLE 1 The role of human prostasomes that support successful fertilization

Functions	Comments	Ref.
Activation of sperm motility	Sperm motility is regulated by Ca^{2+} signaling. The sperm motility depends on the transfer of Ca^{2+} signaling tools form prostasomes.	[32–34]
Immunosuppressive activity	Spermatozoa are considerd as invading pathogens in female reproductive tract. Prostasomes have a variety of immunomodulatory capacities, such as regulating neutrophils and NK cell activity, and interfering with the complement system.	[36–39]
Antioxidant capacity	Reactive oxygen species (ROS) are one of the major causes of idiopathic male infertility. Prostasomes inhibit the ROS production by lipid transfer to the plasma membrane of polymorphonuclear	[42]
Antibacterial activity	Prostasomes contain a C-terminal fragment of chromogranin B, which forms a similar structure of antibacterial protein. Prostasomes also have antibacterial activity to deform bacterial membrane.	[43–45]
Inhibition of capacitation	Prostasomes inhibit prematurely capacitation via cholesterol transfer.	[48, 49]
Supporting acrosome reaction	The progesterone secreted by cumulus cells is one of the major factors that activate the acrosome reaction. Prostasomes make sperm cell more sensitive to the effect of progesterone, and confer progesterone receptors to spermatozoa.	[50]

These examples show that the interaction between prostasomes and spermatozoa is complicated and essential for fertilization at appropriate time and place (Table 1).

4 Prostasomes derived from prostate cancer

EVs have long been considered as disposal vehicles to eliminate waste out of the cells. However, in 2007, Valadi et al. reported the presence of miRNA and mRNA inside EVs and showed potential functionality of these nucleic acids in recipient cells. Further, in 2010, three independent groups reported that miRNA in EVs can be transferred to recipient cells and have functions within these cells [5, 43]. Since then, especially in the cancer field, a number of reports showed that EVs transfer their components from one cell to another and affect cancer progression. In addition, several recent articles have reported that a reduction in cancer-derived EV transfer has the potential to become a new therapeutic strategy for inhibiting cancer proliferation and dissemination [52]. At the same time, EVs harbor various kinds of proteins, nucleic acids, and lipids and are present in many kinds of biological fluids, such as blood, saliva, urine and semen. EVs reflect and carry dynamic changes in diseases, including cancer, allowing us to easily access molecular information about the status of diseases [53]. Therefore, EVs have great potential as useful clinical biomarkers. In this section, we show how prostate cancer-derived prostasomes contribute to the progression of prostate cancer and discuss the development of prostasome-targeting therapy toward clinical application. Furthermore, we also summarize the effectiveness of prostasomes as cancer biomarkers (Fig. 2).

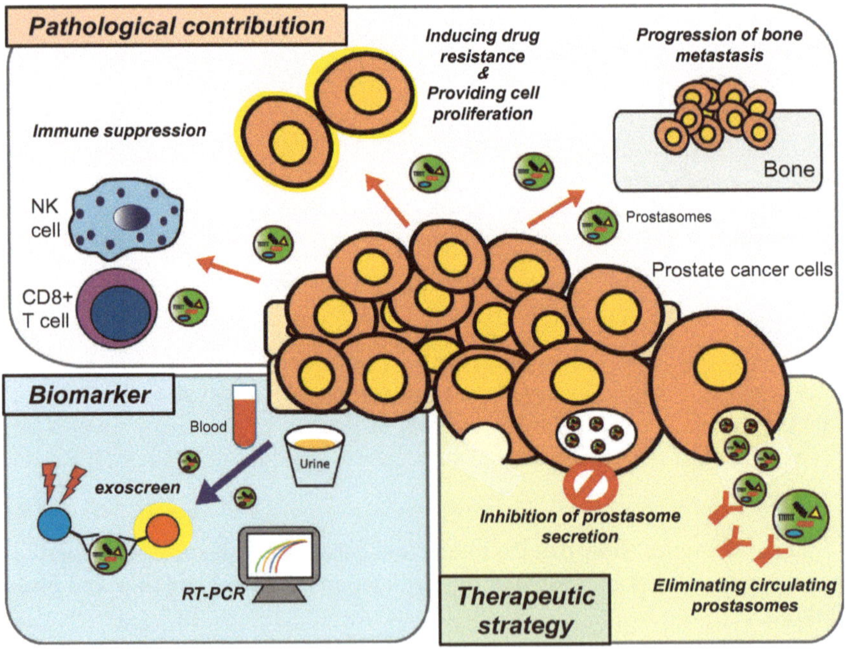

FIG. 2 Pathological contribution of prostasomes derived from prostate cancer and clinical application of the prostasomes. Through the transfer of the content, prostasomes mainly promote tumorigenesis. Tumor-derived prostasomes can contribute to creating an immunosuppressive microenvironment by impairing the function of immune cells. Tumor-derived prostasomes confer proliferation or drug resistance to surrounding prostate cancer cells. In addition, prostasomes dictate bone microenvironments and promote bone metastasis. On the other hand, prostasomes can be useful biomarkers. Tumor-derived prostasomes contain cancer-specific contents and chronologically reflect cancer status. Furthermore, the prevention of intracellular communication via prostasomes could become a new prostate cancer therapeutic strategy.

5 Pathological contribution of prostasomes in prostate cancer

Accumulating evidence indicates that EVs play decisive roles in various cancer phases. Kosaka et al. reported the competition between prostate cancer cells and surrounding normal prostatic epithelial cells through EVs [54]. They further revealed that normal prostatic epithelial cell-derived prostasomes contained tumor suppressor miRNA, miR-143, and attenuated progression of prostate cancer cells in vitro and in vivo [54]. Accordingly, they proposed that tumor suppressor miRNAs in EVs are actively secreted from normal epithelial cells to suppress cancer cells, thereby restoring the tissue to a healthy state during tumor initiation. On the other hand, during tumor progression, EVs derived from various kinds of cancers modulate cancer microenvironment and support tumor proliferation and dissemination via several strategies, such as initiation of a premetastatic niche [55], induction of angiogenesis [56], activation of cancer-associated fibroblasts [57], destruction of the blood-brain barrier [58] or peritoneum [59], and induction of drug resistance [60]. In addition, Ono et al. reported that bone marrow mesenchymal stem cells surrounding cancer secrete EVs, which play an important role in inducing breast cancer cells into a dormant state and, thus, contribute to long-term recurrence [61].

Prostate cancer is the most-frequently diagnosed male tumor and the third leading cause of cancer-related death in the United States [62]. Although the ten-year survival rate of patients with localized prostate cancer is >98%, in patients with metastatic disease, the 5-year relative survival decreases markedly to 30% [62]. Androgen deprivation is the mainstay of metastatic prostate cancer treatment. Despite initial response, almost all patients progress to castration-resistant prostate cancer (CRPC), and the patients will die of the disease within several years. The mechanism of prostate cancer progression is not completely known; therefore, further research is required to improve patients' survival. Previous reports have shown that prostasomes derived from prostate cancer contribute to malignancy by modulating its microenvironment [63–65]. Hosseini-Beheshti et al. reported that prostasomes derived from prostate cancer cells enhance the proliferation and migration of surrounding prostate cancer cells, leading to prostate cancer progression [63]. Lundholm et al. reported a role of prostasomes in immune suppression via NK and CD8+ T cells, which are among the most-important immune cells in cancer immunity. Prostasomes can selectively induce downregulation of an activating cytotoxicity receptor NKG2D in NK and CD8+ T cells in vitro. Consistent with this, they also revealed that circulating EVs in CRPC patients selectively downregulated the expression level of NKG2D on NK cells and CD8+ T cells and suggested the role of prostasomes in impairing the cytotoxic function of lymphocytes and promoting tumor immune escape [64]. While docetaxel offers an improvement in the overall survival of CRPC patients, unfortunately, a relapse is almost inevitable. Corcoran et al. reported that prostasomes play a role as prominent vehicles for the dissemination of cancer drug resistance [65]. In this study, they showed that prostasomes derived from docetaxel-resistant prostate cancer conferred docetaxel-resistance to secondary prostate cancer cells, which may partly be due to the transfer of MDR-1/P-gp via prostasomes [65].

Bone metastases dominate the clinical picture of advanced stage prostate cancer and mostly define morbidity in patients with prostate cancer [66]. In bone metastatic lesions, cancer cells interact with various kinds of cells, including osteoblasts, osteoclasts, and mesenchymal stem cells. Prostate cancer cells frequently induce osteoblastic-type bone metastasis, and several articles have focused on the relationship between prostate cancer cells and osteoblasts. Ito et al. first reported the intracellular communication between prostate cancer cells and osteoblast cells thorough prostasomes. They found that prostasomes derived from human hormone-refractory prostate cancer cells contained Ets-1 protein and induced the differentiation of murine preosteoblastic cells [67]. Ye et al. reported that miR-141-3p in prostasomes promoted osteoblast proliferation and OPG/RANKL expression in vitro by targeting DLC1. In addition, they also showed that prostasomes derived from prostate cancer cells had an osteotropic function in vivo [68]. Recently, Hashimoto et al. reported that prostate cancer-secreted miR-940 reduced the expression level of ARHGAP1 and FAM134A in mesenchymal stem cells, inducing osteogenic differentiation [69]. Furthermore, they also revealed that in an in vivo model, osteoblastic bone metastases were induced by the transfer of miR-940-containing prostasomes [69]. In addition, Karlsson et al. reported that the prostasomes derived from the murine prostate cancer cell line, TRAMPC-1, decreased the differentiation of murine osteoclast precursors to mature multinucleated osteoclasts, promoting osteoblastic bone metastasis [70]. Although the mechanism of prostate cancer progression is quite complex, and further studies are needed, these studies indicate the decisive role of EVs in prostate cancer progression (Table 2).

6 Clinical application of prostasomes for prostate cancer treatment

As we have shown, prostasomes derived from prostate cancer cells confer phenotypical changes to their microenvironment to provide growth benefits to primary or metastatic sites. Recently, many researchers have been focusing on these characteristics of EVs and progressing toward clinical application in cancer treatments. From another aspect, prostasomes contain various kinds of molecules, such as microRNAs and proteins, and reflect the disease condition. In this section, we summarize the current state of EV research regarding the clinical application of prostasomes for prostate cancer diagnosis and treatment.

6.1 Strategy for prostate cancer treatment targeting prostasomes

Intercellular communication via EVs contributes to cancer progression through the transfer of their cargo [53]. Therefore, a reduction in cancer-derived EVs

TABLE 2 The present roles of prostasomes in prostate cancer

Secretion cells	Recipient cells	EV components	Functions	Ref.
Prostate epithelial cells (PNT2)	Prostate cancer cells (PC3M)	miR-143	Inhibit proliferation	[54]
Prostate cancer cells (DU145)	Prostate cancer cells (LNCaP)/ prostate epithelial cells (RWPE-1)	Unknown	Reduce apoptosis, increase proliferation and migration	[63]
Prostate cancer cells (22RV1)	Natural killer cells/CD8+ cells (sorted from PBMCs)	Unknown	Downregulate NKG2D expression and promote immune suppression	[64]
Docetaxel resistant prostate cancer cells (DU145, 22RV1)	Docetaxel sensitive prostate cancer cells (DU145, 22RV1)	MDR-1/P-gp	Confer docetaxel resistance	[65]
Prostate cancer cells (PC3, DU145)	Murine pre-osteoclastic cells (MC3T3-E1)	Ets-1	Induce differentiation	[67]
Prostate cancer cells (MDA PCa 2b)	Osteoblasts (hFOB1.19)	miR-141-3p	Regulate the microenvironment of bone metastasis via activating the p38MAPK pathway	[68]
Prostate cancer cells (C4, C4–2, C4–2B)	Mesenchymal stem cells (immortalized human mesenchymal stem cell line)	miR-940	Downregulate ARHGAP1 and FAM134A and facilitate the osteogenic differentiation	[69]
Murine prostate cancer cells (TRAMP-C1)	Murine osteoclast precursors (RAW264.7, primary mouse bone marrow cells)	Unknown	Decrease fusion and differentiation	[70]

may provide a therapeutic value for inhibiting cancer proliferation and dissemination. Three potential EV therapeutic strategies (inhibition of EV secretion, elimination of circulating EVs, and disruption of EV absorption) have been proposed [52]. Among them, our group has reported the potential of two strategies for EV therapy. Nishida-Aoki et al. reported a novel strategy for therapeutic antibody treatment to target cancer-derived EVs. They showed that in a human breast cancer xenograft model, treatment with antibodies against human-specific CD9 and CD63 enriched on the EV surface, which significantly decreased metastasis, although no obvious effects on primary tumor growth were observed [71]. In another study, Kosaka et al. reported the effectiveness of inhibiting EV production in vitro and in vivo. They revealed that by attenuation of nSMase2, which is required for the synthesis of ceramide, EV secretion and transfer of miR-210-3p, are inhibited, and angiogenesis and metastasis in breast cancer xenograft model are suppressed [56]. However, anti-CD9 and anti-CD63 antibodies cannot selectively identify cancer-derived EVs in humans since CD9 and CD63 are ubiquitous EV markers. In addition, nSMase2 has been reported to be expressed in normal neural cells [72], and the downregulation of nSMase2 does not inhibit EV secretion in prostate cancer [73]. Therefore, to establish cancer-specific EVs, especially for prostasome target therapy, the identification of the prostate cancer-specific genes or molecules that are essential for intracellular communication via prostasomes is required. In 2017, Datta et al. screened two drug libraries to examine the effects of various compounds on prostasome biogenesis [74]. Quantitative high-throughput screened targeting of the GFP signal in a conditioned medium of CD63-GFP-expressing prostate cancer cells treated with compounds was employed. Manumycin-A, a natural microbial metabolite, was identified as an inhibitor of prostasome biogenesis and secretion by prostate cancer but not normal prostatic epithelial cells [74]. In addition, the same group has recently reported several additional potent inhibitors and activators of prostasome secretion using the same high-throughput screening system [75]. These studies showed the potential to target the prostate cancer-specific prostasome biogenesis pathway.

On the other hand, several recent reports have shown the heterogeneity of EVs. In 2016, Kowal et al. reported that different EV fractions have different EV protein markers [29] and, in 2017, Tkach et al. showed that different subpopulations of EVs have different effects on recipient cells [30]. Therefore, the elucidation of the biogenesis details for EV subpopulation with more oncogenic cargo that affects the recipient cells is the most effective strategy. The biogenesis of EVs, including prostasomes, is complicated, and further exploratory research is required for clinical applications. However, the astonishing advances in the EV field promise that these challenges will be overcome.

6.2 Prostasome-associated prostate cancer biomarker

The primary standard of prostate cancer diagnosis is based on the histopathological examination of prostate biopsies. The indications for prostate biopsies mainly rely on digital rectal examination (DRE) and serum PSA. DRE is a subjective examination, and the accuracy depends on the experience of the examiner [76]. In addition, serum PSA lacks the specificity to discriminate benign prostate diseases, such as benign hyperplasia (BPH) and prostatitis. Thus, new biomarkers that can more specifically detect prostate cancer are needed. In addition, prostate cancer is remarkably heterogeneous with tumors ranging from indolent to very aggressive [77]; therefore, the biomarkers that provide information about the severity of the disease and can predict high or low risk of future metastasis are especially required.

Prostasomes reflect the condition of prostate cancer, and they could be a novel biomarker for prostate cancer. However, the detection of prostasomes in bodily fluids have several obstacles since bodily fluids contain EVs from many other sources, and EVs may also reflect the disease condition. Therefore, in this section, we use the term EVs as a general term for all type of vesicles, including prostasomes in the extracellular fluid [9], to introduce the biomarkers for prostate cancer.

Due to the anatomical localization of the prostate and its accessibility, urine appears to be an ideal substrate to detect prostate cancer. Nilsson et al. first showed the presence of two known prostate cancer biomarkers, PCA-3 and TMPRSS2:ERG, in EVs isolated from the urine of prostate cancer patients to show the potential for diagnosis and monitoring cancer patient statuses [78]. Dijkstra et al. reported that digital rectal examination (DRE) enhances the performance of biomarker analysis in urinary EVs. They showed that the expression levels of PCA-3 and TMPRESS2:ERG in EVs were significantly higher when urine was collected after a prostate massage and proposed that the preferential timing of urine collection is after DRE [79]. After that, several studies have reported the effectiveness of urinary EVs as diagnostic biomarkers [80–84]. Recently, McKiernan et al. developed an FDA-approved noninvasive urine exosome gene expression assay, the ExoDx Prostate IntelliScore urine exosome assay. The urine gene expression assay uses a urine sample and utilizes a three EV gene assay signature (PCA3, ERG and SPDEF) to contrast an ExoDx Prostate IntelliScore. They reported that the urine gene expression assay discriminated the high-grade ($\geq$Gleason score 7) from the low-grade (Gleason score 6) cancer and benign cases and can reduce the total number of unnecessary biopsies [85].

EVs isolated from the plasma have been reported to contain prostate cancer-specific proteins. Although the loss of PTEN is commonly associated with an increased risk of aggressive and metastatic prostate cancer, PTEN was found only in plasma-derived EVs from patients with prostate cancer but not from normal subjects [86]. In another study, the anti-apoptotic protein survivin was found to be significantly increased in plasma-derived EVs from patients with prostate cancer compared with patients with BPH or healthy controls [87]. In addition, Tavoosidana et al. developed a highly sensitive and specific method, a so-called proximity-ligation assay, for the detection of EVs derived from plasma. In the assay, they simultaneously targeted multiple proteins on the surface of the same

EVs. After capture by an anti-CD13 monoclonal antibody, the coinciding presence of four EV proteins was probed with DNA-conjugated antibodies to generate amplifiable reporters. With this assay, prostate cancer patients, especially those with high Gleason score (8/9 and 7), were shown to have elevated levels of EVs in plasma samples compared to those with low Gleason score (≤6) and healthy controls [88].

miRNA within serum or plasma prostasomes has a diagnostic and/or prognostic potential for prostate cancer. Although increased levels of miR-141 and miR-375 in serum samples from prostate cancer patients were repeatedly reported [89–91] these miRNAs are also upregulated in serum EVs and are reported as diagnostic biomarkers for prostate cancer [89, 92]. In another study, Bhagirath et al. showed that miR-1246 derived from serum EVs could be a valuable biomarker for diagnosing aggressive prostate cancer. miR-1246 is a prostate cancer tumor-suppressor miRNA, and this miRNA is released in EVs, leading to its high levels in EVs and low levels in prostate cancer cells [93]. In addition, Huang et al. described miR-375 and miR-1290 in serum EVs as potential prognostic biomarkers for patients with metastatic prostate cancer. They performed RNA sequence analysis and found that these miRNAs were significantly associated with poor prognosis in CRPC patients [94].

In addition, although CD9 is a typical EV marker, the expression level of CD9 in EVs is also reported to be a diagnostic and prognostic biomarker for prostate cancer. Soekmadji et al. reported that the number of CD9-positive EVs in plasma is upregulated in prostate cancer patients compared to that of BPH [95]. In another study, they also reported that higher CD9-positive EVs were evident in circulating tumor cell-positive patients compared to negative patients, suggesting that the number of CD9-positive EVs could also be a diagnostic marker for advanced prostate cancer [96].

Therapeutic strategies for CRPC have dramatically changed since 2014, with the advent of novel drugs, such as enzalutamide and abiraterone acetate. While these drugs represent breakthroughs for the treatment of CRPC, unfortunately, approximately 20–40% of patients have no response to these agents [97]. AR-V7, which is a V7 splice variant of androgen receptor, lacks a ligand-binding domain leaving a constitutively active N-terminus [98]. The expression of AR-V7 represents a relatively common cause of a transition to a CRPC phenotype and a cause of resistance to enzalutamide or abiraterone acetate. Del et al. revealed the role of AR-V7 in RNA extracted from EVs in plasma as a predictive biomarker of resistance to hormonal therapy in CRPC patients [99].

As we have shown, the potential of EVs for the development as prostate cancer biomarkers is guaranteed (Table 3). Thinking of clinical application, developing a method for EV isolation and detection without complicated procedure is desirable. Yoshioka et al. established a novel EV-detection method that might facilitate the clinical utilization of EVs by targeting membrane proteins [100]. This method, which is called the ExoScreen assay, can detect circulating EVs in

TABLE 3 Potential use of extracellular vesicles (EVs) in body fluids as prostate cancer biomarkers

Role	Markers	Source	Isolation method	Type assay	Ref.
Diagnostic	PCA-3 and TMPRSS2:ERG	Urine	Ultracentrifugation	RT-PCR	[78, 79]
Diagnostic	lncRNA-21p	Urine	Urine Exosome RNA Isolation Kit	RT-PCR	[80]
Diagnostic	ERG and PCA3 (EXO106 score)	Urine	Urine exosome clinical sample concentrator kit	RT-PCR	[81]
Diagnostic	PSA, CD9 and CD63	Urine	Ultracentrifugation	TR-FIA (time-resolved fluorescence immunoassay)	[82]
Diagnostic	let-7c, miR-21 and miR-375	Urine	Ultracentrifugation	RT-PCR	[83]
Diagnostic	miR-196a-5p and miR-501-3p	Urine	Ultracentrifugation	RNA sequence, RT-PCR	[84]
Diagnostic	PCA3, ERG and SPDEF (ExoDx Prostate IntelliScore)	Urine	EXOPRO Urine Clinical Sample Concentrator Kit, ultracentrifugation	RT-PCR	[85]
Diagnostic	PTEN	Plasma	Ultracentrifugation	Westernblot	[86]
Diagnostic	survivin	Plasma	Ultracentrifugation	ELISA, Westernblot	[87]
Diagnositc	CD13, mAb78, mAb8H10 and human coagulation factor III/tissue	Plasma	Ultracentrifugation and gel chromatography	Proximity ligation assay	[88]
Diagnostic	miR-141 and miR-375	Serum	ExoMiR	RT-PCR	[89]
Diagnostic	miR-141	Serum	ExoQuick	RT-PCR	[92]
Diagnostic	miR-1246	Serum	Total Exosome Isolation reagent, and Plasma/Serum Exosome Purification Kit	RT-PCR	[93]
Prognostic	miR-1290 and miR-375	Plasma	ExoQuick	RNA sequence, RT-PCR	[94]
Diagnostic and Prognostic	CD9	Plasma	Ultracentrifugation	TR-FIA	[95, 96]
Prognostic	AR-V7	Plasma	exoRNeasy kit	RT-PCR	[99]

the conditioned medium or serum based on an amplified luminescent proximity homogeneous assay using photosensitizer beads and two specific antigens residing on EVs. ExoScreen requires only a small amount of a sample (at least 5 μl), eliminates any purification steps and works as a high-throughput analysis. In this study, they reported that EVs derived from plasma could be used to diagnose colorectal cancer more effectively than conventional tumor markers, such as carcinoembryonic antigen (CEA) and carbohydrate antigen 19-9 (CA19-9). Such a method to target membrane protein enables the detection of EVs in a high-throughput manner.

On the other hand, compared to membrane proteins, capturing or purifying the contents packaged in EVs is still unconventional. However, EVs are composed of a lipid bilayer, which can protect their contents from degradation. Notably, RNAs in EVs are stable because EVs protect them from RNases, which are abundant in human body fluids. This protection provides several advantages, such as stability, storage, and reproducibility, in using the EVs as biomarkers. In addition, there are several technologies that can detect nucleic acids, such as quantitative reverse transcription polymerase chain reaction (qRT-PCR) and microarray RNA analysis; therefore, their expression levels could be objectively evaluated. Considering these advantages, nucleic acids in EVs are especially an ideal resource for prostate cancer biomarkers. To stratify the requirements for clinical applications of nucleic acids in EVs, the establishment of high-throughput methods should be pursued. For instance, a combination of EV capture by cancer-specific protein molecules on the membrane and detection of cancer-specific miRNAs by qRT-PCR might be a new standard strategy for EV biomarker in the near future.

7 Conclusion

In this review, we summarize the physiological and pathological roles of prostasomes in humans. Prostasomes have several specific functions to induce fertility within female reproductive tract. On the other hand, prostasomes derive from prostate cancer and contribute to tumor progression, and they could be a therapeutic target. Furthermore, EVs including prostasomes are extremely promising candidates as biomarkers for the diagnosis and/or prognosis biomarkers for prostate cancer. The current expansion of EV research is promising and leading to a growing understanding of the distinctive characteristics and functions of prostasomes.

Acknowledgments

This work was supported by the Practical Research for Innovative Cancer Control (17ck0106366h001) from Japan Agency for Medical Research and Development, AMED.

We thank Dr. Takahiro Kimura (Department of Urology, Jikei University School of Medicine) and Ari Miura (Boston University) for critical discussion regarding this manuscript.

Conflict of interest

None declared.

References

[1] Raposo G, Stoorvogel W. Extracellular vesicles: exosomes, microvesicles, and friends. J Cell Biol 2013;200:373–83.

[2] Valadi H, Ekstrom K, Bossios A, et al. Exosome-mediated transfer of mRNAs and microRNAs is a novel mechanism of genetic exchange between cells. Nat Cell Biol 2007;9:654–9.

[3] Pegtel DM, Cosmopoulos K, Thorley-Lawson DA, et al. Functional delivery of viral miRNAs via exosomes. Proc Natl Acad Sci U S A 2010;107:6328–33.

[4] Kosaka N, Iguchi H, Yoshioka Y, et al. Secretory mechanisms and intercellular transfer of microRNAs in living cells. J Biol Chem 2010;285:17442–52.

[5] Zhang Y, Liu D, Chen X, et al. Secreted monocytic miR-150 enhances targeted endothelial cell migration. Mol Cell 2010;39:133–44.

[6] Yanez-Mo M, Siljander PR, Andreu Z, et al. Biological properties of extracellular vesicles and their physiological functions. J Extracell Vesicles 2015;4:27066.

[7] Cocucci E, Racchetti G, Meldolesi J. Shedding microvesicles: artefacts no more. Trends Cell Biol 2009;19:43–51.

[8] Bergsmedh A, Szeles A, Henriksson M, et al. Horizontal transfer of oncogenes by uptake of apoptotic bodies. Proc Natl Acad Sci U S A 2001;98:6407–11.

[9] Gould SJ, Raposo G. As we wait: coping with an imperfect nomenclature for extracellular vesicles. J Extracell Vesicles 2013;2:. 20389.

[10] Aalberts M, Stout TA, Stoorvogel W. Prostasomes: extracellular vesicles from the prostate. Reproduction 2014;147:R1–14.

[11] Stegmayr B, Ronquist G. Promotive effect on human sperm progressive motility by prostasomes. Urol Res 1982;10:253–7.

[12] Urabe F, Kosaka N, Kimura T, et al. Extracellular vesicles: toward a clinical application in urological cancer treatment. Int J Urol 2018;25:533–43.

[13] Ronquist G, Brody I, Gottfries A, et al. An Mg^{2+} and Ca^{2+}-stimulated adenosine triphosphatase in human prostatic fluid: part I. Andrologia 1978;10:261–72.

[14] Ronquist G, Brody I, Gottfries A, et al. An Mg^{2+} and Ca^{2+}-stimulated adenosine triphosphatase in human prostatic fluid—part II. Andrologia 1978;10:427–33.

[15] Ronquist G, Brody I. The prostasome: its secretion and function in man. Biochim Biophys Acta 1985;822:203–18.

[16] Ronquist GK, Larsson A, Stavreus-Evers A, et al. Prostasomes are heterogeneous regarding size and appearance but affiliated to one DNA-containing exosome family. Prostate 2012;72:1736–45.

[17] Brody I, Ronquist G, Gottfries A. Ultrastructural localization of the prostasome—an organelle in human seminal plasma. Ups J Med Sci 1983;88:63–80.

[18] Sahlen GE, Egevad L, Ahlander A, et al. Ultrastructure of the secretion of prostasomes from benign and malignant epithelial cells in the prostate. Prostate 2002;53:192–9.

[19] Zijlstra C, Stoorvogel W. Prostasomes as a source of diagnostic biomarkers for prostate cancer. J Clin Invest 2016;126:1144–51.

[20] Utleg AG, Yi EC, Xie T, et al. Proteomic analysis of human prostasomes. Prostate 2003;56:150–61.

[21] Poliakov A, Spilman M, Dokland T, et al. Structural heterogeneity and protein composition of exosome-like vesicles (prostasomes) in human semen. Prostate 2009;69:159–67.

[22] Aalberts M, van Dissel-Emiliani FM, van Adrichem NP, et al. Identification of distinct populations of prostasomes that differentially express prostate stem cell antigen, annexin A1, and GLIPR2 in humans. Biol Reprod 2012;86:82.
[23] Ronquist KG, Ronquist G, Carlsson L, et al. Human prostasomes contain chromosomal DNA. Prostate 2009;69:737–43.
[24] Ronquist GK, Larsson A, Ronquist G, et al. Prostasomal DNA characterization and transfer into human sperm. Mol Reprod Dev 2011;78:467–76.
[25] Arienti G, Carlini E, Polci A, et al. Fatty acid pattern of human prostasome lipid. Arch Biochem Biophys 1998;358:391–5.
[26] Carlini E, Palmerini CA, Cosmi EV, et al. Fusion of sperm with prostasomes: effects on membrane fluidity. Arch Biochem Biophys 1997;343:6–12.
[27] Arvidson G, Ronquist G, Wikander G, et al. Human prostasome membranes exhibit very high cholesterol/phospholipid ratios yielding high molecular ordering. Biochim Biophys Acta 1989;984:167–73.
[28] Cross NL. Human seminal plasma prevents sperm from becoming acrosomally responsive to the agonist, progesterone: cholesterol is the major inhibitor. Biol Reprod 1996;54:138–45.
[29] Kowal J, Arras G, Colombo M, et al. Proteomic comparison defines novel markers to characterize heterogeneous populations of extracellular vesicle subtypes. Proc Natl Acad Sci U S A 2016;113:E968–77.
[30] Tkach M, Kowal J, Zucchetti AE, et al. Qualitative differences in T-cell activation by dendritic cell-derived extracellular vesicle subtypes. EMBO J 2017;36:3012–28.
[31] Saez F, Frenette G, Sullivan R. Epididymosomes and prostasomes: their roles in posttesticular maturation of the sperm cells. J Androl 2003;24:149–54.
[32] Fabiani R, Johansson L, Lundkvist O, et al. Promotive effect by prostasomes on normal human spermatozoa exhibiting no forward motility due to buffer washings. Eur J Obstet Gynecol Reprod Biol 1994;57:181–8.
[33] Arienti G, Carlini E, Nicolucci A, et al. The motility of human spermatozoa as influenced by prostasomes at various pH levels. Biol Cell 1999;91:51–4.
[34] Park KH, Kim BJ, Kang J, et al. Ca^{2+} signaling tools acquired from prostasomes are required for progesterone-induced sperm motility. Sci Signal 2011;4:ra31.
[35] Matthijs A, Engel B, Woelders H. Neutrophil recruitment and phagocytosis of boar spermatozoa after artificial insemination of sows, and the effects of inseminate volume, sperm dose and specific additives in the extender. Reproduction 2003;125:357–67.
[36] Skibinski G, Kelly RW, Harkiss D, et al. Immunosuppression by human seminal plasma—extracellular organelles (prostasomes) modulate activity of phagocytic cells. Am J Reprod Immunol 1992;28:97–103.
[37] Tarazona R, Delgado E, Guarnizo MC, et al. Human prostasomes express CD48 and interfere with NK cell function. Immunobiology 2011;216:41–6.
[38] Rooney IA, Atkinson JP, Krul ES, et al. Physiologic relevance of the membrane attack complex inhibitory protein CD59 in human seminal plasma: CD59 is present on extracellular organelles (prostasomes), binds cell membranes, and inhibits complement-mediated lysis. J Exp Med 1993;177:1409–20.
[39] Kitamura M, Namiki M, Matsumiya K, et al. Membrane cofactor protein (CD46) in seminal plasma is a prostasome-bound form with complement regulatory activity and measles virus neutralizing activity. Immunology 1995;84:626–32.
[40] Iwasaki A, Gagnon C. Formation of reactive oxygen species in spermatozoa of infertile patients. Fertil Steril 1992;57:409–16.

[41] Ronquist G. Prostasomes are mediators of intercellular communication: from basic research to clinical implications. J Intern Med 2012;271:400–13.
[42] Saez F, Motta C, Boucher D, et al. Prostasomes inhibit the NADPH oxidase activity of human neutrophils. Mol Hum Reprod 2000;6:883–91.
[43] Strub JM, Garcia-Sablone P, Lonning K, et al. Processing of chromogranin B in bovine adrenal medulla. Identification of secretolytin, the endogenous C-terminal fragment of residues 614-626 with antibacterial activity. Eur J Biochem 1995;229:356–68.
[44] Stridsberg M, Fabiani R, Lukinius A, et al. Prostasomes are neuroendocrine-like vesicles in human semen. Prostate 1996;29:287–95.
[45] Carlsson L, Pahlson C, Bergquist M, et al. Antibacterial activity of human prostasomes. Prostate 2000;44:279–86.
[46] Bailey JL. Factors regulating sperm capacitation. Syst Biol Reprod Med 2010;56:334–48.
[47] Fraser LR. The "switching on" of mammalian spermatozoa: molecular events involved in promotion and regulation of capacitation. Mol Reprod Dev 2010;77:197–208.
[48] Cross NL. Role of cholesterol in sperm capacitation. Biol Reprod 1998;59:7–11.
[49] Pons-Rejraji H, Artonne C, Sion B, et al. Prostasomes: inhibitors of capacitation and modulators of cellular signalling in human sperm. Int J Androl 2011;34:568–80.
[50] Osman RA, Andria ML, Jones AD, et al. Steroid induced exocytosis: the human sperm acrosome reaction. Biochem Biophys Res Commun 1989;160:828–33.
[51] Palmerini CA, Saccardi C, Carlini E, et al. Fusion of prostasomes to human spermatozoa stimulates the acrosome reaction. Fertil Steril 2003;80:1181–4.
[52] Kosaka N, Yoshioka Y, Fujita Y, et al. Versatile roles of extracellular vesicles in cancer. J Clin Invest 2016;126:1163–72.
[53] Urabe F, Kosaka N, Yoshioka Y, et al. The small vesicular culprits: the investigation of extracellular vesicles as new targets for cancer treatment. Clin Transl Med 2017;6:45.
[54] Kosaka N, Iguchi H, Yoshioka Y, et al. Competitive interactions of cancer cells and normal cells via secretory microRNAs. J Biol Chem 2012;287:1397–405.
[55] Peinado H, Aleckovic M, Lavotshkin S, et al. Melanoma exosomes educate bone marrow progenitor cells toward a pro-metastatic phenotype through MET. Nat Med 2012;18:883–91.
[56] Kosaka N, Iguchi H, Hagiwara K, et al. Neutral sphingomyelinase 2 (nSMase2)-dependent exosomal transfer of angiogenic microRNAs regulate cancer cell metastasis. J Biol Chem 2013;288:10849–59.
[57] Pang W, Su J, Wang Y, et al. Pancreatic cancer-secreted miR-155 implicates in the conversion from normal fibroblasts to cancer-associated fibroblasts. Cancer Sci 2015;106:1362–9.
[58] Tominaga N, Kosaka N, Ono M, et al. Brain metastatic cancer cells release microRNA-181c-containing extracellular vesicles capable of destructing blood-brain barrier. Nat Commun 2015;6:6716.
[59] Yokoi A, Yoshioka Y, Yamamoto Y, et al. Malignant extracellular vesicles carrying MMP1 mRNA facilitate peritoneal dissemination in ovarian cancer. Nat Commun 2017;8:14470.
[60] Wei F, Ma C, Zhou T, et al. Exosomes derived from gemcitabine-resistant cells transfer malignant phenotypic traits via delivery of miRNA-222-3p. Mol Cancer 2017;16:132.
[61] Ono M, Kosaka N, Tominaga N, et al. Exosomes from bone marrow mesenchymal stem cells contain a microRNA that promotes dormancy in metastatic breast cancer cells. Sci Signal 2014;7:ra63.
[62] Siegel RL, Miller KD, Jemal A. Cancer Statistics, 2017. CA Cancer J Clin 2017;67:7–30.
[63] Hosseini-Beheshti E, Choi W, Weiswald LB, et al. Exosomes confer pro-survival signals to alter the phenotype of prostate cells in their surrounding environment. Oncotarget 2016;7:14639–58.

[64] Lundholm M, Schroder M, Nagaeva O, et al. Prostate tumor-derived exosomes down-regulate NKG2D expression on natural killer cells and CD8+ T cells: mechanism of immune evasion. PLoS One 2014;9:e108925.

[65] Corcoran C, Rani S, O'Brien K, et al. Docetaxel-resistance in prostate cancer: evaluating associated phenotypic changes and potential for resistance transfer via exosomes. PLoS One 2012;7:e50999.

[66] Ottewell PD, Wang N, Meek J, et al. Castration-induced bone loss triggers growth of disseminated prostate cancer cells in bone. Endocr Relat Cancer 2014;21:769–81.

[67] Itoh T, Ito Y, Ohtsuki Y, et al. Microvesicles released from hormone-refractory prostate cancer cells facilitate mouse pre-osteoblast differentiation. J Mol Histol 2012;43:509–15.

[68] Ye Y, Li SL, Ma YY, et al. Exosomal miR-141-3p regulates osteoblast activity to promote the osteoblastic metastasis of prostate cancer. Oncotarget 2017;8:94834–49.

[69] Hashimoto K, Ochi H, Sunamura S, et al. Cancer-secreted hsa-miR-940 induces an osteoblastic phenotype in the bone metastatic microenvironment via targeting ARHGAP1 and FAM134A. Proc Natl Acad Sci U S A 2018;115:2204–9.

[70] Karlsson T, Lundholm M, Widmark A, et al. Tumor cell-derived exosomes from the prostate cancer cell line TRAMP-C1 impair osteoclast formation and differentiation. PLoS One 2016;11:e0166284.

[71] Nishida-Aoki N, Tominaga N, Takeshita F, et al. Disruption of circulating extracellular vesicles as a novel therapeutic strategy against cancer metastasis. Mol Ther 2017;25:181–91.

[72] Yuyama K, Sun H, Mitsutake S, et al. Sphingolipid-modulated exosome secretion promotes clearance of amyloid-beta by microglia. J Biol Chem 2012;287:10977–89.

[73] Phuyal S, Hessvik NP, Skotland T, et al. Regulation of exosome release by glycosphingolipids and flotillins. FEBS J 2014;281:2214–27.

[74] Datta A, Kim H, Lal M, et al. Manumycin A suppresses exosome biogenesis and secretion via targeted inhibition of Ras/Raf/ERK1/2 signaling and hnRNP H1 in castration-resistant prostate cancer cells. Cancer Lett 2017;408:73–81.

[75] Datta A, Kim H, McGee L, et al. High-throughput screening identified selective inhibitors of exosome biogenesis and secretion: a drug repurposing strategy for advanced cancer. Sci Rep 2018;8:8161.

[76] Varenhorst E, Berglund K, Lofman O, et al. Inter-observer variation in assessment of the prostate by digital rectal examination. Br J Urol 1993;72:173–6.

[77] Boyd LK, Mao X, Lu YJ. The complexity of prostate cancer: genomic alterations and heterogeneity. Nat Rev Urol 2012;9:652–64.

[78] Nilsson J, Skog J, Nordstrand A, et al. Prostate cancer-derived urine exosomes: a novel approach to biomarkers for prostate cancer. Br J Cancer 2009;100:1603–7.

[79] Dijkstra S, Birker IL, Smit FP, et al. Prostate cancer biomarker profiles in urinary sediments and exosomes. J Urol 2014;191:1132–8.

[80] Isin M, Uysaler E, Ozgur E, et al. Exosomal lncRNA-p21 levels may help to distinguish prostate cancer from benign disease. Front Genet 2015;6:168.

[81] Donovan MJ, Noerholm M, Bentink S, et al. A molecular signature of PCA3 and ERG exosomal RNA from non-DRE urine is predictive of initial prostate biopsy result. Prostate Cancer Prostatic Dis 2015;18:370–5.

[82] Duijvesz D, Versluis CY, van der Fels CA, et al. Immuno-based detection of extracellular vesicles in urine as diagnostic marker for prostate cancer. Int J Cancer 2015;137:2869–78.

[83] Foj L, Ferrer F, Serra M, et al. Exosomal and non-exosomal urinary miRNAs in prostate cancer detection and prognosis. Prostate 2017;77:573–83.

[84] Rodriguez M, Bajo-Santos C, Hessvik NP, et al. Identification of non-invasive miRNAs biomarkers for prostate cancer by deep sequencing analysis of urinary exosomes. Mol Cancer 2017;16:156.

[85] McKiernan J, Donovan MJ, O'Neill V, et al. A novel urine exosome gene expression assay to predict high-grade prostate cancer at initial biopsy. JAMA Oncol 2016;2:882–9.

[86] Gabriel K, Ingram A, Austin R, et al. Regulation of the tumor suppressor PTEN through exosomes: a diagnostic potential for prostate cancer. PLoS One 2013;8:e70047.

[87] Khan S, Jutzy JM, Valenzuela MM, et al. Plasma-derived exosomal survivin, a plausible biomarker for early detection of prostate cancer. PLoS One 2012;7:e46737.

[88] Tavoosidana G, Ronquist G, Darmanis S, et al. Multiple recognition assay reveals prostasomes as promising plasma biomarkers for prostate cancer. Proc Natl Acad Sci U S A 2011;108:8809–14.

[89] Bryant RJ, Pawlowski T, Catto JW, et al. Changes in circulating microRNA levels associated with prostate cancer. Br J Cancer 2012;106:768–74.

[90] Yaman Agaoglu F, Kovancilar M, Dizdar Y, et al. Investigation of miR-21, miR-141, and miR-221 in blood circulation of patients with prostate cancer. Tumour Biol 2011;32:583–8.

[91] Haldrup C, Kosaka N, Ochiya T, et al. Profiling of circulating microRNAs for prostate cancer biomarker discovery. Drug Deliv Transl Res 2014;4:19–30.

[92] Li Z, Ma YY, Wang J, et al. Exosomal microRNA-141 is upregulated in the serum of prostate cancer patients. Onco Targets Ther 2016;9:139–48.

[93] Bhagirath D, Yang TL, Bucay N, et al. microRNA-1246 is an exosomal biomarker for aggressive prostate cancer. Cancer Res 2018;78:1833–44.

[94] Huang X, Yuan T, Liang M, et al. Exosomal miR-1290 and miR-375 as prognostic markers in castration-resistant prostate cancer. Eur Urol 2015;67:33–41.

[95] Soekmadji C, Riches JD, Russell PJ, et al. Modulation of paracrine signaling by CD9 positive small extracellular vesicles mediates cellular growth of androgen deprived prostate cancer. Oncotarget 2017;8:52237–55.

[96] Soekmadji C, Corcoran NM, Oleinikova I, et al. Extracellular vesicles for personalized therapy decision support in advanced metastatic cancers and its potential impact for prostate cancer. Prostate 2017;77:1416–23.

[97] Antonarakis ES, Lu C, Wang H, et al. AR-V7 and resistance to enzalutamide and abiraterone in prostate cancer. N Engl J Med 2014;371:1028–38.

[98] Chan SC, Dehm SM. Constitutive activity of the androgen receptor. Adv Pharmacol 2014;70:327–66.

[99] Del Re M, Biasco E, Crucitta S, et al. The detection of androgen receptor splice variant 7 in plasma-derived exosomal RNA strongly predicts resistance to hormonal therapy in metastatic prostate cancer patients. Eur Urol 2017;71:680–7.

[100] Yoshioka Y, Kosaka N, Konishi Y, et al. Ultra-sensitive liquid biopsy of circulating extracellular vesicles using ExoScreen. Nat Commun 2014;5:3591.

Chapter 6

The function and therapeutic use of exosomes in bacterial infections

Yong Cheng, Jeffery S. Schorey
Department of Biological Sciences, Eck Institute for Global Health, Center for Rare and Neglected Diseases, University of Notre Dame, Notre Dame, IN, United States

1 Introduction

Virulence factors released by bacterial pathogens are known to target host cells or components of the innate and specific immune response. The targets can vary for different bacteria species and even between different strains of the same species but in general function to either limit or inhibit the activation of the immune response. Examples include proteins which block complement activation (e.g. SCIN-B by *Staphylococcus aureus*) [1] or inactivate chemotactic factors (e.g. beta-hemolysin by *S. aureus*) [2]. Bacteria can also release factors that can induce cell lysis or block phagocytosis (e.g. M-protein by *Streptococcus pyogenes*) [3]. These approaches to immune subversion can be overcome through mechanisms such as production and release of reactive oxygen and nitrogen species by macrophages and neutrophils. Activation of phagocytes and subsequent killing of ingested bacteria is another mechanism to control a bacterial infection. The activation of phagocytes is facilitated by release of bacterial products (known as PAMPs) which functions through pattern recognition receptors. Activation is further induced by cytokines such as IFN-γ which is primarily released from activated Th1 cells. Stimulation of T and B cells requires exposure to bacterial antigens. For T cells this requires antigen uptake by antigen presenting cells and presentation on MHC (major histocompatibility complex) class II or CD1 molecules for T cell receptor binding. For extracellular bacterial pathogens such as *S. aureus*, the release of bacterial PAMPs (pathogen-associated molecular patterns) and antigens occurs during bacterial growth in tissue or from lysis of bacteria within tissue. For intracellular pathogens, the mechanism of their exposure to phagocytes, B cells, T cells or other immune cells is less clear as the PAMPs and antigens are sequestered within the host cell. Recent studies have

Exosomes. https://doi.org/10.1016/B978-0-12-816053-4.00006-7

suggested that one mechanism for delivery of these bacterial components to the immune system is through release of extracellular vesicles, such as exosomes. However, studies have also suggested that extracellular vesicles (EVs) can carry virulence factors to promote bacterial survival and replication. Therefore during an infection the content of these EVs (both host and bacteria derived) can have a significant effect on who has the advantage in the battle between our immune system and the pathogen. In addition to modulating the host immune response, EVs in the context of bacterial infections can be used therapeutically as disease biomarkers, vaccines and drug delivery systems, which will be discussed at the end of this chapter (Fig. 1).

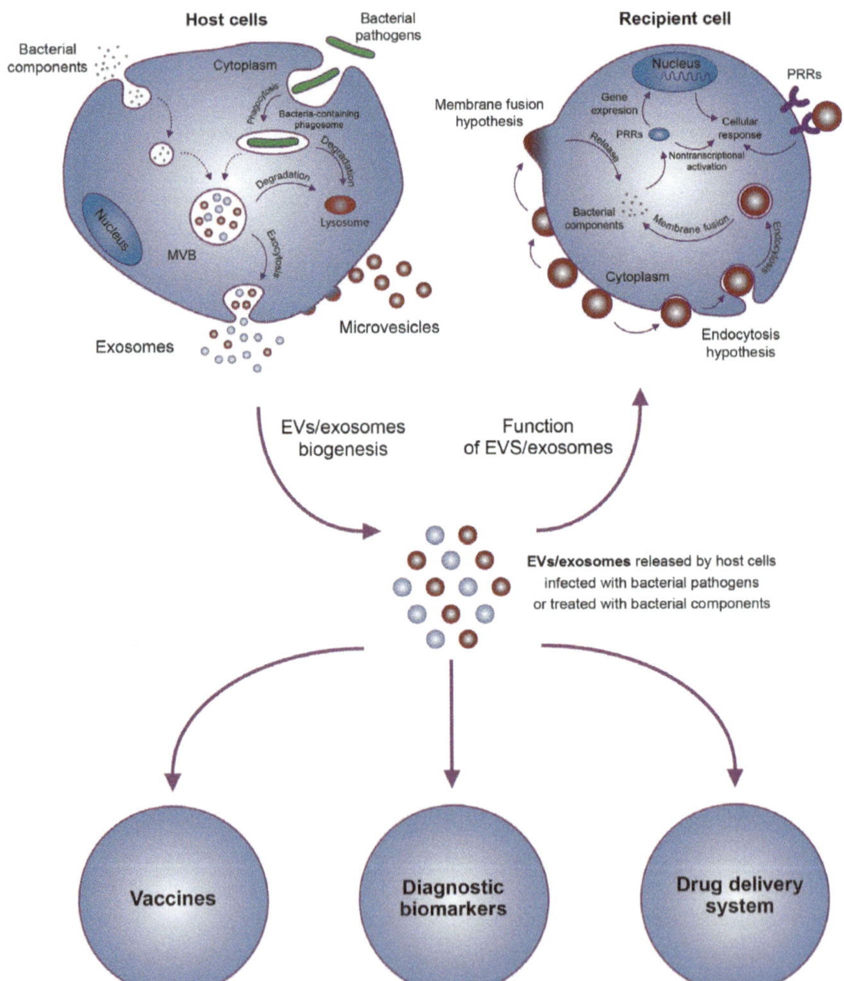

FIG. 1 General mechanisms of EV production and uptake by recipient cells and their therapeutic potential.

2 Extracellular vesicles

In the context of a bacterial infection, EVs can be both host- and bacteria-derived and are broadly define as membrane-bound vesicles released from cells. EVs from bacteria include outer membrane vesicles from gram-negative and membrane vesicles from gram-positive bacteria. A number of excellent reviews have summarized the latest in the field of bacteria-derived vesicles and therefore will not be addressed in this review [4–6]. These vesicles likely play an important role during the course of an extracellular bacterial infection; however, their role in the context of an intracellular pathogen is less clear, as mechanisms to transport the vesicles outside the host cell remains to be defined. EVs differ in their biogenesis and composition and include three main categories; apoptotic bodies, exosomes and microvesicles. All three vesicle populations are enclosed by a lipid bilayer, but vary in size and composition. In contrast to microvesicles, which are generated by budding from the plasma membrane [7], exosomes are derived from the endolysosomal pathway and have a unique lipid and protein makeup. Exosomes have been the most studied in regards to infectious agents; however, the purity of the exosomes was not consistently established and the vesicle population may contain both exosomes and microvesicles since they partially overlap in size and density. Nevertheless, we will use the terminology as defined in the original paper when discussing the results.

3 Exosomes

Exosomes are formed through the fusion of multivesicular bodies (MVBs) with the plasma membrane and release of intra-luminal vesicles (ILVs) as exosomes [8]. Exosomes are 30–150 nm lipid bilayer vesicles with a density of 1.13–1.19 g/mL. Biophysically, exosomes are equivalent to cytoplasm enclosed in a lipid bilayer with the external domains of transmembrane proteins exposed to the extracellular environment. Exosomes have been shown to be released from cells of hematopoietic origin, including B cells [9], mastocytes [10], dendritic cells (DCs) [11, 12], platelets [13], and macrophages [14] as well as cells of non-hematopoietic origin such as neurons and epithelial cells [15–17]. The exosomes can act locally at the site of organ or can circulate through various bodily fluids including blood and lymph resulting in a systemic response [18]. Exosomes were first identified in the culture media of reticulocytes [19, 20]; however, over the past three decades the study of exosomes has extended to encompass most cell types as well as their isolation from different organisms including unicellular eukaryotes. This conservation of exosome production and function through evolution suggest a strong selection for this mechanism of cellular communication. The advantage of using exosomes for cell-to-cell communication stems from their compositional complexity which allows for more control over the communication process. Moreover, the presence of signaling lipids, proteins and various species of RNA within a single structure can lead to rapid and profound changes in the target cell, allowing for a swift response to cellular perturbations. These changes may be induced under physiological or pathological conditions. Although the complexity of exosomes has clear benefits to the organisms that produce them, it has made the study of their function challenging as exosome or pool of exosomes contain a diverse array of components many of which can induced a cellular response in the target cell. Moreover, the tools we have to modulate exosome production and composition in vitro and in vivo are limited, hindering our ability to define exosome function in normal and diseased states. Nevertheless, we have garnered important insight into exosome biogenesis, composition and function over the past decade; a decade that has seen a significant increase in publications on exosomes and other extracellular vesicles.

3.1 Exosomes biogenesis

An important mechanism for removing plasma membrane receptors is through their endocytosis and trafficking to an MVB. The subsequently fusion of the MVB with a lysosome results in the degradation of the endocytosed receptor [21]. However, not all MVBs fuse with lysosomes as a sub-population can fuse with the plasma membrane resulting in the release of the ILVs extracellularly. At present it is unclear if the MVBs that fuse with the lysosome consist of a different subset relative to those that fuse with the plasma membrane or if a particular MVB, under the proper conditions, could follow either path. Data with the trafficking of MHC molecules suggest that there may indeed be separate populations [22]. Despite these recent studies, there remains major gaps in our understanding of MVB biogenesis and exosome release. Several models have been suggested a mechanism for ILV formation. Initial studies in yeast demonstrated a role for the Endosomal Sorting Complex Required for Transport proteins (ESCRT) [23]. Although the ESCRT machinery has primarily been studied for its role in the endosomal sorting and protein degradation, additional studies have imply a role for the ESCRT proteins in membrane invagination [24, 25]. Through its ubiquitin interacting domain, ESCRT-0 clusters ubiquitinate proteins for delivery into MVBs [26]. And then, ESCRT-0 recruits ESCRT-1 to the endosomal membrane, which subsequently recruits the remaining members of the ESCRT machinery; ESCRT-II and ESCRT-III respectively [27, 28]. Through the formation of polymeric filaments mediated by ESCRT-III, membrane invagination results in IVL formation [29] (for a recent review see [30]). In support of the ESCRT being involved in ILV formation, proteomic studies of exosomes have identified the presence of ESCRT proteins in exosomes and knockdown of key components of ESCRT machinery can abrogate ILV formation and exosome release [31]; however, this is likely cell-type specific [32, 33]. Although this general model for MVB biogenesis is supported by published data, it is unclear if this constitutes the primary mechanism of MVB formation as there are published studies which show ESCRT independent mechanisms for MVB biogenesis and exosome release. In support of an ESCRT-independent mechanism Stuffers et al. found that depleting specific subunits from the four ESCRTs complexes did not completely inhibit MVB formation [32]. Studies by van Niel et al. found the tetraspanin, CD63, which is a protein found at high concentrations on exosomes, can promote cargo sorting and ILV formation [34]. Additionally, CD81 has been demonstrated to meditate cargo sorting of tetraspanin ligands such as Rac GTPase, although knockdown of this tetraspanin does not appear to alter MVB morphology or exosome secretion [35]. Altogether, the published data suggest that there are multiple mechanisms that can promote MVB biogenesis and exosome production as well as cargo loading. These mechanisms can differ between cell types and potentially even between different subpopulations of MVBs within a cell. In support of the latter, Buschow and colleagues have shown that within immature DCs, the MHC molecules which are targeted to MVBs low in cholesterol but enriched for lysobisphosphatidic are destined for lysosomal degradation. In contrast, in mature DCs, the MHC molecules which are sorted into MVBs enriched in CD9 and cholesterol are targeted for fusion with the plasma membrane [22].

The fusion of the MVB with the plasma membrane is mediated by the cytoskeleton, fusion machinery such as the SNARE and molecular switches (i.e. small molecular weight GTPases) [36]. Rab GTPases are members of the Ras GTPase superfamily and are known to regulate four steps in membrane trafficking: vesicle formation, vesicle trafficking, tethering, and fusion with target organelles. Presently, nearly 70 different Rab GTPases have been identified in mammalian cells [37] and a number of them have been found in exosomes: including Rab5, Rab11, Rab27, and Rab35. Some of these Rab effectors have been experimentally shown to function in exosome release. Initial studies supported a role for Rab11 in MVB fusion with the plasma membrane in the K562 erythroleukaemic cell line [38]. Additional studies have implicated Rab35 in mediating MVB docking to the plasma membranc in ncuralgia cells and depletion of Rab35 resulted in a significant loss in exosomes release by these cells [39]. More recent studies have demonstrated a role for Rab27a and Rab27b in MVB biogenesis. Data suggest that Rab27a and Rab27b have different but sometimes redundant functions in the MVB biogenesis pathway with Rab27a playing a more prominent role in mediating MVB docking to the plasma membrane [40]. While the Rab-GTPases have been involved in MVB trafficking and fusion, it is important to note that their specific role in the process is still being defined and will likely be cell-type dependent as well as dependent on the physiological/pathological state of the cell.

4 Extracellular vesicles, pathogens and infectious diseases

Exosomes and other EVs have been isolated and characterized in the context of all known pathogen classes including viruses, parasites, fungi and bacteria. However, the composition and activity of these EVs vary significantly between the different taxa and even between pathogens in the same genus. Moreover, various factors such as the animal model used, the experimental design, the cell types chosen for the infection and which recipient cells are targeted, will affect the results observed.

During a bacterial infection, the host immune system is exposed to intact bacteria and to microbial components, and both are key to controlling the infection as well as the subversion of the immune system by the pathogen. Many of the bacterial components known to be involved in the activation/subversion of the immune response are secreted or released from the bacteria during an infection. Defining these bacterial factors as well as how they disseminate during an infection are necessary for our understanding of the disease and how our immune system responds to the infection. EVs are a newly described mechanism for dissemination of these bacterial components (Table 1). Much of our knowledge regarding exosome/EV production and function following a bacterial infection stems from work on mycobacteria, which will be discussed first, followed by an analysis of exosomes/EVs in the context of other bacteria.

5 Composition and function of exosomes and EVs during mycobacterial infections

Russell and colleagues observed that *Mycobacterium tuberculosis* (*M.tb*) components, such as LAM and PIM, are transported from the phagosome to MVBs during a macrophage infection. These mycobacterial components were also present in extracellular vesicles released by infected macrophages, and their

TABLE 1 Exosome production by host cells post bacterial infection

Bacterial pathogen	Host sources	References
Chlamydia trachomatis	Fibroblast	[41]
Chlamydia pneumoniae	ECV304 cells	[42]
Mycobacterium tuberculosis	Macrophages, plasma	[14, 43–49]
Mycobacterium bovis BCG	Macrophages, plasma, BALF	[48, 50]
Mycobacterium smegmatis	Macrophages	[51]
Mycobacterium avium	Macrophages	[51–53]
Salmonella typhimurium	Macrophages	[14, 54]
Mycoplasma spp.	Tumor cells	[55]
Bacillus anthracis	Retinal pigment epithelial cells	[56]
Pseudomonas pneumoniae	Type-I alveolar epithelial cells, BALF	[57]

content could be detected inside neighboring uninfected cells [44]. These vesicles have markers of a late endosomal/lysosomal compartment and are exocytosed in a calcium-dependent manner [45].

Additional studies have expanded on these original observations (Fig. 2). We established that *Mycobacterium avium*-infected macrophages release vesicles that can stimulate a pro-inflammatory response in non-infected or "bystander" macrophages [52]. Similar results were reported by Wang and colleagues [53]. Additional studies found more exosome released by macrophages infected with *M. avium* and *M. smegmatis* compared to uninfected cells, as well as increased levels of the host protein HSP70 in exosomes released from infected macrophages. Anand and colleagues went on to show that HSP70 was an activator of macrophages in vitro [51]. Exosomes released from *M.tb*- or *Mycobacterium bovis* (*M. bovis*) BCG-infected macrophages were also shown to be pro-inflammatory [52]. The mycobacterial 19-kDa lipoprotein present on exosomes that are released from *M. tuberculosis*-infected cells was later shown to be a major driver of this

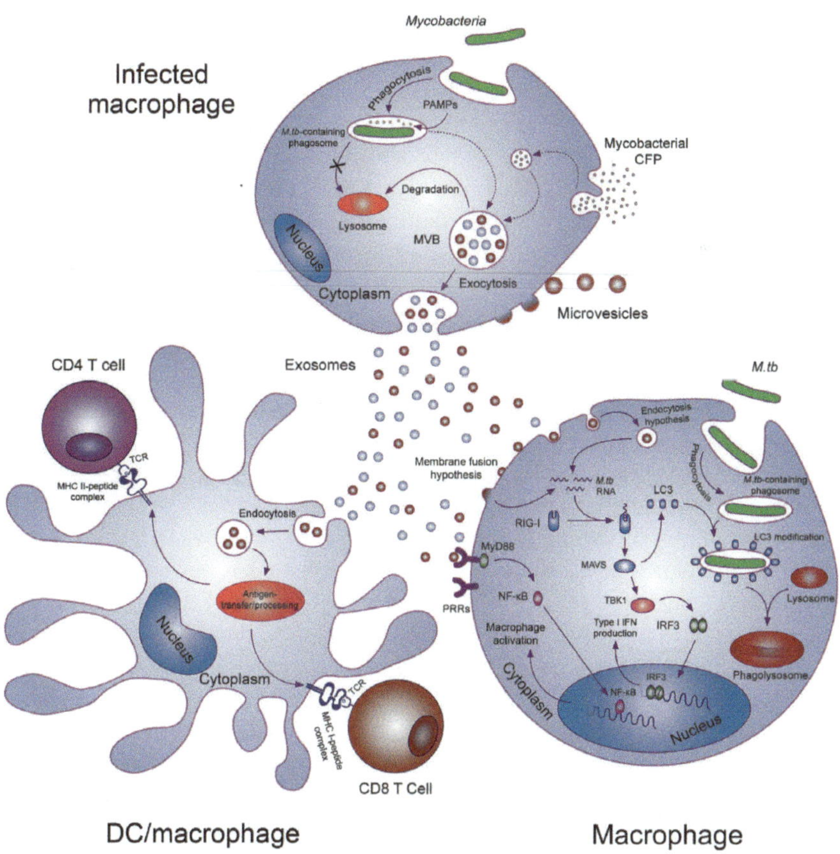

FIG. 2 Stimulation of both innate and adaptive immune responses by exosomes released from Mycobacterium-infected macrophages.

inflammatory response, which was mediated through the TLR/MyD88 pathway [58]. Other types of EVs released from infected macrophages may also stimulate a pro-inflammatory response in target cells [59, 60]. Moreover, EVs isolated from the bronchoalveolar lavage fluid (BALF) of *M. bovis* BCG-infected mice contained mycobacterial components, including the 19 kDa lipoprotein, and were pro-inflammatory ex vivo. In addition, exosomes from *M. bovis* BCG- or *M.tb*-infected macrophages could stimulate a pro-inflammatory response in vivo, as intranasal injection of mice induced TNF-α and IL-12 production, as well as recruitment of macrophages and neutrophils to the lung [52].

Macrophages treated with exosomes from *M.tb*-infected macrophages secrete chemokines that induce naïve macrophage and T cell migration in vitro [48]. Together, these results suggest that exosomes from mycobacterial-infected cells can promote both recruitment and activation of immune cells in vitro and in vivo, and may play a role in promoting the innate immune response upon a mycobacterial infection. However, the mycobacterial components present on/in exosomes could also function to suppress the immune response [47]. Further studies are needed to define the receptors and signaling responses induced upon exosome-macrophage interaction and how these interactions/responses change as the exosome composition is modified during an infection. In addition to their effect on macrophage activation, exosomes from *M.tb*-infected cells have been shown to activate endothelial cells through increased expression of adhesion molecules like ICAM-1 [61]. Exosomes were also found to induce apoptosis in Jurkat T cells via a member of the PE family [62]. These results indicate that the target of these exosome's extend to both immune and non-immune cells.

5.1 *M.tb* infection and characterization of exosome and EV cargo

The presence of the mycobacterial 19 kDa lipoprotein on exosomes released from *M.tb* infected macrophages suggested that other mycobacterial components may also be released through exosomes and other EVs. Indeed, a number of studies have indicated the presence of several mycobacterial proteins, including key antigens that induce an immune response in mice and humans [14, 43, 63, 64]. Importantly, mycobacterial proteins have been detected on EVs isolated from infected mice and from TB patient serum [14, 63, 64]. Host protein content on exosomes also differ between infected and uninfected macrophages as Diaz et al. found 41 proteins to be significantly changed in exosomes released from *M.tb* infected and uninfected THP-1 cells [65]. Hare, et al. also found the proteome to differ between EVs released from *M. tuberculosis*-infected and uninfected macrophages and many of these differentially incorporated proteins are associated with immune function [66]. Subsequent studies extended their findings to include differential incorporation of the glyco-initiating oligosaccharyltransferase complex and glucosidase into EVs [67]. Their data suggest that target cells may undergo differential glycosylation upon uptake of the different EV populations, although this has not been directly tested.

As exosomes and other EVs are known to contain various species of RNA, including miRNAs, studies have addressed how RNA composition differ between vesicles released from mycobacterial-infected compared to uninfected cells. Singh et al. showed approximately 100 transcripts to be enriched or unique to exosomes released from *M.tb* infected murine macrophages, many of which are known to be involved in regulating an immune response [68]. In this same study they also found numerous miRNA to be differentially expressed in exosomes released from infected and uninfected macrophages. Altogether, the data suggest that exosomes released from *M.tb* infected macrophages contain RNA species that induce a more pro-inflammatory response in target cells compared to exosomes from uninfected cells, which matches the previous activation studies described above. Differential expression of miRNA in exosomes released from human monocyte-derived macrophages infected with *M. bovis* BCG compared to uninfected cells was also observed, although the miRNA described appeared unique to each study [69]. One of the more surprising results from the Singh et al. study was the identification of *M.tb* RNA in exosomes released from infected macrophages [68]. At present, the host machinery required for the transport of the mycobacterial RNA into MVBs/exosomes remains undefined; however, a recent study has shown that its release from the *M.tb* requires the expression of the mycobacterial SecA2 secretion system [70]. Interestingly, this same study found the mycobacterial RNA present in the EVs to activate the RIG-I/MAVS RNA sensing pathway in recipient macrophages, leading to increased production of type I interferons (IFNs) [70]. These EVs also promoted macrophage activation leading to increased killing of phagocytosed *M.tb* and again this was dependent on the presence of the mycobacterial RNA and activation of the RIG-I/MAVS pathway. The presence of microbial RNA in EVs suggest a previously unrecognized mechanism of modifying the host immune response; activation of nucleic acid sensors, a mechanism usually associated with controlling viral infections. However, it is important to note that induction of type I IFNs during a *M.tb* infection promotes bacterial survival and disease pathogenesis [71, 72]. Except type I IFN production, the EVs from *M.tb* infected macrophages were also found to promote a noncanonical autophagy pathway and killing of the intracellular *M.tb* via the RIG-I RNA sensor pathway [72]. Similar induction of autophagy and killing of intracellular *M.tb* was found when macrophages were treated with EVs released from *M.tb* infected neutrophils [73]. The dissemination of bacterial RNA via EVs suggest another route of immune surveillance and warrants additional studies in mycobacterial and other bacterial pathogens.

5.2 Modulating of the adaptive immune response by EVs during a mycobacterial infection

Exosomes released from *M.tb*- or *M. bovis* BCG-infected cells, or from *M.tb* culture filtrate protein (CFP)-treated macrophages, can also activate antigen-specific $CD4^+$ and $CD8^+$ T cells in vivo and promote the activation and maturation of bone marrow-derived dendritic cells (BMDCs) [49, 50]. These exosomes induce a Th1 immune response, as defined by antigen-specific T cell production of IFN-γ. Moreover, vaccination of mice with exosomes released from macrophages treated with CFP protect mice against a low-dose aerosolized *M.tb* inoculation; equivalent to BCG vaccinated mice [49]. The release of mycobacterial antigens from infected macrophages is not limited to exosomes. Ramachandra and colleagues observed that infection with *M.tb* or *M. bovis* BCG resulted in increased in both exosome and microvesicle release, and both vesicles could stimulate an antigen-specific T cell response [46]. Together, these results suggest that exosomes and likely other EVs can be a source of antigen for stimulating an acquired immune response. However, other mechanisms for antigen delivery during a mycobacterial infection have been proposed, including necrotic cells, apoptotic bodies and release of free antigen [74–76]. It is likely that M.tb antigens are provided for MHC presentation through multiple routes. Unfortunately, our ability to test the relative importance of exosomes/EVs in antigen delivery is limited, due to the lack of molecular tools to block exosome production in macrophages and dendritic cells without affecting other aspects of vesicular transport and without blocking exosome production by other cell types.

6 Composition and function of exosomes and EVs during other bacterial infections

Although less is known compared to mycobacterial infections, a number of studies have demonstrated the presence of bacterial components in exosomes/EVs and their potential to elicit a host response to the bacterial infection (Fig. 3). Exosomes from *Salmonella*-infected macrophages were found to promote a pro-inflammatory response by inducing TNF-α production by human monocytes [52]. Additionally, exosomes from *S. Typhimurium* infected macrophages stimulate the production of RANTES, IL-1ra, MIP-2 and G-CSF among other cytokines and growth factors and this was dependent on TLR4 expression in the target macrophage [54]. The TLR4 dependence is likely due to the presence of LPS on these exosomes, a known TLR4 ligand present on *Salmonella* and other gram-negative bacteria. The study by Hui et al. also found that *Salmonella*-infected macrophages produce CD63+ and CD9+ subpopulations of exosomes, which fits recent data suggesting that exosomes are heterogeneous in their protein composition even when released from a single cell line [54]. The effect of these different EV populations on the immune response is not known. The pro-inflammatory nature of EVs stemming from gram-negative bacterial infection is not limited to these in vitro studies as experiments characterizing the inflammatory response associated with acute lung injury and acute respiratory distress syndrome found EVs isolated form the bronchoalveolar lavage fluid (BALF) of *Pseudomonas pneumoniae* Infected mice stimulated production of TNF-α, IL-6 and IL-1β [57]. Interestingly, EVs isolated form the BALF following lung oxidative stress were also pro-inflammatory but these EVs were primarily released from type-I alveolar epithelial cells while the active EVs following *P. pneumoniae* infection were primarily from alveolar

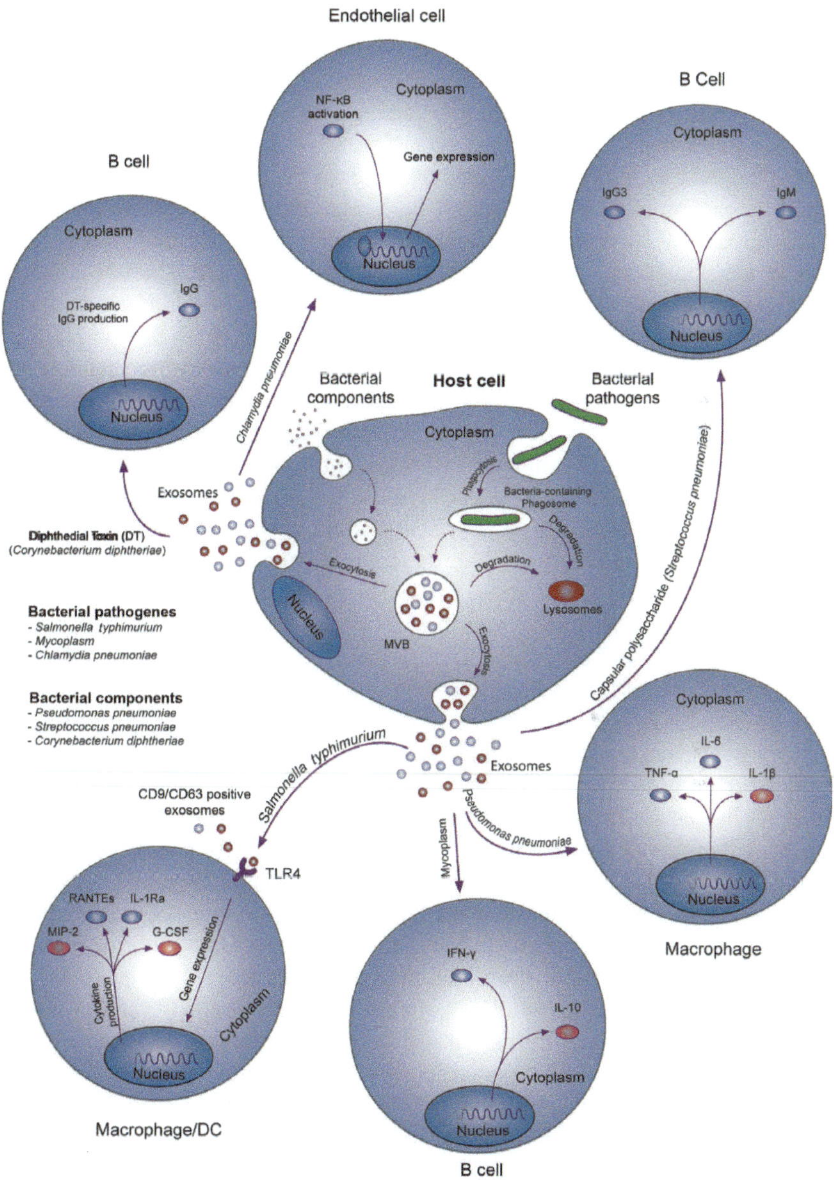

FIG. 3 Responses of different recipient cells to exosomes released from host cells infected with bacterial pathogens.

macrophages. This suggest that multiple cell types are capable of producing pro-inflammatory EVs and these EVs may be associated with tissue damage under various pathological conditions. Targeting these pro-inflammatory EVs may have benefit for treating a number of disease where hyper-inflammation is a significant cause of the disease pathology.

Exosomes from cells infected with *Mycoplasma* induce a mixed cytokine response, including production of both IFN-γ and IL-10 from B cells. However, these exosomes appear to be primarily inhibitory, at least in the context of T cell activation [55]. Exosomes can also be carriers of toxins, as shown by Abrami and colleagues, who found lethal factor (LF), a well-characterized toxin produced by *Bacillus anthracis*, to be packaged into intraluminal vesicles and released on exosomes when expressed in a human epithelial cell line [56]. The LF packaged in exosomes is protected from neutralizing extracellular factors such as antibodies and can be taken up by target cells. Diphtheria toxin has also been found on DC-derived exosomes when the DC were pulsed with the intact diphtheria toxoid [77]. Interestingly these exosomes were found to induce an antibody response specific to the diphtheria toxoid indicating that foreign proteins packaged in exosomes can activate B cells. This is likely occurring as exosomes gain access to the lymph nodes but this remains to be tested experimentally. Colino and Snapper extended this work and demonstrated that exosome-induced antibody response against the capsular polysaccharide of *Streptococcus pneumoniae* type 14 could protect mice from a lethal infection with this *S. pneumoniae* strain [78]. Ettelaie and colleagues reported that "microparticles" released from *Chlamydia pneumoniae*-infected cells contain TF, a blood coagulation protein, which has also been associated with cell proliferation, migration, and apoptosis. The TF positive microparticles activate NF-κB, the transcription factor that partially regulates TF expression in endothelial cells. *C. pneumoniae* elementary bodies were also proposed to be released in microparticles, suggesting a potential role in the dissemination of the infection through the vascular system. These findings have implications in control of the infection by the host, but also potential cardiovascular consequences in relation to inflammatory conditions such as atherosclerosis [42]. Although the *C. pneumoniae* vesicles were referred to as microparticles, the procedures used for isolation would enrich for exosomes. Several cytotoxic and secreted proteins were also associated with host vesicles released from *Chlamydia trachomatis*, which may function in the discrimination of virulence factors [41].

Together, the published studies indicate that packaging of various bacterial components can occur during an active cellular infection or when endocytosed by immune cells such as macrophages and dendritic cells. However, the mechanism by which these components are trafficked to EVs or the specificity of this transport remains an open question. To begin answering this question Smith and colleagues looked at what host factors were required for transport of mycobacterial proteins to exosomes. They found that mycobacterial proteins either released during an infection or endocytosed by macrophages required mono-ubiquitination for trafficking to MVBs and into the intraluminal vesicles [79, 80]. However, whether ubiquitination is a general mechanism for transport of bacterial proteins to MVBs or if other mechanisms are at play is unclear. Moreover, how other bacterial components such as lipids and nucleic acids are transported to exosomes or other EVs remains undefined.

7 EVs as vaccines against bacterial infections

The studies on exosomes as vaccines was inspired by the discovery that exosomes from EBV-transformed B lymphocytes carry MHC II complex and may stimulate antigen-specific MHC II-restricted T lymphocyte activation in the absence of antigen-presenting cells (APCs) [9]. In the past two decades, a number of innovative studies have focused on developing exosome-based vaccines for treatment of cancers in animal models and more recently in human patients. Exosomes have been successfully produced under the regulations of good manufacturing practice (GMP) and three exosome-based vaccine candidates have been evaluated in phase I clinical trials for late-stage cancer patients. One candidate is currently undergoing a phase II clinical trial for non-small cell lung cancer patients [81–84]. In the field of bacterial infections, EVs from pathogen-infected host cells are also shedding light on the development of novel cell-free antibacterial vaccines. An increasing body of evidence highlights EV's distinct advantage as vaccines: (i) EVs are released by bacteria-infected host cells and inherit some properties of parental cells. This includes vesicles from APCs that may carry MHC I/II-antigen complex, free antigens, adjuvant molecules and other regulatory host factors. (ii) As endogenous components, EVs are more stable than most exogenous vaccines. (iii) Different from traditional vaccines, EVs are detected in almost all body fluids. This suggest that EV-carried antigens could reach distal organs or regions that are likely missed by traditional vaccines but nevertheless are important for protective immune responses.

As we indicated above, APC-derived EVs are particularly attractive in the study on EV-based antibacterial vaccines. Colino and colleagues found that exosomes released by mouse bone marrow-derived dendritic cells (BMDCs) pretreated with diphtheria toxin (DT) or capsular polysaccharide of *Streptococcus pneumoniae* type 14 (Cps14) induced Ag (DT/Cps14)-specific Ig responses in naïve recipient mice [77, 78]. Interestingly, exosomes from DT-treated BMDCs also stimulated secondary anti-DT Ig responses in mice primed with free DT. Different from free DT that predominantly induces a type 2 anti-DT response, exosomes from DT-treated BMDCs preferentially stimulate a type 1 anti-DT response [77]. Mouse survival assay further indicated exosomes from Cps14-pulsed BMDCs protected mice from a lethal infection with live *S. pneumoniae* type 14 but not the strains lacking Cps14 [77].

M.tb is currently the number one cause of death by an infectious organism accounting for approximately 1.7 million deaths annually world-wide [85]. *M. bovis* BCG is currently the only licensed anti-TB vaccine but its efficacy waves over time, and has a limited protection in adolescents and adults. Exosomes provide a novel and alternative strategy for anti-TB vaccination. Exosomes from macrophages infected with *M. bovis* BCG or pretreated with *M.tb* CFP induced Ag-specific $CD4^+$ and $CD8^+$ T cell response in naïve mice [49, 50]. Similar to exosomes from DT-treated BMDCs, exosomes from CFP-treated macrophages boosted secondary T cell response in mice vaccinated with *M. bovis*

BCG, providing evidence for exosomes as an enhancer vaccine in combination with primary *M. bovis* BCG vaccination. Unlike *M. bovis* BCG, exosomes from CFP-treated macrophages showed a bias activation of Th1 immune response in vaccinated mice, which is necessary for controlling TB in patients [49, 86]. Additionally, studies show that *M.tb* infection was significantly restricted in mice immunized with exosomes from CFP-treated macrophages in comparison with those receiving exosome from untreated macrophages or a single dose of *M. bovis* BCG. As a booster vaccine, exosomes show a dose-dependent anti-mycobacterial efficacy comparable to or higher than *M. bovis* BCG [49].

A limitation in the published approach to exosome-based vaccines is the use of in vitro antigen-pulsed APCs as the source of exosomes since this approach is not applicable to a large-scale GMP production of exosomes. Furthermore, the published studies indicate that the trafficking of antigens into exosomes depends on the antigens containing a "signal" for delivery to MVBs and intraluminal vesicles. Therefore, a simple and effective approach is needed to target antigens of interest into exosomes regardless of an intrinsic sorting "signal" and is practical for a large-scale production of exosomes. A possible strategy is to engineer a transgenic mammalian cell line that constitutively releases exosomes carrying antigens of choice. Published studies found that a monoubiquitin modification plays a critical role for mycobacterial and host protein trafficking into host-derived exosomes [80]. We took advantage of this information and generated a transgenic HEK 293T cell line in which a monoubiquitin-tag was introduced as a general signal for protein trafficking into exosomes [87]. To optimize this approach, four distinct versions of monoubiquitin modification were tested and the results indicate that an C-terminal wild-type monoubiquitin modification was the most effective in trafficking GFP, *M.tb* Ag85B-EAST6 fusion protein and host Her2 protein into exosomes. However, different from exosomes released from CFP-treated or *M. bovis* BCG-infected macrophages, the exosomes from engineered HEK 293T cells failed to induce Ag85B- or ESAT6-specific T cell activation in the absence of adjuvants. This suggest that the engineered exosomes lacked a mycobacterial component present on exosomes from CFP-treated/BCG-infected exosomes that can function as an adjuvant upon vaccination. A similar strategy has been tested using the membrane C1C2-domain of Lactadherin or HIV Nef protein as an exosome-targeting signal sequence [88–92]. However, the immunogenicity of these engineered Nef exosomes remains to be defined.

8 EVs as a drug delivery system

Exosomes can engage recipient cells to activate cell surface receptors as well as release their contents into host cells. Published studies indicate that exosomes have some specificity in regards to recipient cells which is dictated by their cell of origin [14, 70]. Unlike typical drug delivery systems, exosomes are thought to have higher therapeutic activity [93]. Exosomes have been extensively studied for their potential as cell-free therapeutic agents and as drug delivery system to regulate the cellular processes of recipient cells. Among them, exosomes used as nucleic acid-based drug delivers are particularly attractive, as nucleic acids, including miRNA, hairpin RNA and mRNA, can be loaded endogenously through exosome-producing parental cells or exogenously by electroporation [94]. For example, exosomes from dendritic cells engineered to express Lamp2b, an exosomal membrane protein, fused to the neuron-specific RVG peptide was found to cross the blood-brain barrier (BBB) and deliver exogenously loaded BACE1-specific siRNA into the brain. As a result, the expression of BACE1 was blocked by 60% at the mRNA level and 62% at the protein level, indicating a potential application of exosomes as a therapeutic technique in Alzheimer's disease [95]. In another study, it was found that cell type-targeting exosomes have significantly high therapeutic activity. Exosomes isolated from HEK 293 cells that were pre-loaded with synthetic let-7a miRNA, a tumor suppressor, and expressed the transmembrane domain of platelet-derived growth factor receptor fused to the GE11 peptide, are specifically targeted to xenograft breast cancer cells via a GE11-EGFR interaction and reduce the malignant growth of cancer cells in *RAG2* $^{-/-}$ mice [96].

The use of EVs as drug delivery systems or for enhancing drug treatment in the context of an infectious disease has been relatively unexplored. However, one recent study has looked at EVs in the context of *M.tb* antibiotic treatment. As described above, *M.tb* releases its RNAs into extracellular environment via a SecA2-dependent pathway and some of the released RNAs is packaged into host-derived exosomes [70, 72]. These exosomes induce production of type I IFNs, and activate a ubiquitin-independent LC3-associated modification of the *M.tb*-containing phagosome in recipient infected macrophages. The increase in phagolysosome maturation results in restricting *M.tb* replication in the host cell. Both the type I IFN production and the noncanonical modification of *M.tb*-containing phagosomes by the LC3 rely on the host RIG-I/MAVS-dependent pathway which is activated by the EV-bound *M.tb* RNA. In the same study, exosomes from *M.tb*-infected BMMs were found to limit *M.tb* survival in the lung and spleen of infected mice when administered intratracheally suggesting exosomes from *M.tb*-infected BMMs may work as pre-exposure or post-exposure antimycobacterial agents to control *M.tb* survival in host cells. Additionally, when co-administered with moxifloxacin, a second line anti-TB antibiotics, *M.tb* burden in the lung and spleen significantly declined when compared with moxifloxacin or exosomes alone. Exosome uptake assay in mice showed that intratracheally or intranasally administered exosomes predominantly target alveolar macrophages and DCs [14, 70]. It suggests that exosomes from *M.tb*-infected macrophages may serve as an alternate mechanism to target anti-TB adjunct agents such as to augment the antimicrobial activity of anti-TB drugs. In the context of anti-TB therapy, exosomes could be used to supplement 2nd line antibiotics to treat multidrug-resistant and extensively drug-resistant TB (M/XDR-TB) that are difficult to treat with antibiotics alone [70]. However, similar to the exosome-based vaccines, there are safety concerns as well as concerns about the practicality of developing therapeutic exosomes that met GMP criteria.

9 Exosomes as a source of diagnostic biomarkers for infectious diseases

The discovery of exosomes as natural carriers of nucleic acids and proteins from parental cells has raised great interest in exosome-based biomarkers. A number of studies have shown quantitative and qualitative differences in exosome composition between healthy individuals and those with underlying diseases, including cancers and renal diseases [81]. Moreover in comparison to exosomes from health individuals, exosomes released from patients, such as those with active TB, not only demonstrate changes in host molecules but also carry bacterial components including proteins, nucleic acids and lipids [64, 97]. This provides a distinct advantage for diagnostics as the presences of bacterial macromolecules allows for higher specificity compared quantitative changes in host molecules alone. Therefore, it is surprising that more work has not been invested in developing exosome-based diagnostic for infectious diseases. This potential has been noted previously as publications have encouraged the development of exosome-based diagnostics for bacterial infections, especially for those diseases difficult to diagnose such as TB [98]. The typical diagnostic methods for TB include mycobacterial culture or PCR-based test using patient's sputum. The sensitivity and specificity of these tests are affected by multiple factors. For example, sputum-based tests fail in TB patients having little to no *M.tb* in their sputum or in patients who are unable to produce sputum. Additionally, *M.tb* is slow growing and it takes 3–4 weeks for a mycobacterial culture test. However, at present the culture test is the only one available for the diagnosis of multidrug- and extensively drug-resistant TB.

To investigate whether EVs could be used as biomarkers for TB diagnostics, the *M.tb* protein profile of EVs released from *M.tb*-infected J774 macrophages or from Balb/c mice were analyzed using LC-MS-MS, These studies resulted in the identification of 41 mycobacterial proteins in exosomes from *M.tb*-infected J774 macrophages and 69 proteins in EVs isolated from bronchoalveolar lavage fluid (BALF) of mice infected with *M.tb* H37Rv [43, 63]. During the course of *M.tb* infection in mice, a dynamic profile of mycobacterial proteins in BALF exosomes was observed suggesting that a panel/cocktail of *M.tb* proteins will be needed as biomarkers for an exosome-based TB diagnostics. A similar result was found in serum exosomes from active TB patients [64]. In the cohort of 41 active TB patients no common *M.tb* protein/peptide was found in all EV samples. Nevertheless, the data indicates that 83% of active TB cases could be positively diagnosed using a biomarker candidate panel of 7 proteins that are unique to active TB patients, In contrast to active TB, no unique *M.tb* protein was identified in serum exosomes from individuals with latent TB, although

M.tb peptides were found in these individuals. Considering the fact that only a small group of latent patients ($n = 9$) were tested in the study, a larger number of this cohort is required to potentially identify a exosome-based biomarkers that distinguishes active and latent TB. In addition to *M.tb* protein biomarkers, there is significant interest in defining potential *M.tb* RNA biomarkers in TB patient EVs. Using Illumina RNA sequencing technique, a number of *M.tb* RNAs were identified in exosomes from *M.tb*-infected RAW264.7 macrophages and mouse BMMs [68, 70], and serum exosomes of active TB patients [97].

Beyond the exosome biomarkers, a key step in exosome-based diagnostics is a simple and cost-effective exosome enrichment approach from patient body fluids. For exosomal protein-based test, the high abundance of host blood proteins such as albumins and globulins significantly interfere with the sensitivity and specificity of tests. There are some well-established approaches to enrich for exosomes in body fluids which could potentially be coupled with exosome-based diagnostics. These include antibody-based capture of exosomes such as the use of anti-CD63 antibody-based exosome isolation reagents that are commercially available. Other approaches include size exclusion. A modification of this approach is the development of a Capto Core 700-based size exclusion chromatography method that allows for the removal of free proteins with molecular weights less than 700 kDa while allowing flow through of EVs [99]. Protein quantification confirmed this method removes over 99% of host blood proteins and LC-MS-MS comparison analysis indicates this method significantly increases the test sensitivity [99, 100]. In addition to these methods, there are other exosome purification approaches including the use of polymer-based precipitation, density-gradient centrifugation and differential ultracentrifugation. However, the disadvantage of these approaches are clear. Polymer-based precipitation nonspecifically pull-down vesicles and a significant amount of free proteins. This a particular problematic issue when blood samples are used. For centrifugation-dependent methods, a well-equipped laboratory is needed which is often difficult to obtain in resource poor setting where TB is endemic. For RNA-based exosome diagnostics, the level of free proteins is less a factor as in protein-based exosome test, but a downstream RNA enrichment/purification approach will be crucial.

10 Summary

It is clear that EVs are intimately involved in the host response to bacterial infections and that bacterial components including proteins, lipids and RNA can reside in EVs. However, what is less clear is whether their release promotes or hinders the immune response. It is likely that both are occurring during the course of an infection and this is dependent on the species/strain of bacteria and when during the course of the infection one is measuring the immune response. What is required to answer this question is a better understanding of which bacterial components are trafficked to EVs, what host factors regulate this process and what effect these bacterial components have on the recipient cells. At present we lack many of the tools needed, including approaches to regulate exosome biogenesis both temporally and spatially. Nevertheless, our current understanding has allowed us to explore the use of exosomes and other EVs in the development of vaccines, drug delivery systems and diagnostic markers to better combat bacterial diseases. The use of EVs therapeutically will continue to grow as we learn more about the production, function and composition of EVs in context of bacterial diseases.

References

[1] Laarman A, Milder F, van Strijp J, Rooijakkers S. Complement inhibition by gram-positive pathogens: molecular mechanisms and therapeutic implications. J Mol Med 2010;88:115–20. https://doi.org/10.1007/s00109-009-0572-y.

[2] Tajima A, Iwase T, Shinji H, Seki K, Mizunoe Y. Inhibition of endothelial interleukin-8 production and neutrophil transmigration by staphylococcus aureus beta-hemolysin. Infect Immun 2009;77:327–34. https://doi.org/10.1128/IAI.00748-08.

[3] Frost HR, Sanderson-Smith M, Walker M, Botteaux A, Smeesters PR. Group A streptococcal M-like proteins: from pathogenesis to vaccine potential. FEMS Microbiol Rev 2018;42:193–204. https://doi.org/10.1093/femsre/fux057.

[4] Acevedo R, Fernández S, Zayas C, Acosta A, Sarmiento ME, Ferro VA, Rosenqvist E, Campa C, Cardoso D, Garcia L, Perez JL. Bacterial outer membrane vesicles and vaccine applications. Front Immunol 2014;5:121https://doi.org/10.3389/fimmu.2014.00121.

[5] Deatherage BL, Cookson BT. Membrane vesicle release in bacteria, eukaryotes, and archaea: a conserved yet underappreciated aspect of microbial life. Infect Immun 2012;80:1948–57. https://doi.org/10.1128/IAI.06014-11.

[6] Kulp A, Kuehn MJ. Biological functions and biogenesis of secreted bacterial outer membrane vesicles. Annu Rev Microbiol 2010;64:163–84. https://doi.org/10.1146/annurev.micro.091208.073413.

[7] D'Souza-Schorey C, Clancy JW. Tumor-derived microvesicles: shedding light on novel microenvironment modulators and prospective cancer biomarkers. Genes Dev 2012;26:1287–99. https://doi.org/10.1101/gad.192351.112.

[8] Schorey JS, Cheng Y, Singh PP, Smith VL. Exosomes and other extracellular vesicles in host-pathogen interactions. EMBO Rep 2015;16:24–43. https://doi.org/10.15252/embr.201439363.

[9] Raposo G, Nijman HW, Stoorvogel W, Liejendekker R, Harding CV, Melief CJ, Geuze HJ. B lymphocytes secrete antigen-presenting vesicles. J Exp Med 1996;183:1161–72.

[10] Raposo G, Tenza D, Mecheri S, Peronet R, Bonnerot C, Desaymard C. Accumulation of major histocompatibility complex class II molecules in mast cell secretory granules and their release upon degranulation. Mol Biol Cell 1997;8:2631–45.

[11] Théry C, Regnault A, Garin J, Wolfers J, Zitvogel L, Ricciardi-Castagnoli P, Raposo G, Amigorena S. Molecular characterization of dendritic cell-derived exosomes. Selective accumulation of the heat shock protein hsc73. J Cell Biol 1999;147:599–610.

[12] Zitvogel L, Regnault A, Lozier A, Wolfers J, Flament C, Tenza D, Ricciardi-Castagnoli P, Raposo G, Amigorena S. Eradication of established murine tumors using a novel cell-free vaccine: dendritic cell derived exosomes. Nat Med 1998;4:594–600. https://doi.org/10.1038/nm0598-594.

[13] Heijnen HF, Schiel AE, Fijnheer R, Geuze HJ, Sixma JJ. Activated platelets release two types of membrane vesicles: microvesicles by surface shedding and exosomes derived from exocytosis of multivesicular bodies and alpha-granules. Blood 1999;94:3791–9.

[14] Bhatnagar S, Shinagawa K, Castellino FJ, Schorey JS. Exosomes released from macrophages infected with intracellular pathogens stimulate a proinflammatory response in vitro and in vivo. Blood 2007;110:3234–44. https://doi.org/10.1182/blood-2007-03-079152.

[15] Fauré J, Lachenal G, Court M, Hirrlinger J, Chatellard-Causse C, Blot B, Grange J, Schoehn G, Goldberg Y, Boyer V, Kirchhoff F, Raposo G, Garin J, Sadoul R. Exosomes are released by cultured cortical neurones. Mol Cell Neurosci 2006;31:642–8. https://doi.org/10.1016/j.mcn.2005.12.003.

[16] Guescini M, Genedani S, Stocchi V, Agnati LF. Astrocytes and Glioblastoma cells release exosomes carrying mtDNA. J Neural Transm (Vienna) 2010;117:1–4. https://doi.org/10.1007/s00702-009-0288-8.

[17] Marzesco A-M, Janich P, Wilsch-Bräuninger M, Dubreuil V, Langenfeld K, Corbeil D, Huttner WB. Release of extracellular membrane particles carrying the stem cell marker prominin-1 (CD133) from neural progenitors and other epithelial cells. J Cell Sci 2005;118:2849–58. https://doi.org/10.1242/jcs.02439.

[18] Zhu M, Li Y, Shi J, Feng W, Nie G, Zhao Y. Exosomes as extrapulmonary signaling conveyors for nanoparticle-induced systemic immune activation. Small 2012;8:404–12. https://doi.org/10.1002/smll.201101708.

[19] Harding C, Heuser J, Stahl P. Endocytosis and intracellular processing of transferrin and colloidal gold-transferrin in rat reticulocytes: demonstration of a pathway for receptor shedding. Eur J Cell Biol 1984;35:256–63.

[20] Johnstone RM, Adam M, Hammond JR, Orr L, Turbide C. Vesicle formation during reticulocyte maturation. Association of plasma membrane activities with released vesicles (exosomes). J Biol Chem 1987;262:9412–20.

[21] Woodman PG, Futter CE. Multivesicular bodies: co-ordinated progression to maturity. Curr Opin Cell Biol 2008;20:408–14. https://doi.org/10.1016/j.ceb.2008.04.001.

[22] Buschow SI, Nolte-'t Hoen ENM, van Niel G, Pols MS, ten Broeke T, Lauwen M, Ossendorp F, Melief CJM, Raposo G, Wubbolts R, Wauben MHM, Stoorvogel W. MHC II in dendritic cells is targeted to lysosomes or T cell-induced exosomes via distinct multivesicular body pathways. Traffic 2009;10:1528–42. https://doi.org/10.1111/j.1600-0854.2009.00963.x.

[23] Hurley JH. The ESCRT complexes. Crit Rev Biochem Mol Biol 2010;45:463–87. https://doi.org/10.3109/10409238.2010.502516.

[24] Davies BA, Lee JRE, Oestreich AJ, Katzmann DJ. Membrane protein targeting to the MVB/lysosome. Chem Rev 2009;109:1575–86. https://doi.org/10.1021/cr800473s.

[25] Metcalf D, Isaacs AM. The role of ESCRT proteins in fusion events involving lysosomes, endosomes and autophagosomes. Biochem Soc Trans 2010;38:1469–73. https://doi.org/10.1042/BST0381469.

[26] Raiborg C, Stenmark H. Hrs and endocytic sorting of ubiquitinated membrane proteins. Cell Struct Funct 2002;27:403–8.

[27] Babst M, Katzmann DJ, Snyder WB, Wendland B, Emr SD. Endosome-associated complex, ESCRT-II, recruits transport machinery for protein sorting at the multivesicular body. Dev Cell 2002;3:283–9.

[28] Katzmann DJ, Babst M, Emr SD. Ubiquitin-dependent sorting into the multivesicular body pathway requires the function of a conserved endosomal protein sorting complex, ESCRT-I. Cell 2001;106:145–55.

[29] Wollert T, Wunder C, Lippincott-Schwartz J, Hurley JH. Membrane scission by the ESCRT-III complex. Nature 2009;458:172–7. https://doi.org/10.1038/nature07836.

[30] Hanson PI, Cashikar A. Multivesicular body morphogenesis. Annu Rev Cell Dev Biol 2012;28:337–62. https://doi.org/10.1146/annurev-cellbio-092910-154152.

[31] Tamai K, Tanaka N, Nakano T, Kakazu E, Kondo Y, Inoue J, Shiina M, Fukushima K, Hoshino T, Sano K, Ueno Y, Shimosegawa T, Sugamura K. Exosome secretion of dendritic cells is regulated by Hrs, an ESCRT-0 protein. Biochem Biophys Res Commun 2010;399:384–90. https://doi.org/10.1016/j.bbrc.2010.07.083.

[32] Stuffers S, Sem Wegner C, Stenmark H, Brech A. Multivesicular endosome biogenesis in the absence of ESCRTs. Traffic 2009;10:925–37. https://doi.org/10.1111/j.1600-0854.2009.00920.x.

[33] Trajkovic K, Hsu C, Chiantia S, Rajendran L, Wenzel D, Wieland F, Schwille P, Brügger B, Simons M. Ceramide triggers budding of exosome vesicles into multivesicular endosomes. Science 2008;319:1244–7. https://doi.org/10.1126/science.1153124.

[34] van Niel G, Charrin S, Simoes S, Romao M, Rochin L, Saftig P, Marks MS, Rubinstein E, Raposo G. The tetraspanin CD63 regulates ESCRT-independent and -dependent endosomal sorting during melanogenesis. Dev Cell 2011;21:708–21. https://doi.org/10.1016/j.devcel.2011.08.019.

[35] Perez-Hernandez D, Gutiérrez-Vázquez C, Jorge I, López-Martín S, Ursa A, Sánchez-Madrid F, Vázquez J, Yáñez-Mó M. The intracellular interactome of tetraspanin-enriched microdomains reveals their function as sorting machineries toward exosomes. J Biol Chem 2013;288:11649–61. https://doi.org/10.1074/jbc.M112.445304.

[36] Colombo M, Raposo G, Théry C. Biogenesis, secretion, and intercellular interactions of exosomes and other extracellular vesicles. Annu Rev Cell Dev Biol 2014;30:255–89. https://doi.org/10.1146/annurev-cellbio-101512-122326.

[37] Schwartz SL, Cao C, Pylypenko O, Rak A, Wandinger-Ness A. Rab GTPases at a glance. J Cell Sci 2007;120:3905–10. https://doi.org/10.1242/jcs.015909.

[38] Savina A, Vidal M, Colombo MI. The exosome pathway in K562 cells is regulated by Rab11. J Cell Sci 2002;115:2505–15.

[39] Hsu C, Morohashi Y, Yoshimura S-I, Manrique-Hoyos N, Jung S, Lauterbach MA, Bakhti M, Grønborg M, Möbius W, Rhee J, Barr FA, Simons M. Regulation of exosome secretion by Rab35 and its GTPase-activating proteins TBC1D10A-C. J Cell Biol 2010;189:223–32. https://doi.org/10.1083/jcb.200911018.

[40] Ostrowski M, Carmo NB, Krumeich S, Fanget I, Raposo G, Savina A, Moita CF, Schauer K, Hume AN, Freitas RP, Goud B, Benaroch P, Hacohen N, Fukuda M, Desnos C, Seabra MC, Darchen F, Amigorena S, Moita LF, Thery C. Rab27a and Rab27b control different steps of the exosome secretion pathway. Nat Cell Biol 2010;12:19–30. Sup. pp 1–13, https://doi.org/10.1038/ncb2000.

[41] Frohlich K, Hua Z, Wang J, Shen L. Isolation of *Chlamydia* trachomatis and membrane vesicles derived from host and bacteria. J Microbiol Methods 2012;91:222–30. https://doi.org/10.1016/j.mimet.2012.08.012.

[42] Ettelaie C, Collier MEW, James NJ, Li C. Induction of tissue factor expression and release as microparticles in ECV304 cell line by *Chlamydia pneumoniae* infection. Atherosclerosis 2007;190:343–51. https://doi.org/10.1016/j.atherosclerosis.2006.04.005.

[43] Giri PK, Kruh NA, Dobos KM, Schorey JS. Proteomic analysis identifies highly antigenic proteins in exosomes from *M. tuberculosis*-infected and culture filtrate protein-treated macrophages. Proteomics 2010;10:3190–202. https://doi.org/10.1002/pmic.200900840.

[44] Beatty WL, Russell DG. Identification of mycobacterial surface proteins released into subcellular compartments of infected macrophages. Infect Immun 2000;68:6997–7002.

[45] Beatty WL, Ullrich HJ, Russell DG. Mycobacterial surface moieties are released from infected macrophages by a constitutive exocytic event. Eur J Cell Biol 2001;80:31–40. https://doi.org/10.1078/0171-9335-00131.

[46] Ramachandra L, Qu Y, Wang Y, Lewis CJ, Cobb BA, Takatsu K, Boom WH, Dubyak GR, Harding CV. *Mycobacterium tuberculosis* synergizes with ATP to induce release of microvesicles and exosomes containing major histocompatibility complex class II molecules capable of antigen presentation. Infect Immun 2010;78:5116–25. https://doi.org/10.1128/IAI.01089-09.

[47] Singh PP, LeMaire C, Tan JC, Zeng E, Schorey JS. Exosomes released from *M. tuberculosis* infected cells can suppress IFN-γ mediated activation of naïve macrophages. PLoS ONE 2011;6:e18564. https://doi.org/10.1371/journal.pone.0018564.

[48] Singh PP, Smith VL, Karakousis PC, Schorey JS. Exosomes isolated from mycobacteria-infected mice or cultured macrophages can recruit and activate immune cells in vitro and in vivo. J Immunol 2012;189:777–85. https://doi.org/10.4049/jimmunol.1103638.

[49] Cheng Y, Schorey JS. Exosomes carrying mycobacterial antigens can protect mice against *Mycobacterium tuberculosis* infection. Eur J Immunol 2013;43:3279–90. https://doi.org/10.1002/eji.201343727.

[50] Giri PK, Schorey JS. Exosomes derived from *M. bovis* BCG infected macrophages activate antigen-specific CD4+ and CD8+ T cells in vitro and in vivo. PLoS ONE 2008;3:e2461. https://doi.org/10.1371/journal.pone.0002461.

[51] Anand PK, Anand E, Bleck CKE, Anes E, Griffiths G. Exosomal Hsp70 induces a pro-inflammatory response to foreign particles including mycobacteria. PLoS ONE 2010;5:e10136. https://doi.org/10.1371/journal.pone.0010136.

[52] Bhatnagar S, Schorey JS. Exosomes released from infected macrophages contain *Mycobacterium avium* glycopeptidolipids and are proinflammatory. J Biol Chem 2007;282:25779–89. https://doi.org/10.1074/jbc.M702277200.

[53] Wang J, Chen C, Xie P, Pan Y, Tan Y, Tang L. Proteomic analysis and immune properties of exosomes released by macrophages infected with mycobacterium avium. Microbes Infect 2014;16:283–91. https://doi.org/10.1016/j.micinf.2013.12.001.

[54] Hui WW, Hercik K, Belsare S, Alugubelly N, Clapp B, Rinaldi C, Edelmann MJ. *Salmonella enterica* serovar typhimurium alters the extracellular proteome of macrophages and leads to the production of proinflammatory exosomes. Infect Immun 2018;86 https://doi.org/10.1128/IAI.00386-17.

[55] Yang C, Chalasani G, Ng Y-H, Robbins PD. Exosomes released from mycoplasma infected tumor cells activate inhibitory B cells. PLoS ONE 2012;7 https://doi.org/10.1371/journal.pone.0036138.

[56] Abrami L, Brandi L, Moayeri M, Brown MJ, Krantz BA, Leppla SH, van der Goot FG. Hijacking multivesicular bodies enables long-term and exosome-mediated long-distance action of anthrax toxin. Cell Rep 2013;5:986–96. https://doi.org/10.1016/j.celrep.2013.10.019.

[57] Lee H, Zhang D, Laskin DL, Jin Y. Functional evidence of pulmonary extracellular vesicles in infectious and noninfectious lung inflammation. J Immunol 2018;201:1500–9. https://doi.org/10.4049/jimmunol.1800264.

[58] Schorey JS, Bhatnagar S. Exosome function: from tumor immunology to pathogen biology. Traffic 2008;9:871–81. https://doi.org/10.1111/j.1600-0854.2008.00734.x.

[59] Athman JJ, Wang Y, McDonald DJ, Boom WH, Harding CV, Wearsch PA. Bacterial membrane vesicles mediate the release of *Mycobacterium tuberculosis* lipoglycans and lipoproteins from infected macrophages. J Immunol 2015;195:1044–53. https://doi.org/10.4049/jimmunol.1402894.

[60] Walters SB, Kieckbusch J, Nagalingam G, Swain A, Latham SL, Grau GER, Britton WJ, Combes V, Saunders BM. Microparticles from mycobacteria-infected macrophages promote inflammation and cellular migration. J Immunol 2013;190:669–77. https://doi.org/10.4049/jimmunol.1201856.

[61] Li L, Cheng Y, Emrich S, Schorey J. Activation of endothelial cells by extracellular vesicles derived from *Mycobacterium tuberculosis* infected macrophages or mice. PLoS ONE 2018;13:e0198337. https://doi.org/10.1371/journal.pone.0198337.

[62] Balaji KN, Goyal G, Narayana Y, Srinivas M, Chaturvedi R, Mohammad S. Apoptosis triggered by Rv1818c, a PE family gene from *Mycobacterium tuberculosis* is regulated by mitochondrial intermediates in T cells. Microbes Infect 2007;9:271–81. https://doi.org/10.1016/j.micinf.2006.11.013.

[63] Kruh-Garcia NA, Schorey JS, Dobos KM. Exosomes: new tuberculosis biomarkers—prospects from the bench to the clinic. In: Understanding tuberculosis—global experiences and innovative approaches to the diagnosis. 2012. https://doi.org/10.5772/30720.

[64] Kruh-Garcia NA, Wolfe LM, Chaisson LH, Worodria WO, Nahid P, Schorey JS, Davis JL, Dobos KM. Detection of *Mycobacterium tuberculosis* peptides in the exosomes of patients with active and latent *M. tuberculosis* infection using MRM-MS. PLoS ONE 2014;9:e103811. https://doi.org/10.1371/journal.pone.0103811.

[65] Diaz G, Wolfe LM, Kruh-Garcia NA, Dobos KM. Changes in the membrane-associated proteins of exosomes released from human macrophages after *Mycobacterium tuberculosis* infection. Sci Rep 2016;6:37975. https://doi.org/10.1038/srep37975.

[66] Hare NJ, Chan B, Chan E, Kaufman KL, Britton WJ, Saunders BM. Microparticles released from *Mycobacterium tuberculosis*-infected human macrophages contain increased levels of the type I interferon inducible proteins including ISG15. Proteomics 2015;15:3020–9. https://doi.org/10.1002/pmic.201400610.

[67] Hare NJ, Lee LY, Loke I, Britton WJ, Saunders BM, Thaysen-Andersen M. *Mycobacterium tuberculosis* infection manipulates the glycosylation machinery and the N-glycoproteome of human macrophages and their microparticles. J Proteome Res 2017;16:247–63. https://doi.org/10.1021/acs.jproteome.6b00685.

[68] Singh PP, Li L, Schorey JS. Exosomal RNA from *Mycobacterium tuberculosis*-infected cells is functional in recipient macrophages. Traffic 2015;16:555–71. https://doi.org/10.1111/tra.12278.

[69] Alipoor SD, Mortaz E, Tabarsi P, Farnia P, Mirsaeidi M, Garssen J, Movassaghi M, Adcock IM. Bovis Bacillus Calmette-Guerin (BCG) infection induces exosomal miRNA release by human macrophages. J Transl Med 2017;15:105. https://doi.org/10.1186/s12967-017-1205-9.

[70] Cheng Y, Schorey JS. Extracellular vesicles deliver *Mycobacterium* RNA to promote host immunity and bacterial killing. EMBO Rep 2019. https://doi.org/10.15252/embr.201846613.

[71] Berry MPR, Graham CM, McNab FW, Xu Z, Bloch SAA, Oni T, Wilkinson KA, Banchereau R, Skinner J, Wilkinson RJ, Quinn C, Blankenship D, Dhawan R, Cush JJ, Mejias A, Ramilo O, Kon OM, Pascual V, Banchereau J, Chaussabel D, O'Garra A. An interferon-inducible neutrophil-driven blood transcriptional signature in human tuberculosis. Nature 2010;466:973–7. https://doi.org/10.1038/nature09247.

[72] Cheng Y, Schorey JS. *Mycobacterium tuberculosis*-induced IFN-β production requires cytosolic DNA and RNA sensing pathways. J Exp Med 2018;215:2919–35. https://doi.org/10.1084/jem.20180508.

[73] Alvarez-Jiménez VD, Leyva-Paredes K, García-Martínez M, Vázquez-Flores L, García-Paredes VG, Campillo-Navarro M, Romo-Cruz I, Rosales-García VH, Castañeda-Casimiro J,

González-Pozos S, Hernández JM, Wong-Baeza C, García-Pérez BE, Ortiz-Navarrete V, Estrada-Parra S, Serafín-López J, Wong-Baeza I, Chacón-Salinas R, Estrada-García I. Extracellular vesicles released from mycobacterium tuberculosis-infected neutrophils promote macrophage autophagy and decrease intracellular mycobacterial survival. Front Immunol 2018;9. https://doi.org/10.3389/fimmu.2018.00272.

[74] Behar SM, Martin CJ, Nunes-Alves C, Divangahi M, Remold HG. Lipids, apoptosis, and cross-presentation: links in the chain of host defense against *Mycobacterium tuberculosis*. Microbes Infect 2011;13:749–56. https://doi.org/10.1016/j.micinf.2011.03.002.

[75] Sridharan H, Upton JW. Programmed necrosis in microbial pathogenesis. Trends Microbiol 2014;22:199–207. https://doi.org/10.1016/j.tim.2014.01.005.

[76] Srivastava S, Ernst JD. Cell-to-cell transfer of *M. tuberculosis* antigens optimizes CD4 T cell priming. Cell Host Microbe 2014;15:741–52. https://doi.org/10.1016/j.chom.2014.05.007.

[77] Colino J, Snapper CM. Exosomes from bone marrow dendritic cells pulsed with diphtheria toxoid preferentially induce type 1 antigen-specific IgG responses in naive recipients in the absence of free antigen. J Immunol 2006;177:3757–62.

[78] Colino J, Snapper CM. Dendritic cell-derived exosomes express a *Streptococcus pneumoniae* capsular polysaccharide type 14 cross-reactive antigen that induces protective immunoglobulin responses against pneumococcal infection in mice. Infect Immun 2007;75:220–30. https://doi.org/10.1128/IAI.01217-06.

[79] Smith VL, Cheng Y, Bryant BR, Schorey JS. Exosomes function in antigen presentation during an in vivo *Mycobacterium tuberculosis* infection. Sci Rep 2017;7:43578. https://doi.org/10.1038/srep43578.

[80] Smith VL, Jackson L, Schorey JS. Ubiquitination as a mechanism to transport soluble mycobacterial and eukaryotic proteins to exosomes. J Immunol 2015;195:2722–30. https://doi.org/10.4049/jimmunol.1403186.

[81] Chaput N, Théry C. Exosomes: immune properties and potential clinical implementations. Semin Immunopathol 2011;33:419–40. https://doi.org/10.1007/s00281-010-0233-9.

[82] Dai S, Wei D, Wu Z, Zhou X, Wei X, Huang H, Li G. Phase I clinical trial of autologous ascites-derived exosomes combined with GM-CSF for colorectal cancer. Mol Ther 2008;16:782–90. https://doi.org/10.1038/mt.2008.1.

[83] Escudier B, Dorval T, Chaput N, André F, Caby M-P, Novault S, Flament C, Leboulaire C, Borg C, Amigorena S, Boccaccio C, Bonnerot C, Dhellin O, Movassagh M, Piperno S, Robert C, Serra V, Valente N, Le Pecq J-B, Spatz A, Lantz O, Tursz T, Angevin E, Zitvogel L. Vaccination of metastatic melanoma patients with autologous dendritic cell (DC) derived-exosomes: results of thefirst phase I clinical trial. J Transl Med 2005;3:10. https://doi.org/10.1186/1479-5876-3-10.

[84] Morse MA, Garst J, Osada T, Khan S, Hobeika A, Clay TM, Valente N, Shreeniwas R, Sutton MA, Delcayre A, Hsu D-H, Le Pecq J-B, Lyerly HK. A phase I study of dexosome immunotherapy in patients with advanced non-small cell lung cancer. J Transl Med 2005;3:9. https://doi.org/10.1186/1479-5876-3-9.

[85] WHO. WHO | Global tuberculosis report, http://www.who.int/tb/publications/global_report/en/; 2018. [accessed 2 March 2019].

[86] O'Garra A, Redford PS, McNab FW, Bloom CI, Wilkinson RJ, Berry MPR. The immune response in tuberculosis. Annu Rev Immunol 2013;31:475–527. https://doi.org/10.1146/annurev-immunol-032712-095939.

[87] Cheng Y, Schorey JS. Targeting soluble proteins to exosomes using a ubiquitin tag. Biotechnol Bioeng 2016;113:1315–24. https://doi.org/10.1002/bit.25884.

[88] Delcayre A, Estelles A, Sperinde J, Roulon T, Paz P, Aguilar B, Villanueva J, Khine S, Le Pecq J-B. Exosome display technology: applications to the development of new diagnostics and therapeutics. Blood Cells Mol Dis 2005;35:158–68. https://doi.org/10.1016/j.bcmd.2005.07.003.

[89] Hartman ZC, Wei J, Glass OK, Guo H, Lei G, Yang X-Y, Osada T, Hobeika A, Delcayre A, Le Pecq J-B, Morse MA, Clay TM, Lyerly HK. Increasing vaccine potency through exosome antigen targeting. Vaccine 2011;29:9361–7. https://doi.org/10.1016/j.vaccine.2011.09.133.

[90] Lattanzi L, Federico M. A strategy of antigen incorporation into exosomes: comparing cross-presentation levels of antigens delivered by engineered exosomes and by lentiviral virus-like particles. Vaccine 2012;30:7229–37. https://doi.org/10.1016/j.vaccine.2012.10.010.

[91] Rountree RB, Mandl SJ, Nachtwey JM, Dalpozzo K, Do L, Lombardo JR, Schoonmaker PL, Brinkmann K, Dirmeier U, Laus R, Delcayre A. Exosome targeting of tumor antigens expressed by cancer vaccines can improve antigen immunogenicity and therapeutic efficacy. Cancer Res 2011;71:5235–44. https://doi.org/10.1158/0008-5472.CAN-10-4076.

[92] Zeelenberg IS, Ostrowski M, Krumeich S, Bobrie A, Jancic C, Boissonnas A, Delcayre A, Le Pecq J-B, Combadière B, Amigorena S, Théry C. Targeting tumor antigens to secreted membrane vesicles in vivo induces efficient antitumor immune responses. Cancer Res 2008;68:1228–35. https://doi.org/10.1158/0008-5472.CAN-07-3163.

[93] Tian Y, Li S, Song J, Ji T, Zhu M, Anderson GJ, Wei J, Nie G. A doxorubicin delivery platform using engineered natural membrane vesicle exosomes for targeted tumor therapy. Biomaterials 2014;35:2383–90. https://doi.org/10.1016/j.biomaterials.2013.11.083.

[94] EL Andaloussi S, Mäger I, Breakefield XO, Wood MJA. Extracellular vesicles: biology and emerging therapeutic opportunities. Nat Rev Drug Discov 2013;12:347–57. https://doi.org/10.1038/nrd3978.

[95] Alvarez-Erviti L, Seow Y, Yin H, Betts C, Lakhal S, Wood MJA. Delivery of siRNA to the mouse brain by systemic injection of targeted exosomes. Nat Biotechnol 2011;29:341–5. https://doi.org/10.1038/nbt.1807.

[96] Ohno S, Takanashi M, Sudo K, Ueda S, Ishikawa A, Matsuyama N, Fujita K, Mizutani T, Ohgi T, Ochiya T, Gotoh N, Kuroda M. Systemically injected exosomes targeted to EGFR deliver antitumor microRNA to breast cancer cells. Mol Ther 2013;21:185–91. https://doi.org/10.1038/mt.2012.180.

[97] Schorey JS, Singh PP, Cheng Y. Exosomal biomarkers diagnostic of tuberculosis. 2017. US20170253916A1.

[98] Sheridan C. Exosome cancer diagnostic reaches market. Nat Biotechnol 2016;34:359–60. https://doi.org/10.1038/nbt0416-359.

[99] Schorey JS, Cheng Y. Particle size purification method and devices. 2019. US20190011342A1.

[100] Diaz G, Bridges C, Lucas M, Cheng Y, Schorey JS, Dobos KM, Kruh-Garcia NA. Protein digestion, ultrafiltration, and size exclusion chromatography to optimize the isolation of exosomes from human blood plasma and serum. J Vis Exp 2018. https://doi.org/10.3791/57467.

Chapter 7

Emerging therapeutic roles of exosomes in HIV-1 infection

Siew-Wai Pang, Sin-Yeang Teow
Department of Medical Sciences, School of Healthcare and Medical Sciences, Sunway University, Petaling Jaya, Malaysia

1 Introduction

Human Immunodeficiency Virus Type 1 (HIV-1), together with HIV-2, are retroviruses responsible for causing HIV infection which leads to acquired immunodeficiency syndrome (AIDS) [1]. The virus is normally transmitted through sexual intercourse but can also be transmitted through sharing of needles and from mother to child via breast-feeding [1]. There are many ways in which HIV-1 can propagate once they are in the host, and exosomes which are 30–100nm vesicles have been shown to play a vital role in the pathogenesis of HIV-1. Membrane-bound exosomes are derived from endosomes and are ubiquitously found in many cell types and from various biological fluids including plasma [2], saliva [3], urine [4], ascitic fluid [5], semen [6], breast milk [7], cerebrospinal fluid (CSF) [8], bronchoalveolar lavage (BAL) [9], and amniotic fluid [10]. These nanometer-sized vesicles can be secreted from immune cells including but not limited to macrophages, B-cells, T-cells, dendritic cells (DCs) and tumour cells [11, 12]. Exosomes secretion is being employed for cell-cell communication, and information carried in exosomes can regulate gene expression, cell proliferation and invasion, and immune regulation in various cell types [4, 11, 12]. These findings were discovered more than 30 years ago and became a major topic of research since after it was found that exosomes from B-cells could transport major histocompatibility complex class II (MHC-II) to T-cells [13]. In 2007, human exosomes from mast cells were found to contain messenger RNA (mRNA) and microRNA (miRNA) that have exerted biological effects in the recipient cells [14, 15]. In this article, we will review the implication of exosomes on the pathogenesis of HIV-1, and how a deeper understanding on exosomes could contribute towards better clinical judgements.

Exosomes. https://doi.org/10.1016/B978-0-12-816053-4.00007-9

2 Exosomes and HIV-1: The chicken or egg?

Interestingly, exosomes have been shown to highly resemble HIV-1 particles in terms of physical and chemical properties [16]. The question of whether exosomes evolved into retrovirus or the other way around then becomes a debate among scientists in the field. The model, Trojan exosome hypothesis proposed that the evolution from exosome to retrovirus happens following the mutation of *gag* genes [17, 18]. This idea is supported by the fact that the virus can hijack pre-existing exosome pathways for viral dissemination to cells lacking Env and receptors for viral entry [19]. In contrast, the second theory suggested that the retroviruses may have evolved to exploit the exosomal pathway for the cell-cell communication which then aggravate the virus pathogenesis [20]. The Trojan exosome hypothesis is however a misnomer because the budding of HIV-1 resembles that of ectosomes, a form of extracellular vesicle biochemically different to exosomes [21]. While both models are on two extreme ends, both are on agreement that exosomes play a vital role in HIV-1 pathogenesis due to the similarity of lipid, protein, carbohydrate and even RNAs composition between the viral particle and exosome.

Physically, exosomes and HIV-1 particles are similar in terms of size and density, and both are surrounded by a lipid bilayer. In terms of compositions, both have lipid composed of cholesterol and glycosphingolipids [22], carbohydrate group (high mannose and complex N-linked glycans) [23] and proteins (MHC, tetraspanins, actin, and TSG101) [24]. The type of RNA contents in both exosomes and HIV-1 particles is also fairly similar. A recent study also showed that the secretion of exosomes by DCs is highly dependent on DC immunereceptor, which is the same receptor used by HIV virions to interact with DCs and $CD4^+$ T-cells. The inhibition of this receptor reduced exosomal secretion, suggesting the similarity in the mechanisms utilized by both HIV-1 and exosomes [25]. Hence, it is believed that HIV-1 particles are generated through the same exosome biogenesis pathway [20]. Immune response related molecules such as MHC-II can also be incorporated into the viral particle, to evade the host immune surveillance [17].

Exosomes are extracellular vesicles (EV) secreted by humans, other animals, and microorganisms including bacteria [26], fungi [27], and protozoans [27]. Other than exosomes, EV can be in the form of apoptotic bodies or microvesicles such as ectosomes. What differentiates exosomes to other form of EV (e.g. ectosome) is how the vesicles are generated and its contents. Exosomes are generally either assembled from the plasma membrane, cytosol, or endocytic compartments without the presence of organelles and serum proteins [28]. Its contents include proteins, lipids, carbohydrates, small molecules, and various types of RNA species, but the molecular mechanisms on how these cargos are recruited remains to be elaborated. What is known currently is that endosomal sorting complex required for transport (ESCRT) is important in the formation of intraluminal vesicles and budding of exosomes, and ESCRT-associated protein,

TSG101 contributes to the secretion of cancer exosomes [15, 29]. Interestingly, exosomes have been reported to be secreted in an ESCRT-independent manner involving tetraspanins [30, 31]. Besides tetraspanins, a group of proteins belonging to the small GTPase family identified as RAB family have been shown to also be involved in several processes of vesicular trafficking including vesicle docking, budding, membrane fusion, and possibly exosomal release. In cancer cells, loss of several members of the RAB family have been shown to quickly deplete exosome secretion, supporting its importance in the exosome secretion pathway [15, 32].

While exosomes and viral particles share many biochemical properties, viral particles have several key differences to exosomes. Structurally, exosomes can vary greatly depending on the originating cell type, whereas HIV particles are generally more uniform regardless of the type of infected cells [33]. Biochemically, the internal cargoes of exosomes are heavily cell type-dependent, whereas HIV-1 particles always carry the same contents. For example, the exosomes of HIV-1-infected cells are enriched with viral RNAs and proteins such as Nef [34].

3 Exosomes as double-edged sword in HIV-1

The composition of exosomes from biological fluids vary greatly from one to another, which can either promote or inhibit viral pathogenesis. Whether viral pathogenesis is inhibited or promoted depends greatly on their cellular origins. For example, exosomes derived from HIV-1-infected cells generally exhibit pathogenesis enhancement [35], while exosomes from other sources such as saliva, tears, breast milk and seminal fluids exhibit protective properties [4]. Fig. 1 summarizes the exosome-associated molecules that could either promote or inhibit HIV-1 pathogenesis.

Blood is where HIV-1 resides and is also the mode of transmission for HIV-1. The blood contains various types of cells and is therefore filled with a plethora of exosomes exhibiting different properties [4]. As discussed previously, cells infected with HIV-1 secrete exosomes that could enhance viral dissemination. These exosomes contain chemokine receptors such as CCR5 and CXCR4 that are delivered to recipient cells to facilitate HIV-1 entry through the binding of viral envelope proteins to these receptors [36]. Nef-containing exosomes secreted from infected cells can also induce T-cell apoptosis [34], downregulation of immune cell molecules such as CD4 and MHC-I [37], and inhibit various RNAs [38]. The dissemination of viral particles is also enhanced by Nef as it promotes exosomal secretion [34]. Another way of immune evasion by the virus is by removing several cell surface molecules such as CD45, CD86 and MHC-II through exosomes. Furthermore, the viral particles have also been known to incorporate a variety of host cell molecules, thereby masking its presence from immune surveillance cells such as natural killer (NK) cells [4, 39]. The findings which demonstrated the viral dissemination via Env- and receptor-independent

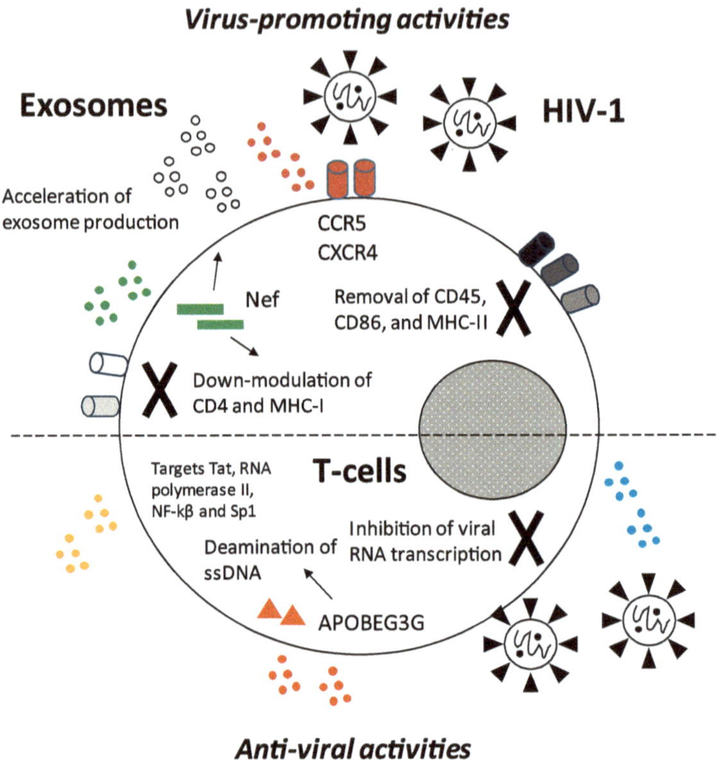

FIG. 1 Double-sword role of exosomes in HIV-1 pathogenesis. Exosomes are in close contact with all cell types. Nef is capable of enhancing the production of HIV-1 associated exosomes, hence promoting the viral pathogenesis. Nef also down-regulates key surface molecules in host immune defense mechanism such as MHC-1 and CD4. Similarly, HIV-1 associated exosomes are able to reduce the expression of CD45, CD86, and MHC-II surface molecules. Meanwhile, exosomes up-regulate the expression of co-receptors CCR5 and CXCR4 to enhance the HIV-1 entry. In contrast, exosomes also exhibit anti-viral activities. They were found to inhibit viral RNA transcription, and interfere with the function of Tat, RNA polymerase II, NF-κβ and Sp1. Exosomes also carry the host-derived APOBEG3G which inhibits HIV-1 replication to the non-infected cells.

pathways also pose a huge challenge to the current development of anti-viral strategies against the fusion of the virus with the recipient cell [19, 40]. In addition, a recent study has strengthened the notion that despite the combination of antiretroviral therapies, extracellular vesicles containing HIV-1 viral products could still continuously be released, which consequently led to neurocognitive immunological disarray [41].

In contrast, exosomes found in breast milk and semen exhibit anti-viral properties. While it is expected for breast milk's exosomes to contain anti-viral properties to provide natural immunity for infants, it is surprising that semen-derived exosomes also exhibit anti-viral properties. It was found that exosomes derived from semen can inhibit the reverse transcription of viral RNA [28].

However, this does not seem to be effective enough to prevent HIV-1 spreading as sexual intercourse is still the main cause of HIV-1 infection [4]. The exact mechanism that is involved in these protective properties is not known, however it is worth investigating. This year in 2018, Jennifer Welch and her team showed that semen exosomes could block the pro-viral transcription of HIV-1 at several transcriptional check points, including the recruitment of transcription factor to the long-tandem repeats (LTR), initiation of transcription, and elongation through biochemical and functional studies. Gel shift assays and qPCR techniques revealed the targets of semen exosomes, which includes the viral protein Tat, human RNA Polymerase II, and human transcription factors NF-kβ and Sp1. In essence, semen exosomes were found to directly inhibit the binding of NF-kβ, RNA Polymerase II, and Sp1 to the LTR, an integral part of the life cycle of HIV-1 [42]. A better understanding on the improvement of protective properties of semen-derived could be useful in preventing the HIV-1 spreading.

Collective studies have also supported the regulation of immune system through exosomes. For example, the function of NK cells has been shown to be indirectly enhanced via exosomes in a fashion unlike that of macrophage activation. It has been documented that the Hepatitis B virus (HBV) [43] and Hepatitis C virus (HCV) [44] could activate NK cells induction, and the depletion of exosomes in the case of HBV markedly reduced the expression of a ligand important for NK cell activation [43]. Exosomes are also involved in the promotion of T-cell activities by inducing cellular maturation and increasing the expression of inflammatory cytokines, which are important in both cellular and humoral immunity [45]. It has always been known that macrophages and DCs are antigen-presenting cells (APCs) that present antigens through MHC I and MHC II to T-cells. Thus, the fact that the surface of exosomes from these two APCs are furnished with MHC I, MHC II, and T-cell co-stimulatory molecules strongly suggest that these exosomes are important elements in the antigen-presenting cascade [45, 46]. To a certain extent, the induction of the immune system through exosomes should provide immunity toward HIV-1.

Apolipoprotein B mRNA-editing enzyme-catalytic polypeptide-like 3G (APOBEC3G) is a host cellular protein that has been shown to protect the recipient cells from HIV-1 infection through exosomal cell-cell transfer [47, 48]. In addition to retroviruses, it has also been shown to protect against herpesvirus, parvovirus, HBV, papillomavirus, and retrotransposons [49]. APOBEC3G is a cytidine deaminase that inhibits HIV-1 replication through DNA-editing and editing-independent activities [50, 51]. The former mechanism involves the deamination of cytosine in single-stranded DNA (−) [49] to uracil during reverse transcription, which can lead to the degradation of the resulting replicon. Additionally, Bishop et al. [52] reported that the reverse transcription inhibition was through restriction of the elongation step rather than degradation [52]. The deamination is catalyzed by conserved zinc-binding domains, however further studies have shown that the deaminase activity is not necessary for the inhibitory effect of APOBEC3G. The loss of restriction capabilities of APOBEC3G becomes apparent only when both the N-terminal and zinc-binding domains are not functioning. This shows that only one domain is enough for HIV-1 inhibitory effects [51]. Of note, the protective properties of APOBEC3G is only apparent in the absence of HIV-1 Vif protein which is one of the HIV-1 accessory proteins. In the presence of Vif, APOBEC3G is polyubiquitinated and directed for proteasomal degradation [53]. The crystal structure of APOBEC3G in complex with ssDNA has been recently elucidated [54]. This is an important publication that provides insights on the mechanisms at the atomic level by which APOBEC3G interact with ssDNA.

4 Exosomes and autophagy

Autophagy, sometimes known as autophagocytosis, comes from the Greek language for 'self-eating'. It is an evolutionary conserved process involved in the maintenance of homeostasis and metabolism at cellular levels through the degradation of proteins with stable half-lives and clearance of cytoplasmic organelles via the lysosomal pathway. In some instances, these unwanted or damaged RNAs and proteins are released into the extracellular environment through exosomes. This will have an impact to the host, which since exosomes can be received by surrounding and neighboring cells. Exosomal secretion can be sometimes dependent on the autophagic system, and the activation of autophagy can either inhibit or promote the release of exosomes [55]. The degradation and recycling of proteins have been evidenced to encourage normal development, cell death, senescence, and even protection against intracellular infections. Cancer, inflammatory, neurodegenerative, and metabolic diseases are possible consequences of a dysregulation in the autophagy pathway [56]. In the case of HIV-1 associated neurodegenerative diseases, many of the autophagy-regulating genes are dysregulated and this in turn may affect exosomal regulation [55].

Since it is known that exosomes contribute significantly to the pathogenesis of HIV-1, it will be important to understand how autophagy can impact exosomal release. Autophagy has been shown to be induced by HIV-1, and the interaction between autophagy and HIV-1 is complicated and depends on the cell type and state of infection [55]. HIV-1 Tat protein can stimulate autophagy in human glial cells, and its interaction with the lysosome associated membrane protein 2A (LAMP2A) indicates that Tat encourages the formation of autophagosomes and lysosomal fusion in neurons [57]. In addition, the increase in autophagy was also observed in postmortem brains from humans inflicted with HIV-1-associated encephalitis [58]. These observations support the notion that autophagy can be induced by HIV-1 components, although the exact mechanisms will likely depend on the cell type and remain to be elucidated [55].

While the exact mechanisms are unclear, autophagy has been shown to provide immunity against HIV-1 to a certain extent. For example, HIV-1 virions were not detected in macrophages with high numbers of autophagosomes, but rather in macrophages with weaker autophagy activities in a study conducted by Espert and co-workers [59]. Sagnier et al. [60] speculated that the restriction could be due to the interaction between Tat and the autophagy protein complex p62/SQSTM1, which directs the Tat for selective autophagy through lysosomal-mediated degradation in CD4+ T-cells. In addition, the degradation of Tat was not only limited to newly infected cells, but also from Tat from infected cells entering uninfected neighboring cells [60]. In another study, Nardacci et al. [61] reported a significant increase in autophagosomes and related markers in peripheral blood mononuclear cells (PBMCs) from HIV-1 infected long term non-progressors as well as elite controllers versus normal progressors [61].

Even though the activation of autophagy pathway provides immunity toward HIV-1 infection through lysosomal-degradation of viral components, HIV-1 has evolved mechanisms against autophagy. Viral proteins such as Nef [62], Vif [63], and Tat [64] have all been shown to inhibit important steps in the autophagic pathway. In dendritic cells, HIV-1 mediates Akt activation, which results in decreased viral antigen presentation to CD4+ T cells by MHC-II [55]. In the event of compromised autophagy, exosome formation and liberation act as alternatives to maintain cellular homeostasis and ameliorate cellular stress. Thus, when autophagosomal degradation is inhibited by HIV-1, the host looks into exosomal transport as an alternative way to remove viral waste [34]. In short, it is possible that HIV-1 deliberately compromises autophagy to activate exosomal release which aids in the spread of viral products to neighboring cells. With this in mind, the induction of autophagy in infected cells may stop the production of exosomes, resulting in the reduction of exosomal-dependent viral spread [55].

As exosomes portray both pathogenic and anti-viral effects towards HIV-1, they play different roles in the treatment of HIV-1 either making the treatment more challenging or partially facilitating the treatment. The pathogenic exosomes or the related proteins can serve as therapeutic targets by current anti-HIV-1 therapeutic strategies whereas the anti-viral exosomes can be potentially developed into a novel antiviral therapeutic modality. The therapeutic potential of exosomes will be discussed in Section 6. Based on the cumulative evidences, there is a great potential in exploring an anti-viral strategy which targets the exosome pathway hijacked by HIV-1 for novel anti-HIV-1 therapeutic development.

5 Exosomes as biomarkers for HIV-1

In the case of cancer patients, exosomes filled with tumor-specific molecules have been found in biofluids and the potential for its use as cancer biomarkers are being evaluated. While the therapeutic and diagnostic potential of exosomes in relation to cancer is moving fast, it is not the case for HIV-1. Exosomes containing HIV-1 RNA and other viral-related molecules can also be potential disease biomarkers aside from traditional antibody and antigen screening test (summarized in Table 1). This is supported by the fact that HIV-1 RNA and protein-loaded exosomes can be found in the blood of seropositive HIV-1 patients.

TABLE 1 Potential molecules within exosomes that can serve as HIV-1 biomarkers

Molecule	Biomarker	Functions	References
MHC II	Diagnostic	Antigen presentation	[4]
Nef	Diagnostic	Viral replication, immune response evasion	[65, 66]
Vif	Diagnostic	Viral replication, inhibition of APOBEC3G	[67]
Vpu	Diagnostic	Viral replication, enhancement of viral release	[68]
Vpr	Diagnostic	Viral replication, immune evasion	[69]
Gag	Diagnostic	Core structural proteins	[70]
Tat	Diagnostic	Enhance viral transcription	[71]
Pol (reverse transcriptase, integrase, protease)	Diagnostic	Viral replication	[72]
Env	Diagnostic	Envelop protein	[72]
Rev	Diagnostic	Gene expression	[72]
miRNA	Diagnostic	RNA silencing and post-transcriptional regulation of gene expression	[73]
HMGB1	Prognostic	DAMP, alarmin	[74]
NF-L	Prognostic	Axonal degeneration marker	[74]
Aβ proteins	Prognostic	Neuroinflammation, oxidative stress, cell signaling	[74]

Together with viral load monitoring, these exosomes can be harvested and be evaluated for the efficacy of anti-retroviral treatment [15].

Accumulative observations have also reported the protective nature of exosomes against HIV-1 infection in breast milk [53, 75] and semen [20, 40]. Cells in vitro were less susceptible to an infection when treated with breast milk and semen, indicating that these exosomes may be used as a prevention method, or even as treatment. In 2015, Madison and his colleagues co-infected the vagina of mice with murine AIDS (mAIDS) and human-derived semen exosomes, and found that viral replication was inhibited, and systemic viral spread and viral load were decreased [21]. While we are now aware of the beneficial protective properties of exosomes, the mechanisms behind its protective attributes are mostly not known [4, 15].

Recently, extracellular vesicles from the urine of HIV-1 positive patients were found to be an attractive non-invasive source for biomarkers. Generally, protein is very scarce in the urine of healthy individuals (0.01%), but in the presence of certain diseases, the protein content and number of extracellular vesicles can elevate [76]. In patients infected with HIV-1, HIV-1 proteins including but not limited to Gag, Pol, Tat, Vif, Vpu, Vpr, and especially Nef [77] can be detected in the extracellular vesicles of urine [78]. All HIV-1 positive urine samples have exosomes containing at least one HIV-1 related proteins while all HIV-1 negative samples have exosomes with no HIV-1 proteins. While HIV-1 virions have been reported in urine, the sensitivity is extremely low. Because of this, the extracellular vesicles in urine is more reliable and can potentially be used for the diagnosis of many diseases [78]. Nef has also been indicated to be consistently present in the extracellular vesicles of transiently transfected cells, systemically circulating in extracellular vesicles, and can be transferred to uninfected cells through exosomes, including uninfectable cells lacking CD4 [65].

Besides urine, neuron-derived exosomes (NDE) are also potential biomarkers for cognitive impairment by HIV-1. Many cells in the nervous system including astrocytes, neurons, microglia, and oligodendrocytes also shed exosomes under normal and pathological conditions, and these neuronal exosomes can be isolated from the brain [79], CSF [8], and even plasma [80]. They can be found in these sites because they can cross the blood brain barrier into other parts of the body. Sun et al. [74] reported that NDE from cognitively impaired patients, regardless of their HIV-1 status will have a marked increase in proteins such as HMGB1, NF-L and Aβ proteins. While it is not entirely specific to HIV-1, NDE is useful in monitoring neuronal health in "real-time" for patients suffering from HIV-1-related cognitive impairment [74].

Exosomes are not only useful as a biomarker for the pathogenesis of HIV-1, it also carries information in relation to immune responses and oxidative stress. Generally, the amount of plasma exosomes was shown to be higher in HIV-1 positive patients in comparison to HIV-1 negative controls. An increase in oxidative stress markers such as cystine, oxidized cys-gly, PRDX1, PRDX2, CAT, and TXN was observed together with the decrease of polyunsaturated fatty acids such as EPA, DHA, and DPA. Immune activation markers such as CD14, HLA-A, HLA-B, and CRP were also detected [81]. Chettimada et al. [81] found that exosomes in HIV-1 patients treated with antiretroviral therapy (ART) carry immune activation and oxidative stress proteins, and these exosomes are capable of immunomodulatory effects on myeloid cells. It is also possible that these exosomes can affect pro-inflammatory and redox responses during the pathogenesis [81].

6 Therapeutic potential of targeting exosomes

6.1 Targeting viral proteins

The potential exosomal pathway targets include the cellular enzymes (lipases, RNAses, and proteases), cytoplasmic proteins (TSG101, MHC-II, tetraspanins, and cyclophilins), and HIV-1 proteins (CCR5, CXCR4, and Nef) [4]. Gag, Pol, Tat, Vif, Vpu, and Vpr are also HIV-1 proteins found in exosomes that are potential therapeutic targets [78] (depicted in Fig. 2). Reports have also found that inhibitors of exosomes can reduce the effectiveness of Env-dependent infection [28, 75], thus the applicability of complementing exosome inhibitors with current anti-viral therapy could be useful. Furthermore, exosomes with anti-viral properties can be purified and potentially be developed as therapy.

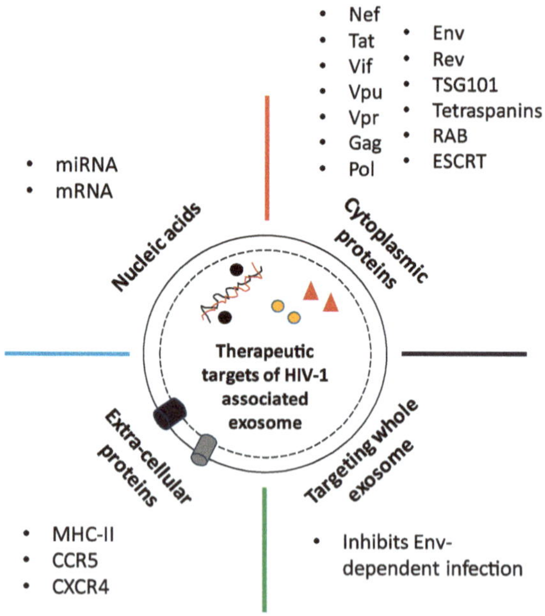

FIG. 2 Therapeutic application using exosomes against HIV-1. There are multiple exosome-associated molecules that could be potentially targeted for antiviral therapies. These include extracellular proteins (e.g. MHC-II, CCR5 and CXCR4), cytoplasmic proteins (e.g. HIV proteins, TSG101, ESCRT, RAB, etc.), nucleic acids (e.g. mRNA and miRNA), and targeting the whole exosomes.

As discussed previously, several cytoplasmic proteins involved in the exosomal pathway are potential therapeutic targets. TSG101, or also known as Tumour Suppressor Gene 101, is a component of cellular ESCRT required for the replication and budding of HIV-1 [82]. MHC-II are cell surface proteins involved in the regulation of immune responses through its ability to interact and process antigens to T-cells [83]. While MHC-II is important in the modulation of immune responses against pathogens, studies have found that a high expression of MHC-II in T-cells is related to an increase in the expression of HIV-1 [84, 85]. The high expression of MHC-II promotes the assembly and budding of HIV-1 particles. Even though MHC-II is a potential therapeutic target, it is important to note that its inhibition can lead to a loss of antigen-presenting abilities, subsequently leading to increased susceptibility to other pathogens [84].

CCR5 and CXCR4 are related G-protein coupled chemokine receptors that HIV-1 exploits to gain entry into the cell. CCR5-deficient individuals have been shown to be highly resistant to CCR5-dependent HIV-1 infection but are not completely resistant due to viral entry via CXCR4 [86]. HIV-1 infected cells are also capable of improving viral dissemination by exporting these receptors to recipient cells through exosomes [4]. With these facts in mind, it is only natural to target these receptors and receptors-containing exosomes [86]. There are two ways in which these receptors can be targeted for prevention of entry. Based on the biology and mechanism of these receptors, therapeutic drugs can be molecules that induce stearic blockade to the binding domain of CCR5/CXCR4. Second, the drugs can be an inducer for the internalization or recycling of these receptors, masking the recipient cells from the virus. The internalization of these two receptors have been shown to be a potential inhibitor of HIV-1 entry [87–89]. That said, the inhibition of CCR5 may have deleterious side effects as it plays a role in the host immune system.

The Nef protein is an accessory protein that is expressed by primate lentiviruses, including HIV-1, HIV-2, and SIV. Nef protein is critical for the survival of the virus as it promotes viral replication and infection through the downregulation of important cell surface molecules. They include MHC-I, MHC-II, CD4 and CD28, all of which are crucial for anti-viral immune responses. This made the development of a vaccine difficult as the immune system cannot eliminate infected cells without the proper functioning of antigen-presenting cells. A study by Geyer et al. [90] identified and described the functional motifs and structures of the Nef protein in detail that can be used as a reference for drug and therapy development [90]. Recently, Dekaban and Dikeakos [91] reviewed the mechanism of a Nef inhibitor, 2c-like compound that blocks the MHC-I-downregulation function of Nef. The restoration of the antigen-presenting function allows for cytotoxic T-cell-mediated killing of infected cells which may potentially act as immune adjuvants to vaccines. However, the author stressed that the current 2c-like Nef inhibitors only restores MHC-I on the cell surface. A second generation 2c-like Nef inhibitor will need to be developed to block the other deleterious functions of Nef [91].

Other than inhibiting Nef protein, Gag, Pol, and Env are also viral proteins found in exosomes that can potentially be targeted for therapeutic applications. HIV-1 Gag is the master orchestrator for viral assembly and is imperative for the formation and release of infectious particles. This makes it an attractive target for drug development. As of current, there are already existing drugs targeting Gag and novel antivirals under development seem to be promising [92]. The Env protein mediates viral entry into the host cell through interaction with the CD4 and CCR5/CXCR4 co-receptor. Because viral entry through this process is complicated, they are vulnerable to inhibition via small molecules. Herschhorn et al. [93] has reviewed several potential Env inhibitors together with the identification of a new compound 18A, which warrants further investigations [93]. Polymerases encoded by HIV-1 which includes reverse transcriptase are necessary for replication. Nucleoside analogs, which targets these polymerases were among the first few to show clinical efficacy and are broadly utilized in the treatment of HIV-1, hepatitis and herpes-simplex virus [94].

Vpr, Vpu, and Vif are viral proteins that are involved in viral replication and are also found in exosomes. Vpr protein can be inhibited by hematoxylin and its derivatives [95]. Vpu is another potential therapeutic target as it is known to downregulate the host protein BST-2, a protein that inhibits the release of viral particles [96]. Vif on the other hand is notoriously known for its ability to inhibit APOBEC3G, a protein that restricts HIV-1 replication through introduction of hypermutations. Pery et al. [97] reported the discovery of a compound that can inhibit the Vif-APOBEC3G interaction while at the same time increasing cellular APOBEC3G levels, resulting in improved attenuation of viral particles [97].

Tat and Rev. are both proteins that are important for gene expression. Without the involvement of Tat in viral transcription, HIV-1 will mostly remain in latency [71]. A recent study by Balachandran et al. [98] revealed the potential of several small molecules in inhibiting the protein expression of not only Tat and Rev., but also Gag and Env. While these molecules show great potential, its biological effects were only demonstrated in vivo [98].

The loss of CD4 T-cells is one of the hallmark features of HIV-1 infection, and HIV-1 infected DCs residing in mucosal tissues have been implied to facilitate the spread and maintenance of HIV-1, with the former through exosomes. Contrary to that, a study demonstrated that DCs exhibited greater exosomal production in the presence of HIV-1 compared to controls. The greater production of exosomes was associated with lesser viral infectivity and better apoptotic rates in CD4 T-cells through the upregulation of pro-apoptotic proteins such as Apaf-1 and Dap-3. The extent to which these proteins affect apoptosis is however still unclear. Through this study, it is now known that DCs pulsed with HIV-1 can significantly improve exosomal secretion [99]. Another group also showed that improved exosomal release by cells can be stimulated through specific receptors associated with the modulation of the immune system [100]. With these results, it is likely that exosomes can act like cytokines by regulating the immune system. Interestingly, Subra also found that mature DCs release more exosomes than immature DCs, indicating that exosomes could be important in inflammatory reactions. All in all, DCs-derived exosomes show immune-modulating properties which helps to fight against HIV-1. However, there are challenges associated with the use of DCs-derived exosomes as a tool for therapy since HIV-1 is notorious for being easily adaptable to its microenvironment [101].

Interestingly, the HIV-1 protein Tat has been recently shown to activate latent HIV-1 in primary, resting CD4+ T-cells by more than 30-fold in HIV-1 mRNA expression. Tang et al. packaged these proteins into exosomes and targeted them towards infected resting CD4+ T cells with minimal toxicity and found that it was a specific HIV-LTR activation instead of a generalized immune stimulation [71]. Tat protein activates the HIV-1 promoter in the long terminal repeat (LTR) region of proviral DNA in which transcription has been initiated and its safety and tolerability by animal models and humans were well documented [102, 103]. That said, it is not likely that Tat will be able to reactivate all latent viruses and will most likely favor proviruses where transcription was previously initiated. In addition, the generation of exosomes containing Tat can be tedious and not be suitable for clinical applications for the time being [71].

6.2 Targeting viral RNAs

In the last 30 years, there have been significant progression in the development of RNA-based gene therapy for HIV-1. These include the use of antisense-ssRNA [104, 105], ribozyme [104], RNA decoy [106], RNA interference (RNAi) [107], and CRISPR involving Cas9 and single-guide RNA [108]. The ability of these RNA species to inhibit active replication and eliminate hidden reservoirs of infection are also well documented [109]. For the RNA-based therapy to be a success, they must be delivered to the target tissues and be stable at adequate doses for an effective amount of time. Knowing how unstable RNA species are, they can easily be cleared out by the host's immune system through RNAse activities. Furthermore, the hydrophilic nature of RNA reduces its affinity to the lipid-based membranes of recipient cells, and the relatively large molecular size of RNA impedes their movement across cell membranes. Consequently, this leads to poor cellular uptake and therefore reduced anti-viral efficacy. These limitations call for more efficient RNA delivery systems with better bioavailability. Traditional drug delivery methods such as the use of viral vectors, liposomes, nano-polymers, and target peptides are commonly used, but even with constant modifications for improvements, they have not met the requirement for an effective gene therapy [104].

Exosomes will be practical replacements for these drug delivery systems as exosomes are naturally occurring and the delivery of RNA such as mRNA and miRNA from one cell to another has been its inherent role for many species and situations [4]. Exosomes that are loaded with the aforementioned RNA species can specifically recognize and target viral proteins or nucleic acid through different mechanisms. Antisense ssRNA can induce RNAse H-dependent degradation of the target RNA through the formation of complimentary base-pairing with HIV-1 RNA or by blocking ribosomal binding [105]. Ribozymes, especially hammerhead and hairpin ribozymes, are catalytically active small RNA molecules which can be engineered to cleave specific sequences of the viral RNA [104]. RNA decoys are also ssRNA with sequences that are like that of viral RNA. As a result, it competes with the viral RNA in binding to replicated-related regulatory proteins such as the ribosome [106]. RNAi on the other hand is a naturally occurring phenomenon [110] that can be used to silence the expression of HIV-1 genes through the assembly of RNA-induced silencing complex (RISC), which leads to the degradation of mRNA in a tightly regulated process. The RISC can be introduced in the form of a double-stranded siRNA or as a short-hairpin RNA [107]. Lastly, CRISPR is a relatively recent technology that is heralded as the future of molecular biology. It is an adaptive immunity system derived from the bacteria *Streptococcus pyogenes* [111] to fight against bacteriophages [112]. It is comprised of the Cas9 protein and a short-guide RNA with a specific sequence designed against the target DNA to be cleaved [108]. Currently, there have been success in permanently inactivating the integrated DNA of HIV-1 through large deletions. Current therapies are unable to effectively suppress the activities of integrated proviral DNA [113].

Exosomes are the ideal nanocarriers of anti-HIV-1 RNA because of several crucial intrinsic properties. The first property is the size of the exosome. The small size allows escape of phagocytosis by the mononuclear phagocyte system, which generally clears particles bigger than 100nm. This allows for the retention of exosomal carriers in vivo. Second, exosomes are endogenously produced from existing cells in the human body and are therefore less cytotoxic and more biocompatible with the immune system. In fact, some exosomes have privileged immune status. Third, exosomes have been shown to be able to cross biological barriers such as the mucosal barrier and the blood-brain barrier that artificial nanocarriers and other drug delivery system cannot. Specifically, semen-derived exosomes have been shown to naturally cross the mucosal barriers [15] while brain endothelial cell-derived exosomes can deliver their cargoes across the blood-brain barrier [114]. Fourth, exosomes are naturally equipped with a high affinity toward recipient cells which can facilitate the internalization of their cargoes through membrane fusion, phagocytosis, or endocytosis [115–117]. Finally, exosomes have a low affinity toward the liver, allowing for a low toxicity profile as the cargoes are not transported for liver processing [118]. Consequently, the cargoes are effectively transported to its intended recipient cell with minimal wastage.

For the past 20 years, there have been success with the use of therapeutic exosomes in vivo, and the use of such for cancer therapy is blooming [119]. While the use of exosomes in HIV-1 therapy remains in its infancy stage, the results and data gathered through therapeutic RNA exosomes in cancer therapy will prove useful in developing a similar system for HIV-1 therapy. Three distinct methods can be utilized to load exosomes with RNA for HIV-1 therapy. The first method involves loading isolated exosomes from parental cells with RNA. The loading process can involve steps such as freeze thawing, sonication, permeabilization, and transfection. This method is the most convenient and direct way to load exosomes with RNA and can yield large amounts of RNA-loaded exosomes within a few isolations. In addition, exosomes from several different cell types can be loaded simultaneously [104, 120]. The second method involves loading the parental cells with exogenous RNA, which will then be encased into the exosomes through inward budding during exosomal biogenesis, followed by the release and isolation of the exosomes. This method is not commonly practiced but have been shown to be useful as a drug delivery system. For example, a study by Qazi and colleagues showed that the elicitation of specific immune response can vary between in vivo and in vitro just by the RNA-loading method used [104, 121]. The third method requires the transfection or infection of parental cells with DNA that can be transcribed into therapeutically-capable RNAs, which will then be encased into exosomes. This method produces RNA with superior integrity but can be expensive and requires specialized equipment. Yeo and colleagues suggested for mesenchymal stem cells (MSCs) to be used as an ex vivo producer of exosomes in a clinically applicable scale as they produce large amounts of exosomes [122]. Another advantage of this method is that a continuous supply of exosomes can be obtained using transfected immortalized MSCs [123], justifying the use of this method compared to the first method of loading isolated exosomes. There is also evidence that genetically-modified macrophages producing such exosomes can help attenuate neuroinflammation for more than 40 days in mouse models because of its ability to continuously produce cargo-bearing exosomes [124]. While all three approaches share the same goal of producing and isolating RNA-loaded exosomes, the decision on which approach to use will depend on the availability of equipment, facilities, type of parental cell, and the overall objective and design of the experiment [104]. While we now have the technology and means to load exosomes with anti-HIV-1 RNA, the next step will be to ensure that the exosomes are efficiently transported to their intended recipient cells. Exosomes in general interact with recipient cells through binding of surface adhesion proteins or other molecules. To maximize the efficiency of exosomal transfer, the surface protein of exosomes can be engineered to have a high affinity with recipient cells. That said, the surface-modifying process of exosomes in targeting the exosomes to CD4 T-cells has yet to be established. Yong et al. [104] has suggested for the DNA of gp-120 fused with exosomal protein Lamp2b (Lamp2b-gp120) be included in a multicistronic plasmid that can simultaneously express anti-HIV short-hairpin RNA to be transfected into parental cells [104].

While RNA-loading and surface-modifying of exosomes show great promises for anti-HIV-1 therapy, there are obstacles and uncertainties that come together in a package. Protein families including RAB [125], tetraspanins [31], and ESCRT [29] are now known to contribute to exosomal cargo assembly.

The precise mechanisms behind the assembly and loading of exosomal cargoes are still in the shadows and need to be elucidated [126]. For these reasons, it is uncertain as to whether the RNA loading through transfection/infection can be efficiently sorted, and the surface proteins expressed from the plasmid can successfully be furnished on the exosomes. More importantly, exosomes are naturally loaded with a myriad of endogenous cargoes depending on the parental cell, which may inhibit the artificial cargo or present toxicity to the recipient cell. The presence of existing endogenous cargoes may also limit the capacity of the exosome to store more cargoes. The choice of parental cell from which exosomes are obtained will also be crucial as this will determine the yield of exosomes [104]. Lastly, if all works, the production of exosomes will need to be scaled-up for clinical applications. Hupfeld et al. [127] and Mitchell et al. [128] have demonstrated the superior exosomal yield of specifically-designed bioreactor systems compared to culture flask methods [127, 128].

7 Lessons from the past

7.1 Understanding exosomal pathway in revising therapeutic strategy against viruses

As studies on exosomes continue to flourish, especially in the field of cancer, the pathways, mechanisms, physical and chemical structures, mediators and signaling cascades are now accessible for researchers studying exosomes in viral pathogenesis. Accumulative findings have demonstrated that exosomes are used both by cancer tumors and viruses as a mean of immune evasion through several methods including but not limited to downregulation of surface molecules [4, 129], and the export of deleterious molecules away from the cell [4, 130]. Furthermore, exosomes have been reported to also improve pathogenesis of cancer and viral spread [4, 130]. With the exosomal pathway elucidated, we see that both viruses and cancer tumors share similarities in terms of exosome biogenesis, and potential therapeutic drugs targeting cancer tumors can be considered and tested in HIV-1 infected cells.

As reported previously, some of the antiretroviral drugs for HIV-1 targets viral entry by inhibiting virus-cell fusion. The fact that HIV-1 can spread through Env negative and receptor-independent pathways through exosomes calls for a better therapeutic strategy, possibly through a combination of exosomal inhibiting drug with Nucleoside Reverse-Transcriptase Inhibitors. From another point, purified exosomes from semen has been found to be inhibitory toward HIV-1 replication [42]. It will be interesting to understand the stimuli that induces exosomal liberation, and perhaps the release of such exosomes can be artificially stimulated in individuals at high risk of an HIV-1 infection.

The identification of the mediators will also allow for their use in therapeutic applications. For example, exosomes can be engineered to contain these specific mediators, or more effective analogs can be designed for better anti-viral properties. Nanoparticle delivery systems resembling semen exosomes can also be considered as they are efficiently transported to their recipient sites and are able to naturally overcome mucosal barriers that other vesicle systems cannot [15].

7.2 Advancement of technologies in exosome detection and characterization

7.2.1 General methods

Exosomes and cell-derived vesicles are abundant in biological fluids and can carry important information on the health and disease status of an individual. Because of this, the detection of vesicles in general have garnered strong clinical, scientific and diagnostic interests. For many years, there have been many techniques developed by researchers worldwide to detect and/or quantify vesicles, but the development of these techniques is challenging because they are normally hindered by the extremely small size of vesicles (less than 100 nm). This is made more challenging when these vesicles have low refractive indexes and vary in shape and sizes [131]. Traditionally, exosomes can be isolated through ultra-centrifugation, however it requires a very long time, expensive equipment, has low yield, low recovery, co-purification with non-exosomal debris, and has low specificity. Isopycnic centrifugation is an improved version of ultra-centrifugation, where exosomes can be recovered with higher purity and yield, however it has the same limitations as ultra-centrifugation in terms of time and expensive laboratory equipment. This makes both the techniques impractical in clinical settings [132]. With these challenges in mind, scientists have managed to come up with several detection methods by working around with other unique properties of vesicles [131]. Several relatively high-throughput vesicle detection methods include flow cytometry, resistive pulse sensing (RPS), and nanoparticle tracking analysis (NTA).

7.2.2 Flow cytometry

Flow cytometry has been used to study vesicles even before the year 2000, but because the methods have been pushed to their limits to detect such small particles, the measurement of false positives or artifact vesicles is difficult to avoid. This method detects and quantify vesicles through the detection of forward and side scatter, similar to how cells are sorted and quantified through a "flow".

With the advent of other technologies such as NTA and RPS, it is then known that flow cytometry is not sensitive enough in detecting the extremely small vesicles as single particles. These findings surprised the exosome and vesicle research community, thus sparking research in the development of better detection methods. At the same time, attempts have been made to standardize the measurements of vesicles. As of now, no gold standard exists for vesicle detection [131]. In 2017 however, Suarez and the team showed that an improved flow cytometry method for the semi-quantitation of extracellular vesicles involving beads coated with antibodies against vesicular proteins such as tetraspanins, MHC molecules, and CD59 can be easily applied in most laboratories with adequate accuracy [133].

7.2.3 Resistive pulse sensing (RPS)

RPS is a method that utilizes the Coulter principle to detect and measure the size of nanoparticles ranging from 50 nm [134, 135]. It looks at the change in current during the analysis, and the bigger the change, the bigger the volume of the vesicle. This allows for accurate size measurements of single vesicles [134].

The limitation for this technology generally depends on the instrument used, and in the case of qNano (Izon Science Ltd., Christchurch, New Zealand), the clogging of the pores in the instrument can hinder the turn-around time of the results, making the qNano unsuitable as a clinical diagnostic tool [131].

7.2.4 Nanoparticle tracking analysis (NTA)

NTA is a technology made commercial by Nanosight (Amesbury, UK) is similar to RPS in that it measures the distribution of vesicles in suspension ranging from around 50 nm. The difference is that vesicles scatter light or emit fluorescence in the presence of laser illumination. The position of the moving vesicles due to Brownian motion are then tracked through a dark-field microscope, and the mean squared velocity of each vesicles measured. The mean squared velocity of the Brownian motion is determined by the diameter of vesicle and thus will provide information on the absolute size distribution of vesicles [132, 134, 136]. NTA can also be used to measure the size of vesicle subgroups through fluorescence labeling [131, 137]. In terms of its use in clinical diagnostics, NTA is convenient to use as it only uses a few minutes for analysis. Real-time feedback from the visualization of samples will also help to identify the presence of cells after vesicle isolation.

The limitation is that each video data generated per measurement can be up to 2GB in size and will therefore require considerably big storage spaces and powerful processing hardware. Skilled operators will also be required to process the data as it involves multiple operations and variables [131].

7.2.5 Immunoassays

Enzyme-linked immunosorbent assay (ELISA) is a method that uses specific capture antibodies to immobilize target antigens followed by detection of fluorescence by a secondary antibody containing a fluorophore/chromophore. ELISA is used to detect exosomes by some researchers through the targeting of vesicle proteins such as tetraspanins, CD9, CD63, and CD81 [138]. The signal given off using this method is proportional to the amount of exosomal protein in the sample. Lateral flow immunoassay (LFIA) on the other hand has a similar basis to ELISA in that it utilizes antibodies to capture antigens. The samples are however first mixed with detection antibodies attached to gold nanoparticles for visualization. In the case of LFIA, the detection signal is inversely proportional to the amount of exosomal protein in the sample. This is because exosomes can be completely covered with antibodies, generating strong stearic hindrances that may prevent its capture by capture antibodies.

With these findings in mind, the use of ELISA or LFIA in clinical diagnosis will need to be carefully considered as both will give completely different outcomes depending on the abundance of exosomal protein. As abundance in different patients can vary greatly, the use of ELISA or LFIA may not be suitable in clinical settings now. Of course, with immunoassays being easily available in most research and clinical laboratories, its potential in clinical diagnostics remain open for exploration [138].

7.2.6 Other methods

In recent years, there are many more advancements in exosome and vesicle detection techniques. These techniques are not commercially available but have the potential to be used in clinical settings for the rapid and accurate measurement of exosomes. They include but not limited to nanowire-on-micropillar from the Liu group [139], acoustic sorting from Lee and Weissleder group [140], alternating current electrohydrodynamic induced nanoshearing by the Trau group [141], on-chip nano biographic imaging by the Ozcan group [142], magnetic resonance by the Lee and Weissleder group [143], Surface Plasmon Resonance by Grigorenko group [144, 145], label-free Frequency-Locked Microtoroid Optical Resonators [146], and others.

7.3 Key challenges of clinical application of exosomes

As mentioned in the previous sections, there are challenges that need overcoming for exosomes to be applied for clinical use. For exosomes to be applied in clinical settings, they must be produced at a large scale. 1mL of culture medium generally yields less than 1 μg of exosomal protein, and the effective dose in most studies is around 10–100 μg. Exosomes are generally prepared by culturing exosome-producing cells in medium for up to a few days, but Riches et al. [147] and Yamashita et al. [73] found that production of exosomes is capped after 12 h of incubation, depending on the cell type [73, 147]. As mentioned previously, the use of a bioreactor may help improve exosomal yield [127]. Additionally, the yield of exosomes has been shown to be increased in the presence of stress such as low pH, anti-cancer drug treatment, and hypoxia [148]. The potentially detrimental effects of these stressed-induced exosomes should however be carefully evaluated as the changes to the composition of the exosomes have been indicated. Furthermore, stressed-induced exosomes may contain elevated levels of apoptotic bodies, which may result in detrimental effects and overestimation of exosome count [73]. Efficient isolation of exosomes can be another approach to be explored. As mentioned previously, ultracentrifugation is a way in which exosomes can be isolated and there are several modifications to the method which may improve isolation yield. Besides that, methods such as size exclusion chromatography [149], aqueous two-phase systems [150], polymer-based precipitation [151], and ultrafiltration [152] can be applied for large-scale exosomal production [73]. It is important to take note however that the differences in exosomal yield, resources, time, and efficiency of these different methods have not been properly compared and evaluated and thus requires further attention.

Another challenge associated with clinical applications of exosome is the collection of uniform and high-quality exosomes. As discussed earlier, there are many methods in which exosomes can be isolated and purified. The method of isolation will greatly affect the physicochemical properties and purity of the exosomes [153]. To be specific, three different ultracentrifugation-based methods can be used for exosome purification including simple pelleting, density cushion method, and density gradient method [153]. The simple pelleting method is easily scalable and has a relatively low running cost, but the exosomes isolated can be highly contaminated with protein. The density cushion method and density gradient method both produces high purity exosomes with the latter being gentler to the exosomes but has low sample capacity. With the lack of efficiency, ultracentrifugation may not be the ideal method for exosomal isolation in clinical settings. Polymer-based precipitations on the other hand is not suitable for use in clinical settings as they are known to cause contamination and severe aggregation [151]. Size-based isolation such as ultrafiltration [152] and size exclusion chromatography [149] also come with their advantages and drawbacks. The former method is very efficient but produces exosomes of low purity and possible damages by shear stress while the chromatography produces exosomes with high purity but has low sample capacity [149, 152].

Another factor to consider is the existence of exosomal sub-populations, which varies in size and composition [154]. Different isolation methods will isolate different sub-populations [155]. This further complicates the use of exosomes for treatment as different sub-populations with different chemical and physical properties may potentially elicit different biological effects. In a nutshell, different isolation methods will produce exosomes with different purity, physical and chemical properties, and composition. For them to be useful in therapeutic applications, the isolation method must be arduously optimized to preserve the properties of exosomes and to ensure the consistency of exosomal contents per every production batch [73].

After successful purification and isolation of exosomes, the next challenge will be to store the exosomes. Accordingly, exosomes can be stored in PBS [156] at −80°C. The addition of trehalose have been shown to protect exosomes from cryodamage [157]. Higher temperatures are associated with fewer exosomes and their contents (80) but is desirable in clinical settings because it does not require specialized equipment. While there is evidence from a patent that exosomes from cardiosphere-derived cells can be lyophilized with minimal loss of biological activity [158], there is no evidence that this will apply to exosomes derived from other sources. The shelf-life of the lyophilized exosomes remain unclear and additional studies will need to be carried out to answer these uncertainties [73].

Now that the strategies on using exosomes in therapeutic applications are discussed, the delivery of purified exosomes into the patient, modified or not, should be discussed. The biological effect of exosomes can only be appreciated upon uptake by intended recipient cells, thus it is important to control the biodistribution of exosomes. Takahashi et al. [159] reported that exosomes that are intravenously injected accumulates rapidly in the spleen, liver and lung. [159]. In the liver and spleen, these exosomes are eventually phagocytosed by macrophages, possibly owing to the fact that phosphatidylserine is contained in the exosomal membrane [160]. Thanks to Wiklander et al. [161], we now know that the route of injection will affect the biodistribution of exosomes. Specifically, intraperitoneal injection led to exosomal accumulation in the gastrointestinal tract and pancreas while subcutaneous injection resulted in lower exosomal accumulation in all measured organs. Generally, most of the injected exosomes end up in macrophages in the reticuloendothelial system regardless of cell origin and administration route [161]. The rapid macrophagic clearance of exosomes from the blood stream will limit systemic administration and calls for research on better delivery and biostability of exosomes in the host.

It can also be challenging when it comes to therapeutic targeting or inhibition of exosomal contents. As discussed previously, Balachandran et al. has identified several small molecules which show great potential in blocking the protein expression of Tat, Rev, Env, and Gag [98]. While these drugs may be effective in a non-exosomal context, it may not be as effective in penetrating the exosomes. It has been shown that the loading of exosomes with cargoes require the participation of several proteins such as RAB [125], ESCRT [29], and tetraspanins [31], but the exact mechanism remains unclear [104]. In addition, because the biophysical and biochemical properties of exosomes vary depending on the cell origin [33], the surface molecules on the exosomes may vary, and this may potentially prevent the entry of small molecules through stearic hindrance. Furthermore, the surface of exosomes generally carries a negative charge [104]. This may render some drugs or molecules ineffective due to electrostatic repulsion.

7.4 Enhancement of exosomal therapeutic efficacy

Since there are challenges associated with producing exosomes for clinical applications, one possible solution is to improve its therapeutic potential. Overexpression of therapeutic molecules or proteins is one simple way to increase efficacy. Zhang et al. and Qu et al. reported that the overexpression of miRNAs increased the therapeutic potential of exosomes [162, 163], and overexpression of proteins in which the miRNA expression profile is increased is also another effective method [164, 165]. Another way is to modify exosomes with functional molecules. Morishita et al. managed to reduce the dose of exosomes needed to induce cancer antigen-specific immune response by up to 100 folds by modifying B16BL6-derived exosomes with an immune adjuvant, CpG DNA [166]. As mentioned previously, hypoxia treatment has also been reported to alter exosomal composition, resulting in improved biological effects [167,

168]. Whilst it is tempting to apply these modifications to clinical applications, changes to the exosome structure and composition may result in undesirable side effects. As such, the safety and effects of exosomal modifications must be evaluated separately [73].

7.5 Future perspectives

We now know that the pathogenesis of HIV-1 can be greatly enhanced by exosomes through the hijacking of the host's biogenesis pathway and can be disseminated through exosomes in receptor and Env-independent pathways. In addition, the anti-viral properties of exosomes derived from breast milk and semen have been demonstrated in several publications, with the mechanism of semen exosomes partly explained [42]. With these contradictions in mind, existing strategies for HIV-1 treatment need to be re-evaluated and this calls for better therapeutics. Since we now know that HIV-1 can be disseminated through exosomes in receptor and Env-independent pathways, a more sensible approach will be to look at exosome inhibitors previously discussed. That said, the specificity of the therapy must be considerably high as exosomes play important roles in other physiological functions. As technology continues to improve every year, the accurate detection and measurement of exosomes in clinical settings without expensive equipment is starting to realize as exosomes carry a wealth of genetic and proteomic information of the host. A simple literature search for methods to detect exosomes and vesicles will yield results ranging from ELISA, LFIA, and flow cytometry to advanced technologies capable of detecting single exosomes. This is because exosomes are not only implicated in HIV-1 infection and other viruses such as HBV [16], HCV [169], and HSV [170], they are also major contributors to the pathogenesis and disease progression for cancer [132]. The implementation of exosome detection and measurement devices in clinical settings will potentially improve and complement existing tests for HIV-1, other viruses, and cancer. Detection aside, purified exosomes containing anti-viral properties can also potentially be used as a prophylaxis for high-risk individuals or as treatment for HIV-1 infected patients. With the recent discovery on the inhibitory mechanism of semen exosomes, exosomes can be engineered to contain mediators or proteins that have similar inhibitory effects.

8 Concluding remarks

In a nutshell, host exosomes could improve the pathogenesis of HIV-1 through intercellular dissemination and immune evasion, while some studies show the anti-viral properties of exosomes derived from breast milk and semen. This goes to show that the properties of said exosome depends on the cellular origin, recipient cell, and its contents. Because HIV-1 can be disseminated via exosomes, a better understanding of exosomes in the pathogenesis of HIV-1 would help in the development of more comprehensive treatment strategies.

Currently, methods for the detection, measurement, and purification of exosomes remain labor intensive and are not suitable for clinical settings. This will however change as it could be seen that increasing articles on newer detection methods for exosomes and vesicles in general are being published. Exosomes with anti-viral properties are also potential therapeutics capable of complementing existing treatment therapies. Finally, even in the field of cancer research, our understanding of the biology of exosomes and the role they play in cancer has only scratched the surface. Far lesser studies have been done on exosomes regarding HIV-1 and this would need to be improved if we are to achieve an HIV-1-free world. As more discoveries are made on exosomes, be it on cancer or HIV-1, this will drive the development of robust and sensitive tests to aid clinical decisions.

References

[1] Mabuka J, Nduati R, Odem-Davis K, Peterson D, Overbaugh J. HIV-specific antibodies capable of ADCC are common in breastmilk and are associated with reduced risk of transmission in women with high viral loads. PLoS Pathog 2012;8(6):e1002739.

[2] Caby MP, Lankar D, Vincendeau-Scherrer C, Raposo G, Bonnerot C. Exosomal-like vesicles are present in human blood plasma. Int Immunol 2005;17(7):879–87.

[3] Michael A, Bajracharya SD, Yuen PST, Zhou H, Star RA, Illei GG, Alevizos I. Exosomes from human saliva as a source of microRNA biomarkers. Oral Dis 2010;16(1):34–8.

[4] Teow SY, Nordin AC, Ali SA, Khoo AS. Exosomes in human immunodeficiency virus type I pathogenesis: threat or opportunity? Adv Virol 2016;2016:9852494.

[5] Andre F, Schartz NEC, Movassagh M, Flament C, Pautier P, Morice P, et al. Malignant effusions and immunogenic tumour-derived exosomes. Lancet 2002;360(9329):295–305.

[6] Sullivan R, Saez F, Girouard J, Frenette G. Role of exosomes in sperm maturation during the transit along the male reproductive tract. Blood Cells Mol Dis 2005;35(1):1–10.

[7] Admyre C, Johansson SM, Qazi KR, Filen J-J, Lahesmaa R, Norman M, et al. Exosomes with immune modulatory features are present in human breast milk. J Immunol 2007;179(3):1969–78.

[8] Street JM, Barran PE, Mackay CL, Weidt S, Balmforth C, Walsh TS, et al. Identification and proteomic profiling of exosomes in human cerebrospinal fluid. J Transl Med 2012;10(1):5.

[9] Admyre C, Grunewald J, Thyberg J, Gripenbäck S, Tornling G, Eklund A, et al. Exosomes with major histocompatibility complex class II and co-stimulatory molecules are present in human BAL fluid. Eur Respir J 2003;22(4):578–83.

[10] Keller S, Rupp C, Stoeck A, Runz S, Fogel M, Lugert S, et al. CD24 is a marker of exosomes secreted into urine and amniotic fluid. Kidney Int 2007;72(9):1095–102.

[11] Teow SY, Peh SC. Exosomes as the promising biomarker of Epstein-Barr virus (EBV)-associated cancers. In: Novel implications of exosomes in diagnosis and treatment of cancer and infectious diseases. vol. 5; 2017. p. 97–114.

[12] Teow SY, Liew K, Khoo ASB, Peh SC. Pathogenic role of exosomes in Epstein-Barr virus (EBV)-associated cancers. Int J Biol Sci 2017;13(10):1276–86.

[13] Raposo G, Nijman HW, Stoorvogel W, Liejendekker R, Harding CV, Melief CJ, Geuze HJ. B lymphocytes secrete antigen-presenting vesicles. J Exp Med 1996;183(3):1161–72.

[14] Lotvall J, Valadi H. Cell to cell signalling via exosomes through esRNA. Cell Adh Migr 2007;1(3):156–8.

[15] Madison MN, Okeoma CM. Exosomes: implications in HIV-1 pathogenesis. Viruses 2015;7(7):4093–118.

[16] Meckes DG, Raab-Traub N. Microvesicles and viral infection. J Virol 2011;85(24): 12844–54.

[17] Gould SJ, Booth AM, Hildreth JEK. The Trojan exosome hypothesis. Proc Natl Acad Sci U S A 2003;100(19):10592–7.

[18] Izquierdo-Useros N, Naranjo-Gómez M, Erkizia I, Puertas MC, Borràs FE, Blanco J, Martinez-Picado J. HIV and mature dendritic cells: Trojan exosomes riding the Trojan horse? PLoS Pathog 2010;6(3):e1000740.

[19] Chow Y, Yu D, Zhang JY, Xie Y, Wei OLC, Chiu C, et al. Gp120-independent infection of CD4- epithelial cells and CD4+ T-cells by HIV-1. J Acquir Immune Defic Syndr 2002;30(1):1–8.

[20] Izquierdo-Useros N, Puertas MC, Borràs FE, Blanco J, Martinez-Picado J. Exosomes and retroviruses: the chicken or the egg? Cell Microbiol 2011;13(1):10–7.

[21] Madison MN, Jones PH, Okeoma CM. Exosomes in human semen restrict HIV-1 transmission by vaginal cells and block intravaginal replication of LP-BM5 murine AIDS virus complex. Virology 2015;482:189–201.

[22] Wubbolts R, Leckie RS, Veenhuizen PTM, Schwarzmann G, Möbius W, Hoernschemeyer J, et al. Proteomic and biochemical analyses of human B cell-derived exosomes: potential implications for their function and multivesicular body formation. J Biol Chem 2003;278(13):10963–72.

[23] Krishnamoorthy L, Bess JW, Preston AB, Nagashima K, Mahal LK. HIV-1 and microvesicles from T cells share a common glycome, arguing for a common origin. Nat Chem Biol 2009;5(4):244–50.

[24] Théry C, Zitvogel L, Amigorena S. Exosomes: composition, biogenesis and function. Nat Rev Immunol 2002;2(8):569–79.

[25] Mfunyi CM, Vaillancourt M, Vitry J, Nsimba Batomene TR, Posvandzic A, Lambert AA, Gilbert C. Exosome release following activation of the dendritic cell immunoreceptor: a potential role in HIV-1 pathogenesis. Virology 2015;484:103–12.

[26] Lee JC, Lee EJ, Lee JH, Jun SH, Choi CW, Kim SI, et al. *Klebsiella pneumoniae* secretes outer membrane vesicles that induce the innate immune response. FEMS Microbiol Lett 2012;331(1):17–24.

[27] Oliveira DL, Nakayasu ES, Joffe LS, Guimarães AJ, Sobreira TJP, Nosanchuk JD, et al. Characterization of yeast extracellular vesicles: evidence for the participation of different pathways of cellular traffic in vesicle biogenesis. PLoS ONE 2010;5(6):e11113.

[28] Madison MN, Roller RJ, Okeoma CM. Human semen contains exosomes with potent anti-HIV-1 activity. Retrovirology 2014;11(1):102.

[29] Colombo M, Moita C, van Niel G, Kowal J, Vigneron J, Benaroch P, et al. Analysis of ESCRT functions in exosome biogenesis, composition and secretion highlights the heterogeneity of extracellular vesicles. J Cell Sci 2013;126(24):5553–65.

[30] Stuffers S, Sem Wegner C, Stenmark H, Brech A. Multivesicular endosome biogenesis in the absence of ESCRTs. Traffic 2009;10(7):925–37.

[31] van Niel G, Charrin S, Simoes S, Romao M, Rochin L, Saftig P, et al. The tetraspanin CD63 regulates ESCRT-independent and -dependent endosomal sorting during melanogenesis. Dev Cell 2011;21(4):708–21.

[32] Ostrowski M, Carmo NB, Krumeich S, Fanget I, Raposo G, Savina A, et al. Rab27a and Rab27b control different steps of the exosome secretion pathway. Nat Cell Biol 2010;12(1):19–30.

[33] Cantin R, Diou J, Bélanger D, Tremblay AM, Gilbert C. Discrimination between exosomes and HIV-1: purification of both vesicles from cell-free supernatants. J Immunol Methods 2008;338(1–2):21–30.

[34] Lenassi M, Cagney G, Liao M, Vaupotič T, Bartholomeeusen K, Cheng Y, et al. HIV Nef is secreted in exosomes and triggers apoptosis in bystander CD4+ T cells. Traffic 2010;11(1):110–22.

[35] Kadiu I, Narayanasamy P, Dash PK, Zhang W, Gendelman HE. Biochemical and biologic characterization of exosomes and microvesicles as facilitators of HIV-1 infection in macrophages. J Immunol 2012;189(2):744–54.

[36] Mack M, Kleinschmidt A, Brühl H, Klier C, Nelson PJ, Cihak J, et al. Transfer of the chemokine receptor CCR5 between cells by membrane- derived microparticles: a mechanism for cellular human immunodeficiency virus 1 infection. Nat Med 2000;6(7):769–75.

[37] Gray LR, Gabuzda D, Cowley D, Ellett A, Chiavaroli L, Wesselingh SL, et al. CD4 and MHC class 1 down-modulation activities of nef alleles from brain-and lymphoid tissue-derived primary HIV-1 isolates. J Neurovirol 2011;17(1):82–91.

[38] Aqil M, Naqvi AR, Bano AS, Jameel S. The HIV-1 Nef protein binds argonaute-2 and functions as a viral suppressor of RNA interference. PLoS ONE 2013;8(9):e74472.

[39] Esser MT, Graham DR, Coren LV, Trubey CM, Bess JW, Arthur LO, et al. Differential incorporation of CD45, CD80 (B7-1), CD86 (B7-2), and major histocompatibility complex class I and II molecules into human immunodeficiency virus type 1 virions and microvesicles: implications for viral pathogenesis and immune regulation. J Virol 2001;75(13):6173–82.

[40] Marras D, Bruggeman LA, Gao F, Tanji N, Mansukhani MM, Cara A, et al. Replication and compartmentalization of HIV-1 in kidney epithelium of patients with HIV-associated nephropathy. Nat Med 2002;8(5):522–6.

[41] Demarino C, Pleet ML, Cowen M, Barclay RA, Akpamagbo Y, Erickson J, et al. Antiretroviral drugs alter the content of extracellular vesicles from HIV-1-infected cells. Sci Rep 2018;8(1):7653.

[42] Welch JL, Kaddour H, Schlievert PM, Stapleton JT, Okeoma CM. Semen exosomes promote transcriptional silencing of HIV-1 by disrupting NF-kB/Sp1/Tat circuitry. J Virol 2018;92:. 00731-18.

[43] Kouwaki T, Fukushima Y, Daito T, Sanada T, Yamamoto N, Mifsud EJ, et al. Extracellular vesicles including exosomes regulate innate immune responses to hepatitis B virus infection. Front Immunol 2016;7:335.

[44] Nakai M, Oshiumi H, Funami K, Okamoto M, Matsumoto M, Seya T, Sakamoto N. Interferon (IFN) and cellular immune response evoked in RNA-pattern sensing during infection with hepatitis C virus (HCV). Sensors (Basel) 2015;15(10):27160–73.

[45] Zhang W, Jiang X, Bao J, Wang Y, Liu H, Tang L. Exosomes in pathogen infections: a bridge to deliver molecules and link functions. Front Immunol 2018;9:90.

[46] Schorey JS, Cheng Y, Singh PP, Smith VL. Exosomes and other extracellular vesicles in host-pathogen interactions. EMBO Rep 2015;16(1):24–43.

[47] Kulkarni R, Prasad A. Exosomes derived from HIV-1 infected DCs mediate viral trans-infection via fibronectin and galectin-3. Sci Rep 2017;7(1):14787.

[48] Wang X, Ao Z, Chen L, Kobinger G, Peng J, Yao X. The cellular antiviral protein APOBEC3G interacts with HIV-1 reverse transcriptase and inhibits its function during viral replication. J Virol 2012;86(7):3777–86.

[49] Stavrou S, Ross SR. APOBEC3 proteins in viral immunity. J Immunol 2015;195(10):4565–70.

[50] Holmes RK, Koning FA, Bishop KN, Malim MH. APOBEC3F can inhibit the accumulation of HIV-1 reverse transcription products in the absence of hypermutation: comparisons with APOBEC3G. J Biol Chem 2007;282(4):2587–95.

[51] Schumacher AJ, Hache G, MacDuff DA, Brown WL, Harris RS. The DNA deaminase activity of human APOBEC3G is required for Ty1, MusD, and human immunodeficiency virus type 1 restriction. J Virol 2008;82(6):2652–60.

[52] Bishop KN, Verma M, Kim EY, Wolinsky SM, Malim MH. APOBEC3G inhibits elongation of HIV-1 reverse transcripts. PLoS Pathog 2008;4(12):e1000231.

[53] Khatua AK, Taylor HE, Hildreth JEK, Popik W. Exosomes packaging APOBEC3G confer human immunodeficiency virus resistance to recipient cells. J Virol 2009;83(2):512–21.

[54] Maiti A, Myint W, Kanai T, Delviks-Frankenberry K, Sierra Rodriguez C, Pathak VK, et al. Crystal structure of the catalytic domain of HIV-1 restriction factor APOBEC3G in complex with ssDNA. Nat Commun 2018;9(1):2460.

[55] Ojha CR, Lapierre J, Rodriguez M, Dever SM, Zadeh MA, Demarino C, et al. Interplay between autophagy, exosomes and HIV-1 associated neurological disorders: new insights for diagnosis and therapeutic applications. Viruses 2017;9(7):E176.

[56] Mizushima N, Komatsu M. Autophagy: renovation of cells and tissues. Cell 2011;147(4):728–41.

[57] Fields J, Dumaop W, Elueteri S, Campos S, Serger E, Trejo M, et al. HIV-1 Tat alters neuronal autophagy by modulating autophagosome fusion to the lysosome: implications for HIV-associated neurocognitive disorders. J Neurosci 2015;35(5):1921–38.

[58] Zhou D, Masliah E, Spector SA. Autophagy is increased in postmortem brains of persons with HIV-1-associated encephalitis. J Infect Dis 2011;203(11):1647–57.

[59] Espert L, Varbanov M, Robert-Hebmann V, Sagnier S, Robbins I, Sanchez F, et al. Differential role of autophagy in CD4 T cells and macrophages during X4 and R5 HIV-1 infection. PLoS ONE 2009;4:e5787.

[60] Sagnier S, Daussy CF, Borel S, Robert-Hebmann V, Faure M, Blanchet FP, et al. Autophagy restricts HIV-1 infection by selectively degrading Tat in $CD4^+$ T lymphocytes. J Virol 2015;89(1):615–25.

[61] Nardacci R, Amendola A, Ciccosanti F, Corazzari M, Esposito V, Vlassi C, et al. Autophagy plays an important role in the containment of HIV-1 in nonprogressor-infected patients. Autophagy 2014;10(7):1167–78.

[62] Campbell GR, Rawat P, Bruckman RS, Spector SA. Human immunodeficiency virus type 1 Nef inhibits autophagy through transcription factor EB sequestration. PLoS Pathog 2015;11(6):e1005018.

[63] Borel S, Robert-Hebmann V, Alfaisal J, Jain A, Faure M, Espert L, et al. HIV-1 viral infectivity factor interacts with microtubule-associated protein light chain 3 and inhibits autophagy. AIDS 2014;29(3):275–86.

[64] Li JCB, Au KY, Fang JW, Yim HC, Chow KH, et al. HIV-1 trans-activator protein dysregulates IFN-γ signaling and contributes to the suppression of autophagy induction. AIDS 2011;25(1):15–25.

[65] McNamara RP, Costantini LM, Myers TA, Schouest B, Maness NJ, Griffith JD, et al. Nef secretion into extracellular vesicles or exosomes is conserved across human and simian immunodeficiency viruses. MBio 2018;9(1). e02344-17.

[66] Das SR, Jameel S. Biology of the HIV Nef protein. Indian J Med Res 2005;121(4):315–32.

[67] Miller JH, Presnyak V, Smith HC. The dimerization domain of HIV-1 viral infectivity factor Vif is required to block virion incorporation of APOBEC3G. Retrovirology 2007;4:81.

[68] Andrew AJ, Miyagi E, Strebel K. Differential effects of human immunodeficiency virus type 1 Vpu on the stability of BST-2/tetherin. J Virol 2011;85(6):2611–9.

[69] Kogan M, Rappaport J. HIV-1 accessory protein Vpr: relevance in the pathogenesis of HIV and potential for therapeutic intervention. Retrovirology 2011;8:25.

[70] Bell NM, Lever AML. HIV Gag polyprotein: processing and early viral particle assembly. Trends Microbiol 2013;21(3):136–44.
[71] Tang X, Lu H, Dooner M, Chapman S, Quesenberry PJ, Ramratnam B. Exosomal Tat protein activates latent HIV-1 in primary, resting CD4+ T lymphocytes. JCI Insight 2018;3(7):e95676.
[72] Nagata S, Imai J, Makino G, Tomita M, Kanai A. Evolutionary analysis of HIV-1 pol proteins reveals representative residues for viral subtype differentiation. Front Microbiol 2017;8:2151.
[73] Yamashita T, Takahashi Y, Takakura Y. Possibility of exosome-based therapeutics and challenges in production of exosomes eligible for therapeutic application. Biol Pharm Bull 2018;41(6):835–42.
[74] Sun B, Dalvi P, Abadjian L, Tang N, Pulliam L. Blood neuron-derived exosomes as biomarkers of cognitive impairment in HIV. AIDS 2017;31(14):F9–17.
[75] Näslund TI, Paquin-Proulx D, Paredes PT, Vallhov H, Sandberg JK, Gabrielsson S. Exosomes from breast milk inhibit HIV-1 infection of dendritic cells and subsequent viral transfer to CD4+ T cells. AIDS 2014;28(2):171–80.
[76] Salih M, Zietse R, Hoorn EJ. Urinary extracellular vesicles and the kidney: biomarkers and beyond. Am J Physiol Renal Physiol 2014;306(11):F1251–9.
[77] Ellwanger JH, Veit TD, Chies JAB. Exosomes in HIV infection: a review and critical look. Infect Genet Evol 2017;53:146–54.
[78] Anyanwu SI, Doherty A, Powell MD, Obialo C, Huang MB, Quarshie A, et al. Detection of HIV-1 and human proteins in urinary extracellular vesicles from HIV+ patients. Adv Virol 2018;2018:7863412.
[79] Banigan MG, Kao PF, Kozubek JA, Winslow AR, Medina J, Costa J, et al. Differential expression of exosomal microRNAs in prefrontal cortices of schizophrenia and bipolar disorder patients. PLoS ONE 2013;8(1):e48814.
[80] Fiandaca MS, Kapogiannis D, Mapstone M, Boxer A, Eitan E, Schwartz JB, et al. Identification of preclinical Alzheimer's disease by a profile of pathogenic proteins in neurally derived blood exosomes: a case-control study. Alzheimers Dement 2015;11(6):600–607.e1.
[81] Chettimada S, Lorenz DR, Misra V, Dillon ST, Reeves RK, Manickam C, et al. Exosome markers associated with immune activation and oxidative stress in HIV patients on antiretroviral therapy. Sci Rep 2018;8(1):7227.
[82] Strickland M, Ehrlich LS, Watanabe S, Khan M, Strub MP, Luan CH, et al. Tsg101 chaperone function revealed by HIV-1 assembly inhibitors. Nat Commun 2017;8(1):1391.
[83] Fridkis-Hareli M. Design of peptide immunotherapies for MHC class-II-associated autoimmune disorders. Clin Dev Immunol 2013;2013:826191.
[84] Finzi A, Brunet A, Xiao Y, Thibodeau J, Cohen EA. Major histocompatibility complex class II molecules promote human immunodeficiency virus type 1 assembly and budding to late endosomal/multivesicular body compartments. J Virol 2006;80(19):9789–97.
[85] Saifuddin M, Spear GT, Chang C, Roebuck KA. Expression of MHC class II in T cells is associated with increased HIV-1 expression. Clin Exp Immunol 2000;121(2):324–31.
[86] Alkhatib G. The biology of CCR5 and CXCR4. Curr Opin HIV AIDS 2009;4(2):96–103.
[87] Alkhatib G, Locati M, Kennedy PE, Murphy PM, Berger EA. HIV-1 coreceptor activity of CCR5 and its inhibition by chemokines: independence from G protein signaling and importance of coreceptor downmodulation. Virology 1997;234(2):340–8.
[88] Amara A, Gall SL, Schwartz O, Salamero J, Montes M, Loetscher P, et al. HIV coreceptor downregulation as antiviral principle: SDF-1alpha-dependent internalization of the chemokine receptor CXCR4 contributes to inhibition of HIV replication. J Exp Med 1997;186(1):139–46.

[89] Signoret N, Oldridge J, Pelchen-Matthews A, Klasse PJ, Tran T, Brass LF, et al. Phorbol esters and SDF-1 induce rapid endocytosis and down modulation of the chemokine receptor CXCR4. J Cell Biol 1997;139(3):651–64.
[90] Geyer M, Fackler OT, Peterlin BM. Structure-function relationships in HIV-1 Nef. EMBO Rep 2001;2(7):580–5.
[91] Dekaban GA, Dikeakos JD. HIV-I Nef inhibitors: a novel class of HIV-specific immune adjuvants in support of a cure. AIDS Res Ther 2017;14:53.
[92] Spearman P. HIV-1 Gag as an antiviral target: development of assembly and maturation inhibitors. Curr Top Med Chem 2016;16(10):1154–66.
[93] Herschhorn A, Gu C, Espy N, Richard J, Finzi A, Sodroski JG. A broad HIV-1 inhibitor blocks envelope glycoprotein transitions critical for entry. Nat Chem Biol 2014;10(10):845–52.
[94] von Kleist M, Metzner P, Marquet R, Schütte C. HIV-1 polymerase inhibition by nucleoside analogs: cellular- and kinetic parameters of efficacy, susceptibility and resistance selection. PLoS Comput Biol 2012;8(1):e1002359.
[95] Hagiwara K, Ishii H, Murakami T, Takeshima SN, Chutiwitoonchai N, Kodama EN, et al. Synthesis of a Vpr-binding derivative for use as a novel HIV-1 inhibitor. PLoS ONE 2015;10(12):e0145573.
[96] Mi Z, Ding J, Zhang Q, Zhao J, Ma L, Yu H, et al. A small molecule compound IMB-LA inhibits HIV-1 infection by preventing viral Vpu from antagonizing the host restriction factor BST-2. Sci Rep 2015;5:18499.
[97] Pery E, Sheehy A, Nebane NM, Brazier AJ, Misra V, Rajendran KS, et al. Identification of a novel HIV-1 inhibitor targeting Vif-dependent degradation of human APOBEC3G protein. J Biol Chem 2015;290(16):10504–17.
[98] Balachandran A, Wong R, Stoilov P, Pan S, Blencowe B, Cheung P, et al. Identification of small molecule modulators of HIV-1 Tat and Rev protein accumulation. Retrovirology 2017;14(1):7.
[99] Wolfson B, Yu JE, Zhou Q. Exosomes may play a crucial role in HIV dendritic cell immunotherapy. Ann Transl Med 2017;5(16):337.
[100] Théry C, Regnault A, Garin J, Wolfers J, Zitvogel L, Ricciardi-Castagnoli P, et al. Molecular characterization of dendritic cell-derived exosomes: selective accumulation of the heat shock protein hsc73. J Cell Biol 1999;147(3):599–610.
[101] Chistiakov DA, Grechko AV, Orekhov AN, Bobryshev YV. An immunoregulatory role of dendritic cell-derived exosomes versus HIV-1 infection: take it easy but be warned. Ann Transl Med 2017;5(17):362.
[102] Chang HC, Samaniego F, Nair BC, Buonaguro L, Ensoli B. HIV-1 Tat protein exits from cells via a leaderless secretory pathway and binds to extracellular matrix-associated heparan sulfate proteoglycans through its basic region. AIDS 1997;11(12):1421–31.
[103] Lin X, Irwin D, Kanazawa S, Huang L, Romeo J, Yen TSB, Peterlin BM. Transcriptional profiles of latent human immunodeficiency virus in infected individuals: effects of Tat on the host and reservoir. J Virol 2003;77(15):8227–36.
[104] Yong Q, Jing M, Yi Z. Therapeutic potential of anti-HIV RNA-loaded exosomes. Biomed Environ Sci 2018;31(3):215–26.
[105] Zeller SJ, Kumar P. RNA-based gene therapy for the treatment and prevention of HIV: from bench to bedside. Yale J Biol Med 2011;84(3):301–9.
[106] Strayer DS, Akkina R, Bunnell BA, Dropulic B, Planelles V, Pomerantz RJ, et al. Current status of gene therapy strategies to treat HIV/AIDS. Mol Ther 2005;11(6):823–42.
[107] Bobbin ML, Burnett JC, Rossi JJ. RNA interference approaches for treatment of HIV-1 infection. Genome Med 2015;7(1):50.

[108] Mali P, Yang L, Esvelt KM, Aach J, Guell M, DiCarlo JE, et al. RNA-guided human genome engineering via Cas9. Science 2013;339(6121):823–6.
[109] Hu W, Kaminski R, Yang F, Zhang Y, Cosentino L, Li F, Luo B, Alvarez-Carbonell D, Garcia-Mesa Y, Karn J, Mo X, Khalili K. RNA-directed gene editing specifically eradicates latent and prevents new HIV-1 infection. Proc Natl Acad Sci U S A 2014;111(31):11461–6.
[110] Swaminathan G, Navas-Martín S, Martín-García J. MicroRNAs and HIV-1 infection: antiviral activities and beyond. J Mol Biol 2014;426(6):1178–97.
[111] Ran FA, Hsu PD, Wright J, Agarwala V, Scott DA, Zhang F. Genome engineering using the CRISPR-Cas9 system. Nat Protoc 2013;8(11):2281–308.
[112] Horvath P, Barrangou R. CRISPR/Cas, the immune system of bacteria and archaea. Science 2010;327(5962):167–70.
[113] Huang Z, Tomitaka A, Raymond A, Nair M. Current application of CRISPR/Cas9 gene-editing technique to eradication of HIV/AIDS. Gene Ther 2017;24(7):377–84.
[114] Yang T, Martin P, Fogarty B, Brown A, Schurman K, Phipps R, et al. Exosome delivered anticancer drugs across the blood-brain barrier for brain cancer therapy in *Danio rerio*. Pharm Res 2015;32(6):2003–14.
[115] Feng D, Zhao WL, Ye YY, Bai XC, Liu RQ, Chang LF, et al. Cellular internalization of exosomes occurs through phagocytosis. Traffic 2010;11(5):675–87.
[116] Morelli AE, Larregina AT, Shufesky WJ, Sullivan MLG, Stolz DB, Papworth GD, et al. Endocytosis, intracellular sorting, and processing of exosomes by dendritic cells. Blood 2004;104(10):3257–66.
[117] Svensson KJ, Christianson HC, Wittrup A, Bourseau-Guilmain E, Lindqvist E, Svensson LM, et al. Exosome uptake depends on ERK1/2-heat shock protein 27 signaling and lipid raft-mediated endocytosis negatively regulated by caveolin-1. J Biol Chem 2013;288(24):17713–24.
[118] Alvarez-Erviti L, Seow Y, Yin H, Betts C, Lakhal S, Wood MJA. Delivery of siRNA to the mouse brain by systemic injection of targeted exosomes. Nat Biotechnol 2011;29(4):341–5.
[119] Zhou Y, Zhou G, Tian C, Jiang W, Jin L, Zhang C, Chen X. Exosome-mediated small RNA delivery for gene therapy. Wiley Interdiscip Rev RNA 2016;7(6):758–71.
[120] Quah BJC, O'Neill HC. The immunogenicity of dendritic cell-derived exosomes. Blood Cells Mol Dis 2005;35(2):94–110.
[121] Qazi KR, Gehrmann U, Jordö ED, Karlsson MCI, Gabrielsson S. Antigen-loaded exosomes alone induce Thl-type memory through a B cell dependent mechanism. Blood 2009;113(12):2673–83.
[122] Yeo RWY, Lai RC, Zhang B, Tan SS, Yin Y, Teh BJ, Lim SK. Mesenchymal stem cell: an efficient mass producer of exosomes for drug delivery. Adv Drug Deliv Rev 2013;65(3):336–41.
[123] Chen TS, Arslan F, Yin Y, Tan SS, Lai RC, Choo ABH, et al. Enabling a robust scalable manufacturing process for therapeutic exosomes through oncogenic immortalization of human ESC-derived MSCs. J Transl Med 2011;9:47.
[124] Haney MJ, Zhao Y, Harrison EB, Mahajan V, Ahmed S, He Z, et al. Specific transfection of inflamed brain by macrophages: a new therapeutic strategy for neurodegenerative diseases. PLoS ONE 2013;8(4):e61852.
[125] Savina A, Fader CM, Damiani MT, Colombo MI. Rab11 promotes docking and fusion of multivesicular bodies in a calcium-dependent manner. Traffic 2005;6(2):131–43.
[126] Janas T, Janas MM, Sapoń K, Janas T. Mechanisms of RNA loading into exosomes. FEBS Lett 2015;589(13):1391–8.
[127] Hupfeld J, Gorr IH, Schwald C, Beaucamp N, Wiechmann K, Kuentzer K, et al. Modulation of mesenchymal stromal cell characteristics by microcarrier culture in bioreactors. Biotechnol Bioeng 2014;111(11):2290–302.

[128] Mitchell JP, Court J, Mason MD, Tabi Z, Clayton A. Increased exosome production from tumour cell cultures using the Integra CELLine culture system. J Immunol Methods 2008;335(1–2):98–105.

[129] Barros FM, Carneiro F, Machado JC, Melo SA. Exosomes and immune response in cancer: friends or foes? Front Immunol 2018;9:730.

[130] Yang C, Robbins PD. The roles of tumor-derived exosomes in cancer pathogenesis. Clin Dev Immunol 2011;2011:842849.

[131] van der Pol E, Coumans F, Varga Z, Krumrey M, Nieuwland R. Innovation in detection of microparticles and exosomes. J Thromb Haemost 2013;1:36–45.

[132] Ko J, Carpenter E, Issadore D. Detection and isolation of circulating exosomes and microvesicles for cancer monitoring and diagnostics using micro-/nano-based devices. Analyst 2016;141(2):450–60.

[133] Suárez H, Gámez-Valero A, Reyes R, López-Martín S, Rodríguez MJ, Carrascosa JL, et al. A bead-assisted flow cytometry method for the semi-quantitative analysis of extracellular vesicles. Sci Rep 2017;7(1):11271.

[134] Ahmad MA. Electrical detection, identification and quantification of exosomes. IEEE Access 2018;6.

[135] Vogel R, Willmott G, Kozak D, Roberts GS, Anderson W, Groenewegen L, et al. Quantitative sizing of nano/microparticles with a tunable elastomeric pore sensor. Anal Chem 2011;83(9):3499–506.

[136] Dragovic RA, Gardiner C, Brooks AS, Tannetta DS, Ferguson DJP, Hole P, et al. Sizing and phenotyping of cellular vesicles using nanoparticle tracking analysis. Nanomed Nanotechnol Biol Med 2011;7(6):780–8.

[137] Braeckmans K, Buyens K, Bouquet W, Vervaet C, Joye P, De Vos F, et al. Sizing nanomatter in biological fluids by fluorescence single particle tracking. Nano Lett 2010;10(11):4435–42.

[138] López-Cobo S, Campos-Silva C, Moyano A, Oliveira-Rodríguez M, Paschen A, Yáñez-Mó M, et al. Immunoassays for scarce tumour-antigens in exosomes: detection of the human NKG2D-ligand, MICA, in tetraspanin-containing nanovesicles from melanoma. J Nanobiotechnol 2018;16(1):47.

[139] Wang Z, Wu HJ, Fine D, Schmulen J, Hu Y, Godin B, et al. Ciliated micropillars for the microfluidic-based isolation of nanoscale lipid vesicles. Lab Chip 2013;13(15):2879–82.

[140] Lee K, Shao H, Weissleder R, Lee H. Acoustic purification of extracellular microvesicles. ACS Nano 2015;9(3):2321–7.

[141] Vaidyanathan R, Naghibosadat M, Rauf S, Korbie D, Carrascosa LG, Shiddiky MJA, Trau M. Detecting exosomes specifically: a multiplexed device based on alternating current electrohydrodynamic induced nanoshearing. Anal Chem 2014;86(22):11125–32.

[142] McLeod E, Dincer TU, Veli M, Ertas YN, Nguyen C, Luo W, et al. High-throughput and label-free single nanoparticle sizing based on time-resolved on-chip microscopy. ACS Nano 2015;9(3):3265–73.

[143] Issadore D, Min C, Liong M, Chung J, Weissleder R, Lee H. Miniature magnetic resonance system for point-of-care diagnostics. Lab Chip 2011;11(13):2282–7.

[144] Im H, Lee K, Weissleder R, Lee H, Castro CM. Novel nanosensing technologies for exosome detection and profiling. Lab Chip 2017;17(17):2892–8.

[145] Kravets VG, Schedin F, Jalil R, Britnell L, Gorbachev RV, Ansell D, et al. Singular phase nano-optics in plasmonic metamaterials for label-free single-molecule detection. Nat Mater 2013;12(4):304–9.

[146] Su J. Label-free single exosome detection using frequency-locked microtoroid optical resonators. ACS Photonics 2015;2(9):1241–5.

[147] Riches A, Campbell E, Borger E, Powis S. Regulation of exosome release from mammary epithelial and breast cancer cells—a new regulatory pathway. Eur J Cancer 2014;50(5):1025–34.
[148] Harmati M, Tarnai Z, Decsi G, Kormondi S, Szegletes Z, Janovak L, et al. Stressors alter intercellular communication and exosome profile of nasopharyngeal carcinoma cells. J Oral Pathol Med 2017;46(4):259–66.
[149] Böing AN, van der Pol E, Grootemaat AE, Coumans FAW, Sturk A, Nieuwland R. Single-step isolation of extracellular vesicles by size-exclusion chromatography. J Extracell Vesicles 2014;3(1):23430.
[150] Shin H, Han C, Labuz JM, Kim J, Kim J, Cho S, et al. High-yield isolation of extracellular vesicles using aqueous two-phase system. Sci Rep 2015;5:13103.
[151] Weng Y, Sui Z, Shan Y, Hu Y, Chen Y, Zhang L, Zhang Y. Effective isolation of exosomes with polyethylene glycol from cell culture supernatant for in-depth proteome profiling. Analyst 2016;141(15):4640–6.
[152] Heinemann ML, Ilmer M, Silva LP, Hawke DH, Recio A, Vorontsova MA, et al. Benchtop isolation and characterization of functional exosomes by sequential filtration. J Chromatogr A 2014;1371:125–35.
[153] Yamashita T, Takahashi Y, Nishikawa M, Takakura Y. Effect of exosome isolation methods on physicochemical properties of exosomes and clearance of exosomes from the blood circulation. Eur J Pharm Biopharm 2016;98:1–8.
[154] Willms E, Johansson HJ, Mäger I, Lee Y, Blomberg KEM, Sadik M, et al. Cells release subpopulations of exosomes with distinct molecular and biological properties. Sci Rep 2016;6:22519.
[155] Tauro BJ, Greening DW, Mathias RA, Ji H, Mathivanan S, Scott AM, Simpson RJ. Comparison of ultracentrifugation, density gradient separation, and immunoaffinity capture methods for isolating human colon cancer cell line LIM1863-derived exosomes. Methods 2012;56(2):293–304.
[156] Witwer KW, Buzás EI, Bemis LT, Bora A, Lässer C, Lötvall J, et al. Standardization of sample collection, isolation and analysis methods in extracellular vesicle research. J Extracell Vesicles 2013;2(1):20360.
[157] Bosch S, De Beaurepaire L, Allard M, Mosser M, Heichette C, Chrétien D, et al. Trehalose prevents aggregation of exosomes and cryodamage. Sci Rep 2016;6:36162.
[158] Kreke, M., Smith, R., Hanscome, P., Peck, K., Ibrahim, A., 2016. Processes for producing stable exosome formulations. US Patent 20160158291A1.
[159] Takahashi Y, Nishikawa M, Shinotsuka H, Matsui Y, Ohara S, Imai T, Takakura Y. Visualization and in vivo tracking of the exosomes of murine melanoma B16-BL6 cells in mice after intravenous injection. J Biotechnol 2013;165(2):77–84.
[160] Imai T, Takahashi Y, Nishikawa M, Kato K, Morishita M, Yamashita T, et al. Macrophage-dependent clearance of systemically administered B16BL6-derived exosomes from the blood circulation in mice. J Extracell Vesicles 2015;4:26238.
[161] Wiklander OPB, Nordin JZ, O'Loughlin A, Gustafsson Y, Corso G, Mäger I, et al. Extracellular vesicle in vivo biodistribution is determined by cell source, route of administration and targeting. J Extracell Vesicles 2015;4:26316.
[162] Qu Y, Zhang Q, Cai X, Li F, Ma Z, Xu M, Lu L. Exosomes derived from miR-181-5p-modified adipose-derived mesenchymal stem cells prevent liver fibrosis via autophagy activation. J Cell Mol Med 2017;21(10):2491–502.
[163] Zhang J, Ma J, Long K, Qiu W, Wang Y, Hu Z, et al. Overexpression of exosomal cardioprotective miRNAs mitigates hypoxia-induced H9c2 cells apoptosis. Int J Mol Sci 2017;18(4):711.

[164] Yang MQ, Du Q, Varley PR, Goswami J, Liang Z, Wang R, et al. Interferon regulatory factor 1 priming of tumour-derived exosomes enhances the antitumour immune response. Br J Cancer 2018;118(1):62–71.

[165] Yu B, Kim HW, Gong M, Wang J, Millard RW, Wang Y, et al. Exosomes secreted from GATA-4 overexpressing mesenchymal stem cells serve as a reservoir of anti-apoptotic microRNAs for cardioprotection. Int J Cardiol 2015;182(C):349–60.

[166] Morishita M, Takahashi Y, Matsumoto A, Nishikawa M, Takakura Y. Exosome-based tumor antigens–adjuvant co-delivery utilizing genetically engineered tumor cell-derived exosomes with immunostimulatory CpG DNA. Biomaterials 2016;111:55–65.

[167] Cui GH, Wu J, Mou FF, Xie WH, Wang FB, Wang QL, et al. Exosomes derived from hypoxia-preconditioned mesenchymal stromal cells ameliorate cognitive decline by rescuing synaptic dysfunction and regulating inflammatory responses in APP/PS1 mice. FASEB J 2018;32(2):654–68.

[168] Lamichhane TN, Leung CA, Douti LY, Jay SM. Ethanol induces enhanced vascularization bioactivity of endothelial cell-derived extracellular vesicles via regulation of microRNAs and long non-coding RNAs. Sci Rep 2017;7(1):13794.

[169] Bukong TN, Momen-Heravi F, Kodys K, Bala S, Szabo G. Exosomes from hepatitis C infected patients transmit HCV infection and contain replication competent viral RNA in complex with Ago2-miR122-HSP90. PLoS Pathog 2014;10(10):e1004424.

[170] Wurdinger T, Gatson NN, Balaj L, Kaur B, Breakefield XO, Pegtel DM. Extracellular vesicles and their convergence with viral pathways. Adv Virol 2012;2012:767694.

Chapter 8

Extracellular vesicles in parasitic disease

Patricia Xander, André Cronemberger-Andrade, and Ana Claudia Torrecilhas
Laboratory of Cellular Immunology and Biochemistry of Fungi and Protozoa, Department of Pharmaceutical Sciences, Federal University of São Paulo (UNIFESP), Diadema, SP, Brazil

1 Extracellular vesicles and parasite

Pathogens shed extracellular vesicles (EVs) in the host microenvironment. EVs released by the pathogens and those produced by the host have been characterized and isolated from diverse organisms and these EVs have been shown to play an important role in intracellular communication between pathogens and hosts [1–4].

Virus, bacteria, fungi, protozoa and helminth release EVs continuously into extracellular environment [1–3]. These EVs measure between 20 and 200 nm in diameter and contain proteins, glycoconjugates, lipids, nucleic acid (RNAs, non-transcribed RNAs and microRNAs). They can be detected in various body fluids (serum, plasma, saliva, urine, breast milk, etc.). EVs modulate the communication and interaction with the host, also delivering virulence factors and antigens. Indeed, EVs isolated from parasites act either as messengers priming host cells for specific interactions, promoting invasion by the pathogen, or activating host innate and acquired immunity [1, 2, 5, 6].

Many studies have shown that EVs released from parasites are able to affect host cells. For example, this interaction changes the composition of exosomes released by macrophages. Pathogens can also stimulate the release of host EVs containing proteins or other components of the parasite as a mechanism for transporting contents of these microorganisms. Moreover, some pathogens have as mechanism of adaptation to the host, the release of EVs, which facilitates their dissemination or escape of the immune system.

In this chapter, we intend to present updated information regarding the role of EVs released by some major Protozoan parasites.

Exosomes. https://doi.org/10.1016/B978-0-12-816053-4.00008-0

2 *Leishmania*

Leishmaniasis are neglected diseases (WHO, 2015) (https://www.who.int/leishmaniasis/en), which are endemic in several countries. The diseases are caused by protozoa belonging to the *Leishmania* genus and can be grouped into different clinical forms: (a) localized cutaneous leishmaniasis (CL), (b) mucocutaneous leishmaniasis (MCL), (c) diffuse cutaneous leishmaniasis (DCL), and visceral leishmaniasis (VL). Currently, it is estimated that 556 million and almost 399 million people are living in areas at risk to contract VL and CL, respectively, located mostly in high-burden countries [7, 7a, 7b].

The cycle of *Leishmania* is digenetic occurring between mammalian hosts and the insect vectors of the Phlebotomine family [8]. During the blood repast, the female sandflies ingest the amastigotes present in infected mammalian hosts. Inside the vectors, the amastigotes become metacyclic promastigotes, and in a new blood uptake the promastigotes are egested by the vector and phagocytosed by host macrophages. Inside the phagocytes, the parasites differentiate into amastigotes and multiply. Rupture of the host cell membrane allow infection of new macrophages [9].

The innate immunity plays an important role in infection by *Leishmania* [10]. Neutrophils, macrophages and dendritic cells can recognize and be activated in response to contact or infection with *Leishmania* [11]. However, macrophages are the major cells involved in immunity and pathogenesis of leishmaniasis. *Leishmania* are phagocytosed by macrophages by binding of surface molecules of the parasites, such as lipophosphoglycan (LPG), to receptors expressed on the surface of macrophages, such as complement 1 and 3 receptors (CR1 and CR3), mannose receptor (MR), Fc receptors (FcR) and fibronectin receptors [12]. Activated macrophages release reactive oxygen species (ROS), and increase the expression of nitric oxide synthase (iNOS), which leads to the production of nitric oxide (NO). ROS and NO are highly toxic to *Leishmania* [10].

Innate immune cells express pattern recognition receptors (PRRs) that recognize pathogen-associated molecular patterns (PAMPs). These receptors include Toll-like receptors (TLRs), cytosolic DNA sensors (CDSs), NOD-like receptors (NLRs), RIG-I-like receptors (RLRs) and C-type lectin receptors (CLRs). Toll receptor 2 (TLR2) and TLR4 recognize *Leishmania* extracellularly, while TLR3 and TLR9 recognize the parasite in the vacuole. The role of each TLRs in *Leishmania* defense or persistency will depend on the species of the parasite [13]. The function of NLRP3 inflammasome are, however, still controversial. Some studies showed that NLRP3 can be activated by *Leishmania* and lead to the production of the pro-inflammatory cytokines IL-1β and IL-18 (reviewed in [10]. On the other hand, *L. donovani* stimulated the expression of A20 by macrophages, a molecule that inhibits NLRP3 inflammasome activation and the production of IL-1β [14].

Adaptative immunity to *Leishmania* can then contribute to the clearance of the parasite or can participate in the pathogenesis processes [15]. Protective immunity to *Leishmania* requires appropriate activation of innate cells. The production of the IL-12, IL-18 and IL-27 cytokines by activated dendritic cells stimulates the differentiation of IFN-γ-producing Th1 $CD4^+$ T cells. The cytokines produced by Th1 (especially IFN-alpha increase the activation of macrophages to produce higher amounts of NO and ROS, thus contributing to the elimination of the parasites) (reviewed in [16]). On the other hand, the Th2 response, with high levels of IL-4 and IL-13, as well as the hyper-inflammation, associated with high production of TNF-α, can contribute to the susceptibility or development of the DCL, respectively [17]. The role of the other Th subsets has not been completely understood. The Th17 T cells have a controversial, species-specific role [18]. The IL-10-producing Treg cells have been associated with parasite latency, treatment resistance and disease relapse in *Leishmania* infection [19].

Although the immune system has mechanisms to eliminate the parasite, *Leishmania* developed several strategies to avoid the immune response. Molecules produced by *Leishmania* can interfere with cell signaling pathways altering the expression of proinflammatory receptors and cytokines leading to parasite survival [20]. GP63 is an abundant major surface protease considered a virulence factor that participates in the parasite evasion by altering complement mediated lysis, interfering with PKC signaling in macrophages, and acting on the interaction of the parasite with host cells and extracellular matrix [21]. The *Leishmania* surface coat also contains a lipophosphoglycan (LPG) that protects the parasite from complement-mediated lysis. However, LPG induced complement activation, promotes parasite phagocytosis with a delay in the formation of phago-lysosomes, allowing the parasite differentiation into the proliferative amastigote form [22].

Components released by the parasite can modulate the function and activation of phagocytic cells and the immune response. Several studies showed that parasite's secretoma contain molecules with potential to modulate the immune cells [23]. In addition to secreted molecules, *Leishmania* also spontaneously releases extracellular vesicles (Fig. 1) that showed biological effects in immunity and disease progression [24, 25].

Silverman et al [26] reported for the first time the presence of EVs budding from the flagellar pocket and plasma membrane of stationary phase *Leishmania infantum* promastigotes. Other studies showed that *Leishmania* EVs contain virulence factors (such as GP63 and LPG) and several molecules involved in the pathogenesis of *Leishmania* [25, 27], suggesting that these EVs can participate in the leishmaniasis infection and disease progression. EVs released by *Leishmania* were detected in the supernatant of promastigotes cultures. They were also detected within the lumen of the sand fly midgut, corroborating the idea that *Leishmania* EVs may contribute to the host infectious process [27]. Cultured macrophages engulf *Leishmania* exosomes through surface binding, fusion with the plasma membrane, or by endocytosis [25, 28]. EVs released by *Leishmania donovani* increased IL-8 production by differentiated THP-1 macrophages [25]. However, human monocytes exposed to *L. donovani* EVs, subsequently infected with the parasite, and then treated with IFN-γ showed a significant decrease in IL-8 and TNF-α levels and an increase in IL-10 levels, suggesting that EVs produced by the parasite show immunosuppressive effect on human monocytes [29].

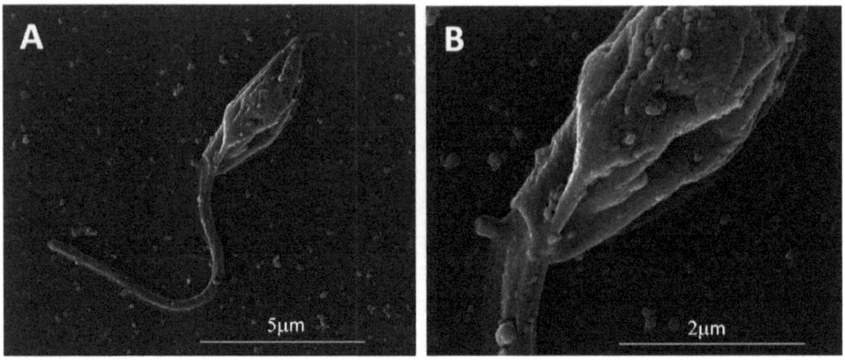

FIG. 1 Scanning electron microscopy (SEM) of *Leishmania (Leishmania) amazonensis* MHOM/BR/1973/M2269 strain showing shedding of EVs. The panels show promastigotes forms incubated in DMEM with 2% glucose and attached to glass coverslips containing poly-lysine, fixed and processed for SEM. Briefly, the samples were fixed in 2.5% glutaraldehyde solution according to established preparation protocols, washed in 0.1 M cacodylate solution, fixed with osmium tetroxide, treated with tannic acid, dehydrated with ethanol, and dried in a CPD 030 critical point dryer. The samples were coated with a gold layer using a sputtering method ('sputtering', © Leica EM 500 SCD, Germany) and then observed in a Field Emission FEI Quanta 250 FEG scanning electron microscope. The bar sizes are indicated in each image.

Experiments using mice models have also contributed to better understand the role of EVs released by *Leishmania* in disease progression. C57Bl/6 mice previously treated with *L. donovani* EVs showed higher parasitic load and IL-10 levels, than untreated animals [25]. Similar results were obtained with BALB/c mice treated with *Leishmania major* EVs. Mice pre-injected twice with 15 μg of *L. major* EVs showed a prominent Th2 polarization and disease exacerbation with higher lesion volume (mm), as compared to control group after infection with *L. major* promastigotes [25]. The impact of *L. major* EVs in the infection site was also evaluated by co-injecting the parasite and EVs in BALB/c footpads. *L. major* EVs significantly enhance lesion development by favoring parasite replication and by enhancing cytokine expression, especially IL-17a (a pro-inflammatory cytokine) [27]. Thus, *Leishmania* EVs seems to be a mechanism used by the parasite to establish the infection, favoring its survival and contributing to disease progression.

The presence of pathogens components has been identified in EVs released by infected cells [30]. Thus, several studies have been conducted to evaluate the release of EVs by cells infected with *Leishmania* [29, 31]. Proteomic analyzes showed the presence of GP63 in EVs released by J774 cells infected with *L. mexicana* promastigotes [31]. It has also been found that EVs released by J774 infected with *L. mexicana* were able to induce the phosphorylation of MAP kinases in naive macrophages [31]. Primary cultures of murine bone marrow-derived macrophages significantly increased the inflammatory cytokines production after treatment with EVs from macrophages infected with *L. amazonensis* promastigotes, which contributes to elimination of the pathogen [32].

In addition to pathogen molecules EVs also contain cell immune molecules such as, MHC-antigenic peptide complexes, unbound antigens, cytokines and co-stimulatory molecules, contributing to their immunomodulatory effects [30]. Therefore, EVs from *Leishmania*-infected cells should be considered relevant in cell-cell communications, affecting immune cells for antigen presentation and immune activation.

3 *Trypanosoma cruzi*

Trypanosoma cruzi, also a protozoan parasite, is the causative agent of Chagas' disease, or American trypanosomiasis, one of the deadliest infectious diseases in Latin and Central America. It is estimated that 6–8 million people are contaminated, and approximately 25 million are at risk of acquiring the disease. More than 5 million people do not know they have the disease (www.who.int/tdr/en/). Chagas Disease is also an emerging disease in U.S and Europe due to migration of chronically infected individuals from endemic countries and no regular screening on blood bank and hospitals in these areas. This protozoan is transmitted to humans by deposition of insect-vector in wounds and mucosa, or orally by foods contaminated with the insect such as observed in acai fruit or sugar cane juice. The parasite is also transmitted through blood transfusion, organ transplantation, and through the placenta during gestation and/or delivery. Vector-mediated transmission begins when the insect-derived metacyclic-trypomastigote form, present in the insect feces, enter mammalian hosts and invade cells by forming a parasitophorous vacuole. Inside the host cell, the parasite transforms into amastigotes that multiply freely in the cytoplasm. In the cell, the amastigotes differentiate into infective trypomastigotes, which rupture the cell plasma membrane, reach the extracellular matrix, and then the bloodstream to invade surrounding cells and tissues [33]. Bloodstream forms are then available to the insect vector, or can be transmitted to other hosts.

In human, Chagas disease is characterized by an acute followed by a chronic phase. The acute phase shows microscopically detectable blood parasitemia and can last up to 8 weeks after the initial contact. Chronic phase, the diagnostic relies on serologic assays as the parasite can only be detected by highly sensitive PCR techniques. Patients with chronic *T. cruzi* infection but without signs or symptoms of Chagas disease are considered to have the indeterminate form. It is estimated that from 20% to 30% of those people, who initially were serologically positive, will progress over a period of years, or decades, to clinically evident cardiomyopathies, or mega gastrointestinal diseases, or both.

T. cruzi employs a highly elaborated array of molecules and strategies to invade a wide range of host cells [34, 35], as well as to escape from host immune defense mechanisms [36]. The invasion and immune resistance are both vital processes required for the successful survival, proliferation, and perpetuation of *T. cruzi* inside the mammalian hosts. Several parasite surface molecules have been implicated in host cell invasion and/or host immunomodulation [37]. From 60% to 80% of trypomastigote surface is composed by glycoconjugates. These are mostly mucin-like glycoproteins, mucin-associated surface proteins (MASP), and members of superfamily gp85/trans-sialidase like glycoproteins [38, 39]. All these are associated to the membrane via glycosylphosphatidyl inositol (GPI) lipids. Glycosyl phospholipids, called GIPLs are also present in the parasite surface [37]. When the parasites enter the host cells, macrophages are activated, and a cascade of events culminates in cytokines production, leukocyte attraction and activation, and increased IL-12- versus IL-10 expression. The production of IL-12 and IFN-gamma by macrophages and natural killer (NK) cells influences macrophage effectors functions. This induces the polarization towards pro-inflammatory responses, in which helps to eliminate *T. cruzi*, versus a the Th2-type response, which favors parasite maintenance [37].

Many studies were conducted to identify parasite components that were able to induce these cytokines and NO production. Isolated mucin-like glycoproteins, the most abundant surface protein of trypomastigotes, were found to activate pro-inflammatory responses, and this have been attributed to the presence of unique unsaturated fatty acids in the GPI anchor [37]. However, it was unclear how these and other surface and GPI-anchored molecules are delivered and how they trigger specific signaling pathways through their cognate receptors. Two possibilities have been envisaged: (i) release as soluble molecules; and/or (ii) in the form of extracellular vesicles. Enzyme-mediated, or spontaneous release of GPI-anchored components has been observed in protozoa [37]. Shedding of antigens is also a well-documented phenomenon in *T. cruzi* [1, 40]. However, these studies did not reveals whether these antigen were released as soluble and/or membrane-bound form.

In parallel, it has been shown that trypomastigotes spontaneously release EVs of approximately 20–200 nm, as demonstrated by electron scanning microscopy (Fig. 2). EVs are released by different isolates of *T. cruzi* in a process

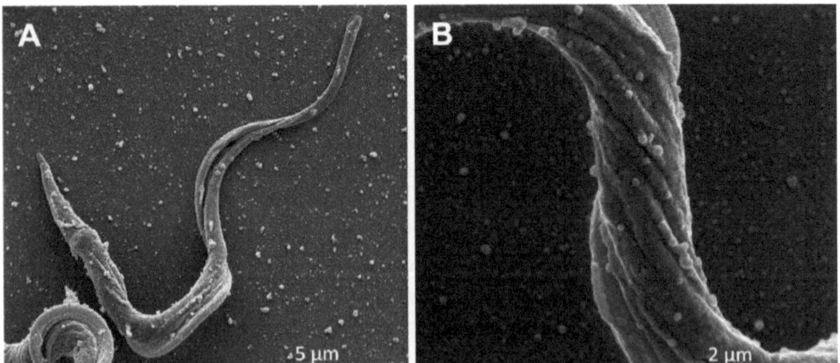

FIG. 2 SEM of *T. cruzi* trypomastigotes from Y strain showing shedding of EVs. Protozoa parasite were pre-incubated in DMEM with 2% glucose and attached to glass coverslips containing poly-lysine, fixed and processed for SEM as described in Fig. 1. The bar sizes are indicated in each image.

that occur without lost in parasite infectivity, suggesting that cell death does not occur during the process of vesicle secretion. The EVs were shown to contain members of TS/gp85 and mucin-like superfamily of glycoproteins and to have potent proinflammatory activity [40, 41]. EVs purified by gel-filtration presented as a major component the mucin-like glycoproteins containing substituted α-galactosyl residues, which are with highly immunogenic, as well as several members of the TS-like glycoproteins. These α-galactosyl-enriched (α-Gal) EVs strongly trigger proinflammatory responses in murine macrophages via TLR2-dependent pathway [42]. The interaction between *T. cruzi* EVs and the host TLR2 leads to inflammation and enhancement of cell invasion, which may favor the long-term maintenance of the parasite within the host (unpublished data). Indeed, BALB/c mice pretreated with these EVs before infection with *T. cruzi* trypomastigotes developed more severe heart pathology with an intense inflammatory reaction and an increased number of intracellular amastigote nests in heart tissue [4].

EVs are released by different isolates of *T. cruzi*. However, the secretome of YuYu strain, which is more infective than Y, contains more EVs than Y strain [41]. The proteomic analysis of these EVs revealed the presence of several members of the TS/gp85 superfamily in both cases. As these proteins are most likely involved in adhesion of the parasite to the host cells, it is possible that differences in infectivity could be due to the presence of more EVs in the more virulent strain. In addition, some members of these family of proteins correspond to trans-sialidase enzymes that catalyzes the transfer of sialic acid from host molecules to parasite mucins and have been shown to participate in host cell adhesion and invasion, parasite escaping from the phagolysosome in the host cell cytoplasm and exit from the mammalian cell. In addition, the presence of TS could protect the parasite from host complement and lytic Ch anti-α-Gal antibodies and modulate host immune and apoptotic response. It has been proposed that the immune evasion mechanism of relies on the ability of the parasite to induce the release of host cell-derived vesicles, which contribute to immune evasion by protecting the parasite from the complement attack, and ultimately resulting in increased parasite infectivity and survival [43].

Noteworthy, TS is differentially expressed by the two major *T. cruzi* phylogenetic lineages, namely *T. cruzi* I and *T. cruzi* II. In the latter, which is commonly found associated to human infection, TS is expressed and shed in much higher quantities. This observation clearly underscores the relevance of active TS, possible in EVs as a virulence factor for the parasite. Other related members of the TS/gp85 superfamily which do not have enzymatic activity, present in EVs could be playing a major role in the interaction with host cells. EVs released by YuYu strain (more virulent) caused a lower infection but higher intracellular proliferation in macrophages than EVs from Y strain. In contrast, YuYu strain-derived EVs caused higher infection of epithelial cells than Y strain-derived EVs [41]. Other qualitative differences in EVs from different *T. cruzi* strains were found, but at the present it is not possible to correlate them with infectivity/virulence differences.

Other parasite stages besides blood trypomastigotes also release specific EVs. For example proliferative epimastigotes, which are non-infective and the infective metacyclic-trypomastigote form of *T. cruzi* formed at the hindgut of the insect vector showed distinct proteins sets in their EVs population [44]. These differences may arise from differences in the metabolism and gene expression of these parasite stages and suggest specific roles for each EVs, probably related to the interaction with the insect vector and/or in the initial steps of the parasite infection. The EVs seems to appear as a consequence of exocytic fusion of multivesicular bodies (MVB) resulting in exosomes, and budding of vesicles directly from the plasma membrane, resulting in microvesicles (MVs, also named ectosomes or plasma membrane-derived vesicles). In conclusion, all parasite *T. cruzi* stages uses EVs as a strategy to release major surface components, which are involved in parasite adhesion, invasion, survival and evasion from host immune response.

4 *Trypanosoma brucei*

The Kinetoplastida *Trypanosoma brucei* (genus *Trypanosoma*) is the causative agent of sleeping sickness or African Trypanosomiasis that affects millions of people in 36 countries in sub-Saharan Africa and Nagana in animals [45]. There are three subspecies group, the *T. b. brucei*, which infect mammals, and *T. b. gambiense* and *T. b. rhodesiense*, which infect humans. These parasites are transmitted the Tsetse fly (*Glossina* spp.). The insect injects trypomastigotes into skin tissue during the blood meal. From the bite, parasites first enter the lymphatic system and then pass into the bloodstream. The parasite persist in blood causing anemia, neurological disorders and death if left untreated [45]. *T. brucei* is an extracellular parasite and divide in the mammalian bloodstream or in the insect gut and salivary glands of the insect vector. The bloodstream trypomastigotes express majorly a surface glycoprotein that form a protective coat around the parasite. The protein, named variant surface glycoprotein (VSG) is encoded by a multigene family and only one member of this family is expressed at each time. During the infection, once the host establish an immune response, another VSG is expressed allowing the parasite to escape the immune defenses and leading to chronic infection. In prolonged infection, the parasite cross blood brain barrier.

The VSGs which are attached to the membrane via GPI-anchors protects the parasite from lysis by the alternative pathway of the complement and against non-specific antibodies [46]. In infected hosts, VSG and parasite DNA containing CqG are recognized by SR-A and TLR9, respectively, promoting macrophage and dendritic cells (DC) activation and production of proinflammatory cytokines (TNF-α, IL-6 and IL-12). These parasite components and production of INF-γ by activated NK- and T-cells induce activated type 1 macrophages (M1). These strong Th1 type of response induces the production of NO, which contributes to parasitic control. At same time it enhances uptake of erythrocytes causing anemia during acute phase of the disease [47]. More recently it has been shown that *T. brucei* EVs can also promote erythrocytes clearance resulting in anemia [48].

EVs of *T. b. gambiense* are released through nanotubes that extend from the flagellar pocket. These EVs were obtained from secreted material as well as from infected rat sera and range from 50 to 100 nm in diameter, which correspond to exosomes size [49]. The proteins identified in the secretome appear to have a direct role in the parasite survival strategies [49]. For example, EVs released by bloodstream forms of *T. b. rhodesiense* were shown to contain the serum resistance-associated protein (SRA) that allow the parasite to survive in human blood. This protein, is not present in *T. b. brucei*, one of the reasons why this subspecies does not infect humans and some primates that present host innate immune factor [48]. It has been shown that EVs derived from *T. brucei* fuse with erythrocytes membranes transferring lipids and VSG glycoproteins. This alteration of the erythrocytes membrane results in recognition and phagocytosis by macrophages in the liver and spleen leading to anemia [48].

Many orthologs of the secretory machinery of eukaryote cells (ESCRT machinery) have been described in *T. brucei* [50] and it was shown that inhibiting Vsp36 (an ESCRT component) compromises EVs secretion is compromised, but not of nanotube derived EVs [51]. Therefore, different mechanisms of secretion may occur for these two types of vesicles that are also biochemically distinct [51].

T. brucei EVs are also related the social motility of parasites, which induces these parasites to migrate away from either damaged cells and from harmful environments. These EVs are internalized by insect form (procyclics of *T. brucei* in the insect host and are key components in the parasite-parasite communication [51]. Taken together, the release of *T. brucei* EVs plays a role in cell-cell communication with the host and among themselves and open new insights to development new potential therapeutic targets or diagnostic markers.

5 *Plasmodium*

Malaria is a disease found in 91 countries in tropical and subtropical regions (WHO, 2017) (https://www.who.int/malaria/en). The World Health Organization reported 219 million cases and 435,000 malaria deaths worldwide. Currently, it is estimated that approximately 3.3 billion people are living in risky areas (WHO, 2017) (https://www.who.int/malaria/en). The etiologic agent of malaria are parasites of *Plasmodium* genus and three species present a significant health threat for humans, *Plasmodium vivax*, *Plasmodium malariae* and *Plasmodium falciparum*. The human infection is initiated with the injection of the sporozoites into the dermis by a feeding female anopheline mosquito. The parasites reach the liver, and invade hepatocyte. Inside the cells, the sporozoites undergo schizogony and thousands of merozoites are released in the blood. In *P. falciparum*, merozoites invade erythrocytes and begin a chronic cycle of asexual schizogony in the bloodstream. The host reticulocytes are infected by *P. vivax*. A proportion of asexually reproducing merozoites are reprogrammed to undergo gametocytogenesis that gives rise to male and female gametocytes. The mosquitos ingest both gametocytes present in peripheral circulation, which undergoes several process and in the midgut of the vector until giving rise to sporozoites that migrate to the salivary glands where they can be injected into the next human host [52].

Immune responses are involved in malaria pathogenesis and disease control, although repeated *Plasmodium* infections do not result in complete immunity, leading populations of endemic regions to develop several infections throughout life. After inoculation in the skin, an innate immune response is triggered [53]. Some innate receptors are activated by *Plasmodium* PAMPs and contribute to dendritic cells functions, such as T and B activation [54]. However, the parasite is able to evade the immune response by using several mechanisms including the inhibition of DCs maturation.

It has been shown that several species of *Plasmodium,* including human and rodent malaria release of EVs in human and mouse serum or plasma during active infection [1, 2, 55]. In the case of malaria, EVs injection was found to control infection by *Plasmodium yoelii*-infected reticulocytes in the mouse model [56]. In mice infected with *Plasmodium berghei* (ANKA strain) EVs isolated from infected erythrocytes induce a potent macrophages stimulation via toll-like receptor (TLR) [57]. The EVs isolates from infected erythrocytes contain parasite protein and RNA and pro-inflammatory effect and stimulate TLR pathways. Exosomes derived from infected reticulocytes contained many host and parasite proteins and have also been found to control host immune response in *P. yoelii* 17X infection of BALB/c mice [56]. These proteins included merozoite surface proteins 1 and 9 and enzymes related to proteolysis and metabolic processes, which induced strong IgG responses. Furthermore, these exosomes combined with CpG-oligodeoxynucleotides induced protection against lethal infections in 85% of the mice tested. In other studies, it was found that *P. falciparum*-infected erythrocytes also release EVs, and that these particles contained host and parasites proteins. These particles acts in the intercellular communication between host and parasite [6]. In addition, EVs released by infected erythrocytes (iRBC) transfer genetic material to cells. The EVs isolated from iRBC also induce gametocytogenesis and confer resistance to the parasites that receive genetic materials. There are many reports showing that *Plasmodium* EVs localize in brain and that can modulate host immune system, originating severe clinical symptoms in malaria. In patients with cerebral pathogenesis there is an increase of EVs. Furthermore, platelet EVs are involved in iRBC cytoadhesion.

The EVs have been considered biomarkers in severe cerebral malaria and its presence is correlated with clinical manifestations such as fever and cerebral dysfunctions such as inflammatory responses [58]. Therefore, EVs derived from reticulocytes or red blood cells could be explored as connection between parasite and host cells and the modulation of pathogeneses of several malaria infections. More relevant, is the fact that exosomes derived from reticulocytes could be explored as a vaccine platform against malaria infections.

6 *Toxoplasma gondii*

Toxoplasmosis is caused by *Toxoplasma gondii*, an obligatory intracellular protozoan that infects nearly one-third of the world's human population [59]. Most individuals infected with the parasite are asymptomatic but the disease can be symptomatic or fatal in immunocompromised patients. The congenital infection can cause severe sequelae or death in fetus in primary infection of pregnant women [60].

The sexual cycle of *Toxoplasma gondii* occurs in the small intestine of the feline family and an asexual cycle is present in infected animals, including humans. The infection in humans initiates by ingestion of oocysts (released in feces of infected cats) or cysts (found in undercooked meat). In the intestinal lumen, the cysts or oocysts release bradyzoites and sporozoites that invade cells, transform in tachyzoites that reach the vasculature or lymphatic system. The tachyzoites can virtually infect all nucleated cells and tissues and are detected in biological fluids during the acute phase. In chronic phase, the tachyzoites transform in slow growing bradyzoites that can persist inside cysts in the host tissues [61].

In healthy individuals, the activation of innate and adaptative response eliminates most parasites. This occur because macrophages and dendritic cells can be activated in response to parasite products during infection. The stimulation of TLRs and other PAMPs receptors induce an innate immune response with the production of pro-inflammatory cytokines, contributing to an appropriated inflammatory milieu followed by the recruitment of neutrophils, monocytes and NK cells. Later on the activation of adaptative immune cells (T and B lymphocytes) ensues. However, the immune pressure on tachyzoites induce their conversion into bradyzoites that evades the host immune response, contributing to the disease progress to the chronic phase [62].

T. gondii express several antigens on the cell surface of the tachyzoites and bradyzoites [63]. These antigens have several roles including adhesion of the parasite to host cells and parasite survival on the cell environment [63]. The surface antigen 1 (SAG1) and SAG2 families are GPI anchored and one of the most important surface antigens [64]. In addition, *T. gondii* secretes antigens that are able to stimulate the antibodies and T-cells responses. Molecules from dense granules, micronemes, and rhoptries, which are structures involved in the parasite entry into the cells, highly immunogenic and elicit a protective response in addition to serves in serologic diagnosis [65–67].

Some cell surface antigens and secreted proteins were detected extracellular vesicles released by *T. gondii* by proteomic approach [68]. Small RNA and miRNA were also detected in EVs released by these parasites [69], suggesting that EVs from *T. gondii* can delivery virulence factors and other molecules contributing to the host-parasite interactions. The recognition of these EVs by sera from chronically infected human or mice also confirmed the presence of the immunogenic molecules expressed in EVs [69].

The effects of *T. gondii*-EVs in immunity were demonstrated by in vitro and in vivo studies. EVs from *T. gondii* stimulate murine macrophages to increase the expression of IL-10, iNOS and TNF-alpha [69]. RAW264.7 cells treated with *T. gondii*-EVs induce the production of IL-12, TNF-alpha, IFN-gamma with a concomitant inhibition of IL-10 expression [70]. In vivo, mice immunized with *T. gondii*-EVs showed a predominance of IgG2a over IgG1 and high levels of both IFN-γ and IL-12, as compared with control groups, suggesting a Th1-type cell immune profile induced by immunizations with EVs [70]. Immunized mice challenged with lethal infection with the parasites had significant increase in the survival time, as compared to control group, demonstrating a partial protective effect in experimental toxoplasmosis [70].

Dendritic cells infected with the parasite and pulsed EVs with *T. gondii* antigens release exosomes with immunomodulatory properties. These EVs induced a specific Th1 immune response in immunized animals [71]. In addition, these EVs led to a protective systemic immunity to *T. gondii* in mice challenged with the parasites [71]. Immunomodulatory effects were also observed by EVs released by THP-1 cells infected with *T. gondii*. These EVs stimulate TNF-α production by uninfected THP-1 cells [72]. Thus, immune cells infected with *T. gondii* release EVs that can stimulate proinflammatory response, contributing to efficient immunity against the parasite. The better understand of EVs released by *T. gondii* or by infected cells in immunity and immune surveillance can contribute to the development of new strategies to protection in pregnant women and people living in risk areas.

7 *Giardia duodenalis*

The enteric protozoan *Giardia duodenalis* is responsible for 280 million worldwide cases per year of diarrhea infections (giardiasis), with the highest prevalence found in developing countries [73]. Giardia infection range from asymptomatic to a severe malabsorption syndrome. Intestine epithelial cells play an important role in protection against invading microbes and express Pathogen Recognition Receptors PRRs that are capable of sense pathogens promoting a first line in immune response. The IL-17 production by CD4 + Th17 cells induce production of NO, antimicrobial peptides, chemokines and transports IgA by intestinal epithelial cells [74].

Cyst of Giardia, which is its infective form are ingested through contaminated food or water. Excystation (release of trophozoites) occurs after the cysts enters in contact with acid pH and intestinal proteases degrading the cyst wall proteins (CWPs) in stomach. Then, cysts release trophozoites (replicative form) in the intestine that attach to the epithelial cells. Trophozoites release through exocytosis encystation-specific vesicles (ESV) containing CWPs in membrane during encysting [75, 76]. The released of these vesicles is associated with the process of encysting. Some authors reported that in response to pH changes and presence of bile the ESVs formation increases in trophozoites [77]. This could be important for parasite survival in the host during infection.

Cysteine-rich proteins called variable surface proteins (VSPs) cover trophozoite surface and they help to evade the host immune system due their constant switching [78, 79]. Some of these proteins have cysteine protease activity that can be secreted, acting as virulence factors [80–83]. Giardia secretes a cathepsin B-like cysteine protease that degrades CXCL8 and attenuates neutrophils chemotaxis [82]. A proteomic study identified several proteins in parasite secretome and it has been shown that some are released as cargo of EVs [84]. EVs from Giardia range from 100 to 250nm in size [84]. Recent studies have shown that EVs released by Giardia contribute to parasite colonization, increasing adhesion to epithelial cells and also induction of immune responses [85]. The vesicle trafficking and biogenesis is not completely understood in *Giardia* as this organisms does not have Golgi complex and a well-defined endosomal system [86]. Orthologs of the subunits of ESCRT complex that is involved in endosomal-lysosomal sorting processes in yeast and mammals were found but it seems that these proteins diverge not only in sequence but also in functional characteristics [87]. These observations highlight to the importance of vesicles in pathogenesis of giardiasis and more work is necessary to clarify the role of EVs in the course of its infection.

8 Concluding remarks

Here we illustrate how extracellular vesicles are related to many aspects of some pathogenic protozoan parasites during the interaction with their host. The acknowledgment that many microbial pathogens secrete membrane and cytosolic materials as exosomes and/or any type of extracellular vesicles is allowing us to better understand the complex mechanisms of parasite interaction with the host. EVs participate in multiple events such as tissue and cell invasion. They also directly affect the host innate and adaptative immune responses, which ultimately promote parasite survival and transmission. Therefore, the release of EVs have also many effects in the pathogenesis. Table 1 summarizes the present knowledge about the biogenesis and characterization of parasite and the main roles during the interaction with host cell-derived EVs. Furthermore, EVs are also released from infected cells, some of them containing parasite components, extending the role of these structures in the disease and healing process. Some major questions, however, remain unanswered. For example, the mechanism by which the different parasite release EVs remains elusive and should more studied as a possible strategy to prevent disease progression. Moreover, the observations that EVs isolate from parasite have a strong potential for immunoprevention and therapy should be further explored, mainly in the case of Neglected Tropical Diseases.

TABLE 1 **Parasite extracellular vesicles-associated distinct glycoproteins, lipids or nucleic acid and their effect on host parasite interaction.**

Parasite	Molecules	Effect on host	Reference
L. donovani *L. infantum* *L. mexicana*	GP63 LPG HSP	Parasite evasion by altering complement mediated lysis and promoting parasite phagocytosis	[25, 88]
T. cruzi	Trans-sialidase Mucin Cruzipain GP85 Parasite antigen	Interaction, invasion and induce host inflammation and induce innate immune response via TLR2 parasites escaping the complement attack (TGF-β-bearing EVs released from monocytes and lymphocytes EVs enhance communication *T. cruzi* and host human cell promoting invasion	[1, 41–43, 89]
T. gondii	Small RNAs/ miRNA	Induce humoral and cellular immune responses	[69]
P. vivax *P. falciparum* *P. berghei* *P. yoelii*	Parasite Protein CSP? ABCA1 transporter	Interaction host cells EVs correlated with clinical manifestations such as fever and cerebral dysfunctions, suggesting a role of EVs in malaria pathogenesis Induced a potent activation of macrophages via TLR Modulator of the presence of phosphatidylserine (PS) in the outer layer of the plasma membrane, and PS is a major component of the surface of MVs Role in modulating immune responses in vivo EVs shown to promote the transference of genetic material between iRBCs (iRBC-derived EVs and induce gametocytogenesis)	[2, 6, 56]
T. brucei	VSG protease	Evasion from the host immune system and increase invasion BBB	[48]
G. duodenalis	Protein	Mechanism to the parasite to adapt to the host changing environment in the course of the infection	[85]

LPG, lipophosphoglycan; *VSG*, variant surface glycoprotein; *CSP*, circumsporozoite protein; *PS*, phosphatidylserine; *MVs*, microvesicles; *iRBCs*, infected red blood cells; *BBB*, blood brain barrier.

References

[1] Campos JH, Soares RP, Ribeiro K, Andrade AC, Batista WL, Torrecilhas AC. Extracellular vesicles: role in inflammatory responses and potential uses in vaccination in cancer and infectious diseases. J Immunol Res 2015;832057. https://doi.org/10.1155/2015/832057.

[2] Marcilla A, Martin-Jaular L, Trelis M, de Menezes-Neto A, Osuna A, Bernal D, Fernandez-Becerra C, Almeida IC, Del Portillo HA. Extracellular vesicles in parasitic diseases. J Extracell Vesicles 2014;3:25040.

[3] Torrecilhas AC, Schumacher RI, Alves MJ, Colli W. Vesicles as carriers of virulence factors in parasitic protozoan diseases. Microbes Infect 2012;14(15):1465–74. https://doi.org/10.1016/j.micinf.2012.07.008.

[4] Trocoli Torrecilhas AC, Tonelli RR, Pavanelli WR, da Silva JS, Schumacher RI, de Souza W, Silva NCE, de Almeida Abrahamsohn I, Colli W, Manso Alves MJ. *Trypanosoma cruzi*: parasite shed vesicles increase heart parasitism and generate an intense inflammatory response. Microbes Infect 2009;11(1):29–39. https://doi.org/10.1016/j.micinf.2008.10.003.

[5] Coakley G, Maizels RM, Buck AH. Exosomes and other extracellular vesicles: the new communicators in parasite infections. Trends Parasitol 2015;31(10):477–89. https://doi.org/10.1016/j.pt.2015.06.009.

[6] Regev-Rudzki N, Wilson DW, Carvalho TG, Sisquella X, Coleman BM, Rug M, Bursac D, Angrisano F, Gee M, Hill AF, Baum J, Cowman AF. Cell-cell communication between malaria-infected red blood cells via exosome-like vesicles. Cell 2013;153(5):1120–33. https://doi.org/10.1016/j.cell.2013.04.029.

[7] Alvar J, Vélez ID, Bern C, Herrero M, Desjeux P, Cano J, Jannin J, den Boer M, Leishmaniasis Control Team WHO. Leishmaniasis worldwide and global estimates of its incidence. PLoS ONE 2012;7(5):e35671. https://doi.org/10.1371/journal.pone.0035671.

[7a] Ross R. Note on the bodies recently described by Leishman and Donovan. Br Med J 1903;2:1261–2.

[7b] Ross R. Further notes of Leishman's bodies. Br Med J 1903;2:1401.

[8] Lainson R, Ryan L, Shaw JJ. Infective stages of *Leishmania* in the sandfly vector and some observations on the mechanism of transmission. Mem Inst Oswaldo Cruz 1987;82(3):421–4.

[9] Chappuis F, Sundar S, Hailu A, Ghalib H, Rijal S, Peeling RW, Alvar J, Boelaert M. Visceral leishmaniasis: what are the needs for diagnosis, treatment and control? Nat Rev Microbiol 2007;5(11):873–82. https://doi.org/10.1038/nrmicro1748.

[10] Gurung P, Kanneganti TD. Innate immunity against *Leishmania* infections. Cell Microbiol 2015;17(9):1286–94. https://doi.org/10.1111/cmi.12484.

[11] Liu D, Uzonna JE. The early interaction of *Leishmania* with macrophages and dendritic cells and its influence on the host immune response. Front Cell Infect Microbiol 2012;2:83. https://doi.org/10.3389/fcimb.2012.00083.

[12] Ueno N, Wilson ME. Receptor-mediated phagocytosis of *Leishmania*: implications for intracellular survival. Trends Parasitol 2012;28(8):335–44. https://doi.org/10.1016/j.pt.2012.05.002.

[13] Faria MS, Reis FC, Lima AP. Toll-like receptors in *Leishmania* infections: guardians or promoters? J Parasitol Res 2012;2012:930257. https://doi.org/10.1155/2012/930257.

[14] Gupta AK, Ghosh K, Palit S, Barua J, Das PK, Ukil A. inhibits inflammasome-dependent macrophage activation by exploiting the negative regulatory proteins A20 and UCP2. FASEB J 2017;31(11):5087–101. https://doi.org/10.1096/fj.201700407R.

[15] Scott P, Novais FO. Cutaneous leishmaniasis: immune responses in protection and pathogenesis. Nat Rev Immunol 2016;16:581–92. https://doi.org/10.1038/nri.2016.72.

[16] Soong L. Subversion and utilization of host innate defense by *Leishmania amazonensis*. Front Immunol 2012;3:58. https://doi.org/10.3389/fimmu.2012.00058.

[17] Silveira FT, Lainson R, De Castro Gomes CM, Laurenti MD, Corbett CE. Immunopathogenic competences of *Leishmania* (V.) *braziliensis* and *L.* (L.) *amazonensis* in American cutaneous leishmaniasis. Parasite Immunol 2009;31(8):423–31. https://doi.org/10.1111/j.1365-3024.2009.01116.x.

[18] Lopez Kostka S, Dinges S, Griewank K, Iwakura Y, Udey MC, von Stebut E. IL-17 promotes progression of cutaneous leishmaniasis in susceptible mice. J Immunol 2009;182(5):3039–46. https://doi.org/10.4049/jimmunol.0713598.

[19] Katara GK, Ansari NA, Verma S, Ramesh V, Salotra P. Foxp3 and IL-10 expression correlates with parasite burden in lesional tissues of post kala azar dermal leishmaniasis (PKDL) patients. PLoS Negl Trop Dis 2011;5(5):e1171. https://doi.org/10.1371/journal.pntd.0001171.

[20] Bhardwaj S, Srivastava N, Sudan R, Saha B. *Leishmania* interferes with host cell signaling to devise a survival strategy. J Biomed Biotechnol 2010;2010:109189. https://doi.org/10.1155/2010/109189.

[21] Isnard A, Shio MT, Olivier M. Impact of *Leishmania* metalloprotease GP63 on macrophage signaling. Front Cell Infect Microbiol 2012;2:72. https://doi.org/10.3389/fcimb.2012.00072.

[22] Forestier CL, Gao Q, Boons GJ. Leishmania lipophosphoglycan: how to establish structure-activity relationships for this highly complex and multifunctional glycoconjugate? Front Cell Infect Microbiol 2014;4:193. https://doi.org/10.3389/fcimb.2014.00193.

[23] Paape D, Aebischer T. Contribution of proteomics of *Leishmania* spp. to the understanding of differentiation, drug resistance mechanisms, vaccine and drug development. J Proteomics 2011;74(9):1614–24. https://doi.org/10.1016/j.jprot.2011.05.005.

[24] Paranaiba LF, Pinheiro LJ, Macedo DH, Menezes-Neto A, Torrecilhas AC, Tafuri WL, Soares RP. An overview on *Leishmania* (Mundinia) *enriettii*: biology, immunopathology, LRV and extracellular vesicles during the host-parasite interaction. Parasitology 2017;1–9. https://doi.org/10.1017/S0031182017001810.

[25] Silverman JM, Clos J, de' Oliveira CC, Shirvani O, Fang Y, Wang C, Foster LJ, Reiner NE. An exosome-based secretion pathway is responsible for protein export from *Leishmania* and communication with macrophages. J Cell Sci 2010;123(Pt. 6):842–52. https://doi.org/10.1242/jcs.056465.

[26] Silverman JM, Chan SK, Robinson DP, Dwyer DM, Nandan D, Foster LJ, Reiner NE. Proteomic analysis of the secretome of *Leishmania donovani*. Genome Biol 2008;9(2):R35. https://doi.org/10.1186/gb-2008-9-2-r35.

[27] Atayde VD, Aslan H, Townsend S, Hassani K, Kamhawi S, Olivier M. Exosome secretion by the parasitic protozoan *Leishmania* within the sand fly midgut. Cell Rep 2015;13(5):957–67. https://doi.org/10.1016/j.celrep.2015.09.058.

[28] Silverman JM, Reiner NE. *Leishmania* exosomes deliver preemptive strikes to create an environment permissive for early infection. Front Cell Infect Microbiol 2011;1:26. https://doi.org/10.3389/fcimb.2011.00026.

[29] Silverman JM, Clos J, Horakova E, Wang AY, Wiesgigl M, Kelly I, Lynn MA, McMaster WR, Foster LJ, Levings MK, Reiner NE. *Leishmania* exosomes modulate innate and adaptive immune responses through effects on monocytes and dendritic cells. J Immunol 2010;185(9):5011–22. https://doi.org/10.4049/jimmunol.1000541.

[30] Schwab A, Meyering SS, Lepene B, Iordanskiy S, van Hoek ML, Hakami RM, Kashanchi F. Extracellular vesicles from infected cells: potential for direct pathogenesis. Front Microbiol 2015;6:1132. https://doi.org/10.3389/fmicb.2015.01132.

[31] Hassani K, Olivier M. Immunomodulatory impact of *Leishmania*-induced macrophage exosomes: a comparative proteomic and functional analysis. PLoS Negl Trop Dis 2013;7(5):e2185. https://doi.org/10.1371/journal.pntd.0002185.

[32] Cronemberger-Andrade A, Aragão-França L, de Araujo CF, Rocha VJ, Borges-Silva MC, Figueira CP, Figueiras CP, Oliveira PR, de Freitas LA, Veras PS, Pontes-de-Carvalho L. Extracellular vesicles from *Leishmania*-infected macrophages confer an anti-infection cytokine-production profile to naïve macrophages. PLoS Negl Trop Dis 2014;8(9):e3161. https://doi.org/10.1371/journal.pntd.0003161.

[33] Burleigh BA, Andrews NW. The mechanisms of *Trypanosoma cruzi* invasion of mammalian cells. Annu Rev Microbiol 1995;49:175–200. https://doi.org/10.1146/annurev.mi.49.100195.001135.

[34] Burleigh BA, Woolsey AM. Cell signalling and *Trypanosoma cruzi* invasion. Cell Microbiol 2002;4(11):701–11.

[35] Sibley LD, Andrews NW. Cell invasion by un-palatable parasites. Traffic 2000;1(2):100–6.

[36] Ouaissi A, Ouaissi M. Molecular basis of *Trypanosoma cruzi* and *Leishmania* interaction with their host(s): exploitation of immune and defense mechanisms by the parasite leading to persistence and chronicity, features reminiscent of immune system evasion strategies in cancer diseases. Arch Immunol Ther Exp (Warsz) 2005;53(2):102–14.

[37] Almeida IC, Gazzinelli RT. Proinflammatory activity of glycosylphosphatidylinositol anchors derived from *Trypanosoma cruzi*: structural and functional analyses. J Leukoc Biol 2001;70(4):467–77.

[38] Almeida IC, Camargo MM, Procópio DO, Silva LS, Mehlert A, Travassos LR, Gazzinelli RT, Ferguson MA. Highly purified glycosylphosphatidylinositols from *Trypanosoma cruzi* are potent proinflammatory agents. EMBO J 2000;19(7):1476–85. https://doi.org/10.1093/emboj/19.7.1476.

[39] Colli W. Trans-sialidase: a unique enzyme activity discovered in the protozoan *Trypanosoma cruzi*. FASEB J 1993;7(13):1257–64.

[40] Gonçalves MF, Umezawa ES, Katzin AM, de Souza W, Alves MJ, Zingales B, Colli W. *Trypanosoma cruzi*: shedding of surface antigens as membrane vesicles. Exp Parasitol 1991;72(1):43–53.

[41] Ribeiro KS, Vasconcellos CI, Soares RP, Mendes MT, Ellis CC, Aguilera-Flores M, de Almeida IC, Schenkman S, Iwai LK, Torrecilhas AC. Proteomic analysis reveals different composition of extracellular vesicles released by two. J Extracell Vesicles 2018;7(1):1463779. https://doi.org/10.1080/20013078.2018.1463779.

[42] Nogueira PM, Ribeiro K, Silveira AC, Campos JH, Martins-Filho OA, Bela SR, Campos MA, Pessoa NL, Colli W, Alves MJ, Soares RP, Torrecilhas AC. Vesicles from different *Trypanosoma cruzi* strains trigger differential innate and chronic immune responses. J Extracell Vesicles 2015;4:28734.

[43] Cestari I, Ansa-Addo E, Deolindo P, Inal JM, Ramirez MI. *Trypanosoma cruzi* immune evasion mediated by host cell-derived microvesicles. J Immunol 2012;188(4):1942–52. https://doi.org/10.4049/jimmunol.1102053.

[44] Bayer-Santos E, Aguilar-Bonavides C, Rodrigues SP, Cordero EM, Marques AF, Varela-Ramirez A, Choi H, Yoshida N, da Silveira JF, Almeida IC. Proteomic analysis of *Trypanosoma cruzi* secretome: characterization of two populations of extracellular vesicles and soluble proteins. J Proteome Res 2013;12(2):883–97. https://doi.org/10.1021/pr300947g.

[45] Büscher P, Cecchi G, Jamonneau V, Priotto G. Human African trypanosomiasis. Lancet 2017;390(10110):2397–409. https://doi.org/10.1016/S0140-6736(17)31510-6.

[46] Borst P, Ulbert S. Control of VSG gene expression sites. Mol Biochem Parasitol 2001;114(1):17–27.

[47] Stijlemans B, Caljon G, Van Den Abbeele J, Van Ginderachter JA, Magez S, De Trez C. Immune evasion strategies of *Trypanosoma brucei* within the mammalian host: progression to pathogenicity. Front Immunol 2016;7:233. https://doi.org/10.3389/fimmu.2016.00233.

[48] Szempruch AJ, Sykes SE, Kieft R, Dennison L, Becker AC, Gartrell A, Martin WJ, Nakayasu ES, Almeida IC, Hajduk SL, Harrington JM. Extracellular vesicles from *Trypanosoma brucei* mediate virulence factor transfer and cause host anemia. Cell 2016;164(1–2):246–57. https://doi.org/10.1016/j.cell.2015.11.051.

[49] Geiger A, Hirtz C, Becue T, Bellard E, Centeno D, Gargani D, Rossignol M, Cuny G, Peltier JB. Exocytosis and protein secretion in *Trypanosoma*. BMC Microbiol 2010;10:20. https://doi.org/10.1186/1471-2180-10-20.

[50] Silverman JS, Muratore KA, Bangs JD. Characterization of the late endosomal ESCRT machinery in *Trypanosoma brucei*. Traffic 2013;14(10):1078–90. https://doi.org/10.1111/tra.12094.

[51] Eliaz D, Kannan S, Shaked H, Arvatz G, Tkacz ID, Binder L, Waldman Ben-Asher H, Okalang U, Chikne V, Cohen-Chalamish S, Michaeli S. Exosome secretion affects social motility in *Trypanosoma brucei*. PLoS Pathog 2017;13(3):e1006245. https://doi.org/10.1371/journal.ppat.1006245.

[52] Cowman AF, Healer J, Marapana D, Marsh K. Malaria: biology and disease. Cell 2016;167(3):610–24. https://doi.org/10.1016/j.cell.2016.07.055.

[53] Yamauchi LM, Coppi A, Snounou G, Sinnis P. *Plasmodium* sporozoites trickle out of the injection site. Cell Microbiol 2007;9(5):1215–22. https://doi.org/10.1111/j.1462-5822.2006.00861.x.

[54] Good MF, Doolan DL. Malaria vaccine design: immunological considerations. Immunity 2010;33(4):555–66. https://doi.org/10.1016/j.immuni.2010.10.005.

[55] Campos FM, Franklin BS, Teixeira-Carvalho A, Filho AL, de Paula SC, Fontes CJ, Brito CF, Carvalho LH. Augmented plasma microparticles during acute *Plasmodium vivax* infection. Malar J 2010;9:327. https://doi.org/10.1186/1475-2875-9-327.

[56] Martin-Jaular L, Nakayasu ES, Ferrer M, Almeida IC, Del Portillo HA. Exosomes from *Plasmodium yoelii*-infected reticulocytes protect mice from lethal infections. PLoS ONE 2011;6(10):e26588. https://doi.org/10.1371/journal.pone.0026588.

[57] Couper KN, Barnes T, Hafalla JC, Combes V, Ryffel B, Secher T, Grau GE, Riley EM, de Souza JB. Parasite-derived plasma microparticles contribute significantly to malaria infection-induced inflammation through potent macrophage stimulation. PLoS Pathog 2010;6(1):e1000744. https://doi.org/10.1371/journal.ppat.1000744.

[58] Martin-Jaular L, Ferrer M, Calvo M, Rosanas-Urgell A, Kalko S, Graewe S, Soria G, Cortadellas N, Ordi J, Planas A, Burns J, Heussler V, del Portillo HA. Strain-specific spleen remodelling in *Plasmodium yoelii* infections in Balb/c mice facilitates adherence and spleen macrophage-clearance escape. Cell Microbiol 2011;13(1):109–22. https://doi.org/10.1111/j.1462-5822.2010.01523.x.

[59] Skorokhod OA, Alessio M, Mordmüller B, Arese P, Schwarzer E. Hemozoin (malarial pigment) inhibits differentiation and maturation of human monocyte-derived dendritic cells: a peroxisome proliferator-activated receptor-gamma-mediated effect. J Immunol 2004;173(6):4066–74.

[60] Montoya JG, Remington JS. Management of *Toxoplasma gondii* infection during pregnancy. Clin Infect Dis 2008;47(4):554–66. https://doi.org/10.1086/590149.

[61] Saadatnia G, Golkar M. A review on human toxoplasmosis. Scand J Infect Dis 2012;44(11):805–14. https://doi.org/10.3109/00365548.2012.693197.

[62] Sasai M, Pradipta A, Yamamoto M. Host immune responses to *Toxoplasma gondii*. Int Immunol 2018;30(3):113–9. https://doi.org/10.1093/intimm/dxy004.

[63] Lekutis C, Ferguson DJ, Grigg ME, Camps M, Boothroyd JC. Surface antigens of *Toxoplasma gondii*: variations on a theme. Int J Parasitol 2001;31(12):1285–92.

[64] Jung C, Lee CY, Grigg ME. The SRS superfamily of *Toxoplasma* surface proteins. Int J Parasitol 2004;34(3):285–96. https://doi.org/10.1016/j.ijpara.2003.12.004.

[65] Carruthers VB. Host cell invasion by the opportunistic pathogen *Toxoplasma gondii*. Acta Trop 2002;81(2):111–22.

[66] Denkers EY, Gazzinelli RT. Regulation and function of T-cell-mediated immunity during *Toxoplasma gondii* infection. Clin Microbiol Rev 1998;11(4):569–88.

[67] Meira CS, Costa-Silva TA, Vidal JE, Ferreira IM, Hiramoto RM, Pereira-Chioccola VL. Use of the serum reactivity against *Toxoplasma gondii* excreted-secreted antigens in cerebral toxoplasmosis diagnosis in human immunodeficiency virus-infected patients. J Med Microbiol 2008;57(Pt. 7):845–50. https://doi.org/10.1099/jmm.0.47687-0.

[68] Wowk PF, Zardo ML, Miot HT, Goldenberg S, Carvalho PC, Mörking PA. Proteomic profiling of extracellular vesicles secreted from *Toxoplasma gondii*. Proteomics 2017;17(15–16):1600477. https://doi.org/10.1002/pmic.201600477.

[69] Silva VO, Maia MM, Torrecilhas AC, Taniwaki NN, Namiyama GM, Oliveira KC, Ribeiro KS, Toledo MDS, Xander P, Pereira-Chioccola VL. Extracellular vesicles isolated from *Toxoplasma gondii* induce host immune response. Parasite Immunol 2018;40(9):e12571. https://doi.org/10.1111/pim.12571.

[70] Li Y, Liu Y, Xiu F, Wang J, Cong H, He S, Shi Y, Wang X, Li X, Zhou H. Characterization of exosomes derived from *Toxoplasma gondii* and their functions in modulating immune responses. Int J Nanomedicine 2018;13:467–77. https://doi.org/10.2147/IJN.S151110.

[71] Aline F, Bout D, Amigorena S, Roingeard P, Dimier-Poisson I. *Toxoplasma gondii* antigen-pulsed-dendritic cell-derived exosomes induce a protective immune response against *T. gondii* infection. Infect Immun 2004;72(7):4127–37. https://doi.org/10.1128/IAI.72.7.4127-4137.2004.

[72] Bhatnagar S, Schorey JS. Exosomes released from infected macrophages contain *Mycobacterium avium* glycopeptidolipids and are proinflammatory. J Biol Chem 2007;282(35):25779–89. https://doi.org/10.1074/jbc.M702277200.

[73] Einarsson E, Ma'ayeh S, Svärd SG. An up-date on Giardia and giardiasis. Curr Opin Microbiol 2016;34:47–52. https://doi.org/10.1016/j.mib.2016.07.019.

[74] Fink MY, Singer SM. The intersection of immune responses, microbiota, and pathogenesis in giardiasis. Trends Parasitol 2017;33(11):901–13. https://doi.org/10.1016/j.pt.2017.08.001.

[75] Benchimol M. The release of secretory vesicle in encysting *Giardia lamblia*. FEMS Microbiol Lett 2004;235(1):81–7. https://doi.org/10.1016/j.femsle.2004.04.014.

[76] Gottig N, Elías EV, Quiroga R, Nores MJ, Solari AJ, Touz MC, Luján HD. Active and passive mechanisms drive secretory granule biogenesis during differentiation of the intestinal parasite *Giardia lamblia*. J Biol Chem 2006;281(26):18156–66. https://doi.org/10.1074/jbc.M602081200.

[77] Deolindo P, Evans-Osses I, Ramirez MI. Microvesicles and exosomes as vehicles between protozoan and host cell communication. Biochem Soc Trans 2013;41(1):252–7. https://doi.org/10.1042/BST20120217.

[78] Papanastasiou P, Hiltpold A, Bommeli C, Köhler P. The release of the variant surface protein of *Giardia* to its soluble isoform is mediated by the selective cleavage of the conserved carboxy-terminal domain. Biochemistry 1996;35(31):10143–8. https://doi.org/10.1021/bi960473b.

[79] Prucca CG, Slavin I, Quiroga R, Elías EV, Rivero FD, Saura A, Carranza PG, Luján HD. Antigenic variation in *Giardia lamblia* is regulated by RNA interference. Nature 2008;456(7223):750–4. https://doi.org/10.1038/nature07585.

[80] Amat CB, Motta JP, Fekete E, Moreau F, Chadee K, Buret AG. Cysteine protease-dependent mucous disruptions and differential mucin gene expression in giardia duodenalis infection. Am J Pathol 2017;187(11):2486–98. https://doi.org/10.1016/j.ajpath.2017.07.009.

[81] Bhargava A, Cotton JA, Dixon BR, Gedamu L, Yates RM, Buret AG. *Giardia duodenalis* surface cysteine proteases induce cleavage of the intestinal epithelial cytoskeletal protein villin via myosin light chain kinase. PLoS One 2015;10(9):e0136102. https://doi.org/10.1371/journal.pone.0136102.

[82] Cotton JA, Bhargava A, Ferraz JG, Yates RM, Beck PL, Buret AG. *Giardia duodenalis* cathepsin B proteases degrade intestinal epithelial interleukin-8 and attenuate interleukin-8-induced neutrophil chemotaxis. Infect Immun 2014;82(7):2772–87. https://doi.org/10.1128/IAI.01771-14.

[83] Ortega-Pierres G, Argüello-García R, Laredo-Cisneros MS, Fonseca-Linán R, Gómez-Mondragón M, Inzunza-Arroyo R, Flores-Benítez D, Raya-Sandino A, Chavez-Munguía B, Ventura-Gallegos JL, Zentella-Dehesa A, Bermúdez-Cruz RM, González-Mariscal L. Giardipain-1, a protease secreted by *Giardia duodenalis* trophozoites, causes junctional, barrier and apoptotic damage in epithelial cell monolayers. Int J Parasitol 2018;48(8):621–39. https://doi.org/10.1016/j.ijpara.2018.01.006.

[84] Ma'ayeh SY, Liu J, Peirasmaki D, Hörnaeus K, Bergström Lind S, Grabherr M, Bergquist J, Svärd SG. Characterization of the *Giardia intestinalis* secretome during interaction with human intestinal epithelial cells: the impact on host cells. PLoS Negl Trop Dis 2017;11(12):e0006120. https://doi.org/10.1371/journal.pntd.0006120.

[85] Evans-Osses I, Mojoli A, Monguió-Tortajada M, Marcilla A, Aran V, Amorim M, Inal J, Borràs FE, Ramirez MI. Microvesicles released from *Giardia intestinalis* disturb host-pathogen response in vitro. Eur J Cell Biol 2017;96(2):131–42. https://doi.org/10.1016/j.ejcb.2017.01.005.

[86] Touz MC, Rivero MR, Miras SL, Bonifacino JS. Lysosomal protein trafficking in *Giardia lamblia*: common and distinct features. Front Biosci (Elite Ed) 2012;4:1898–909.

[87] Saha N, Dutta S, Datta SP, Sarkar S. The minimal ESCRT machinery of *Giardia lamblia* has altered inter-subunit interactions within the ESCRT-II and ESCRT-III complexes. Eur J Cell Biol 2018;97(1):44–62. https://doi.org/10.1016/j.ejcb.2017.11.004.

[88] Hassani K, Antoniak E, Jardim A, Olivier M. Temperature-induced protein secretion by *Leishmania mexicana* modulates macrophage signalling and function. PLoS One 2011;6(5):e18724. https://doi.org/10.1371/journal.pone.0018724.

[89] Ramirez MI, de Ruiz RC, Araya JE, Da Silveira JF, Yoshida N. Involvement of the stage-specific 82-kilodalton adhesion molecule of *Trypanosoma cruzi* metacyclic trypomastigotes in host cell invasion. Infect Immun 1993;61(9):3636–41.

Chapter 9

Exosomes as intercellular communication messengers for cardiovascular and cerebrovascular diseases

Antonia Teona Deftu[a,b,#], Beatrice Mihaela Radu[a,b,#], Dragos Cretoiu[c,d], Alexandru Florian Deftu[a,b], Sanda Maria Cretoiu[c], Junjie Xiao[e,f]

[a]*Department of Anatomy, Animal Physiology and Biophysics, Faculty of Biology, University of Bucharest, Bucharest, Romania,* [b]*Life, Environmental and Earth Sciences Division, Research Institute of the University of Bucharest (ICUB), Bucharest, Romania,* [c]*Department of Cell and Molecular Biology and Histology, Carol Davila University of Medicine and Pharmacy, Bucharest, Romania,* [d]*Alessandrescu-Rusescu National Institute of Mother and Child Health, Fetal Medicine Excellence Research Center, Bucharest, Romania,* [e]*Department of Cardiology, The First Affiliated Hospital of Nanjing Medical University, Nanjing, People's Republic of China,* [f]*Cardiac Regeneration and Ageing Lab, Experimental Center of Life Sciences, School of Life Science, Shanghai University, Shanghai, People's Republic of China*

1 Introduction

The secretome of the cell is considered to contain all the proteins secreted by the cell and includes paracrine substances, exosomes and microvesicles [1]. Although there is still no consensus on the nomenclature of the vesicles released by a cell, considerable efforts are being made to quench this desideratum [2]. Reunited under the generic name of extracellular vesicles (a term introduced by introduced by International Society of Extracellular Vesicles), exosomes and microvesicles, along with apoptotic bodies, can influence by cargo molecules the nearby or remote cells [3]. At present, several public online databases are available such as Vesiclepedia (www.microvesicles.org), EVpedia (www.evpedia.info) and ExoCarta (www.exocarta.org) [4].

[#] AT Deftu and BM Radu had equal contributions.

Exosomes. https://doi.org/10.1016/B978-0-12-816053-4.00009-2

Exosomes mediate intercellularly, communication in normal and pathological conditions and they are found in body fluids such as plasma, malignant ascites, urine, amniotic fluid, sputum, seminal fluid, breast milk saliva and pericardial fluid [5–8].

Exosomes are cell-derived small extracellular membrane vesicles, with 50–100nm in diameter, that are actively secreted and released both in physiological and pathological conditions. Often, there is no clear distinction between exosomes and microvesicles [9]. Exosomes originate in the cytoplasmic multivesicular bodies, which fuses with plasma lemma to release the exosomes. Microvesicles, also named ectosomes, are larger vesicles than exosomes which result from direct budding of the plasma membrane [10]. Exosomes also differ from microvesicles in chemical composition, being enriched in lipids, particularly cholesterol (while microvesicles abound in GM1 ganglioside, also present in small quantities in exosomes) and transmembrane proteins—tetraspanins such as CD9, CD63, and CD81 (considered as exosome markers) [11].

Exosomes are important cargos for proteins (e.g., cytokine receptors, tumor-associated antigens, oncoproteins, and antigen presentation molecules), coding RNA (mRNA), non-coding RNA (miRNAs), or even DNA, and ensure communication at cellular, tissue or organ level [12–16]. Due to their small size and protection have given by exosomes, miRNAs can escape from protease and nuclease degradation in the serum or plasma [17].

Although not much is known about the role of exosomes in normal conditions, their roles in pathological conditions have been much more extensively studied [18]. Exosomes were proposed as an effective means to detect a variety of disease states since they have specific genomic and proteomic signatures characteristic of the cells which they are derived from [19]. Exosomes-based biomarkers are highly relevant for the prognosis of cardiovascular diseases [20, 21] and are widely recognized as key players in the prevention, repair or progression of cardiovascular disease [22].

In this chapter, we aim to update the knowledge about exosomes in various cardiovascular and cerebrovascular pathologies. We are reviewing the methods used for isolation and analysis of circulating exosomes, and we provide an extensive overview of the current knowledge about exosomes in the clinical practice as biomarkers of various cardiovascular diseases.

2 Exosomes biogenesis and release

The biogenesis and release of exosomes is an autoregulated process. Thus, during exosome biogenesis, the message send/released to the target cells depends on exosomes content, which, in turn, is influenced by the cell of origin and by the local microenvironment [23]. Exosomes that ensure the intercellular communication in physiological and pathological states of the cardiovascular system and cerebrovascular system (Fig. 1) derive from and act on various cell subtypes, including cardiomyocytes, endothelial cells, mesenchymal stem/stromal cells, erythrocytes, brain cells derived exosomes and others.

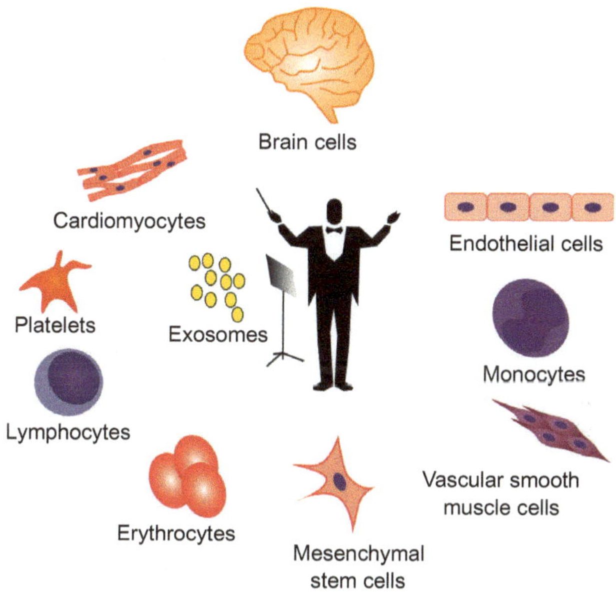

FIG. 1 Exosomes ensure the intercellular communication in the cardiovascular system and cerebrovascular system.

Studies have also demonstrated that exosomes released by different types of stem cells, including mesenchymal stem cells, hematopoietic stem cells, cardiac progenitor stem cells, cardiosphere cells, induced pluripotent cells, are directed to endothelial cells, cardiomyocytes, fibroblasts and other types of cells, and play a cardioprotective role, by increasing cardiac angiogenesis, diminishing oxidative stress and inflammation, and reducing apoptosis [24].

2.1 Cardiomyocytes-derived exosomes

Cardiomyocytes release extracellular vesicles (e.g., exosomes or microvesicles) in physiological and physiopathological conditions. To date, stress conditions, such as hypoxia, inflammation or injury trigger in cardiac cells the secretion of extracellular vesicles that contribute to heart regeneration by their content (i.e., angiogenic, anti-apoptotic, mitogenic and growth factors) [25].

There is important crosstalk between cardiomyocytes and endothelial cells that is ensured by means of the cardiomyocytes-derived exosomes that are uptaken by endothelial cells. In this cardio-endothelial communication system, exosomes contain functional glucose transporters and glycolytic enzymes that were demonstrated to be transferred to the endothelial cells [26]. Interestingly, the biogenesis of cardiomyocytes-derived exosomes can be stimulated by glucose starvation, and the exposure of endothelial cells to exosomes delivered in these stress conditions induced angiogenesis [27].

The extensive crosstalk between different cardiac cells (e.g., cardiomyocytes—endothelial cells—stem cells—fibroblasts—smooth muscle cells) is essential in maintaining cardiac homeostasis. Beside the active communication between cardiomyocytes and endothelial cells, cardiomyocytes-derived exosomes were also demonstrated to be transferred to fibroblasts [28, 29].

Cardiomyocytes progenitor cells also release exosomes that were described to be efficiently uptaken by endothelial cells and to stimulate the endothelial cell migration and angiogenesis, by the activation of the ERK/Akt-signaling pathways [30].

2.2 Endothelial cell-derived exosomes

In vitro and in vivo studies demonstrated that exosomes from circulating endothelial progenitor cells transfer cargo mRNA and activate angiogenesis in recipient endothelial cells [31, 32]. On the other hand, exosomes derived from various cell types have the potential to deliver their content that induces pro- (e.g., transfer of microRNAs, proteins, lipids, transcription factors or activation of signaling pathways) or antiangiogenic (LDL receptor-mediated endocytosis, CD36-dependent uptake of extracellular vesicles, induction of oxidative stress) signaling in endothelial cells. This is a dynamic process, where exosomes adapt their composition in response to local microenvironmental changes and modulate angiogenesis [23].

The circulating levels of endothelial cell-derived microvesicles were unchanged for type 2 diabetes patients [33, 34], while they were significantly increased for patients with metabolic syndrome [35].

Pharmacological therapy in patients with vascular complications, diabetes or metabolic syndrome might increase or decrease the levels of endothelially derived microvesicles [36, 37].

Cardiac endothelial cells (CEC) are important players in cardiac physiology and pathophysiology, with a distinct transcriptome in comparison with other endothelial cells (e.g., renal, cerebral, or pulmonary endothelial cells) [38]. CEC-derived exosomes express the marker integrin $\alpha v\beta 6$ and determine the activation of the latent transforming growth factor (TGF)-β into its active form in B cells [38].

2.3 Mesenchymal stem/stromal cells derived exosomes

Mesenchymal stem cells (MSCs) are multipotent progenitors that can be isolated from different types of tissue (e.g., bone marrow, adipose, muscle bone, brain, spleen, liver, kidney, lung, thymus, and pancreas). MSCs-derived exosomes have been described to play multiple roles, including tissue repair, anti-inflammatory effects, immunomodulation, and neuroprotection. Interestingly, the MSCs-derived exosomes are more stable than MSCs *per se*, and when in vivo injected exosomes are more prone not to be rejected by the immune system due to the lack of aneuploidy in comparison with the origin cells [39]. However, human MSCs-derived exosomes were demonstrated to exert *in vitro* immunomodulatory effects against peripheral blood mononuclear cells (e.g., T lymphocytes), increasing their apoptosis but without any effects on the proliferation [40].

It has been demonstrated that MSCs were able to produce larger quantities of exosomes in comparison with other cell subtypes, including myoblasts, human acute monocytic leukemia cells or human embryonic kidney cells [41].

2.4 Platelet-derived exosomes

Platelet-derived exosomes have also been described [42]. Interestingly, platelets host a complex transcriptome, including approximately 9500 mRNAs, different classes of non-coding RNAs (e.g., microRNAs), and proteins (involved in adhesion, aggregation, granule secretion, RNA transfer, mRNA translation, and immune regulation processes [43–49]. Platelet-derived exosomes have similar size, protein markers, and ratio of the amount of CD81 marker-to-exosome count, in comparison with exosomes derived from other cell subtypes [42, 50]. Extracellular calcium influences the content of the platelet-derived exosomes, by changing the total exosomal content and the levels per exosome of platelet glycoprotein VI, CXCL7, or high-mobility group box 1 protein [50].

In patients with sepsis, platelet-derived exosomes were shown to contribute to myocardial dysfunction [51]. Platelets play an essential role in recovery after the blood vessels damage and in thrombosis. To detail, upon vessel injury, platelets interact with the endothelial cells and release their cargo (e.g., adhesive molecules, including von Willebrand factor and fibrinogen, growth factors and inflammatory and angiogenic mediators) [52]. In mice, platelet-derived exosomes were demonstrated to increase the protein ubiquitination and to increase the proteasome degradation of CD36 and thus to contribute to the inhibition of the athero-thrombotic processes [53]. Moreover, platelet-derived exosomes were described to the apoptosis of endothelial cells apoptosis [54].

2.5 Leukocyte-derived exosomes

Leukocyte-derived microparticles express different protein markers in correlation with the cellular origin, such as leukocyte (CD45), neutrophil (CD15, CD64, CD66b, CD66e), monocyte (CD11a, CD14, CD18), lymphocyte (CD2, CD3, CD4, CD8, CD19, CD20) [55]. If stimulated by starvation or by exposure to endotoxin or calcium ionophore A23187, monocytes (i.e., monocyte cell line THP-1) release exosomes expressing the following markers, Tsg 101, negative phospholipids, CD18, CD14 and the active tissue factor (TF) [56]. Interestingly, if endothelial cells were exposed to these monocytes-derived microvesicles, endothelial apoptosis and nuclear fragmentation were triggered, and procoagulation effects were activated [56].

An analysis of the protein markers between the cultured leukocytes (e.g., T cells, monocytes, NK cells, and B cells) and their derived exosomes revealed that some cell surface proteins are not transferred to the surface of the extracellular vesicles, including CD3, CD14, CD16, and CD19 [57]. On the other hand, other cell surface proteins, like CD9, CD63, and CD81 are transferred from parent cells to the extracellular vesicles for the majority of cultured leukocytes [57].

2.6 Erythrocytes-derived exosomes

Erythrocytes release both exosomes and ectosomes depending on their phase of development. In particular, erythrocytes-derived exosomes (EDE) are released only during the development of red blood cells in the bone marrow [58]. EDE were identified in the blood and had been characterized to bind to monocytes inducing secretion of proinflammatory cytokines (e.g., TNF-α) and to boost the mitogen-induced CD4+ and CD8+ T-cell proliferation [59]. The proinflammatory effect (i.e., production of TNF, IL-6, and IL-8) induced by erythrocyte-derived exosomes is influenced by the storage duration of the red blood cells [60]. In particular, EDE may contribute to the negative effects of blood storage that prevents blood transfusion. Mechanisms involve the altered NO scavenging by haemoglobin inside the red blood cells [61], and the abnormal vasoregulation [62].

The production of vesicles by red blood cells may be triggered by various stimuli, including increased cytosolic Ca^{2+} [63], ATP depletion, increased K^+ leakage [64], and activation of various intracellular signaling cascades (e.g., GPCR, PI3K-Akt, Jak-STAT, and Raf-MEK pathways) [65].

EDE also play an essential role in various blood and cardiovascular pathologies. To date, EDE contribute to the cell-to-cell communication in blood infective diseases (e.g., malaria) [66]. In patients with sickle cell disease, erythrocytes release spectrin-free haemoglobin-containing spicules as rods and microspheres, and rods containing polymerized haemoglobin may eventually become chains of microvesicles [67]. The spicules released by sickled erythrocytes were demonstrated to the altered blood clotting probably due to the reorganization of the phospholipids inside their spectrin-free membrane [68]. High levels of erythrocyte-derived microvesicles CD235a-positive were detected in the plasma of obese patients with type 2 diabetes, without a direct correlation with the plasma haemoglobin A1c levels [69]. Erythrocyte-derived microvesicles were also significantly increased in patients with metabolic syndrome, with diabetes as comorbidity (⅓ of the patients) [70].

2.7 Vascular smooth muscle cells derived exosomes

Vascular smooth muscle cells (VSMCs) are also producing exosomes with potential proangiogenic activity. VSMCs-derived exosomes have a diameter of 60–150 nm and express protein markers like flotillin-1, CD81, and syntenin-1 [71]. The protein profile of the human VSMC-derived exosomes analyzed by nanoLC-MS/MS-based proteomics revealed the presence of 459 proteins [72]. Another extensive proteomic study identified 349 proteins in the VSMC-derived exosomes, most of them being an extracellular matrix (ECM) or ECM-related proteins and cell adhesion molecules, which implies that VSMC-derived exosomes are involved in focal adhesion and to intercellular signaling [73].

VSMC-derived exosomes play an important role in vascular pathologies, including restenosis, atherosclerosis, calcification, and coagulation [74, 75]. In VSMCs-overexpressing Krüppel-like factor 5 (KLF5) was demonstrated that VSMCs-derived exosomes transfer miR-155 from VSMCs to endothelial cells, thus disturbing the integrity of the endothelial barrier and stimulating the atherosclerosis progression [76].

2.8 Brain cells derived exosomes

Beside the extracerebral sources of exosomes, all brain cells (i.e., neurons, astrocytes, cerebral endothelial cells, microglia, oligodendrocytes and others) can synthesize and release exosomes [77–80]. To date, cultured glioblastoma cells [81], brain microvascular endothelial cells [82, 83], primary cultured rat microglia [84], PC12 neuronal differentiated cells [85] and astrocytes [80] secrete exosomes.

Activation of different cellular components of the neurovascular unit by physiological or pathological stimuli determines the release of exosomes. In HIV infection the communication between brain microvascular endothelial cells and macrophages is ensured by exosomes derived from TLR3-activated brain microvascular endothelial cells [86]. The angiogenesis process in brain cancer was also mediated extracellular vesicles that contribute to the communication between brain microvascular endothelial cells and glioblastoma [87]. In neuronal stress conditions (e.g., oxidative stress, hypoxia, ischemia or hypoglycemia), the release of exosomes by astrocytes with a high content of prion proteins ensured the neuronal survival [88]. In spinal cord regeneration, upon treatment with retinoic acid receptor β (RARβ) agonist, phosphatase and tensin homolog were inactivated in the neuronal cytoplasm being further transferred by exosomes to astrocytes and thus contributing to axonal regeneration [89].

In this context, brain remodeling is strongly influenced by the exosome-mediated intercellular communication, and the vascular component plays an essential role.

3 Exosomes characterization and methods of study and analysis

3.1 Exosomes size

It is generally stated that exosomes size varies between 40 and 100 nm in diameter. The size of exosomes varies between geometric (hydrated or desiccated), hydrodynamic and volumetric estimations. Various methods or cell subtypes were used for exosomes characterization and results indicated differences in exosomes size and shape [90]. The size of exosomes might be obtained employing several techniques, including atomic force microscopy, dynamic light scattering, nanoparticle tracking analysis, endothelial colony forming cells, scanning electron microscopy, transmission electron microscopy, and freeze-fracture transmission electron microscopy.

The size of exosomes might also be affected by the method of drying or desiccating. For example, the hydrodynamic diameter of mesenchymal stem cell derived exosomes was approximately ~98 nm by nanoparticle tracking analysis, while it was approximately 55.5 nm by transmission electron microscopy measurements [91].

The method used for isolating exosomes is also a factor that contributes to size variability [92–94]. Interestingly, the number of circulating exosomes and their size distribution might be influenced by the age, the gender and the health status of the patients [95, 96].

3.2 Exosomes separation

Exosomes are generally separated from conditioned culture media and body fluids (e.g., blood, urine, saliva, cerebrospinal fluid, joint fluid, breast milk, ascitic fluid, etc.) Fig. 2. Although multiple methods of exosome separation have been developed, there is no reliable protocol for exosome purification and analysis [97].

The most employed methods for exosome purification include differential centrifugation (low speeds, approx. 1500 × g), high-speed ultracentrifugation (> 100,000 × g), ultrafiltration, solvent precipitation, microfluidic techniques, immunoaffinity isolation, density gradient and size-exclusion chromatography [92, 98–102].

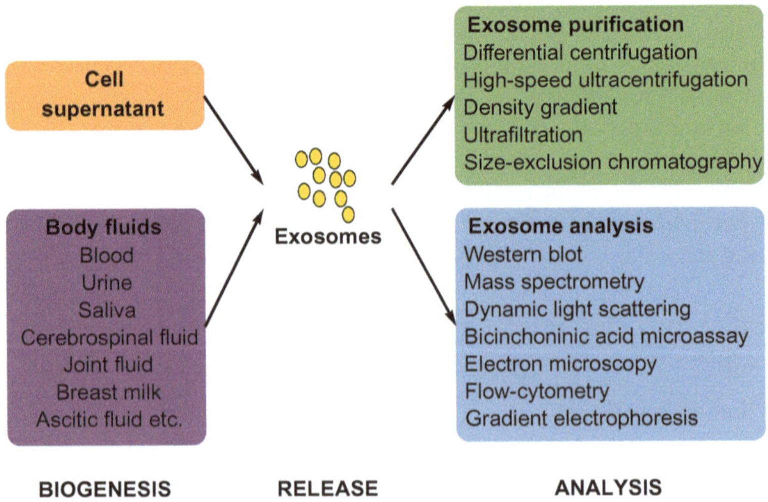

FIG. 2 Overview of the exosomes biogenesis and analysis.

Immunoaffinity capture was considered to be the best method for exosomes separation compared to ultracentrifugation, density gradient separation, due to the enrichment in exosome markers and exosome-associated proteins [92].

In particular, isolation of neuron-derived exosomes is challenging. Methods of fetal neuronal exosomes separation from maternal blood have also been developed, followed by subsequent neuronal markers analysis to determine the fetal origin (e.g., CD81, neuron-specific enolase, neurofilament light chain, L1 cell adhesion molecule), or to exclude the placental origin (type 1 pregnancy-specific β-1-glycoprotein) [50]. Recently, exosomes have been separated from the in vivo brain tissue [103].

Exosome isolation is clinically relevant especially due to the accessibility and diversity of body fluids in which exosomes are present. However, there are difficulties related to the lack of standardized clinical protocols for exosomes isolation [104]. The protocols of extracellular vesicles separation in clinical practice should be adapted to the required output parameters, e.g., the final concentration of extracellular vesicles, the purity of the extracellular vesicles, selection of extracellular vesicle subtype, and the time/cost benefits [102].

3.3 Exosomes analysis

Several methods have been employed for exosome analysis, including Western blot analysis, mass spectrometry [105], dynamic light scattering, bicinchoninic acid microassay [97], electron microscopy [106], flow-cytometry [107, 108], and gradient electrophoresis [109].

The exosome's size of tens of nanometer does not allow it to be visualized by conventional optical microscopy, making electron microscopy necessary to characterize its morphology [110]. By transmission electron microscopy, collapsed extracellular vesicles appear with characteristic cup-shaped or doughnut morphology, while intact exosomes are spherical (diameter from 30 nm to 1 μm), as seen by cryo-electron microscopy [108]. The atomic force microscopy is also a method used for the characterization of extracellular vesicles and exosomes, in particular, primarily to analyze their size and topology [111].

Exosomes are detergent-resistant due to their rich content in cholesterol, sphingomyelin, and ganglioside GM3, or lipids [109]. For some exosomes the flotation density was measured in a sucrose gradient at values of 1.10–1.18 g/mL [107].

Cardiac exosomes have specific protein markers, including proteins from the endosomal sorting complexes required for transport machinery, tetraspanins (e.g., CD9, CD63, CD81), and heat shock protein 70 [112]. Depending on their origin, leukocyte-derived microparticles have various markers [55].

Exosomal protein content is influenced by the cell source and stimulus for exosome formation. Western blot analyses provided evidence that cardiac exosomes retained their protein cargo under different physiological/pathological conditions. Moreover, mass spectrometry demonstrated that the protein content of cardiac exosomes differed significantly from other types of exosomes, and contained cytosolic, sarcomeric, and mitochondrial proteins [105].

In the case of erythrocyte-derived exosomes, their detection was strongly influenced by the storage conditions of the red blood cells concentrates. Flow-cytometry and dynamic light scattering were unable to detail erythrocyte-derived exosomes until a particular concentration was reached at approximately 42–43 days of storage of the red blood cells concentrates [113].

4 Exosomes and coronary artery disease

4.1 Exosomes as biomarkers in coronary artery disease

Although there is expanding evidence about the importance of extracellular vesicles, and in particular of exosomes, in cardiovascular disease, their utilization as markers for a specific disease is still limited by different technological obstacles (e.g., most studies are done *in vitro*).

Coronary artery disease (CAD), the most common type of heart disease, is a result of the existence of inflammation and cholesterol-containing deposits (plaque) in the arteries. It is known that plaque formation is accompanied by the presence of extracellular vesicles originating in leukocytes, erythrocytes, smooth muscle cells, and endothelial cells [114]. By contrast, extracellular vesicles circulating in plasma are platelet-derived and show no smooth muscle cell origin. It was shown that extracellular vesicles are active participants in endothelial dysfunction, intravascular calcification, unstable plaque progression, and thrombus formation after rupture [115].

For assessing CADs, to predict or stratify the risk of mortality, various methods were used, starting with Killip classification, The Canadian Cardiovascular Society Angina Grading Scale, and finishing with prognostic biomarkers including high sensitivity cardiac troponin, high-sensitivity C-reactive protein, and creatine kinase MB. Unfortunately, nowadays there are no specific markers included in the guidelines. Studies showed that PTX3 appears to be more specific than C-reactive protein for predicting cardiovascular disease [116]. Recent studies showed that miRNAs identified in plasma are transported by circulating proteins and lipoproteins and also by exosomes [117] which can be determined as useful biomarkers because they can reflect disease stage or progressions [118]. Several miRNAs were shown to be altered in coronary heart diseases (miR-17, miR-19a, miR-21, miR-92a, miR-126, miR-146a, miR-222, and miR-223) [119].

More on more studies pointed that miRNAs may serve as prognostic and therapeutic targets for cardiovascular diseases [120]. In 2017, Karakas *et al.* performed the most extensive study, so far, evaluating the prognostic value of circulating miRNAs in cardiovascular disease. They showed that miR-132, miR-140-3p, and miR-210 were able to predict cardiovascular death reliably. Three miRNAs miR-223, miR-197, and miR-126 reached statistical significance;

however none was predictive for both early and late cardiovascular events [121]. Another very recent study showed that miR-142–3p and miR-17-5p might be useful as future biomarkers in assessing the risk in CAD [122]. Moreover, in patients with CADs significantly reduced levels of miR-126 were found in comparison with healthy controls [119], and it was suggested that in the future the development of extracellular vesicle-based miR-126 therapeutic strategy might be of use to treat CADs.

4.2 Exosomes in the recovery after coronary artery disease

The formation of the atherosclerotic plaque on the arteries of the heart is followed in evolution by the different degrees of obstruction of the vessel leading to myocardial ischemia. In the worst scenario, the plaque can rupture causing thrombotic occlusion of the artery resulting in acute myocardial infarction. The most appropriate treatment in these cases is to try reperfusion of infarcted myocardial territory. However, this is accompanied by so-called reperfusion ischemia syndrome responsible for additional injury [123]. Many researchers aim to reduce the effects of reperfusion in order to improve the activity of the affected myocardial area. Among these attempts are listed those that address factors released by stem cells [124] or even a by a new cell type described in the myocardium, namely telocytes [21, 125, 126]. These factors might act by paracrine or autocrine effect, and they can be transported with the aid of exosomes [127]. Sahoo *et al.* demonstrated that exosome secreted by CD34+ stem cells might exhibit independent angiogenic activity both *in vitro* and in vivo [128]. Exosomes secreted by mesenchymal stem cells confer cardioprotective effects if they are perfused in rats, directly in the myocardium or intravenously before reperfusion [76, 129, 130]. It has been proposed that exosomes have cardioprotective effects in Langendorff-perfused rat heart [131].

Exosomes have been proposed as an alternative to cell therapies since for example, exosomes secreted by human cardiosphere-derived cells possess regenerative and cardioprotective properties by inhibiting apoptosis and promoting proliferation of cardiomyocytes and angiogenesis [132].

5 Exosomes in myocardial ischemia and infarction

5.1 Exosomes as biomarkers in myocardial ischemia/infarction

5.1.1 Exosomes as biomarkers in myocardial ischemia/infarction

Biomarkers are well known as useful diagnostic tools in many diseases. In cardiovascular medicine, the discovery of such biomarkers, especially for early diagnosis of the pre-necrotic stage occurred prior to the infarction, will be of great value [133]. Experiments performed *in vitro* showed that exosomes released from cardiomyocytes subjected to hypoxia, carry cytokine TNFα and the heat shock protein HSP60 [134, 135]. These findings introduce the possibility that some markers might be released from cardiomyocytes prior to necrosis when the cells are in hypoxia, idea supported by the increased plasmatic levels of exosomes after coronary artery bypass surgery [136]. There are many studies aiming to identify an acute myocardial infarction (AMI) cardiomyocyte-specific miRNA, although some authors suggested that not all changes in miRs levels indicate a cardiomyocyte origin since other cell types are involved in AMI, e.g., endothelial cells [137]. In this sense, recently, a panel of miRs which can be used to differentiate Takotsubo cardiomyopathy from AMI was described: miR-1, miR-16, miR-26a, and miR-133a differentiated TTC from healthy subjects and STEMI patients [138]. Results showed that cardiac-specific miR-1 and miR-133a were up-regulated in ST segment elevation myocardial infarction (STEMI) patients compared with healthy controls [138].

Exosomal miRNA-133a can be considered among the biomarkers for **myocardial infarction** [139, 140]. miR-1, miR-133a, miR-208a, and miR-499 levels were documented in mice as being significantly reduced in infarcted and adjacent infarcted cardiac areas [140]. Additionally, the capture of miR-133a by adjacent infarcted cardiac areas can contribute to hypertrophy inhibition [139]. In cardiac injuries, such as AMI, the above-mentioned cardiomyocyte produced miRNAs (miR-1, miR-133a/b, miR-208a, and miR-499) are found at dramatically increased levels in the blood, indicating myocardial damage [141, 142]. This suggests a potential value these miRNAs as future diagnostic biomarkers for early diagnosis of AMI [143, 144].

5.2 Exosomes in the recovery after myocardial ischemia/infarction

5.2.1 Exosomes involved in the endogenous recovery mechanisms after myocardial ischemia/infarction

Several cell subtypes, including mesenchymal stem cells, cardiac fibroblasts, cardiac progenitor cells, and hematopoietic stem cells have been described to release exosomes that contribute to cardiac regeneration upon myocardial ischemia or infarction.

To date, mesenchymal stem cells derived exosomes and their microRNA profile were described to be involved in the cardiac tissue regeneration upon **ischemic heart injury**, by several mechanisms including, reduction of collagen production, protection against apoptosis, and increased angiogenesis [91]. Moreover, in the murine model of acute ischemia/reperfusion injury, cardiac progenitor-derived exosomes were demonstrated to protect ischemic myocardium by inhibiting the cardiomyocyte apoptosis [145]. In particular, cardiac progenitor-derived exosomal miR-21 was shown to inhibit apoptosis in cardiomyocytes by targeting the programmed cell death 4 (PDCD4) gene [146]. Human cardiac progenitor cells derived exosomes were also demonstrated to inhibit cardiomyocyte apoptosis and to improve cardiac function upon myocardial infarction when injected in the area surrounding the infarction of rat hearts one hour after in vivo coronary artery occlusion [147].

Cardiac fibroblasts were also described to play an important role in the cardiac repair process upon **infarction**. To date, myocardial infarction is associated with acute inflammation processes that trigger the production of inflammatory mediators, the activation of the matrix-degrading activity, and the proliferation and subsequent migration of cardiac fibroblasts [148]. At the site of the myocardial damage, fibroblasts migrate and transdifferentiate into myofibroblasts in response to local inflammatory mediators (e.g., TGF-β), and thus contribute to the repair process [149]. The proliferative phase of the myocardial repair is characterized by the multiplication of cardiac fibroblasts in response to proliferative agents (e.g., fibroblast growth factor 2, platelet-derived growth factor, angiotensin II, vasopressin, endothelin I, cardiotrophin, tryptase, and chymase) and their migration to the injury site [150–155]. In the recovery upon myocardial infarction, a complex intercellular communication is involved by means of exosomes, that implies the cross-talk between cardiomyocytes—endothelial cells—stem cells—fibroblasts—smooth muscle cells [156].

It is known that miR-302–367 promote embryonic cardiomyocyte proliferation and therefore, studies were performed about its role in myocardial infarction [157]. Myocardial regeneration after myocardial infarction is directly influenced by transient activation of miR-302–367, miR-17-92, miR-590, miR-199a, miR-17-92 cluster, miR-199a-214 cluster, miR-34a, and miR-15 family [158]. Moreover, miR-31a-5p has been reported to control postnatal cardiomyocyte proliferation in vivo by targeting RhoBTB1, suggesting that it might have therapeutic implications promoting cardiac regeneration in diseased myocardium [159].

5.2.2 Exosomes used in the therapeutic approaches after myocardial ischemia/infarction

The therapeutic potential of exosomes in the heart recovery after myocardial ischemia and infarction was tested by *in vitro*, ex vivo, or in vivo approaches delivering them to the injury site. Moreover, a large variety of engineering methodology has been employed in order to encapsulate and deliver exosomes.

In vitro assays demonstrated that extracellular vesicles derived from AC10 ventricular cardiomyocyte cell line were more prone to be internalized by endothelial cells compared to fibroblasts, and the paracrine crosstalk between cardiomyocytes-endothelial cells-fibroblasts was proposed to contribute in cardiac remodeling upon ischemic injury [160].

In vivo injection of exosomes derived from mesenchymal stem cells proved to be efficient in the myocardial recovery after ischaemia/reperfusion injury by triggering autophagy in cardiomyocytes via AMPK/mTOR and Akt/mTOR signaling pathways [161]. Moreover, injection of mesenchymal stem cell-derived exosomes into the areas bordering the myocardial infarction was demonstrated to exert local effects, by stimulating angiogenesis and reducing the inflammation response [162]. Additionally, dendritic cells derived exosomes injection after myocardial infarction improved the cardiac function by activating CD4(+) T lymphocytes [163].

Cardiac stem cells exosomes conjugated with cardiac homing peptide (CHP; CSTSMLKAC) have been successfully delivered within heart sections or in neonatal rat cardiomyocytes and demonstrated their efficacy in the treatment of myocardial infarction [164]. Additionally, exosomal enriched membrane protein (Lamp2b) fused with ischemic myocardium-targeting peptide CSTSMLKAC (IMTP) had an efficient anti-apoptotic effect upon *in vitro* internalization by hypoxia-injured H9C2 cells, while IMTP-exosomes injected systematically into the mice experiencing myocardial infarction were able to target ischemic myocardium preferentially and to reduce cardiac inflammation and apoptosis within the ischemic area [165].

Recent technologies advances enabled researchers to obtain human cardiac muscle patches engineered from cardiomyocytes, smooth muscle cells, and endothelial cells differentiated from human induced pluripotent stem cells, that were successfully tested *in vitro* or transplanted in mice models of myocardial infarction. Experiments demonstrated that human cardiac muscle patches release exosomes that exert cardiomyocyte cytoprotective effects and stimulate the angiogenic activity in endothelial cells [166].

6 Exosomes in heart failure

6.1 Exosomes as biomarkers in heart failure

Many CADs progress up to chronic heart failure (HF), and the number of patients developing HF is increasing each year. A challenge in HF management is to identify a reliable approach for the prognosis of the disease. HF development is determined by complex processes of cardiac remodeling and vascular dysfunction [167]. Exosome analysis might represent a useful tool for evaluating the development of HF in patients and multiple studies reported their role in regulating this pathological condition [168, 169].

A novel paracrine regulator mediating intercellular communications is represented by exosomes. The cross-talk between cardiomyocytes and cardiac fibroblasts are mediated by cardiac fibroblast-derived microRNA-enriched exosomes, a process leading to cardiomyocyte hypertrophy and contributing to HF [170].

Using exosomes as biomarkers have some advantages, though there are still limitations in risk stratification among HF patients. A classical way to evaluate the extent of HF is The New York Heart Association (NYHA) Functional Classification, which remains arguably the most important prognostic method in routine clinical, though this is an assessment with limitations such as the challenge of consistently classifying patients in class II or III. Furthermore, the accuracy of the results might be significantly influenced if relying on patients' subjective statements instead of the objective condition [171].

The plasma endothelium-derived extracellular vesicles levels were found to be significantly increased in HF patients with NYHA functional class I or more. Analysis proved that high extracellular vesicles level was associated with a significantly increased probability of cardiovascular events [172].

6.1.1 *Proteins from exosomes in heart failure*

Several studies show that degeneration of endothelial function and vascular integrity, imbalanced angiogenesis, and inflammation critically contribute to the progression of HF. Vascular endothelial dysfunction its been shown to be closely related to the levels of circulating extracellular vesicles derived from endothelial cells [21]. Patients with chronic HF have high levels of extracellular vesicles CD144+/CD31+/AnnexinV+ and CD31+/AnnexinV+ levels. A good clinical prediction for HF may be obtained by the detection of endothelial extracellular vesicles (i.e., CD144+/CD31+/AnnexinV+ and CD31+/AnnexinV+ extracellular vesicles) with the NYHA Functional Classification [169]. Other studies on patients with chronic HF reported a correlation between mortality and an elevated ratio of CD31+/Annexin V+ endothelial-derived apoptotic extracellular vesicles to mononuclear progenitor cells [173]. The detection of these extracellular vesicles HF patients may be considered as a good potential biomarker associated with a high risk of developing this pathology [21].

The European Society of Cardiology and the American Heart Association/American Colleges of Cardiology recommend natriuretic peptides, pentraxin-3, galectin-3, and cardiac-specific troponin as HF biomarkers with high predictive value [174].

One of the most studied and validated biomarkers used in chronic HF is the B-type natriuretic peptide, that is employed for the assessment of severity and diagnosis of the pathology. However, its prognostic value is under debate due to the fact that some clinical trials reported the lack of correlation between the levels of B-type natriuretic peptide and the death rate or the clinical outcome [175].

Heat shock protein 20 (Hsp20) is responsive to both acute and chronic progress of hyperglycemia in a mouse heart, assuming that low levels of Hsp20 could affect the evolution of diabetic cardiomyopathy and the propagation towards HF [176].

The circulating exosomes containing AT1R, delivered from cardiomyocytes over pressure overload, may play important roles in regulating the blood pressure in deleterious circumstances such as hypertension and HF [21].

SCL8A1 is involved in $Ca2+$ homeostasis in both mitochondria and cardiomyocytes and has been implicated in the development of arrhythmias and HF [177].

Guescini *et al.* showed that both of the plasma exosome particles numbers and exosomal mitochondrial DNA (mtDNA) copy number were increased in chronic HF patients. Exosomes originating from astrocytes, glioblastoma cells, and C2C12 cells had been found to contain mtDNA [178, 179].

Plasma-derived exosomes from HF patients and healthy controls exposed to THP-1 cells or Raji were incorporated by cells. Subsequently, the internalized exosomes induce the secretion of IL-1 and IL-8 and increase the expression of proteins in the TLR9-NF-κB (toll-like receptor 9-nuclear factor-κB) pathway. The increasing of the inflammatory cytokines is very related to mtDNA copy number, and not precisely related to the source of exosomal mtDNA, patients or control. Chloroquine, an inhibitor of TLR9, could block the TLR9-NF-κB pathway and reduce the expression of the proteins in this pathway. A recent study reported that THP-1 cells expressed TLR2 and TLR4, a fact related the exosome-induced inflammation. Excluding the TLR2/4 pathway, exosomes could also activate TLR9-NF-κB pathway through mtDNA and contribute to inflammation. In chronic HF patients, with increased plasma exosome numbers, more exosomal mtDNA enters the target immune cells and promotes the secretion of cytokines by activating TLR9-NF-κB pathway, which contributes to the process of chronic inflammation observed in HF patients [180].

Peripartum cardiomyopathy (PPCM) HF related has a high mortality risk, since lacks of specific diagnosis biomarkers [181]. Though specific biomarkers of PPCM are not discovered yet, routine blood work is still useful for the early diagnosis of PPCM. Inflammatory markers as IL-6, TNF-α, and CRP are highly expressed. Marker of apoptosis, Fas/Apo-1 is also elevated. Brain natriuretic peptide (BNP) and N-terminal portion of proBNP (pro-NT BNP) are sensitive biomarkers to HF but not specific for PPCM. High level of pro-NT BNP is associated with worse prognosis [182]. Additionally, evidence suggests that specific exosomes-associated miRNAs are differentially regulated in the failing heart [183, 184], suggesting their potential complementary role in the diagnosis of HF [21].

6.1.2 miRNAs from exosomes in heart failure

Previous reports have indicated miRNAs as biomarkers in multiple human cardiovascular diseases [185]. A number of dysregulated miRNAs are recurrent in the HF literature and confirmed to be altered in various forms of diseased and damaged human cardiac tissue [186].

Clinical studies have shown that microRNA-27a, one of the predicted microRNAs targeting nuclear factor erythroid 2-related factor 2 (Nrf2), was significantly upregulated in the human HF [187], and suggested that microRNAs may be involved in the dysregulation of Nrf2 during the onset and progression of HF. Recent studies have demonstrated that microRNAs can be actively transported by either binding to RNA-binding proteins resistant to nuclease or entrapped in exosomes to be implicated in signaling and intercellular communication [188, 189]. MicroRNA-enriched exosome-mediated intercellular communication between cardiac fibroblasts and cardiomyocytes contributes to the dysregulation of the Nrf2/ARE signaling pathway. The studies demonstrated that microRNA potentially targeting Nrf2 mRNA were highly expressed in the post-myocardial infarction (post-MI) heart and abundant in exosomes released from cardiac fibroblasts, which may have the potential to communicate with cardiomyocytes. This mechanism may contribute to the inhibition of Nrf2 expression induced by myocardial infarction and oxidative stress in chronic HF. It is still unclear if Nrf2-related microRNAs participate in the dysregulation of Nrf2 in the infarcted heart and HF [190].

Some studies suggested that exosomes based circulating p53-responsive miRNAs (miRNA-34a, miRNA-192, and miRNA-194) are highly related to the development of HF [191], increase significantly in the HF patients after acute AMI development [184], and may be regarded as predictors of ischemic HF that develops after AMI. Interestingly, the three miRNAs were determined to be abundant in the exosome fraction in HF patients [192]. Knockdown of the miRNAs mentioned above, enhanced cell viability after doxorubicin treatment, while overexpression of these miRNAs decreased cell survival *in vitro*, suggesting that higher level of these miRNAs may cause heart malfunction [21]. Remarkably, the miR-34a and miR-194 levels were related to left ventricle end-diastolic dimension after MI, indicating that the two might be useful in diagnosing HF development in MI patients [193].

Several studies proved that circulating miRNA-92 is in close relation with cardiovascular diseases [182, 194], and its concentration in the serum, was determined to be in higher level in the exosomal fraction of the HF patients compared to control [Goren *et al.* 2012]. A significantly higher plasma miRNA expression level has miR-210, in patients with NYHA class III and IV HF than in those with class II HF and controls [195].

MiRNA-146a is reported to be a contingent tool for diagnosing PPCM associated HF. A study carried out by Halkein *et al.* showed that patients with acute PPCM have significantly higher levels of circulating miRNA-146a, meanwhile, in patients with acute PPCM after following the usual therapy of HF, the level of circulating exosomal miRNA-146a was reduced, implying to consider miRNA-146a a promising biomarker for PPCM associated acute HF [181].

miR-1 [139, 196], miR-21 [197], miR-24, miR-29b [198], miR-133a and b [197], miR-199 [197], miR-208 [199, 200], miR-214 [197] and miR-499 [201] are consistently affected in human HF of varying origin and degree. This panel of miRNAs may serve as a preliminary template for the miRNA-mediated heart–skeletal muscle communication during HF [202]. MiR-21 is up-regulated with various degrees of diagnosed HF [203].

More experiments are needed to describe the miRNA profile of human cardiomyocyte-derived extracellular vesicles, but consistent with what is found in myocardial tissue and the circulation with HF, cardiomyocytes cultured from human progenitor cells robustly express miR-1 and miR-499 [30]. Interestingly, circulating miR-1 levels are diminished in direct relationship with the severity of HF in aged patients [204]. An interaction between aging and aging with HF appears to manifest in the expression of circulating miRNAs [202]. Of the miRNAs enriched in both cardiac muscle and the circulation during HF, myomiRs (namely miR-1, miR-133, miR-208, and miR-499 [205], are shown to affect developed skeletal muscle strongly. Of these myomiRs, miR-1 and miR-133a are elevated in the circulation within various forms of cardiovascular disease [140] and remain elevated for more than three months post-MI [206]. It was also reported that circulating myomiRs (including miR-1) are upregulated in the most advanced HF patients [207]. Inhibition of miR-1 and miR-133a are involved in mature skeletal muscle hypertrophy [208], probably through anti-repressive targeting of the insulin-like growth factor-1 (IGF-1) growth signaling axis [209, 210]. Thus, increased circulating miR-1 and miR-133a could facilitate skeletal muscle atrophy in the wake of a traumatic cardiovascular event. Interestingly, once HF is compensated with a left ventricular assist device, myomiR levels in the myocardium (specifically miR-1, miR-133a, and miR-133b) are decreased [211].

More recently, a panel of circulating miRNAs sensitive enough to differentiate between chronic HF with preserved versus reduced ejection fraction was identified and circulating miR-221, and miR-328 levels increased the discriminatory power of circulating B-type natriuretic peptide for assessing HF [212].

Several studies have identified markers akin to an acute exhaustive exercise in patients with chronic HF. As an example, one study showed that serum miR-21, miR-378, and miR-940 levels were significantly increased shortly after acute exercise [213].

In patients with relatively stable chronic HF, the levels of miR-548 family members (i.e., miR-548c, miR-548i) in peripheral blood mononuclear cells were decreased compared to controls [214], while levels of miR-22, miR-92b, miR-320 appeared to be increased [203].

In a post-infarct mouse HF model, a group investigated whether post-infarction exosomes treatment, synthesized by human embryonic stem cell-derived cardiovascular progenitors (hESC-Pg) can accord similar benefits to administered hESC-Pg. The data revealed that the exosomes were as effective as the stem cells in improving cardiac function, which was mediated by paracrine factors that stimulate cell division and viability [215].

In veterinary patients, changes in various miRNA concentrations have been observed in canine ventricular and atrial muscles after chronic HF evolution followed by experimental ventricular pacing, covering downregulation of cfa-miR-1, cfa-miR-26a, cfa-miR-26b, cfa-miR-29a, cfa-miR-30a, cfa-miR-133a, cfa- miR-133b, cfa-miR-208a and cfa-miR-218, and the upregulation of cfa-miR-21 and cfa-miR-146b [216]. Another study show that cfa-miR-9, cfa-miR-181c, cfa-miR-495, and cfa-miR-599 expressions differ with disease development and progression of HF, but also reveal modifications as a function of age [217].

6.2 Exosomes in the recovery after heart failure

Spironolactone and digoxin are involved in traditional medication for HF [193]. Angiotensin-converting enzyme inhibitors (ACEI) and β blockers have been recognized to be some of the fundamental medications for HF for many years. It is recommended, and it decreases mortality for any tolerated HF patients. Studies showed that, despite these benefits, numerous side effects on the fetus had been observed, as prematurity, bony deformity, oligohydramnios, limb contractures, intrauterine growth delay, neonatal death and pulmonary hypoplasia [218]. A better alternative is hydralazine which usually replaces ACEI. Treatment for dilated HF is similar to current treatment for PPCM [219].

Studies revealed that a specific molecular pattern in children with HF contributes to age-related variations in response to therapy, including multitudinous effects of phosphodiesterase treatment, adrenergic receptor system adjustment, the fibrotic gene profile, and miRNA expression level [220, 221].

Nrf2-related microRNAs could be considered as potential therapeutic targets in chronic HF or post-MI states [190].

A previous study showed that recovery of normal cardiac activity, improvement of coronary blood flow, an increase of expression levels of Sirtuin 1 and B-cell leukemia/lymphoma 2 after miRNA-195 inhibition, prove that decreasing miRNA-195 levels can delay the establishment or development of diabetic cardiomyopathy towards complete HF [222].

Several guidelines for the cardiac condition diagnosis and therapies for patients with HF have been published by European Society of Cardiology and the American Heart Association/American Colleges of Cardiology [223, 224].

7 Exosome in cardiac hypertrophy

7.1 Exosomes as biomarkers in cardiac hypertrophy

Cardiac hypertrophy is due to extensive cardiac stress and often is associated with cardiac remodeling. It should be emphasized that cardiac hypertrophy might contribute to heart failure. Analysis of the cardiac fibroblast-derived exosomes content revealed the abundance of several miRNA, e.g., miR-21_3p (miR-21), that are inducing cardiac hypertrophy. Additionally, proteome analysis indicated some miR-21 targets, including sorbin, SH3 domain-containing protein 2 (SORBS2), PDZ and LIM domain 5 (PDLIM5), and cardiomyocyte hypertrophy was associated with silencing SORBS2 or PDLIM5 [170].

Human and animal studies have demonstrated that oxidative stress is among the key factors contributing to cardiac hypertrophy [225]. To date, in patients affected by Friedreich's ataxia was described an association between oxidative stress and cardiac hypertrophy, and the administration of idebenone (i.e., a short chain analogue of coenzyme Q10) reduced the cardiac hypertrophy in approximately 50% of the patients [226]. Additionally, isoproterenol-induced cardiac hypertrophy in Wistar rats was demonstrated to be associated with an increase of lipid peroxides in plasma and myocardial tissue, while fosfobion treatment had a protective effect against free radicals production [227].

Another biomarker in cardiac hypertrophy is the posttranslational alteration of SERCA2a, the cardiac isoform of sarcoplasmic reticulum calcium ATPase pump, that can be diminished by the small ubiquitin-like modifier type 1 (SUMO-1) [228].

In rats with cardiac hypertrophy induced by right renal artery, the ligature was demonstrated that cardiomyocyte-derived exosomes containing heat shock protein 90 and interleukin-6 migrate to cardiac fibroblasts, thus contributing to the biphasic activation of signal transducer and activator of transcription 3 (STAT-3) signaling in cardiac fibroblasts that triggers an increased collagen synthesis [229].

7.2 Exosomes in the recovery after cardiac hypertrophy

Cardiac hypertrophy due to myocardial infarction was demonstrated to be reduced by the capture of exosomal miRNA-133a in normal cells surrounding the infarcted areas [139]. Stimulating oxidative stress might also reduce cardiac hypertrophy. Thus, oxidative stress induced in cardiac progenitor cells by stimulation with H_2O_2, determines the release of exosomes containing high levels of miRNA-21, that finally prevent cardiomyocytes apoptosis [146].

It has been proposed as a therapeutic approach in cardiac hypertrophy the targeting of angiotensin II-induced exosome release from cardiac fibroblasts, by the activation of angiotensin II receptor types 1 and 2 [230].

In vitro delivery of miR-200a in adipocyte-derived exosomes to cardiomyocytes, induced their hypertrophy, while delivery of the antagomir to miR-200a reversed cardiomyocytes hypertrophy. Moreover, in vivo ablation of peroxisome proliferator-activated receptor gamma (PPAR-γ) reduced cardiac hypertrophy [231].

Overexpression of heat shock protein 20 (Hsp20) in transgenic mice reduced the streptozotocin-induced cardiac dysfunction and hypertrophy. In these transgenic mice, cardiomyocytes-derived exosomes had a high content of Hsp20, p-Akt, survivin, and superoxide dismutase 1 and ensured protection against in vivo streptozotocin-induced heart remodeling [232].

8 Exosome in cardiac arrhythmia

8.1 Exosomes as biomarkers in arrhythmia

The role played by exosomes in cardiac arrhythmia is mostly unknown. However, some studies indicated that exosomes rich in miR-1 and miR-133, commonly identified as being secreted by cardiomyocytes in patients affected coronary artery disease or in rat models of heart ischemia, are also playing an important role in cardiac arrhythmia, being able modulate action potential and cardiac conduction via the Ca^{2+}/calmodulin-dependent protein kinase II signaling [233].

Additionally, some exosomal miRs (e.g., miR-328) were suggested to have proarrhythmogenic properties [234, 235]. Studies indicated that ischemia was associated with miR-328 upregulation in exosomes and suggested that its proarrhythmogenic effects might be correlated with the targeting of L-type calcium channels [234]. It should be emphasized that the targeting of L-type calcium channels by exosomal miRs is very important in arrhythmia, as clinical and experimental atrial fibrillation was demonstrated to be linked with a decrease of L-type Ca^{2+} currents and the reduction of the action potential duration [236–238].

8.2 Exosomes in the recovery after arrhythmia

Some studies demonstrated the positive role of stem cells upon arrhythmia, while others proved the induction of arrhythmogenicity upon stem cells transplantation. The administration of adult human mesenchymal stem cells into the canine left ventricular myocardium normalized the pacemaking without rejection of the xenograft for six weeks [239]. Interestingly, incubation of neonatal rat cardiomyocytes with exosomes derived from human mesenchymal stem cells did not increase the action potential duration (APD). On the other hand, engraftment of human adult mesenchymal stem cells should be treated with caution due to the proarrhythmogenic induced effects, such as slowing the conduction or reducing the repolarization rate [240]. An integrated experimental and simulation study demonstrated that exosomes released by human mesenchymal stem cells affect the human cardiac contractility and arrhythmogenicity mainly by paracrine signaling, and to a lesser extent by heterocellular coupling [241].

9 Exosome in cerebrovascular diseases

9.1 Exosomes as biomarkers in cerebrovascular diseases

Exosomes content can be used to identify valuable biomarkers in cerebrovascular diseases (see Table 1).

In the serum of acute ischemic stroke patients were identified increased levels of exosomes containing brain-specific miR-9 and miR-124, that had a positive correlation with scores in the National Institutes of Health Stroke Scale, the infarct volumes and the IL-6 serum levels [242], being promising biomarkers for acute ischemic stroke. Moreover, in the serum of neonates affected by acute hypoxic–ischemic encephalopathy was detected the neuronal exosome synaptopodin as a promising biomarker, its levels being associated with short-term neurologic outcomes [243].

In patients with hypoxia associated pathologies (i.e., stroke, myocardial infarction, preeclampsia, tumor hypoxia, and others) were identified circulating exosomes as biomarkers associated with the state of the disease [249].

In glioblastoma multiforme patients, exosomes obtained from tumors indicate the hypoxic status of the U87MG glioma cells and their content is rich in hypoxia-regulated mRNAs and proteins, such as matrix metalloproteinases, IL-8, PDGFs, caveolin 1, and lysyl oxidase. Moreover, exosomes released by glioma cells modulate the activity of endothelial cells and thus contribute to tumor development [244].

TABLE 1 Exosomes as biomarkers in human cerebrovascular diseases.

Cerebrovascular disease	Type of exosomes	Biomarkers	Reference
Acute ischemic stroke	Serum exosomes	Brain-specific miR-9 and miR-124	[242]
Acute hypoxic-ischemic encephalopathy	Serum exosomes	Synaptopodin	[243]
Malignant brain tumor glioblastoma multiforme	Cell supernatant	Hypoxia regulated proteins	[244]
Alzheimer's disease Frontotemporal dementia	Plasma (astrocyte- and neuron-derived exosomes)	• β-site amyloid precursor protein-cleaving enzyme 1 (BACE-1) • γ-secretase • soluble Aβ42 • soluble amyloid precursor protein β (sAPPβ) • soluble amyloid precursor protein α (sAPPα), glial-derived neurotrophic factor (GDNF) • P-T181-tau • P-S396-tau	[80]
Parkinson's disease	Erythrocyte-derived exosomes	α-Synuclein	[245]
Traumatic brain injury	Brain derived extracellular vesicles Cerebrospinal fluid (brain-derived exosomes)	↓ miR-212 ↑ miR-21, miR-146, miR-7a, and miR-7b Nucleotide-binding and oligomerization domain-like receptor protein-1 inflammasome αII-spectrin breakdown products, GFAP, and its breakdown products, and ubiquitin carboxy-terminal hydrolase L1	[246] [247, 248]

In Alzheimer's disease, astrocyte-derived exosomes have a higher protein content (e.g. β-site amyloid precursor protein-cleaving enzyme 1, γ-secretase, soluble Aβ42, soluble amyloid precursor protein β, soluble amyloid precursor protein α, glial-derived neurotrophic factor, P-T181-tau, and P-S396-tau) than the neuronal-derived exosomes in the plasma of the patients [80].

The initiation/propagation of Parkinson's disease is influenced by the passage through the blood-brain barrier of the erythrocyte-derived exosomes containing α-synuclein [245].

In traumatic brain injury, several biomarkers can be identified in the body fluids exosomes, especially blood and cerebrospinal fluid. miRNAs in brain-derived exosomes have been shown to mediate traumatic brain injury, their expression being downregulated (miR-212) or upregulated (miR-21, miR-146, miR-7a, and miR-7b) [246]. In rat brain with induced traumatic brain injury was also demonstrated that phosphorylation of connexin 43 stimulated the release of brain-derived exosomes and the activation of ERK signaling is required [250]. In patients affected by spinal cord injury or by traumatic brain injury, components of the nucleotide-binding and oligomerization domain-like receptor protein-1 inflammasome were identified as a biomarker in exosomes in cerebrospinal fluid [247]. Additionally, an extensive proteomic study in traumatic brain injury patients indicated several biomarkers in the cerebrospinal spinal, including αII-spectrin breakdown products, GFAP, and its breakdown products, and ubiquitin carboxy-terminal hydrolase L1 [248].

9.2 Exosomes as cargo in the brain

Exosomes are essential for the intercellular communication, and the current knowledge is considering them as cargo for the transfer of exosomal protein and RNA from various sources (e.g., cells and body fluids) to the brain tissue. On the other hand, exosomes transfer was also evidenced in the opposite direction upon brain injury. Thus, it was shown that brain-derived exosomes could pass through the blood-brain barrier (BBB) (Fig. 3) and their presence was evidenced in peripheral blood or the cerebrospinal fluid [78, 251, 252]. *In vitro* studies on brain microvascular endothelial cells monolayers, demonstrated the passage of exosomes through BBB in stroke-like and TNF-α induced inflammation conditions but not in control conditions [253].

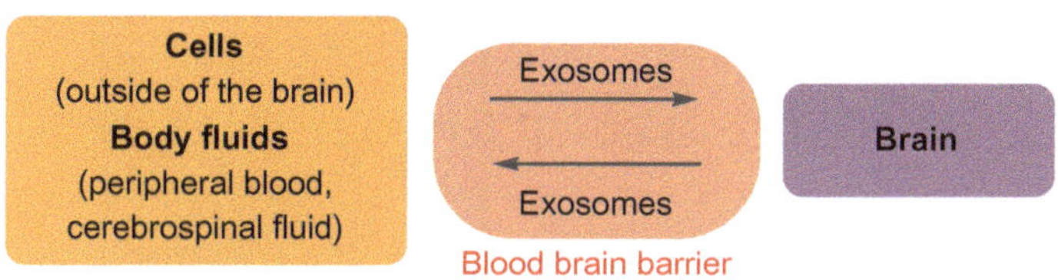

FIG. 3 Exosomes transfer through the blood-brain barrier.

Brain vascular integrity is regulated by neuronal miR-132-containing exosomes that are transferred to endothelial cells [254]. Moreover, MSCs-derived exosomes exert *in vitro* positive effects by promoting the axonal growth of cortical neurons [255].

Interestingly, in early human neurodevelopment, exosomes also serve as cargo from the maternal blood [50].

9.3 Exosomes in the treatment of cerebrovascular diseases

After a stroke, the intravenous administration of exosomes released from mesenchymal stem cells was tested in several animal models and could represent a translational treatment strategy for brain repair [79, 252, 256]. To date, systemic administration of MSCs-derived exosomes in adult male Wistar rats, previously subjected to 2h of middle cerebral artery occlusion, improved the functional recovery upon stroke, by stimulating neurite remodeling, neurogenesis, and angiogenesis [256].

Additionally, systemic administration of exosomes derived from pluripotent mesenchymal stromal cells or microglia was demonstrated to induce functional recovery upon traumatic brain injury (Fig. 4). To detail, transplantation of multipotent mesenchymal stromal cells into Wistar rats with induced traumatic brain injury improved their functional sensorimotor recovery, promoted angiogenesis and neurogenesis, and reduced inflammation [257]. Systemic administration of exosomes from human bone marrow-derived mesenchymal stem cells in rats after traumatic brain injury determined an increased generation of new cells (i.e., endothelial cells and neurons) in the lesion site, the reduction of neuroinflammation, and the improvement of the functional recovery [255]. Moreover, exosomes derived from stem cells of human exfoliated deciduous teeth were injected into rat models of traumatic brain injury, and these exosomes were able to reduce neuroinflammation by changing the polarization of microglia (i.e., preventing the M1 microglia phenotype and promoting the M2 microglia phenotype) and to recover the motor function [258]. In an *in vitro* model was demonstrated that miR-124-3p from microglia-derived exosomes was transferred to neurons, and exosomes inhibited the neuronal inflammation, promoted neurite outgrowth in brain extracts from mice with induced traumatic brain injury [165].

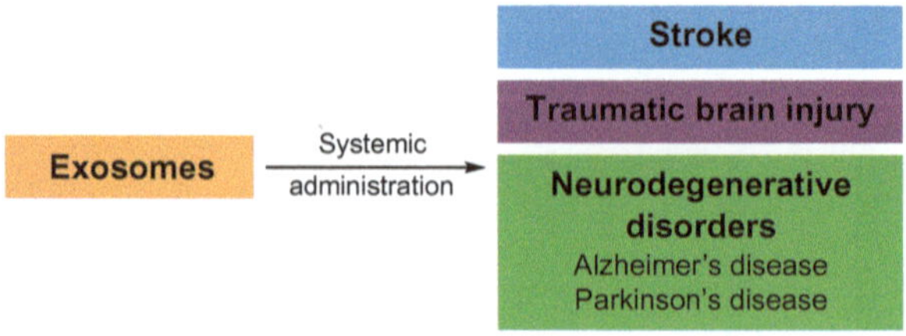

FIG. 4 Systemic administration of exosome for cerebrovascular conditions.

Neurodegenerative disorders are also considered as cerebrovascular diseases. In vivo studies on mice with induced Alzheimer's disease demonstrated that preventing exosomes secretion by the intraperitoneal injection with GW4869, an inhibitor of neutral sphingomyelinase 2, determined a reduction in the amyloid plaque formation [259]. Exosomes-based therapeutic approaches have also been tempted in Parkinson's disease. Thus, exosomes derived from dental pulp stem cells were demonstrated to prevent apoptosis in human dopaminergic neurons [260]. Exosomes loaded with catalase were uptaken *in vitro* by neurons and were able to exert in vivo neuroprotective effects upon intranasal administration in mice with induced Parkinson's disease [261].

10 Concluding observations

Exosomes are cargo vesicles of interest in the future because they reflect the pathophysiological state of their cells of origin. Donor cell exosomes contain RNA messages capable of influencing recipient cells found either locally or at long distance. It was even hypothesized that they might carry epigenetic modulators capable of influencing recipient cells. Also, exosomes could be used as biomarkers of evolution and progression of cardiovascular disease, but to the same extent as a means of treating or regenerating various degrees of cardiac damage. Messages carried by exosomes may be protective or destructive on the cardiovascular apparatus, depending on the cell from which they originate. Some of the transported miRNAs, protected from degradation by RNAses and proteinases, were suggested to be attractive biomarker candidates in the cardiovascular system, e.g., sera miR-192, miR-194, and miR-34a present in patients with AMI. These markers can predict the next year's evolution towards heart failure of ventricular remodeling. In the future, some other biomarkers will be discovered and exosomes will become more and more useful as biomarkers, diagnostic, and prognostic factors.

Although *in vitro* and in vivo studies have indicated positive effects of using stem cells cardiac recovery upon injury, the engraftment stem cells in cardiovascular pathologies should be considered with caution. Many studies are indicating that the results are mediated by paracrine actions. Nowadays exosomes are regarded as natural nanoparticles capable of being delivered to a specific target location. Depending on their origin, such as stem cells or cardiac progenitors, somatic cells or different body fluids, exosomes will play an important role in the processes of repair or regeneration in the cardiovascular system. Moreover, artificially engineered exosomes were produced by manipulating their cargo, with a specific destination for ischemic heart disease.

Prospects related to the use of exosomes as drug delivery systems are very promising, but many preclinical and clinical studies are needed before their use in practice.

Acknowledgements

This work was supported by the grants from National Natural Science Foundation of China (81722008, 91639101 and 81570362 to JJ Xiao), Innovation Program of Shanghai Municipal Education Commission (2017-01-07-00-09-E00042 to JJ Xiao), the grant from Science and Technology Commission of Shanghai Municipality (17010500100 to JJ Xiao), and the development fund for Shanghai talents (to JJ Xiao).

Assoc. Prof. BM Radu, Assist. Prof. AF Deftu and Research assistant AT Deftu were funded from Competitiveness Operational Programme 2014–2020 project P_37_675 (contract no. 146/2016), Priority Axis 1, Action 1.1.4, co-financed by the European Funds for Regional Development and Romanian Government funds, and the contents of this publication do not necessarily reflect the official position of the European Union or Romanian Government.

Competing financial interests

The authors declare no competing financial interests.

References

[1] Bruno S, Deregibus MC, Camussi G. The secretome of mesenchymal stromal cells: role of extracellular vesicles in immunomodulation. Immunol Lett 2015;168(2):154–8.

[2] Kalra H, Simpson RJ, Ji H, Aikawa E, Altevogt P, Askenase P, et al. Vesiclepedia: a compendium for extracellular vesicles with continuous community annotation. PLoS Biol 2012;10(12):e1001450.

[3] Camussi G, Deregibus MC, Bruno S, Cantaluppi V, Biancone L. Exosomes/microvesicles as a mechanism of cell-to-cell communication. Kidney Int 2010;78(9):838–48.

[4] Yanez-Mo M, Siljander PR, Andreu Z, Zavec AB, Borras FE, Buzas EI, et al. Biological properties of extracellular vesicles and their physiological functions. J Extracell vesicles 2015;4:27066.

[5] Keller S, Ridinger J, Rupp AK, Janssen JW, Altevogt P. Body fluid derived exosomes as a novel template for clinical diagnostics. J Transl Med 2011;9:86.

[6] Machida T, Tomofuji T, Ekuni D, Maruyama T, Yoneda T, Kawabata Y, et al. MicroRNAs in salivary exosome as potential biomarkers of aging. Int J Mol Sci 2015;16(9):21294–309.

[7] Beltrami C, Besnier M, Shantikumar S, Shearn AI, Rajakaruna C, Laftah A, et al. Human pericardial fluid contains exosomes enriched with cardiovascular-expressed microRNAs and promotes therapeutic angiogenesis. Mol ther 2017;25(3):679–93.

[8] Barcelo M, Mata A, Bassas L, Larriba S. Exosomal microRNAs in seminal plasma are markers of the origin of azoospermia and can predict the presence of sperm in testicular tissue. Hum Reprod 2018;33(6):1087–98.

[9] Evans-Osses I, Reichembach LH, Ramirez MI. Exosomes or microvesicles? Two kinds of extracellular vesicles with different routes to modify protozoan-host cell interaction. Parasitol Res 2015;114(10):3567–75.

[10] Cocucci E, Meldolesi J. Ectosomes and exosomes: shedding the confusion between extracellular vesicles. Trends Cell Biol 2015;25(6):364–72.

[11] Kalra H, Drummen GP, Mathivanan S. Focus on extracellular vesicles: introducing the next small big thing. Int J Mol Sci 2016;17(2):170.

[12] Thery C, Zitvogel L, Amigorena S. Exosomes: composition, biogenesis and function. Nat Rev Immunol 2002;2(8):569–79.

[13] Rink J, Ghigo E, Kalaidzidis Y, Zerial M. Rab conversion as a mechanism of progression from early to late endosomes. Cell 2005;122(5):735–49.

[14] Balaj L, Lessard R, Dai L, Cho YJ, Pomeroy SL, Breakefield XO, et al. Tumour microvesicles contain retrotransposon elements and amplified oncogene sequences. Nat Commun 2011;2:180.

[15] Buzas EI, Gyorgy B, Nagy G, Falus A, Gay S. Emerging role of extracellular vesicles in inflammatory diseases. Nat Rev Rheumatol 2014;10(6):356–64.

[16] Abels ER, Breakefield XO. Introduction to extracellular vesicles: biogenesis, rna cargo selection, content, release, and uptake. Cell Mol Neurobiol 2016;36(3):301–12.

[17] Henderson MC, Azorsa DO. The genomic and proteomic content of cancer cell-derived exosomes. Front Oncol 2012;2:38.

[18] Gangoda L, Boukouris S, Liem M, Kalra H, Mathivanan S. Extracellular vesicles including exosomes are mediators of signal transduction: are they protective or pathogenic? Proteomics 2015;15(2-3):260–71.

[19] Raimondo F, Morosi L, Chinello C, Magni F, Pitto M. Advances in membranous vesicle and exosome proteomics improving biological understanding and biomarker discovery. Proteomics 2011;11(4):709–20.

[20] Cosme J, Liu PP, Gramolini AO. The cardiovascular exosome: current perspectives and potential. Proteomics 2013;13(10-11):1654–9.

[21] Bei Y, Yu P, Cretoiu D, Cretoiu SM, Xiao J. Exosomes-based biomarkers for the prognosis of cardiovascular diseases. Adv Exp Med Biol 2017;998:71–88.

[22] Vanhaverbeke M, Gal D, Holvoet P. Functional role of cardiovascular exosomes in myocardial injury and atherosclerosis. Exosomes in cardiovascular diseases advances in experimental medicine and biology. In: Exosomes in Cardiovascular Diseases. vol. 998. Singapore: Springer; 2017. p. 45–58.

[23] Todorova D, Simoncini S, Lacroix R, Sabatier F, Dignat-George F. Extracellular vesicles in angiogenesis. Circ Res 2017;120(10):1658–73.

[24] Ailawadi S, Wang X, Gu H, Fan GC. Pathologic function and therapeutic potential of exosomes in cardiovascular disease. Biochim Biophys Acta 2015;1852(1):1–11.

[25] Chistiakov DA, Orekhov AN, Bobryshev YV. Cardiac extracellular vesicles in normal and infarcted heart. Int J Mol Sci 2016;17(1):63–81.

[26] Garcia NA, Moncayo-Arlandi J, Sepulveda P, Diez-Juan A. Cardiomyocyte exosomes regulate glycolytic flux in endothelium by direct transfer of GLUT transporters and glycolytic enzymes. Cardiovasc Res 2016;109(3):397–408.

[27] Garcia NA, Ontoria-Oviedo I, Gonzalez-King H, Diez-Juan A, Sepulveda P. Glucose starvation in cardiomyocytes enhances exosome secretion and promotes angiogenesis in endothelial cells. PLoS One 2015;10(9):e0138849.

[28] Kakkar R, Lee RT. Intramyocardial fibroblast myocyte communication. Circ Res 2010;106(1):47–57.

[29] Cervio E, Barile L, Moccetti T, Vassalli G. Exosomes for intramyocardial intercellular communication. Stem Cells Int 2015;2015:482171.

[30] Vrijsen KR, Maring JA, Chamuleau SA, Verhage V, Mol EA, Deddens JC, et al. Exosomes from cardiomyocyte progenitor cells and mesenchymal stem cells stimulate angiogenesis via EMMPRIN. Adv Healthc Mater 2016;5(19):2555–65.

[31] Deregibus MC, Cantaluppi V, Calogero R, Lo Iacono M, Tetta C, Biancone L, et al. Endothelial progenitor cell derived microvesicles activate an angiogenic program in endothelial cells by a horizontal transfer of mRNA. Blood 2007;110(7):2440–8.

[32] Cantaluppi V, Biancone L, Figliolini F, Beltramo S, Medica D, Deregibus MC, et al. Microvesicles derived from endothelial progenitor cells enhance neoangiogenesis of human pancreatic islets. Cell Transplant 2012;21(6):1305–20.

[33] Tsimerman G, Roguin A, Bachar A, Melamed E, Brenner B, Aharon A. Involvement of microparticles in diabetic vascular complications. Thromb Haemost 2011;106(2):310–21.

[34] Sabatier F, Darmon P, Hugel B, Combes V, Sanmarco M, Velut JG, et al. Type 1 and type 2 diabetic patients display different patterns of cellular microparticles. Diabetes 2002;51(9):2840–5.

[35] Helal O, Defoort C, Robert S, Marin C, Lesavre N, Lopez-Miranda J, et al. Increased levels of microparticles originating from endothelial cells, platelets and erythrocytes in subjects with metabolic syndrome: relationship with oxidative stress. Nutr Metab Cardiovasc Dis 2011;21(9):665–71.

[36] Esposito K, Ciotola M, Giugliano D. Pioglitazone reduces endothelial microparticles in the metabolic syndrome. Arterioscler Thromb Vasc Biol 2006;26(8):1926.

[37] Mobarrez F, Egberg N, Antovic J, Broijersen A, Jorneskog G, Wallen H. Release of endothelial microparticles in vivo during atorvastatin treatment; a randomized double-blind placebo-controlled study. Thromb Res 2012;129(1):95–7.

[38] Lother A, Bergemann S, Deng L, Moser M, Bode C, Hein L. Cardiac endothelial cell transcriptome. Arterioscler Thromb Vasc Biol 2018;38(3):566–74.

[39] Yu B, Zhang X, Li X. Exosomes derived from mesenchymal stem cells. Int J Mol Sci 2014;15(3):4142–57.

[40] Chen W, Huang Y, Han J, Yu L, Li Y, Lu Z, et al. Immunomodulatory effects of mesenchymal stromal cells-derived exosome. Immunol Res 2016;64(4):831–40.

[41] Yeo RW, Lai RC, Zhang B, Tan SS, Yin Y, Teh BJ, et al. Mesenchymal stem cell: an efficient mass producer of exosomes for drug delivery. Adv Drug Deliv Rev 2013;65(3):336–41.

[42] Heijnen HF, Schiel AE, Fijnheer R, Geuze HJ, Sixma JJ. Activated platelets release two types of membrane vesicles: microvesicles by surface shedding and exosomes derived from exocytosis of multivesicular bodies and alpha-granules. Blood 1999;94(11):3791–9.

[43] Landry P, Plante I, Ouellet DL, Perron MP, Rousseau G, Provost P. Existence of a microRNA pathway in anucleate platelets. Nat Struct Mol Biol 2009;16(9):961–6.

[44] Rowley JW, Oler AJ, Tolley ND, Hunter BN, Low EN, Nix DA, et al. Genome-wide RNA-seq analysis of human and mouse platelet transcriptomes. Blood 2011;118(14):e101–11.

[45] Bray PF, McKenzie SE, Edelstein LC, Nagalla S, Delgrosso K, Ertel A, et al. The complex transcriptional landscape of the anucleate human platelet. BMC Genomics 2013;14:1.

[46] Schubert S, Weyrich AS, Rowley JW. A tour through the transcriptional landscape of platelets. Blood 2014;124(4):493–502.

[47] Rondina MT, Weyrich AS. Regulation of the genetic code in megakaryocytes and platelets. J Thromb Haemost 2015;13(Suppl 1):S26–32.

[48] Alhasan AA, Izuogu OG, Al-Balool HH, Steyn JS, Evans A, Colzani M, et al. Circular RNA enrichment in platelets is a signature of transcriptome degradation. Blood 2016;127(9):e1–11.

[49] Ple H, Landry P, Benham A, Coarfa C, Gunaratne PH, Provost P. The repertoire and features of human platelet microRNAs. PLoS One 2012;7(12):e50746.

[50] Goetzl L, Darbinian N, Goetzl EJ. Novel window on early human neurodevelopment via fetal exosomes in maternal blood. Ann Clin Transl Neurol 2016;3(5):381–5.

[51] Azevedo LC, Janiszewski M, Pontieri V, Pedro Mde A, Bassi E, Tucci PJ, et al. Platelet-derived exosomes from septic shock patients induce myocardial dysfunction. Crit Care 2007;11(6):R120.

[52] Heijnen H, van der Sluijs P. Platelet secretory behaviour: as diverse as the granules... or not? J Thromb Haemost 2015;13(12):2141–51.

[53] Srikanthan S, Li W, Silverstein RL, McIntyre TM. Exosome poly-ubiquitin inhibits platelet activation, downregulates CD36 and inhibits pro-atherothombotic cellular functions. J Thromb Haemost 2014;12(11):1906–17.

[54] Gambim MH, do Carmo Ade O, Marti L, Verissimo-Filho S, Lopes LR, Janiszewski M. Platelet-derived exosomes induce endothelial cell apoptosis through peroxynitrite generation: experimental evidence for a novel mechanism of septic vascular dysfunction. Crit Care 2007;11(5):R107.

[55] Angelillo-Scherrer A. Leukocyte-derived microparticles in vascular homeostasis. Circ Res 2012;110(2):356–69.

[56] Aharon A, Tamari T, Brenner B. Monocyte-derived microparticles and exosomes induce procoagulant and apoptotic effects on endothelial cells. Thromb Haemost 2008;100(5):878–85.

[57] Niu C, Wang X, Zhao M, Cai T, Liu P, Li J, et al. Macrophage foam cell-derived extracellular vesicles promote vascular smooth muscle cell migration and adhesion. J Am Heart Assoc 2016;5(10). pii: e004099.

[58] Kuo WP, Tigges JC, Toxavidis V, Ghiran I. Red blood cells: a source of extracellular vesicles. Methods Mol Biol 2017;1660:15–22.

[59] Danesh A, Inglis HC, Jackman RP, Wu S, Deng X, Muench MO, et al. Exosomes from red blood cell units bind to monocytes and induce proinflammatory cytokines, boosting T-cell responses in vitro. Blood 2014;123(5):687–96.

[60] Straat M, Boing AN, Tuip-De Boer A, Nieuwland R, Juffermans NP. Extracellular vesicles from red blood cell products induce a strong pro-inflammatory host response, dependent on both numbers and storage duration. Transfus Med Hemother 2016;43(4):302–5.

[61] Azarov I, Liu C, Reynolds H, Tsekouras Z, Lee JS, Gladwin MT, et al. Mechanisms of slower nitric oxide uptake by red blood cells and other hemoglobin-containing vesicles. J Biol Chem 2011;286(38):33567–79.

[62] Said AS, Doctor A. Influence of red blood cell-derived microparticles upon vasoregulation. Blood transfusion = Trasfusione del sangue 2017;15(6):522–34.

[63] Bevers EM, Comfurius P, Dekkers DW, Zwaal RF. Lipid translocation across the plasma membrane of mammalian cells. Biochim Biophys Acta 1999;1439(3):317–30.

[64] Burger P, Kostova E, Bloem E, Hilarius-Stokman P, Meijer AB, van den Berg TK, et al. Potassium leakage primes stored erythrocytes for phosphatidylserine exposure and shedding of pro-coagulant vesicles. Br J Haematol 2013;160(3):377–86.

[65] Kostova EB, Beuger BM, Klei TR, Halonen P, Lieftink C, Beijersbergen R, et al. Identification of signalling cascades involved in red blood cell shrinkage and vesiculation. Biosci Rep 2015;35(2). pii: e00187.

[66] Regev-Rudzki N, Wilson DW, Carvalho TG, Sisquella X, Coleman BM, Rug M, et al. Cell-cell communication between malaria-infected red blood cells via exosome-like vesicles. Cell 2013;153(5):1120–33.

[67] Allan D, Limbrick AR, Thomas P, Westerman MP. Release of spectrin-free spicules on reoxygenation of sickled erythrocytes. Nature 1982;295(5850):612–3.

[68] Westerman MP, Cole ER, Wu K. The effect of spicules obtained from sickle red cells on clotting activity. Br J Haematol 1984;56(4):557–62.

[69] Alkhatatbeh MJ, Enjeti AK, Acharya S, Thorne RF, Lincz LF. The origin of circulating CD36 in type 2 diabetes. Nutrition & diabetes 2013;3:e59.

[70] Agouni A, Lagrue-Lak-Hal AH, Ducluzeau PH, Mostefai HA, Draunet-Busson C, Leftheriotis G, et al. Endothelial dysfunction caused by circulating microparticles from patients with metabolic syndrome. Am J Pathol 2008;173(4):1210–9.

[71] Boyer M, Baggett A, Scalia R, Eguchi S, Rizzo V. Characterization of exosomes isolated from cultured vascular endothelial and smooth muscle cells. FASEB J 2017;31(1_supplement). 837.7-.7.

[72] Qiu H, Shi S, Wang S, Peng H, Ding SJ, Wang L. Proteomic profiling exosomes from vascular smooth muscle cell. Proteomics Clin Appl 2018;12(5):e1700097.

[73] Comelli L, Rocchiccioli S, Smirni S, Salvetti A, Signore G, Citti L, et al. Characterization of secreted vesicles from vascular smooth muscle cells. Mol BioSyst 2014;10(5):1146–52.

[74] Kapustin AN, Chatrou ML, Drozdov I, Zheng Y, Davidson SM, Soong D, et al. Vascular smooth muscle cell calcification is mediated by regulated exosome secretion. Circ Res 2015;116(8):1312–23.

[75] Kapustin AN, Shanahan CM. Emerging roles for vascular smooth muscle cell exosomes in calcification and coagulation. J Physiol 2016;594(11):2905–14.

[76] Zheng B, Yin WN, Suzuki T, Zhang XH, Zhang Y, Song LL, et al. Exosome-mediated miR-155 transfer from smooth muscle cells to endothelial cells induces endothelial injury and promotes atherosclerosis. Mol ther 2017;25(6):1279–94.

[77] Faure J, Lachenal G, Court M, Hirrlinger J, Chatellard-Causse C, Blot B, et al. Exosomes are released by cultured cortical neurones. Mol Cell Neurosci 2006;31(4):642–8.

[78] Fruhbeis C, Frohlich D, Kuo WP, Kramer-Albers EM. Extracellular vesicles as mediators of neuron-glia communication. Front Cell Neurosci 2013;7:182.

[79] Zhang ZG, Chopp M. Exosomes in stroke pathogenesis and therapy. J Clin Invest 2016;126(4):1190–7.

[80] Goetzl EJ, Mustapic M, Kapogiannis D, Eitan E, Lobach IV, Goetzl L, et al. Cargo proteins of plasma astrocyte-derived exosomes in Alzheimer's disease. FASEB J 2016;30(11):3853–9.

[81] Skog J, Wurdinger T, van Rijn S, Meijer DH, Gainche L, Sena-Esteves M, et al. Glioblastoma microvesicles transport RNA and proteins that promote tumour growth and provide diagnostic biomarkers. Nat Cell Biol 2008;10(12):1470–6.

[82] Haqqani AS, Delaney CE, Tremblay TL, Sodja C, Sandhu JK, Stanimirovic DB. Method for isolation and molecular characterization of extracellular microvesicles released from brain endothelial cells. Fluids Barriers CNS 2013;10(1):4.

[83] Yamamoto S, Niida S, Azuma E, Yanagibashi T, Muramatsu M, Huang TT, et al. Inflammation-induced endothelial cell-derived extracellular vesicles modulate the cellular status of pericytes. Sci Rep 2015;5:8505.

[84] Hooper C, Sainz-Fuertes R, Lynham S, Hye A, Killick R, Warley A, et al. Wnt3a induces exosome secretion from primary cultured rat microglia. BMC Neurosci 2012;13:144.

[85] Bahrini I, Song JH, Diez D, Hanayama R. Neuronal exosomes facilitate synaptic pruning by up-regulating complement factors in microglia. Sci Rep 2015;5:7989.

[86] Sun L, Wang X, Zhou Y, Zhou RH, Ho WZ, Li JL. Exosomes contribute to the transmission of anti-HIV activity from TLR3-activated brain microvascular endothelial cells to macrophages. Antivir Res 2016;134:167–71.

[87] Giusti I, Delle Monache S, Di Francesco M, Sanita P, D'Ascenzo S, Gravina GL, et al. From glioblastoma to endothelial cells through extracellular vesicles: messages for angiogenesis. Tumour Biol 2016;37(9):12743–53.

[88] Guitart K, Loers G, Buck F, Bork U, Schachner M, Kleene R. Improvement of neuronal cell survival by astrocyte-derived exosomes under hypoxic and ischemic conditions depends on prion protein. Glia 2016;64(6):896–910.
[89] Goncalves MB, Malmqvist T, Clarke E, Hubens CJ, Grist J, Hobbs C, et al. Neuronal RAR-beta signaling modulates PTEN activity directly in neurons and via exosome transfer in astrocytes to prevent glial scar formation and induce spinal cord regeneration. J Neurosci 2015;35(47):15731–45.
[90] Chernyshev VS, Rachamadugu R, Tseng YH, Belnap DM, Jia Y, Branch KJ, et al. Size and shape characterization of hydrated and desiccated exosomes. Anal Bioanal Chem 2015;407(12):3285–301.
[91] Ferguson SW, Wang J, Lee CJ, Liu M, Neelamegham S, Canty JM, et al. The microRNA regulatory landscape of MSC-derived exosomes: a systems view. Sci Rep 2018;8(1):1419.
[92] Tauro BJ, Greening DW, Mathias RA, Ji H, Mathivanan S, Scott AM, et al. Comparison of ultracentrifugation, density gradient separation, and immunoaffinity capture methods for isolating human colon cancer cell line LIM1863-derived exosomes. Methods 2012;56(2):293–304.
[93] Alvarez ML, Khosroheidari M, Kanchi Ravi R, DiStefano JK. Comparison of protein, microRNA, and mRNA yields using different methods of urinary exosome isolation for the discovery of kidney disease biomarkers. Kidney Int 2012;82(9):1024–32.
[94] Rekker K, Saare M, Roost AM, Kubo AL, Zarovni N, Chiesi A, et al. Comparison of serum exosome isolation methods for microRNA profiling. Clin Biochem 2014;47(1-2):135–8.
[95] Rautou PE, Mackman N. Microvesicles as risk markers for venous thrombosis. Expert Rev Hematol 2013;6(1):91–101.
[96] Gustafson CM, Shepherd AJ, Miller VM, Jayachandran M. Age- and sex-specific differences in blood-borne microvesicles from apparently healthy humans. Biol Sex Differ 2015;6:10.
[97] Malik ZA, Liu TT, Knowlton AA. Cardiac myocyte exosome isolation. Methods Mol Biol 2016;1448:237–48.
[98] Chen C, Skog J, Hsu CH, Lessard RT, Balaj L, Wurdinger T, et al. Microfluidic isolation and transcriptome analysis of serum microvesicles. Lab Chip 2010;10(4):505–11.
[99] Szczepanski MJ, Szajnik M, Welsh A, Whiteside TL, Boyiadzis M. Blast-derived microvesicles in sera from patients with acute myeloid leukemia suppress natural killer cell function via membrane-associated transforming growth factor-beta1. Haematologica 2011;96(9):1302–9.
[100] Taylor DD, Zacharias W, Gercel-Taylor C. Exosome isolation for proteomic analyses and RNA profiling. Methods Mol Biol 2011;728:235–46.
[101] Hong CS, Muller L, Whiteside TL, Boyiadzis M. Plasma exosomes as markers of therapeutic response in patients with acute myeloid leukemia. Front Immunol 2014;5:160.
[102] Saenz-Cuesta M, Arbelaiz A, Oregi A, Irizar H, Osorio-Querejeta I, Munoz-Culla M, et al. Methods for extracellular vesicles isolation in a hospital setting. Front Immunol 2015;6:50.
[103] Levy E. Exosomes in the diseased brain: first insights from in vivo studies. Front Neurosci 2017;11:142.
[104] van der Meel R, Krawczyk-Durka M, van Solinge WW, Schiffelers RM. Toward routine detection of extracellular vesicles in clinical samples. Int J Lab Hematol 2014;36(3):244–53.
[105] Malik ZA, Kott KS, Poe AJ, Kuo T, Chen L, Ferrara KW, et al. Cardiac myocyte exosomes: stability, HSP60, and proteomics. Am J Physiol Heart Circ Physiol 2013;304(7):H954–65.
[106] van Weering JR, Brown E, Sharp TH, Mantell J, Cullen PJ, Verkade P. Intracellular membrane traffic at high resolution. Methods Cell Biol 2010;96:619–48.
[107] Admyre C, Johansson SM, Qazi KR, Filen JJ, Lahesmaa R, Norman M, et al. Exosomes with immune modulatory features are present in human breast milk. J Immunol 2007;179(3):1969–78.

[108] Arraud N, Linares R, Tan S, Gounou C, Pasquet JM, Mornet S, et al. Extracellular vesicles from blood plasma: determination of their morphology, size, phenotype and concentration. J Thromb Haemost 2014;12(5):614–27.

[109] Wubbolts R, Leckie RS, Veenhuizen PT, Schwarzmann G, Mobius W, Hoernschemeyer J, et al. Proteomic and biochemical analyses of human B cell-derived exosomes. Potential implications for their function and multivesicular body formation. J Biol Chem 2003;278(13):10963–72.

[110] van der Pol E, Hoekstra AG, Sturk A, Otto C, van Leeuwen TG, Nieuwland R. Optical and non-optical methods for detection and characterization of microparticles and exosomes. J Thromb Haemost 2010;8(12):2596–607.

[111] Ashcroft BA, de Sonneville J, Yuana Y, Osanto S, Bertina R, Kuil ME, et al. Determination of the size distribution of blood microparticles directly in plasma using atomic force microscopy and microfluidics. Biomed Microdevices 2012;14(4):641–9.

[112] Davidson SM, Takov K, Yellon DM. Exosomes and cardiovascular protection. Cardiovasc Drugs Ther 2017;31(1):77–86.

[113] Almizraq RJ, Seghatchian J, Holovati JL, Acker JP. Extracellular vesicle characteristics in stored red blood cell concentrates are influenced by the method of detection. Transfus Apher Sci 2017;56(2):254–60.

[114]. Leroyer AS, Isobe H, Leseche G, Castier Y, Wassef M, Mallat Z, et al. Cellular origins and thrombogenic activity of microparticles isolated from human atherosclerotic plaques. J Am Coll Cardiol 2007;49(7):772-7.

[115] Jansen F, Li Q, Pfeifer A, Werner N. Endothelial- and immune cell-derived extracellular vesicles in the regulation of cardiovascular health and disease. JACC Basic to Transl Sci 2017;2(6):790–807.

[116] Yayan J. Emerging families of biomarkers for coronary artery disease: inflammatory mediators. Vasc Health Risk Manag 2013;9:435–56.

[117] Kosaka N, Iguchi H, Yoshioka Y, Takeshita F, Matsuki Y, Ochiya T. Secretory mechanisms and intercellular transfer of microRNAs in living cells. J Biol Chem 2010;285(23):17442–52.

[118] Cretoiu SM. Circulating microRNAs in cardiovascular diseases: recent progress and challenges. J Hypertens Res 2016;2(1):15–8.

[119] Fichtlscherer S, De Rosa S, Fox H, Schwietz T, Fischer A, Liebetrau C, et al. Circulating microRNAs in patients with coronary artery disease. Circ Res 2010;107(5):677–84.

[120] Bei Y, Tao L, Cretoiu D, Cretoiu SM, Xiao J. MicroRNAs mediate beneficial effects of exercise in heart. Adv Exp Med Biol 2017;1000:261–80.

[121] Karakas M, Schulte C, Appelbaum S, Ojeda F, Lackner KJ, Munzel T, et al. Circulating microRNAs strongly predict cardiovascular death in patients with coronary artery disease-results from the large AtheroGene study. Eur Heart J 2017;38(7):516–23.

[122] Zhong Z, Hou J, Zhang Q, Zhong W, Li B, Li C, et al. Circulating microRNA expression profiling and bioinformatics analysis of dysregulated microRNAs of patients with coronary artery disease. Medicine 2018;97(27):e11428.

[123] Hausenloy DJ, Garcia-Dorado D, Botker HE, Davidson SM, Downey J, Engel FB, et al. Novel targets and future strategies for acute cardioprotection: position paper of the european society of cardiology working group on cellular biology of the heart. Cardiovasc Res 2017;113(6):564–85.

[124] Davidson SM, Yellon DM. Exosomes and cardioprotection—a critical analysis. Mol Asp Med 2018;60:104–14.

[125] Marini M, Ibba-Manneschi L, Manetti M. Cardiac telocyte-derived exosomes and their possible implications in cardiovascular pathophysiology. Adv Exp Med Biol 2017;998:237–54.

[126] Cretoiu D, Radu BM, Banciu A, Banciu DD, Cretoiu SM. Telocytes heterogeneity: from cellular morphology to functional evidence. Semin Cell Dev Biol 2017;64:26–39.

[127] Yellon DM, Davidson SM. Exosomes: nanoparticles involved in cardioprotection? Circ Res 2014;114(2):325–32.

[128] Sahoo S, Klychko E, Thorne T, Misener S, Schultz KM, Millay M, et al. Exosomes from human CD34(+) stem cells mediate their proangiogenic paracrine activity. Circ Res 2011;109(7):724–8.

[129] Lai RC, Arslan F, Lee MM, Sze NS, Choo A, Chen TS, et al. Exosome secreted by MSC reduces myocardial ischemia/reperfusion injury. Stem Cell Res 2010;4(3):214–22.

[130] Arslan F, Lai RC, Smeets MB, Akeroyd L, Choo A, Aguor EN, et al. Mesenchymal stem cell-derived exosomes increase ATP levels, decrease oxidative stress and activate PI3K/Akt pathway to enhance myocardial viability and prevent adverse remodeling after myocardial ischemia/reperfusion injury. Stem Cell Res 2013;10(3):301–12.

[131] Bell R, Beeuwkes R, Botker HE, Davidson S, Downey J, Garcia-Dorado D, et al. Trials, tribulations and speculation! Report from the 7th biennial hatter cardiovascular institute workshop. Basic Res Cardiol 2012;107(6):300.

[132] Ibrahim AG, Cheng K, Marban E. Exosomes as critical agents of cardiac regeneration triggered by cell therapy. Stem Cell Rep 2014;2(5):606–19.

[133] Xiao J, Cretoiu D, Lei Z, Das S, Li X. Genetic and epigenetic regulation networks: governing from cardiovascular development to remodeling. Biomed Res Int 2017;2017:4135956.

[134] Yu X, Deng L, Wang D, Li N, Chen X, Cheng X, et al. Mechanism of TNF-alpha autocrine effects in hypoxic cardiomyocytes: initiated by hypoxia inducible factor 1alpha, presented by exosomes. J Mol Cell Cardiol 2012;53(6):848–57.

[135] Gupta S, Knowlton AA. HSP60 trafficking in adult cardiac myocytes: role of the exosomal pathway. Am J Physiol Heart Circ Physiol 2007;292(6):H3052–6.

[136] Emanueli C, Shearn AI, Laftah A, Fiorentino F, Reeves BC, Beltrami C, et al. Coronary artery-bypass-graft surgery increases the plasma concentration of exosomes carrying a cargo of cardiac MicroRNAs: an example of exosome trafficking out of the human heart with potential for cardiac biomarker discovery. PLoS One 2016;11(4):e0154274.

[137] Meder B, Keller A, Vogel B, Haas J, Sedaghat-Hamedani F, Kayvanpour E, et al. MicroRNA signatures in total peripheral blood as novel biomarkers for acute myocardial infarction. Basic Res Cardiol 2011;106(1):13–23.

[138] Jaguszewski M, Osipova J, Ghadri JR, Napp LC, Widera C, Franke J, et al. A signature of circulating microRNAs differentiates takotsubo cardiomyopathy from acute myocardial infarction. Eur Heart J 2014;35(15):999–1006.

[139] Care A, Catalucci D, Felicetti F, Bonci D, Addario A, Gallo P, et al. MicroRNA-133 controls cardiac hypertrophy. Nat Med 2007;13(5):613–8.

[140] Kuwabara Y, Ono K, Horie T, Nishi H, Nagao K, Kinoshita M, et al. Increased microRNA-1 and microRNA-133a levels in serum of patients with cardiovascular disease indicate myocardial damage. Circ Cardiovasc Genet 2011;4(4):446–54.

[141] Corsten MF, Dennert R, Jochems S, Kuznetsova T, Devaux Y, Hofstra L, et al. Circulating microRNA-208b and MicroRNA-499 reflect myocardial damage in cardiovascular disease. Circ Cardiovasc Genet 2010;3(6):499–506.

[142] Oliveira-Carvalho V, Carvalho VO, Bocchi EA. The emerging role of miR-208a in the heart. DNA Cell Biol 2013;32(1):8–12.

[143] Cheng C, Wang Q, You W, Chen M, Xia J. MiRNAs as biomarkers of myocardial infarction: a meta-analysis. PLoS One 2014;9(2):e88566.

[144] Bei Y, Chen T, Banciu DD, Cretoiu D, Xiao J. Circulating exosomes in cardiovascular diseases. Adv Exp Med Biol 2017;998:255–69.
[145] Chen L, Wang Y, Pan Y, Zhang L, Shen C, Qin G, et al. Cardiac progenitor-derived exosomes protect ischemic myocardium from acute ischemia/reperfusion injury. Biochem Biophys Res Commun 2013;431(3):566–71.
[146] Xiao J, Pan Y, Li XH, Yang XY, Feng YL, Tan HH, et al. Cardiac progenitor cell-derived exosomes prevent cardiomyocytes apoptosis through exosomal miR-21 by targeting PDCD4. Cell Death Dis 2016;7(6):e2277.
[147] Barile L, Lionetti V, Cervio E, Matteucci M, Gherghiceanu M, Popescu LM, et al. Extracellular vesicles from human cardiac progenitor cells inhibit cardiomyocyte apoptosis and improve cardiac function after myocardial infarction. Cardiovasc Res 2014;103(4):530–41.
[148] Chistiakov DA, Orekhov AN, Bobryshev YV. The role of cardiac fibroblasts in post-myocardial heart tissue repair. Exp Mol Pathol 2016;101(2):231–40.
[149] Meng XM, Nikolic-Paterson DJ, Lan HY. TGF-beta: the master regulator of fibrosis. Nat Rev Nephrol 2016;12(6):325–38.
[150] Booz GW, Baker KM. Molecular signalling mechanisms controlling growth and function of cardiac fibroblasts. Cardiovasc Res 1995;30(4):537–43.
[151] Dostal DE, Booz GW, Baker KM. Angiotensin II signalling pathways in cardiac fibroblasts: conventional versus novel mechanisms in mediating cardiac growth and function. Mol Cell Biochem 1996;157(1-2):15–21.
[152] Piacentini L, Gray M, Honbo NY, Chentoufi J, Bergman M, Karliner JS. Endothelin-1 stimulates cardiac fibroblast proliferation through activation of protein kinase C. J Mol Cell Cardiol 2000;32(4):565–76.
[153] Virag JI, Murry CE. Myofibroblast and endothelial cell proliferation during murine myocardial infarct repair. Am J Pathol 2003;163(6):2433–40.
[154] Virag JA, Rolle ML, Reece J, Hardouin S, Feigl EO, Murry CE. Fibroblast growth factor-2 regulates myocardial infarct repair: effects on cell proliferation, scar contraction, and ventricular function. Am J Pathol 2007;171(5):1431–40.
[155] Zymek P, Bujak M, Chatila K, Cieslak A, Thakker G, Entman ML, et al. The role of platelet-derived growth factor signaling in healing myocardial infarcts. J Am Coll Cardiol 2006;48(11):2315–23.
[156] Yuan MJ, Maghsoudi T, Wang T. Exosomes mediate the intercellular communication after myocardial infarction. Int J Med Sci 2016;13(2):113–6.
[157] Tian Y, Liu Y, Wang T, Zhou N, Kong J, Chen L, et al. A microRNA-Hippo pathway that promotes cardiomyocyte proliferation and cardiac regeneration in mice. Sci Transl Med 2015;7(279):279ra38.
[158] Wang J, Martin JF. Macro advances in microRNAs and myocardial regeneration. Curr Opin Cardiol 2014;29(3):207–13.
[159] Xiao J, Liu H, Cretoiu D, Toader DO, Suciu N, Shi J, et al. miR-31a-5p promotes postnatal cardiomyocyte proliferation by targeting RhoBTB1. Exp Mol Med 2017;49(10):e386.
[160] Ontoria-Oviedo I, Dorronsoro A, Sanchez R, Ciria M, Gomez-Ferrer M, Buigues M, et al. Extracellular vesicles secreted by hypoxic AC10 cardiomyocytes modulate fibroblast cell motility. Front Cardiovas Med 2018;5:152.
[161] Liu L, Jin X, Hu CF, Li R, Zhou Z, Shen CX. Exosomes derived from mesenchymal stem cells rescue myocardial ischaemia/reperfusion injury by inducing cardiomyocyte autophagy via AMPK and Akt pathways. Cell Physiol Biochem 2017;43(1):52–68.
[162] Teng X, Chen L, Chen W, Yang J, Yang Z, Shen Z. Mesenchymal stem cell-derived exosomes improve the microenvironment of infarcted myocardium contributing to angiogenesis and anti-inflammation. Cell Physiol Biochem 2015;37(6):2415–24.

[163] Liu H, Gao W, Yuan J, Wu C, Yao K, Zhang L, et al. Exosomes derived from dendritic cells improve cardiac function via activation of CD4(+) T lymphocytes after myocardial infarction. J Mol Cell Cardiol 2016;91:123–33.
[164] Vandergriff A, Huang K, Shen D, Hu S, Hensley MT, Caranasos TG, et al. Targeting regenerative exosomes to myocardial infarction using cardiac homing peptide. Theranostics 2018;8(7):1869–78.
[165] Huang S, Ge X, Yu J, Han Z, Yin Z, Li Y, et al. Increased miR-124-3p in microglial exosomes following traumatic brain injury inhibits neuronal inflammation and contributes to neurite outgrowth via their transfer into neurons. FASEB J 2018;32(1):512–28.
[166] Gao L, Gregorich ZR, Zhu W, Mattapally S, Oduk Y, Lou X, et al. Large cardiac muscle patches engineered from human induced-pluripotent stem cell-derived cardiac cells improve recovery from myocardial infarction in swine. Circulation 2018;137(16):1712–30.
[167] Shen L, Wang H, Bei Y, Cretoiu D, Cretoiu SM, Xiao J. Formation of new cardiomyocytes in exercise. Adv Exp Med Biol 2017;999:91–102.
[168] Waldenström A, Ronquist G. Role of exosomes in myocardial remodeling. Circ Res 2014;114(2):315–24.
[169] Berezin AE, Kremzer AA, Samura TA, Martovitskaya YV, Malinovskiy YV, Oleshko SV, et al. Predictive value of apoptotic microparticles to mononuclear progenitor cells ratio in advanced chronic heart failure patients. J Cardiol 2015;65(5):403–11.
[170] Bang C, Batkai S, Dangwal S, Gupta SK, Foinquinos A, Holzmann A, et al. Cardiac fibroblast-derived microRNA passenger strand-enriched exosomes mediate cardiomyocyte hypertrophy. J Clin Invest 2014;124(5):2136–46.
[171] Raphael C, Briscoe C, Davies J, Ian Whinnett Z, Manisty C, Sutton R, et al. Limitations of the New York Heart Association functional classification system and self-reported walking distances in chronic heart failure. Heart 2007;93(4):476–82.
[172] Nozaki T, Sugiyama S, Sugamura K, Ohba K, Matsuzawa Y, Konishi M, et al. Prognostic value of endothelial microparticles in patients with heart failure. Eur J Heart Fail 2010;12(11):1223–8.
[173] Berezin AE, Kremzer AA, Martovitskaya YV, Samura TA, Berezina TA. The predictive role of circulating microparticles in patients with chronic heart failure. BBA Clin 2015;3:18–24.
[174] Dubin R, Li Y, Ix JH, Shlipak MG, Whooley M, Peralta C. Associations of pentraxin-3 with cardiovascular events, incident heart failure and mortality among persons with coronary heart disease: data from the heart and soul study. Am Heart J 2012;163:274–9.
[175] Schneider HG, Lam L, Lokuge A, Krum H, Naughton M, De Villiers SP, et al. B-Type natriuretic peptide testing, clinical outcomes, and health services use in emergency department patients with dyspnea a randomized trial. Ann Intern Med 2009;150:365–71.
[176] Salem ESB, Fan GC. Pathological effects of exosomes in mediating diabetic cardiomyopathy. Adv Exp Med Biol 2017;998:113–38.
[177] Khananshvili D. The SLC8 gene family of sodium–calcium exchangers (NCX)—structure, function, and regulation in health and disease. Mol Aspects Med 2013;34:220–35.
[178] Guescini M, Guidolin D, Luciana V, Casadei L, Gioacchini A, Tibollo P, et al. C2C12 myoblasts release micro-vesicles containing mtDNA and proteins involved in signal transduction. Exp Cell Res 2010;316:1977–84.
[179] Guescini M, Genedani S, Stocchi V, Agnati LF. Astrocytes and glioblastoma cells release exosomes carrying mtDNA. J Neural Transm 2010;117(1):1–4.
[180] Ye W, Tang X, Yang Z, Liu C, Zhang X, Jin J, et al. Plasma-derived exosomes contribute to inflammation via the TLR9-NF-kappaB pathway in chronic heart failure patients. Mol Immunol 2017;87:114–21.

[181] Halkein J, Tabruyn SP, Ricke-Hoch M, Haghikia A, Nguyen NQ, Scherr M, et al. MicroRNA-146a is a therapeutic target and biomarker for peripartum cardiomyopathy. J Clin Invest 2013;123(5):2143–54.

[182] Natarelli L, Schober A. MicroRNAs and the response to injury in atherosclerosis. Hamostaseologie 2015;35(2):142–50.

[183] Divakaran V, Mann DL. The emerging role of microRNAs in cardiac remodeling and heart failure. Circ Res 2008;103(10):1072–83.

[184] Matsumoto S, Sakata Y, Suna S, Nakatani D, Usami M, Hara M, et al. Circulating p53-responsive microRNAs are predictive indicators of heart failure after acute myocardial infarction. Circ Res 2013;113(3):322–6.

[185] Wu T, Chen Y, Du Y, Tao J, Zhou Z, Yang Z. Serum exosomal MiR-92b-5p as a potential biomarker for acute heart failure caused by dilated cardiomyopathy. Cell Physiol Biochem 2018;46(5):1939–50.

[186] Melman YF, Shah R, Das S. MicroRNAs in heart failure: is the picture becoming less miRky? Circ Heart Fail 2014;7(1):203–14.

[187] Matkovich SJ, Van Booven DJ, Youker KA, Torre-Amione G, Diwan A, Eschenbacher WH, et al. Reciprocal regulation of myocardial microRNAs and messenger RNA in human cardiomyopathy and reversal of the microRNA signature by biomechanical support. Circulation 2009;119(9):1263–71.

[188] Turchinovich A, Weiz L, Langheinz A, Burwinkel B. Characterization of extracellular circulating microRNA. Nucleic Acids Res 2011;39(16):7223–33.

[189] Valadi H, Ekstrom K, Bossios A, Sjostrand M, Lee JJ, Lotvall JO. Exosome-mediated transfer of mRNAs and microRNAs is a novel mechanism of genetic exchange between cells. Nat Cell Biol 2007;9(6):654–9.

[190] Tian C, Gao L, Zimmerman MC, Zucker IH. Myocardial infarction-induced microRNA-enriched exosomes contribute to cardiac Nrf2 dysregulation in chronic heart failure. Am J Physiol Heart Circ Physiol 2018;314(5):H928–39.

[191] O'Neill S, Bohl M, Gregersen S, Hermansen K, O'Driscoll L. Blood-based biomarkers for metabolic syndrome. Trends Endocrinol Metab 2016;27(6):363–74.

[192] Evans S, Mann DL. Circulating p53-responsive microRNAs as predictive biomarkers in heart failure after acute myocardial infarction: the long and arduous road from scientific discovery to clinical utility. Circ Res 2013;113(3):242–4.

[193] Lindenfeld J, Albert NM, Boehmer JP, Collins SP, Ezekowitz JA, Givertz MM, et al. HFSA 2010 comprehensive heart failure practice guideline. J Card Fail 2010;16(6):e1–194.

[194] Jansen F, Yang X, Proebsting S, Hoelscher M, Przybilla D, Baumann K, et al. MicroRNA expression in circulating microvesicles predicts cardiovascular events in patients with coronary artery disease. J Am Heart Assoc 2014;3(6). e001249-e.

[195] Endo K, Naito Y, Ji X, Nakanishi M, Noguchi T, Goto Y, et al. MicroRNA 210 as a biomarker for congestive heart failure. Biol Pharm Bull 2013;36(1):48–54.

[196] Thum T, Galuppo P, Wolf C, Fiedler J, Kneitz S, van Laake LW, et al. MicroRNAs in the human heart: a clue to fetal gene reprogramming in heart failure. Circulation 2007;116(3):258–67.

[197] van Rooij E, Sutherland LB, Liu N, Williams AH, McAnally J, Gerard RD, et al. A signature pattern of stress-responsive microRNAs that can evoke cardiac hypertrophy and heart failure. Proc Natl Acad Sci U S A 2006;103(48):18255–60.

[198] van Rooij E, Sutherland LB, Thatcher JE, DiMaio JM, Naseem RH, Marshall WS, et al. Dysregulation of microRNAs after myocardial infarction reveals a role of miR-29 in cardiac fibrosis. Proc Natl Acad Sci U S A 2008;105(35):13027–32.

[199] Bostjancic E, Zidar N, Stajer D, Glavac D. MicroRNAs miR-1, miR-133a, miR-133b and miR-208 are dysregulated in human myocardial infarction. Cardiology 2010;115(3):163–9.
[200] Satoh M, Minami Y, Takahashi Y, Tabuchi T, Nakamura M. Expression of microRNA-208 is associated with adverse clinical outcomes in human dilated cardiomyopathy. J Card Fail 2010;16(5):404–10.
[201] Ikeda S, Kong SW, Lu J, Bisping E, Zhang H, Allen PD, et al. Altered microRNA expression in human heart disease. Physiol Genomics 2007;31(3):367–73.
[202] Murach KA, McCarthy JJ. MicroRNAs, heart failure, and aging: potential interactions with skeletal muscle. Heart Fail Rev 2017;22(2):209–18.
[203] Goren Y, Kushnir M, Zafrir B, Tabak S, Lewis BS, Amir O. Serum levels of microRNAs in patients with heart failure. Eur J Heart Fail 2012;14(2):147–54.
[204] Sygitowicz G, Tomaniak M, Blaszczyk O, Koltowski L, Filipiak KJ, Sitkiewicz D. Circulating microribonucleic acids miR-1, miR-21 and miR-208a in patients with symptomatic heart failure: preliminary results. Arch Cardiovasc Dis 2015;108(12):634–42.
[205] van Rooij E, Liu N, Olson EN. MicroRNAs flex their muscles. Trends Genet 2008;24(4):159–66.
[206] Zile MR, Mehurg SM, Arroyo JE, Stroud RE, DeSantis SM, Spinale FG. Relationship between the temporal profile of plasma microRNA and left ventricular remodeling in patients after myocardial infarction. Circ Cardiovasc Genet 2011;4:614–9.
[207] Akat KM, Moore-McGriff D, Morozov P, Brown M, Gogakos T, Correa Da Rosa J, et al. Comparative RNA-sequencing analysis of myocardial and circulating small RNAs in human heart failure and their utility as biomarkers. Proc Natl Acad Sci U S A 2014;111(30):11151–6.
[208] McCarthy JJ, Esser KA. MicroRNA-1 and microRNA-133a expression are decreased during skeletal muscle hypertrophy. J Appl Physiol 2007;102(1):306–13.
[209] Elia L, Contu R, Quintavalle M, Varrone F, Chimenti C, Russo MA, et al. Reciprocal regulation of microRNA-1 and insulin-like growth factor-1 signal transduction cascade in cardiac and skeletal muscle in physiological and pathological conditions. Circulation 2009;120(23):2377–85.
[210] Huang MB, Xu H, Xie SJ, Zhou H, Qu LH. Insulin-like growth factor-1 receptor is regulated by microRNA-133 during skeletal myogenesis. PLoS One 2011;6(12):e29173.
[211] Schipper ME, van Kuik J, de Jonge N, Dullens HF, de Weger RA. Changes in regulatory microRNA expression in myocardium of heart failure patients on left ventricular assist device support. J Heart Lung Transplant 2008;27(12):1282–5.
[212] Watson CJ, Gupta SK, O'Connell E, Thum S, Glezeva N, Fendrich J, et al. MicroRNA signatures differentiate preserved from reduced ejection fraction heart failure. Eur J Heart Fail 2015;17(4):405–15.
[213] Xu T, Zhou Q, Che L, Das S, Wang L, Jiang J, et al. Circulating miR-21, miR-378, and miR-940 increase in response to an acute exhaustive exercise in chronic heart failure patients. Oncotarget 2016;7(11):12414–25.
[214] Gupta MK, Halley C, Duan Z-H, Lappe J, Viterna J, Jana S, et al. miRNA-548c: a specific signature in circulating PBMCs from dilated cardiomyopathy patients. J Mol Cell Cardiol 2013;62:131–41.
[215] Kervadec A, Bellamy V, El Harane N, Arakelian L, Vanneaux V, Cacciapuoti I, et al. Cardiovascular progenitor-derived extracellular vesicles recapitulate the beneficial effects of their parent cells in the treatment of chronic heart failure. J Heart Lung Transplant 2016;35(6):795–807.
[216] Chen Y, Wakili R, Xiao J, Wu CT, Luo X, Clauss S, et al. Detailed characterization of microRNA changes in a canine heart failure model: relationship to arrhythmogenic structural remodeling. J Mol Cell Cardiol 2014;77:113–24.

[217] Yang VK, Loughran KA, Meola DM, Juhr CM, Thane KE, Davis AM, et al. Circulating exosome microRNA associated with heart failure secondary to myxomatous mitral valve disease in a naturally occurring canine model. J Extracell vesicles 2017;6(1):1350088.

[218] Shotan A, Widerhorn J, Hurst A, Elkayam U. Risks of angiotensin-converting enzyme inhibition during pregnancy: experimental and clinical evidence, potential mechanisms, and recommendations for use. Am J Med 1994;96(5):451–6.

[219] Modi KA, Illum S, Jariatul K, Caldito G, Reddy PC. Poor outcome of indigent patients with peripartum cardiomyopathy in the United States. Am J Obstet Gynecol 2009;201(2). 171.e1-5.

[220] Jiang X, Sucharov J, Stauffer BL, Miyamoto SD, Sucharov CC. Exosomes from pediatric dilated cardiomyopathy patients modulate a pathological response in cardiomyocytes. Am J Physiol Heart Circ Physiol 2017;312(4):H818–26.

[221] Stauffer BL, Russell G, Nunley K, Miyamoto SD, Sucharov CC. miRNA expression in pediatric failing human heart. J Mol Cell Cardiol 2013;57:43–6.

[222] Zheng D, Ma J, Yu Y, Li M, Ni R, Wang G, et al. Silencing of miR-195 reduces diabetic cardiomyopathy in C57BL/6 mice. Diabetologia 2015;58(8):1949–58.

[223] Hunt SA, Abraham WT, Chin MH, Feldman AM, Francis GS, Ganiats TG, et al. ACC/AHA 2005 guideline update for the diagnosis and management of chronic heart failure in the adult: a report of the American College of Cardiology/American Heart Association task force on practice guidelines (writing committee to update the 2001 guidelines for the evaluation and management of heart failure): developed in collaboration with the American College of Chest Physicians and the international society for heart and lung transplantation: endorsed by the Heart Rhythm Society. Circulation 2005;112(12):e154–235.

[224] McMurray JJ, Adamopoulos S, Anker SD, Auricchio A, Bohm M, Dickstein K, et al. ESC Guidelines for the diagnosis and treatment of acute and chronic heart failure 2012: the task force for the diagnosis and treatment of acute and chronic heart failure 2012 of the European Society of Cardiology. Developed in collaboration with the Heart Failure Association (HFA) of the ESC. Eur Heart J 2012;33(14):1787–847.

[225] Maulik SK, Kumar S. Oxidative stress and cardiac hypertrophy: a review. Toxicol Mech Methods 2012;22(5):359–66.

[226] Hausse AO, Aggoun Y, Bonnet D, Sidi D, Munnich A, Rötig A, et al. Idebenone and reduced cardiac hypertrophy in Friedreich's ataxia. Heart 2002;87(4):346–9.

[227] Balta N, Dumitru IF, Stoian G, Petec G, Dinischiotu A. Influence of isoproterenol-induced cardiac hypertrophy on oxydative myocardial stress. Rom J Physiol 1995;32(1-4): 149–54.

[228] Lee A, Jeong D, Mitsuyama S, Oh JG, Liang L, Ikeda Y, et al. The role of SUMO-1 in cardiac oxidative stress and hypertrophy. Antioxid Redox Signal 2014;21(14):1986–2001.

[229] Datta R, Bansal T, Rana S, Datta K, Datta Chaudhuri R, Chawla-Sarkar M, et al. Myocyte-derived Hsp90 modulates collagen upregulation via biphasic activation of STAT-3 in fibroblasts during cardiac hypertrophy. J Mol Cell Biol 2017;37(6). pii: e00611-16.

[230] Lyu L, Wang H, Li B, Qin Q, Qi L, Nagarkatti M, et al. A critical role of cardiac fibroblast-derived exosomes in activating renin angiotensin system in cardiomyocytes. J Mol Cell Cardiol 2015;89(Pt B):268–79.

[231] Fang X, Stroud MJ, Ouyang K, Fang L, Zhang J, Dalton ND, et al. Adipocyte-specific loss of PPARgamma attenuates cardiac hypertrophy. JCI insight 2016;1(16):e89908.

[232] Wang X, Gu H, Huang W, Peng J, Li Y, Yang L, et al. Hsp20-Mediated activation of exosome biogenesis in cardiomyocytes improves cardiac function and angiogenesis in diabetic mice. Diabetes 2016;65(10):3111–28.

[233] Belevych AE, Sansom SE, Terentyeva R, Ho H-T, Nishijima Y, Martin MM, et al. MicroRNA-1 and -133 Increase arrhythmogenesis in heart failure by dissociating phosphatase activity from RyR2 complex. PLoS One 2011;6(12):e28324.
[234] Lu Y, Zhang Y, Wang N, Pan Z, Gao X, Zhang F, et al. MicroRNA-328 contributes to adverse electrical remodeling in atrial fibrillation. Circulation 2010;122(23):2378–87.
[235] Neppl RL, Wang D-Z. The myriad essential roles of microRNAs in cardiovascular homeostasis and disease. Genes & diseases 2014;1(1):18–39.
[236] Lai LP, Su MJ, Lin JL, Lin FY, Tsai CH, Chen YS, et al. Down-regulation of L-type calcium channel and sarcoplasmic reticular Ca(2+)-ATPase mRNA in human atrial fibrillation without significant change in the mRNA of ryanodine receptor, calsequestrin and phospholamban: an insight into the mechanism of atrial electrical remodeling. J Am Coll Cardiol 1999;33(5):1231–7.
[237] Schotten U, Ausma J, Stellbrink C, Sabatschus I, Vogel M, Frechen D, et al. Cellular mechanisms of depressed atrial contractility in patients with chronic atrial fibrillation. Circulation 2001;103(5):691–8.
[238] Mancarella S, Yue Y, Karnabi E, Qu Y, El-Sherif N, Boutjdir M. Impaired Ca2+ homeostasis is associated with atrial fibrillation in the alpha1D L-type Ca2+ channel KO mouse. Am J Physiol Heart Circ Physiol 2008;295(5):H2017–24.
[239] Plotnikov AN, Shlapakova I, Szabolcs MJ, Danilo Jr. P, Lorell BH, Potapova IA, et al. Xenografted adult human mesenchymal stem cells provide a platform for sustained biological pacemaker function in canine heart. Circulation 2007;116(7):706–13.
[240] Askar SF, Ramkisoensing AA, Atsma DE, Schalij MJ, de Vries AA, Pijnappels DA. Engraftment patterns of human adult mesenchymal stem cells expose electrotonic and paracrine proarrhythmic mechanisms in myocardial cell cultures. Circ Arrhythm Electrophysiol 2013;6(2):380–91.
[241] Mayourian J, Cashman TJ, Ceholski DK, Johnson BV, Sachs D, Kaji DA, et al. Experimental and computational insight into human mesenchymal stem cell paracrine signaling and heterocellular coupling effects on cardiac contractility and arrhythmogenicity. Circ Res 2017;121(4):411–23.
[242] Ji Q, Ji Y, Peng J, Zhou X, Chen X, Zhao H, et al. Increased brain-specific MiR-9 and MiR-124 in the serum exosomes of acute ischemic stroke patients. PLoS One 2016;11(9):e0163645.
[243] Goetzl L, Merabova N, Darbinian N, Martirosyan D, Poletto E, Fugarolas K, et al. Diagnostic potential of neural exosome cargo as biomarkers for acute brain injury. Ann Clin Transl Neurol 2017;5(1):4–10.
[244] Kucharzewska P, Christianson HC, Welch JE, Svensson KJ, Fredlund E, Ringner M, et al. Exosomes reflect the hypoxic status of glioma cells and mediate hypoxia-dependent activation of vascular cells during tumor development. Proc Natl Acad Sci U S A 2013;110(18):7312–7.
[245] Matsumoto J, Stewart T, Sheng L, Li N, Bullock K, Song N, et al. Transmission of alpha-synuclein-containing erythrocyte-derived extracellular vesicles across the blood-brain barrier via adsorptive mediated transcytosis: another mechanism for initiation and progression of Parkinson's disease? Acta Neuropathol Commun 2017;5(1):71.
[246] Harrison EB, Hochfelder CG, Lamberty BG, Meays BM, Morsey BM, Kelso ML, et al. Traumatic brain injury increases levels of miR-21 in extracellular vesicles: implications for neuroinflammation. FEBS Open Bio 2016;6(8):835–46.
[247] de Rivero Vaccari JP, Brand 3rd F, Adamczak S, Lee SW, Perez-Barcena J, Wang MY, et al. Exosome-mediated inflammasome signaling after central nervous system injury. J Neurochem 2016;136(Suppl 1):39–48.

[248] Manek R, Moghieb A, Yang Z, Kumar D, Kobessiy F, Sarkis GA, et al. Protein biomarkers and neuroproteomics characterization of microvesicles/exosomes from human cerebrospinal fluid following traumatic brain injury. Mol Neurobiol 2018;55(7):6112–28.

[249] Belting M, Christianson HC. Role of exosomes and microvesicles in hypoxia-associated tumour development and cardiovascular disease. J Intern Med 2015;278(3):251–63.

[250] Chen W, Guo Y, Yang W, Chen L, Ren D, Wu C, et al. Phosphorylation of connexin 43 induced by traumatic brain injury promotes exosome release. J Neurophysiol 2018;119(1):305–11.

[251] Zhuang X, Xiang X, Grizzle W, Sun D, Zhang S, Axtell RC, et al. Treatment of brain inflammatory diseases by delivering exosome encapsulated anti-inflammatory drugs from the nasal region to the brain. Mol ther 2011;19(10):1769–79.

[252] Otero-Ortega L, Laso-Garcia F, Gomez-de Frutos M, Fuentes B, Diekhorst L, Diez-Tejedor E, et al. Role of exosomes as a treatment and potential biomarker for stroke. Transl Stroke Res 2019;10:241–49.

[253] Chen CC, Liu L, Ma F, Wong CW, Guo XE, Chacko JV, et al. Elucidation of exosome migration across the blood-brain barrier model in vitro. Cell Mol Bioeng 2016;9(4):509–29.

[254] Xu B, Zhang Y, Du XF, Li J, Zi HX, Bu JW, et al. Neurons secrete miR-132-containing exosomes to regulate brain vascular integrity. Cell Res 2017;27(7):882–97.

[255] Zhang Y, Chopp M, Liu XS, Katakowski M, Wang X, Tian X, et al. Exosomes derived from mesenchymal stromal cells promote axonal growth of cortical neurons. Mol Neurobiol 2017;54(4):2659–73.

[256] Xin H, Li Y, Cui Y, Yang JJ, Zhang ZG, Chopp M. Systemic administration of exosomes released from mesenchymal stromal cells promote functional recovery and neurovascular plasticity after stroke in rats. J Cereb Blood Flow Metab 2013;33(11):1711–5.

[257] Zhang Y, Chopp M, Meng Y, Katakowski M, Xin H, Mahmood A, et al. Effect of exosomes derived from multipluripotent mesenchymal stromal cells on functional recovery and neurovascular plasticity in rats after traumatic brain injury. J Neurosurg 2015;122(4):856–67.

[258] Li Y, Yang YY, Ren JL, Xu F, Chen FM, Li A. Exosomes secreted by stem cells from human exfoliated deciduous teeth contribute to functional recovery after traumatic brain injury by shifting microglia M1/M2 polarization in rats. Stem Cell Res Ther 2017;8(1):198.

[259] Dinkins MB, Dasgupta S, Wang G, Zhu G, Bieberich E. Exosome reduction in vivo is associated with lower amyloid plaque load in the 5XFAD mouse model of Alzheimer's disease. Neurobiol Aging 2014;35(8):1792–800.

[260] Jarmalaviciute A, Tunaitis V, Pivoraite U, Venalis A, Pivoriunas A. Exosomes from dental pulp stem cells rescue human dopaminergic neurons from 6-hydroxy-dopamine-induced apoptosis. Cytotherapy 2015;17(7):932–9.

[261] Haney MJ, Klyachko NL, Zhao Y, Gupta R, Plotnikova EG, He Z, et al. Exosomes as drug delivery vehicles for Parkinson's disease therapy. J Control Release 2015;207:18–30.

Chapter 10

Exosomes in cutaneous biology and dermatologic disease

Jeffrey D. McBride, Divya Aickara, Evangelos Badiavas
Department of Dermatology and Cutaneous Surgery, University of Miami, Miami, FL, United States

1 Keratinocytes, the major cells of the upper skin, secrete exosomes that modulate pigment production by melanocytes

Exosomes play a role in one of the most fundamental aspects of basic skin biology – skin pigmentation. Skin pigmentation relies on communication between keratinocytes and melanocytes. Skin pigmentation is a primary way to protect skin cells and their DNA from damage induced by ultraviolet light. In the outermost layer of the skin, keratinocytes with melanocytes form the epidermal-melanin unit, which can be activated by ultraviolet radiation [1]. This radiation, partly in the form of UVA and UVB, activates signaling pathways that induce secretion of growth factors and extracellular proteins that lead to production of melanin, the major pigment in the skin. Cells secrete extracellular vesicles (exosomes and microvesicles), which can transfer proteins, lipids, and RNAs to regulate functions of the recipient cells [1]. Keratinocytes secrete exosomes that enhance melanin synthesis via increasing expression and activity of melanosomal proteins [1]. The functions of these exosomes are dependent on the phototype (the lightness or darkness of the skin, and how easily the skin typically burns in the sun), and this effect is modulated by ultraviolet B [1]. Exosomes carrying selected microRNAs (miRNAs) are targeted to melanocytes and modulate skin pigmentation via alterating of gene and protein expression in melanocytes (Fig. 1) [1]. When keratinocytes are co-cultured with melanocytes, the multivesicular bodies (containing CD63+-intraluminal vesicles that are destined to be extruded as CD63+-exosomes) are redistributed to areas in which keratinocytes make contact with melanocytes [1]. Keratinocyte exosomes are taken up by melanocytes shortly after secretion into the extracellular environment [1]. Exosomes from keratinocytes derived from phototype V skin (darker pigmented individuals) significantly increased

Exosomes. https://doi.org/10.1016/B978-0-12-816053-4.00010-9

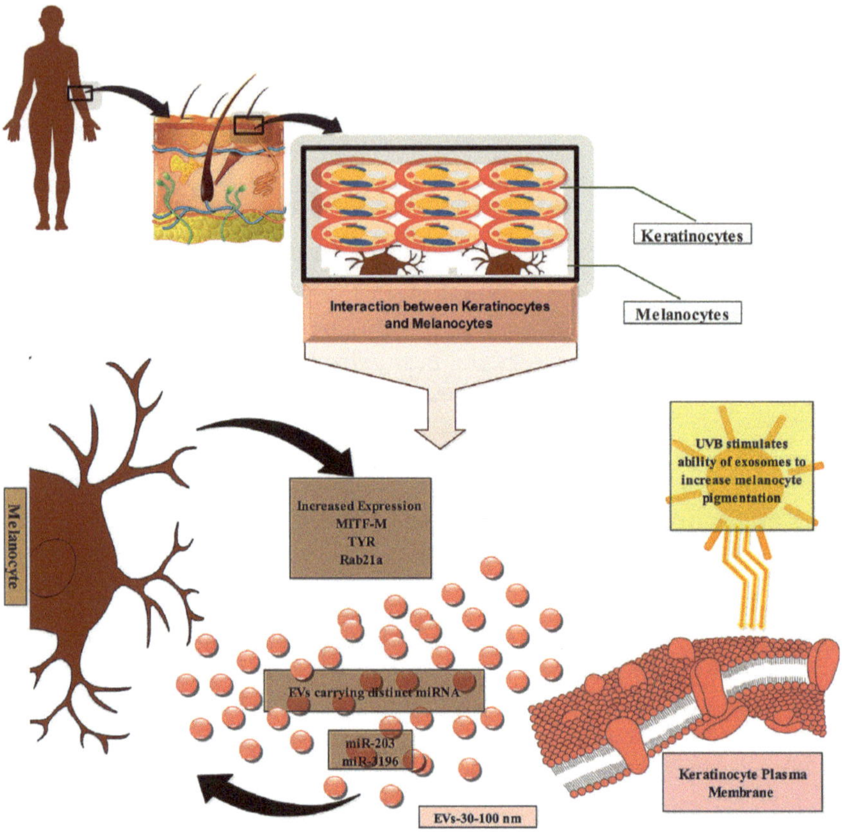

FIG. 1 Ultraviolet B stimulates keratinocytes to secrete exosomes that stimulate melanin production by melanocytes. UVB stimulates keratinocytes to produced microRNAs miR-203 and miR-3196 that ultimately stimulate melanogenesis via upregulated MITF, TRY, and Rab21a. *Credit: Illustration created by Divya Aickara. Based on information in article by Lo Cicero A, et al. Exosomes released by keratinocytes modulate melanocyte pigmentation. Nat Commun 2015;6:7506.*

the melanin content in melanocytes derived from type I-II/light-complected or Caucasian individuals [1]. Thus, secreted exosomes from high phototype-derived keratinocytes may have intrinsic properties and cargo that stimulate melanin production [1]. When Caucasian-derived keratinocyte exosomes were used to stimulate melanocytes from the same donor, there was no increase in the tyrosinase enzyme (the main enzyme in the melanin production pathway); however, when exosomes derived from UVB-stimulated keratinocytes were used to stimulate melanocytes from the same donor, there was a significant increase in tyrosinase expression and microphthalmia-transcription factor expression (which is the master regulator of melanogenesis), suggesting UVB-stimulated keratinocyte exosomes are powerful inducers of pigmented production in Caucasians [1]. The same effects were seen when UVB-induced keratinocyte exosomes were applied to human-constructed epidermis in vitro [1]. Exosomes carry miRNAs that regulate pigmentation, and the miRNA content is altered after exposure to UVB, specifically miR-3196 [1]. Additionally, miR-203, a key regulator of melanogenesis in melanoma cells via upregulation of TYR protein, was found to be highly expressed in exosomes derived from dark-complected individuals [1]. miRNAs likely depend on contextual molecules, such as proteins or lipids, that supplement or modulate the overall effect of the differentially expressed keratinocytes [1]. Keratinocyte exosomal miRNA likely fine-tunes melanocyte pigmentation via both miR-3196 and MITF-dependent and miR-203 and MITF-independent signaling pathways. Thus, skin pigmentation relies on exosome relay between keratinocytes, and environmental and genetic factors likely play a role in the exosomal cargo that ultimately impacts skin pigmentation.

Ultraviolet light irradiation induced pigmentation in the skin via intercellular transfer of melanin from melanocytes to keratinocytes [2]. Wäster et al. showed a novel cell response after UVA irradiation, resulting in transfer of lysosomal-derived extracellular vesicles from melanocytes to keratinocytes [2]. UVA induces an immediate shedding of extracellular vesicles from the plasma membrane of melanocytes, which is likely following UVA-induced plasma membrane damage [2]. Many of these vesicles were positive for CD63 and other canonical exosome markers [2]. These EVs are preferentially endocytosed by keratinocytes [2]. Melanosome transfer from melanocytes to keratinocytes is equally stimulated by IVA and UVB and relies on a functional cytoskeleton [2]. UVA-induced plasma membrane damage in melanocytes is rapidly repaired via calcium-dependent lysosomal exocytosis [2]. Furthermore, the melanocyte-derived exosomes enhanced keratinocyte proliferation [2].

One group discovered that extracellular vesicles containing fibronectin can protect melanocytes against UV radiation-induced cytotoxicity [3]. UV irradiation activates skin melanocytes to secrete melanosomes as a mechanism to protect against the UV-induced damage [3]. It appears that melanocytes can use their own extracellular vesicles, coupled with an extracellular matrix component, to protect each other from UV radiation [3]. Using proteome analysis, fibronectin was identified [3]. Interestingly, in patients with melasma, the extracellular space around melanocytes contained more fibronectin compared with normal skin, correlating with the model that extracellular fibronectin is involved in promoting even pathologic survival of melanocytes [3]. Because melanocytes cycle slower, express higher levels of anti-apoptotic Bcl-2, and therefore have a high need to protect their genomic stability, there is a high need for a biologic mechanism that protects melanocytes from UV-induced genomic damage [3]. In this study, analysis of EV-associated proteins showed fibronectin to be a major component with the highest centrality based on degree, betweenness, closeness and radiality among membrane proteins [3]. Confocal microscopy analysis demonstrated that CD81 (exosome markers) colocalized with fibronectin [3].

2 Keratinocyte and immune system exosomes modulate fibroblast expression of proteins that modify the extracellular matrix

Keratinocytes secrete proteins associated with exosomes that ultimately regulate expression of proteins, such as matrix metalloproteinases, that are known to modify the extracellular matrix. Stratifin (also known as 14-3-3σ) is a protein that was known to be "externalized" by keratinocytes and stimulates matrix metalloprotein-1 in fibroblasts, but the mechanism of "externalization" was unknown, until it was discovered to be associated with exosomes [4]. It was found that stratifin (14-3-3σ) exited in purified exosomes from keratinocytes and these stratifin-containing exosomes were able to stimulate expression of MMP-1 from fibroblasts, related to activation of the p38/MAPK pathway (Fig. 2) [4]. This concept opens the possibility that keratinocytes utilize their exosomes to induce fibroblasts to remodel the extracellular matrix, which is important in skin homeostasis and will likely be important in understanding the pathogenesis of fibrotic diseases in dermatology. For example, it is unknown if a lack of keratinocyte exosome transfer to fibroblasts may result in reduced MMP production, resulting in a skewed homeostasis toward fibrosis. In general, in various organs, the epithelial layers likely modulate the underlying connective tissues via these mechanisms. Furthermore, the immune system also secretes exosomes with 14-3-3 protein that modulate MMP expression-1 (Fig. 3) [5]. Thus, both the epithelium and the immune system can influence the major dermal cells, fibroblast, to modulate expression of MMPs that can remodel the surrounding collagen. These concepts demonstrate the complexity of multiple cell-to-cell communication pathways that affect expression of extracellular matrix remodeling proteins. These concepts suggest that defects in the epithelium or immune system could result in effects on the status of the fibroblasts and fibrosis in the dermis.

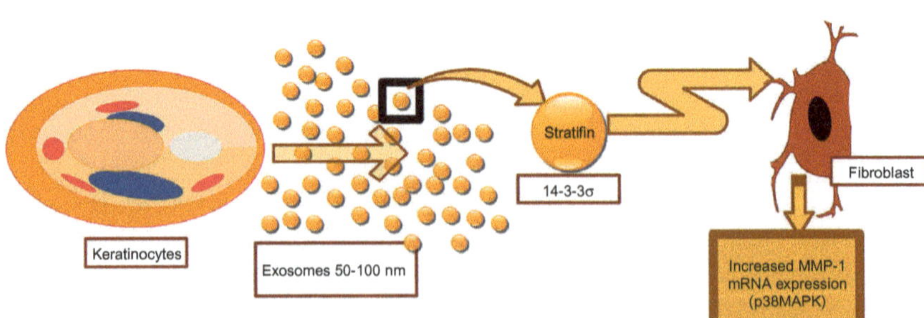

FIG. 2 Keratinocytes secrete exosome-associated stratifin to increase matrix metalloproteinase expression. Activation of p38/MAPK pathways results in secretion of MMP-1 from fibroblasts. *Credit: Illustration created by Divya Aickara. Based on information in article by Chavez-Munoz C, et al. Primary human keratinocytes externalize stratifin protein via exosomes. J Cell Biochem 2008;104(6):2165–73.*

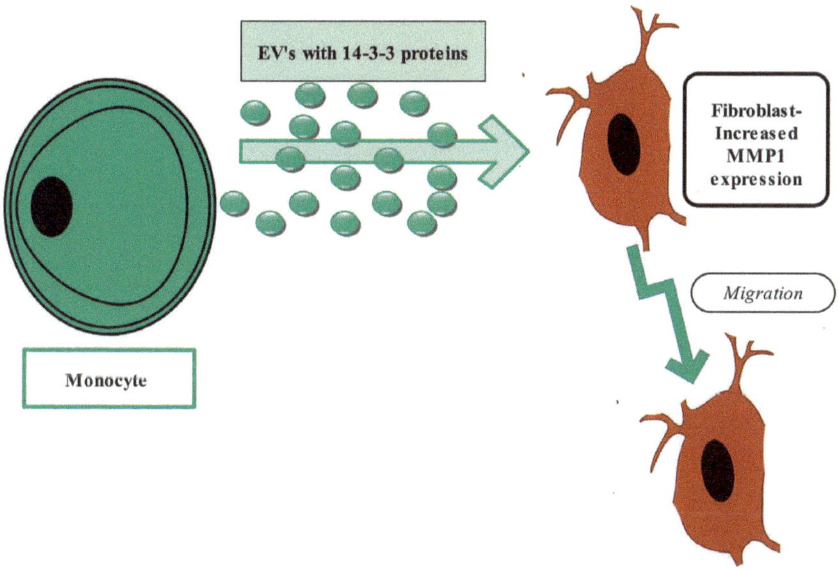

FIG. 3 Monocytes secrete exosomes with 14-3-3 proteins that induce matrix metalloproteinase expression. Increased fibroblast MMP-1 expression promotes extracellular matrix remodeling and fibroblast migration. *Credit: Illustration created by Divya Aickara. Based on information in article by Medina A, Ghahary A. Transdifferentiated circulating monocytes release exosomes containing 14-3-3 proteins with matrix metalloproteinase-1 stimulating effect for dermal fibroblasts. Wound Repair Regen 2010;18(2):245–53.*

3 Keratinocyte exosomes modulate the immune system

The interaction of the epithelium with microbiota utilizes exosomes to relay information to the immune system. The skin can be considered a large organ of the immune system, with cells each have different functions contributing to overall optimal immune system function. *Staphylococcus aureus* are common bacteria present on the healthy skin. Infections with *S. aureus* are common among atopic dermatitis (eczema) patients, as their impaired skin barrier, increased inflammation, and poor skin hydration makes the skin more prone to infection. Exosomes from the skin play vital roles in the function of the immune system and response to infections. In both atopic dermatitis and psoriasis, for example, infection and colonization with *S. aureus* induces the onset of inflammation, partly through secretion of its enterotoxin B. The major cell type of the epidermis, keratinocytes make up approximately 90% of the epidermis and act as non-professional antigen presenting cells and promote superantigen-induced proliferation of T cells [6]. Previous work suggests that keratinocytes support the proliferation of T cells after interaction with antigens [7]. In an experiment involving a transwell co-culture system, human keratinocytes can induce T cells to proliferation via indirect contact—via exosomes [6]. Exosomes

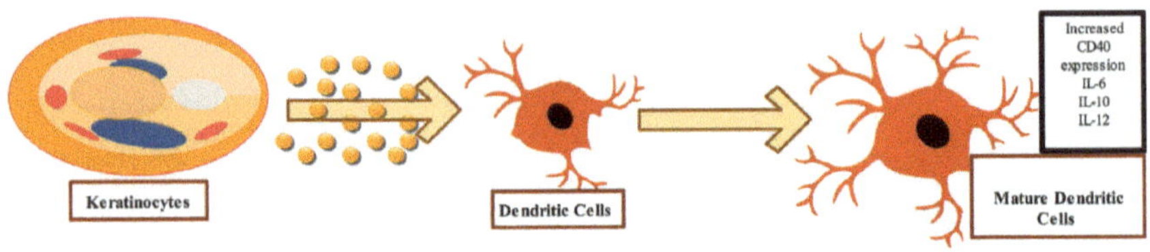

FIG. 4 Keratinocytes secrete exosomes which stimulate the development of mature dendritic cells. Keratinocyte exosomes induced increased expression of CD40 and interleukins IL-6, IL-10, and IL-12 by mature dendritic cells. *Credit: Illustration created by Divya Aickara. Based on information in article by Kotzerke K, et al. Immunostimulatory activity of murine keratinocyte-derived exosomes. Exp Dermatol 2013;22(10):650–5.*

can transfer antigens to recipient T cells; keratinocyte exosomes contained MHC I and MHC II molecules and interact with T cells as a part of antigen presentation [6]. Following stimulation with interferon-γ, keratinocytes (previously loaded with staphylococcal aureus enterotoxin B) secreted exosomes to induce proliferation of CD4 and CD8 T cells [6]. Thus, keratinocytes use exosomes in order to prime T cells to respond to *S. aureus* infection. Exosomes are an important part of the communication system between the external environment and the immune system's response. In another study, a murine keratinocyte cell line (MPEK cells) produced exosomes under both inflammatory and non-inflammatory conditions and that these exosomes were absorbed by bone marrow dendritic cells (likely via direct fusion or other mechanisms—see previous chapters in this book for a discussion of these mechanisms of uptake), inducing a mature phenotype via increased CD40 expression and cytokine production, including IL-6, IL-10 and IL-12 (Fig. 4) [8]. When keratinocytes were exposed to the antigen ovalbumin, they transferred it to their exosomes; but interestingly these exosomes did not induce specific immune responses to T cells via the bone marrow dendritic cells [8]. Thus, this report suggests keratinocytes, via exosomes, may activate broad responses in T cells, independent of the stimulatory antigen. Keratinocyte-derived exosomes were able to affect bone marrow dendritic cell phenotypes and interleukin production, indicating that this crosstalk between keratinocytes and dendritic cells using exosomes are important for immune system activation [8].

4 The dermis fibroblasts, specifically the dermal papilla cells, secrete exosomes that stimulate hair follicle growth phase, known as the anagen phase

Hair follicle density is critical to skin-bacteria homeostasis, temperature regulation, sexual signaling, and even human cosmetics and self-esteem. Exosomes secreted by dermal papilla cells were found to be important in the development of hair follicles [9]. Dermal papilla cells play a crucial role in the regulation of hair follicle growth, formation and cycling [9]. It was hypothesized that the dermal papilla regulated hair follicle growth via paracrine mechanisms, in part via exosomes [9]. In this study, hair follicles at different stages of cycling were injected with exosomes derived from dermal papilla cells, which expressed CD9, CD63, and TSG101 [9]. The effects of the exosomes on outer root sheath cells proliferation, migration and cell cycle status were evaluated [9]. Dermal papilla cells-derived exosomes accelerated the onset of hair follicle transition from telogen (resting phase) to anagen phase (the follicle growth phase) and delayed catagen (the follicle breakdown phase) in mice [9]. Exosomes upregulated both beta-catenin and Shh levels in the hair follicle and increased outer root sheet proliferation and migration [9]. Thus, it is likely that the developing hair follicle is stimulated from exosomes from the dermal papilla that stimulate the Wnt/beta-catenin signaling pathways and the sonic hedgehog pathway [9]. Thus, exosomes from fibroblasts stimulate the growth of hair follicles, critical for skin homeostasis.

5 Disorders of fibrosis have dysregulated expression of fibroblasts exosomes

Systemic sclerosis is a chronic, autoimmune disease characterized by tissue fibrosis of the skin and internal organs. One study sought to determine a putative role of exosomes in systemic sclerosis and to elucidate the effect of exosomes on wound healing [10]. Common markers for exosomes (CD9, CD63, and CD81) were increased in dermal fibroblasts of systemic sclerosis compared to fibroblasts from normal subjects [10]. Exosomes derived from fibroblasts of systemic sclerosis patients stimulated the production of type I collagen in normal fibroblasts [10]. It appeared that the content of exosomes (specifically mRNAs) from systemic sclerosis patients were altered [10]. Interestingly, the sera of patients with systemic sclerosis had significantly decreased levels of exosomes [10]. The authors hypothesized that the vascular abnormalities in systemic sclerosis may be the result of these decreased exosomes [10]. This decrease in exosomes in the serum of systemic sclerosis patients may contribute to delayed wound healing due to down-regulation of collagen, higher likelihood of ulcers and pitting scars [10]. Still, not much is known regarding the role of exosomes in modulation of fibrotic diseases such as systemic sclerosis. Nevertheless, fibroblast exosome release is likely critical for the status of connective tissue diseases, such as systemic sclerosis.

6 Mast cell exosomes, containing phospholipase activity, stimulate Langerhans cell presentation of lipid antigens to T cells

Psoriasis is considered a chronic inflammatory disease of the skin, but also can involve other organs, and is associated with a T helper 17 response [11]. It is associated with infiltration of a variety of cell types, including CD4+ and CD+ T cells, neutrophils, NK cells, mast cells, innate lymphoid cells and monocytes/macrophages [11]. The role of peptide-based antigens has remained unclear; it has been hypothesized that non-peptide antigens, such as lipids, may play a role in the pathogenesis of psoriasis. In addition, activation and degranulation of mast cells, and their proinflammatory cytokines, may contribute to the pathology of psoriasis in skin [11]. Langerhans cells are antigen presenting cells in the skin, and in psoriatic lesions expressive high levels of CD1a, which present lipid antigens (such as fatty acids, wax esters and squalene, to T cells [11]. In psoriasis, interferon-alpha-induced mast cells release of exosomes transferred phospholipase A_2 (an enzyme that generates lipid antigens and has been linked to lipid-specific T cell inflammatory skin reactions) to CD1a-expressing cells (Fig. 5) [11]. There was a larger proportion of exosome-responsive CD1a-reactive T cells in lesional psoriatic skin than in peripheral blood and nonlesional skin of psoriasis patients [11]. These exosomes resulted in an increase in lipid antigen presentation to T-cells, inducing production of IL-17a and IL-22 [11]. Finding phospholipase A_2 activity in the exosomes explained how mast cells contribute to generation of lipid antigens for ultimate presentation by Langerhans cells to T cells, which until this discovery has been unexplained [11]. Thus, this model suggests that mast cell exosomes are upstream in the pathogenesis of psoriasis [11].

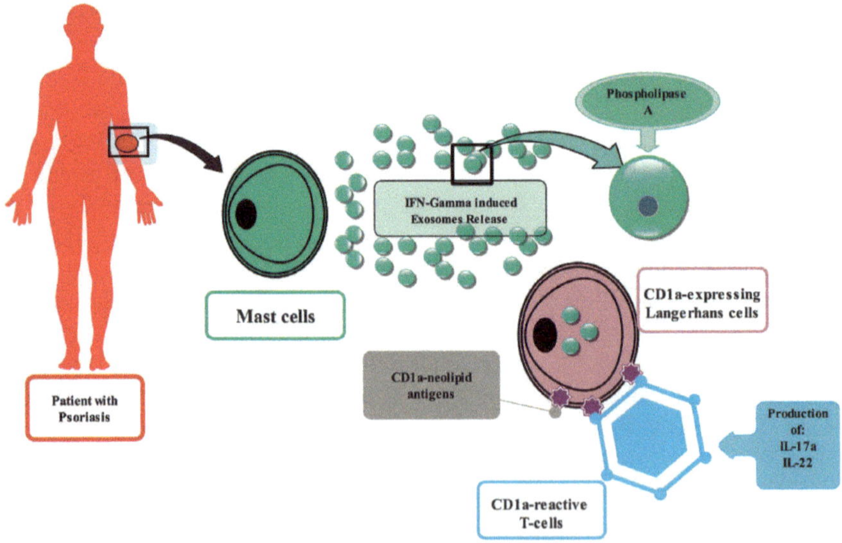

FIG. 5 Expression of exosomes carrying phospholipase A activity promotes lipid antigen expression to T cells in psoriasis. IFN-gamma-induced release of exosomes carrying phospholipase by mast cells stimulate neolipid antigen presentation by CD1a-Langerhans cells to T cells, contributing to the pathogenesis of psoriasis via IL-17 and IL-22 expression by T cells. *Credit: Illustration created by Divya Aickara. Based on information in article by Cheung KL, et al. Psoriatic T cells recognize neolipid antigens generated by mast cell phospholipase delivered by exosomes and presented by CD1a. J Exp Med 2016;213(11):2399–2412.*

7 Exosomes in the fluid of bullous pemphigoid patients potentiates pathogenic inflammation

Bullous pemphigoid is an autoimmune blistering disease in which antibodies, produced by B cells, are directed against basement membrane hemidesmosomal antigens, BP180 and BP230, leading to formation of subepidermal blisters. Within the blister, there is fluid containing numerous inflammatory cells, including eosinophils and neutrophils, and CD9+, CD63+, and CD81+ exosomes [12]. These blister-fluid derived exosomes, most likely from the infiltrating granulocytes, were internalized by primary keratinocytes, inducing the expression of pro-inflammatory cytokines and chemokines and activating ERK1/2 and STAT3 signaling pathways, constituting a proinflammatory amplification loop (Fig. 6) [12]. In mass spectrometry analysis of blister fluid exosomes, immune and inflammatory molecules, such as s100-A8, matrix metalloproteinase 1, heat shock protein, and HLA class II histocompatibility antigen were enriched compared to control exosomes [12]. Neutrophil markers, including MPO, and eosinophils biomarkers, including eosinophil peroxidase and eosinophil cationic protein were present in blister fluid exosomes but control exosomes [12]. Overall, the proteomic profile of bullous pemphigoid blister-fluid-derived exosomes suggest they may contribute, or at least reflect, the inflammatory pathogenesis of bullous pemphigoid [12]. Overall, exosomes in blistering disease contribute to maintenance of pathogenic inflammation.

8 Skin repair

8.1 Stem cell exosomes stimulate skin repair

While skin trauma and breakdown are common, many cutaneous wounds are difficult to heal; when they do heal, they result in scarring that impairs cosmetic or functional outcomes. Conventional methods to accelerate healing include skin grafting, special biomaterials or dressings, application of growth factors, gene therapy and even laser therapy. Injection of growth factors alone may result in degradation by proteinase in the extracellular fluids. Goals of skin repair include shortening healing time and reducing scarring. There has been a focus on exosomes as regenerative tools in that they can shield growth factors from degradation and help transport proteins from one cell to the other. The role of exosomes in cutaneous wound healing has been an area of active exploration, because they are considered highly stable and of low immunogenicity. Furthermore, their ability to home to the site of tissue damage is crucial to potential repair of skin. In terms of pharmaceutical potential, they may be easier to standardize dosing and concentration [13]. In dermatology, exosomes are considered promising potential regenerative tools for cutaneous wound healing.

Accumulating evidence suggests mesenchymal stem cells are beneficial for repair and regeneration of the skin. Bone marrow-derived mesenchymal stem cells (BM-MSCs) dose-dependently enhanced proliferation and migration of

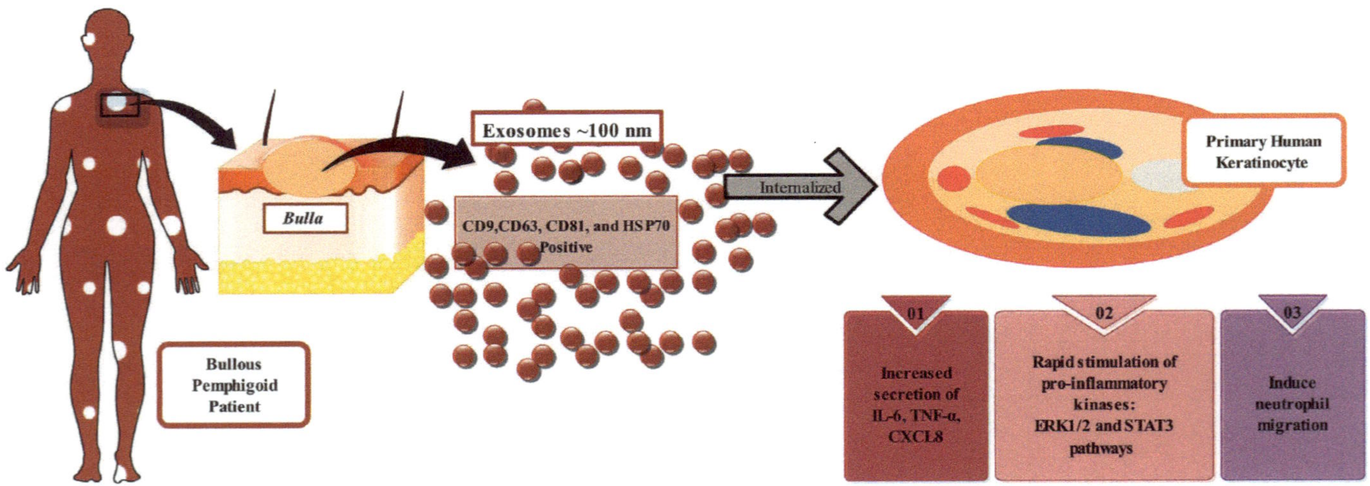

FIG. 6 Blister fluid exosomes in bullous pemphigoid are absorbed by keratinocytes to further promote proinflammatory activities. CD9, CD63, CD81 and HSP70+ exosomes are internalized by keratinocytes and result in increased secretion of IL-6, TNF-alpha, and CXCL8, stimulate pro-inflammatory kinases, ERK1/2 and STAT3 pathways, and induce migration of neutrophils. *Credit: Illustration created by Divya Aickara. Based on information in article by Fang H, et al. Proinflammatory role of blister fluid-derived exosomes in bullous pemphigoid. J Pathol 2018;245(1):114–125.*

fibroblasts derived from normal donors and chronic wound patients [14]. BM-MSC-exosomes activated several signaling pathways in fibroblasts, including the Akt, ERK, and STAT3 mediated pathways and induced the expression of numerous growth factors, including hepatocyte growth factor, insulin-like growth factor-1, nerve growth factor, and stromal-derived growth factor-1 [14]. Furthermore, BM-MSCs have been found to contain active transcription factors (STAT3) [14], and helps facilitate transport of hydrophobic/lipidated glycoprotein growth factors (ex. Wnt3a) [15] and basement membrane proteins (ex. Collagen VII) [16]. BM-MSC-exosomes transfer of Wnt3a stimulated fibroblast proliferation, migration and endothelial angiogenesis [15]. BM-MSC-extracellular vesicle-mediated transfer of collagen VII protein and *COL7A* mRNA stimulated proliferation of recessive dystrophic epidermolysis bullosa (RDEB) fibroblasts (that lack collagen VII production) and induced them to make new collagen VII protein [16]. When cells were treated with BM-MSC extracellular vesicles, then washed and allowed to secreted new proteins in serum-free conditioned media, there was a dose-dependent accumulation of collagen VII in the media, suggesting BM-MSC extracellular vesicles can induce production of collagen VII in RDEB fibroblasts (Fig. 7) [16]. Excessive scar formation caused by myofibroblasts is an important therapeutic target in

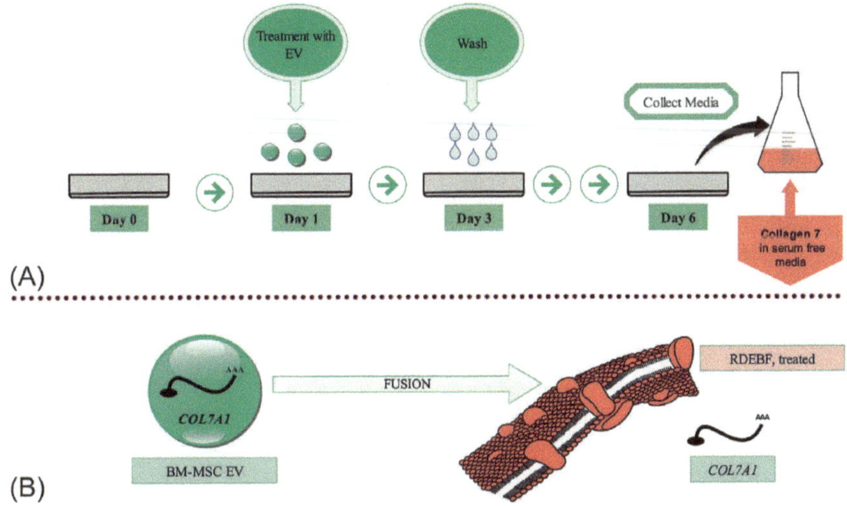

FIG. 7 Bone marrow-derived mesenchymal stem cell extracellular vesicles induce collagen VII production in fibroblasts from recessive dystrophic epidermolysis bullosa patient. (A) Treatment of RDEB fibroblasts at Day 1, for 48h followed by extensive washing, resulted in detectable production of collagen VII by day 6 in the serum-free conditioned media. (B) COL7A1 mRNA was detected in BM-MSC extracellular vesicles; when these vesicles fuse with RDEB fibroblasts, the mRNA can be internalized into the cell, making it available for use in production of new collagen VII protein. *Credit: Illustration created by Divya Aickara. Based on information in article by McBride JD, et al. Dual mechanism of type VII collagen transfer by bone marrow mesenchymal stem cell extracellular vesicles to recessive dystrophic epidermolysis bullosa fibroblasts. Biochimie 2018.*

scar prevention and treatment [17]. In a murine full-thickness cutaneous wound model, Umbilical cord-derived mesenchymal stem cells reduced scar formation and myofibroblast accumulation, in a manner largely dependent on exosomes and their microRNA contents, specifically miR-21, -23a, -125b, and 145, which suppressed myofibroblast formation via inhibition of the TGF-beta2/SMAD2 pathway [17].

Exosomes derived from human adipose mesenchymal stem cells (AD-MSC-exosomes) have also been shown to accelerate cutaneous wound healing via fibroblast reprogramming [13]. Because fat tissue is an active endocrine organ, supports the skin's normal physiologic state, provides nutrients and support, and contain stem cells and growth factors, there has been a focus on harnessing the potential of fat tissue to promote skin regeneration [13]. After AD-MSC-exosomes treatment, fibroblasts were stimulated to proliferation, migration, and produce collagens I and III and elastin in a dose-dependent manner [13]. Expression of various genes were enhanced: N-cadherin, cyclin-1, PCNA and collagen I and III [13]. In vivo, the AD-MSCs homed to the wound environment and secreted exosomes, with histologic evidence of increased collagens I and III in the wound in the early stages of wound healing [13]. Interestingly, in the late stage of acute wound healing, it appears AD-MSC-exosomes paradoxically inhibit collagen expression to reduce the scarring process. Human umbilical cord-MSCs secrete exosomes that can induce Wnt/beta-catenin signaling to accelerate skin repair [18]. However, in high cell density environments, umbilical cord-exosomes, via 14-3-3ζ protein, inhibited Wnt/beta-catenin signaling via YAP phosphorylation, ultimately to attenuate the Wnt signal, possibly contributing to decreased scarring of the wound repair in the remodeling phase [18].

Angiogenesis, the formation of new vascular sprouts and vessels from pre-existing vessels, is an important process in cutaneous wound healing. Angiogenesis requires cooperation between endothelial cells and their surrounding environment, resulting in proliferation, migration, invasion, sprouting and three-dimensional organization of endothelial cells to form new blood vessels [19]. AD-MSCs have been a candidate endogenous regulator of endothelial cells during angiogenesis, as they have been thought to create environments prime for neovessel formation via paracrine mechanisms [19, 20]. Both BM-MSC-exosomes and AD-MSC-exosomes were absorbed by endothelial cells (HUVECs) and promoted angiogenesis both in vitro and in vivo [14, 19]. In the AD-MSC-exosomes, miR-125a was enriched and repressed the expression of the delta-like 4 (DLL4), an angiogenic inhibitor, by targeting its 3′ untranslated region, promoting formation of endothelial vessel tip cells [19]. Overall, exosomes, especially from stem cells, induce a complex program that results in improved cutaneous repair in a variety of models.

9 Intercellular adhesion molecules associated with exosomes, such as desmogleins, secreted from malignant cells can modulate the extracellular environment to promote tumor progression

Interestingly, some components of the intercellular adhesion molecules have been shown to modulate expression of exosomes from certain types of skin cancer. Accumulating evidence has shown that cancer-derived extracellular vesicles, secreted into bodily fluids, can play an important role in modulating the tumor microenvironment, pathogenesis and metastasis of cancer [21]. Squamous cell carcinoma-derived extracellular vesicles were enriched with the C-terminal fragment of desmoglein 2 (Dsg2) [21]. When Dsg2 was overexpressed, extracellular vesicles release was increased by the squamous cell carcinoma cells [21]. When fibroblasts were co-cultured with SCC cells expressing Dsg2-green fluorescent protein (GFP), detection of GFP was found inside of the fibroblasts [21]. Dsg2 enhanced the mitogenicity of extracellular vesicles to enhance fibroblast cell growth [21]. SCC extracellular vesicles activated Erk1/2 and Akt signaling and stimulated fibroblast proliferation [21]. In head and neck SCCs, Dsg2 was highly upregulated; extracellular vesicles in SCC patients' sera showed enrichment of Dsg2 C-terminal fragment and growth factors, such as epidermal growth factor [21]. In SCC cells, Dsg2 down-regulated caveolin-1 levels, suggesting that Dsg2 may enhance endocytosis of Cavolin-1 [21].

10 Exosomes are critical in the pathogenesis of malignant melanoma

Exosome have been shown to be critical in the pathogenesis of melanoma cells. CD147, originally identified as a cell-surface protein on the immunoglobulin superfamily, is highly expressed on the surface of various tumor cells and is a marker of malignant tumors on the cell surface of melanoma [22]. CD147 expressed on malignant cells can induce tumor cell invasion by stimulating the production of matrix metalloproteinases (ex. MMP2) by fibroblasts and promotes angiogenesis [22]. Cleaved CD147 fragments exist on the surface of melanoma extracellular vesicles [22]. Matrix metalloproteinases are a family of enzymes that degrade different components of the extracellular matrix, and a number of studies indicate that they play an important role in remodeling of the extracellular environment [22]. It was found that malignant melanoma cells could stimulate the production of MMPs in fibroblasts without direct contact, giving a hint that exosomes may be involved in transporting a stimulating factor to the fibroblasts (ex. CD147 via extracellular exosomes) [22]. Melanoma is a highly aggressive form of cancer that accounts for 4% of cancers among men and women and approximately 80% of skin-cancer related deaths in the

US [23]. microRNAs (miRNAs) are a class of 22–25 nucleotide RNAs that modulate gene expression post-transcription [23]. A noninvasive screening tool to identify patients with a predisposition to melanoma is currently lacking, and blood holds several key advantages as a biomarker specimen, in that sampling and processing is much simpler than skin, or alternatively, as an extreme, performing whole body PET-CT scans to look for metastatic melanoma in all at-risk patients. In plasma-derived exosomal miRNAs from clinically affected familial melanoma patients (CDKN2A/p16 gene carriers), versus unaffected control subjects, miR-17, miR-19a, miR-21, miR-126, and miR-149 were significantly higher in patients with metastatic melanoma compared to those without [23]. These miRNAs were involved in regulation of nuclear hormone receptors, cell-cell adhesion molecules, tyrosine kinase receptors, lymphocyte transcription factors, p63 expression, zinc finger proteins, and tumor necrosis factor receptor superfamily members, and many others [23]. The miRNAs that are differentially expressed in metastatic melanoma patients may be used as predictive biomarkers to monitor onset, remission and relapse, and therapeutic responses [23]. The remarkable stability of exosomal miRNAs makes them great candidates to monitor disease progression in a variety of cancers [23].

While metastatic cancers can produce exosomes that condition pre-metastatic niches in remote microenvironments to promote metastasis, there is evidence that before melanoma metastasizes, it secretes exosomes that suppress metastasis [24]. These anti-metastatic exosomes stimulate an innate immune response through the expansion of a certain subset of monocytes, called Ly6C^{low} patrolling monocytes in the bone marrow [24]. These monocytes in turn can clear cancer cells from the pre-metastatic niche via NK cell and macrophage recruitment [24]. Very interestingly, these events depended on induction of the Nr4a1 transcription factor and pigment epithelium-derived factor (PEDF) on the outer surface of exosomes [24]. Exosomes isolated from patients with primary melanomas (but no metastases) have an ability to suppress lung metastases. Thus, exosomes from melanoma cells may initially suppress metastases indirectly via stimulation of the bone marrow and immune system to enhance surveillance and cancer cell clearance at potential sites of metastases [24].

11 Exosomes in the circulation signal presence of metastatic squamous cell carcinomas, especially in recessive dystrophic epidermolysis bullosa patients

Squamous cell carcinomas utilize exosomes to promote tumor progression. Transforming growth factor beta (TGFbeta) signaling promotes cancer growth and fibrosis but can act as a tumor suppressor or promoter depending on context [25]. Oral squamous cell carcinomas (SCCs) frequently have loss of function mutations in TFG-beta type II receptor [25]. Heterogeneity of TGFbeta activity in the leading edge of SCCs, which can influence drug resistance [25]. Exosomes from oral SCC-derived fibroblasts contained TGFbeta type

II receptor [25]. SCC transfer of TGF-beta type II receptor allowed keratinocytes devoid of the receptor to become responsive to TGF-beta, suggesting that stromal communication between tumor cells and non-tumor cells can alter the sensitivity to TGF beta signaling [25]. Patients with recessive dystrophic epidermolysis bullosa (RDEB) carry loss-of-function mutations in *COL7A1*, which is essential for maintaining the basement membrane in the skin and mucous membranes. Due to the constant and repeated cycles of wounding, inflammation, infection, and attempted wound healing, patients with recessive dystrophic epidermolysis bullosa are susceptible to developing invasive and aggressive squamous cell carcinomas. It was hypothesized that these squamous cell carcinoma tumors may emit exosomes containing detectable signatures in the circulation that may signal the presence of invasive squamous cell carcinomas [26]. Tumors that have metastasized to internal tissues are not easily detected or screened for on every clinical visit. One research group investigated the feasibility of using SCC-derived extracellular vesicles as a "liquid biopsy" to detect a tumor marker gene *Ct-SLCO1B3*, which encodes a member of the organic anion transporting polypeptide superfamily whose expression is typically in the liver, but had been reported to be overexpressed by RDEB-SCCs [26]. *Ct-SLCO1B3* was detected in extracellular vesicles from seven difference RDEB-SCC cell lines, but not in exosomes from normal keratinocytes or keratinocytes of RDEB patients [26].

12 Exosomes in circulation of patients with melanoma signal metastases and worsened prognosis

Accumulating evidence suggests that exosomes can regulate tumor metastasis; exosomes derived from dermatologic tumors are no exception. Tumor-derived exosomes are emerging as mediators of tumorigenesis. Exosomes from highly metastatic melanomas increased the metastatic behavior of the primary melanoma by activating tyrosine kinase MET in bone marrow progenitors [27]. Melanoma-derived exosomes induced vascular leakage at the pre-metastatic sites and reprogrammed bone marrow progenitors to support pathogenic blood vessel formation [27]. MET was elevated in the bone marrow progenitors of patients with metastatic melanoma and reduction of MET expression in exosomes diminished the pro-metastatic behavior of the bone marrow progenitors in the context of malignant melanoma [27]. In this same work, Peinado et al. identified an exosome-specific melanoma signature with both prognostic and therapeutic potential, comprising TRYP2, VLA-4, HSP70, an HSP90 isoform and the MET oncoprotein, with many of the genes relating to extracellular matrix remodeling and inflammation [27]. MET was higher in circulating exosomes in both stage 3 and stage 4 melanoma compared to control subjects or stage 1 disease [27]. In patients with Stage 4 melanoma, the cumulative survival probabilities of patients was significantly lower if the total amount of protein in isolated circulating exosomes per milliliter of plasma was greater than 50 micrograms per milliliter (versus less than 50 micrograms per milliliter), with all patients dying by 11months in the former group (versus greater than 50% surviving after 28months in the latter group). Thus, exosome content in the circulation of patients with metastatic melanoma, both in terms of total protein in exosomes and specific gene products, may help predict survival [27].

13 Conclusion

The role of exosomes in dermatology are beginning to be understood as critical for our understanding of fundamental biological processes in the skin. Furthermore, exosomes play key roles in the pathogenesis of atopic dermatitis, bullous pemphigoid, psoriasis, and skin cancer. As therapeutics, stem cell exosomes can stimulate skin repair and regeneration. With all that has been discovered to date, it is clear that the field of dermatology will benefit from exosome research, allowing developing better biomarkers for disease detection and prognosis and therapeutic tools to manage dermatologic disease.

References

[1] Lo Cicero A, et al. Exosomes released by keratinocytes modulate melanocyte pigmentation. Nat Commun 2015;6:7506.

[2] Waster P, et al. Extracellular vesicles are transferred from melanocytes to keratinocytes after UVA irradiation. Sci Rep 2016;6:27890.

[3] Bin BH, et al. Fibronectin-containing extracellular vesicles protect melanocytes against ultraviolet radiation-induced cytotoxicity. J Invest Dermatol 2016;136(5):957–66.

[4] Chavez-Munoz C, et al. Primary human keratinocytes externalize stratifin protein via exosomes. J Cell Biochem 2008;104(6):2165–73.

[5] Medina A, Ghahary A. Transdifferentiated circulating monocytes release exosomes containing 14-3-3 proteins with matrix metalloproteinase-1 stimulating effect for dermal fibroblasts. Wound Repair Regen 2010;18(2):245–53.

[6] Cai XW, et al. A novel noncontact communication between human keratinocytes and T cells: exosomes derived from keratinocytes support superantigen-induced proliferation of resting T cells. Mol Med Rep 2017;16(5):7032–8.

[7] Nickoloff BJ, et al. Accessory cell function of keratinocytes for superantigens. Dependence on lymphocyte function-associated antigen-1/intercellular adhesion molecule-1 interaction. J Immunol 1993;150(6):2148–59.

[8] Kotzerke K, et al. Immunostimulatory activity of murine keratinocyte-derived exosomes. Exp Dermatol 2013;22(10):650–5.

[9] Zhou L, et al. Regulation of hair follicle development by exosomes derived from dermal papilla cells. Biochem Biophys Res Commun 2018;500(2):325–32.

[10] Nakamura K, et al. Altered expression of CD63 and exosomes in scleroderma dermal fibroblasts. J Dermatol Sci 2016;84(1):30–9.

[11] Cheung KL, et al. Psoriatic T cells recognize neolipid antigens generated by mast cell phospholipase delivered by exosomes and presented by CD1a. J Exp Med 2016;213(11):2399–412.

[12] Fang H, et al. Proinflammatory role of blister fluid-derived exosomes in bullous pemphigoid. J Pathol 2018;245(1):114–25.

[13] Hu L, et al. Exosomes derived from human adipose mensenchymal stem cells accelerates cutaneous wound healing via optimizing the characteristics of fibroblasts. Sci Rep 2016;6:32993.

[14] Shabbir A, et al. Mesenchymal stem cell exosomes induce proliferation and migration of normal and chronic wound fibroblasts, and enhance angiogenesis in vitro. Stem Cells Dev 2015;24(14):1635–47.
[15] McBride JD, et al. Bone marrow mesenchymal stem cell-derived CD63(+) exosomes transport Wnt3a exteriorly and enhance dermal fibroblast proliferation, migration, and angiogenesis in vitro. Stem Cells Dev 2017;26(19):1384–98.
[16] McBride JD, et al. Dual mechanism of type VII collagen transfer by bone marrow mesenchymal stem cell extracellular vesicles to recessive dystrophic epidermolysis bullosa fibroblasts. Biochimie 2018;155:50–8.
[17] Fang S, et al. Umbilical cord-derived mesenchymal stem cell-derived exosomal microRNAs suppress myofibroblast differentiation by inhibiting the transforming growth factor-beta/SMAD2 pathway during wound healing. Stem Cells Transl Med 2016;5(10):1425–39.
[18] Zhang B, et al. HucMSC exosome-delivered 14-3-3zeta orchestrates self-control of the Wnt response via modulation of YAP during cutaneous regeneration. Stem Cells 2016;34(10):2485–500.
[19] Liang X, et al. Exosomes secreted by mesenchymal stem cells promote endothelial cell angiogenesis by transferring miR-125a. J Cell Sci 2016;129(11):2182–9.
[20] Bronckaers A, et al. Mesenchymal stem/stromal cells as a pharmacological and therapeutic approach to accelerate angiogenesis. Pharmacol Ther 2014;143(2):181–96.
[21] Overmiller AM, et al. Desmoglein 2 modulates extracellular vesicle release from squamous cell carcinoma keratinocytes. FASEB J 2017;31(8):3412–24.
[22] Hatanaka M, et al. Cleaved CD147 shed from the surface of malignant melanoma cells activates MMP2 produced by fibroblasts. Anticancer Res 2014;34(12):7091–6.
[23] Pfeffer SR, et al. Detection of exosomal miRNAs in the plasma of melanoma patients. J Clin Med 2015;4(12):2012–27.
[24] Plebanek MP, et al. Pre-metastatic cancer exosomes induce immune surveillance by patrolling monocytes at the metastatic niche. Nat Commun 2017;8(1):1319.
[25] Languino LR, et al. Exosome-mediated transfer from the tumor microenvironment increases TGFbeta signaling in squamous cell carcinoma. Am J Transl Res 2016;8(5):2432–7.
[26] Sun Y, et al. Extracellular vesicles as biomarkers for the detection of a tumor marker gene in epidermolysis bullosa-associated squamous cell carcinoma. J Invest Dermatol 2018;138(5):1197–200.
[27] Peinado H, et al. Melanoma exosomes educate bone marrow progenitor cells toward a pro-metastatic phenotype through MET. Nat Med 2012;18(6):883–91.

Chapter 11

Exosomes in nephrology

Robert W. Hunter, James W. Dear, and Matthew A. Bailey
University/BHF Centre for Cardiovascular Science, Queen's Medical Research Institute, University of Edinburgh, Edinburgh, United Kingdom

1 Introduction

1.1 Aim

In this chapter, we aim to provide an up-to-date insight into exosomes in the kidney. We will review the role of exosomes in normal physiology and in disease.

We will begin by considering urinary exosomes, the main source of clinical and experimental exosomal material from the kidney and urinary tract. We will discuss the potential for urinary exosomes to provide biomarkers of nephrological disease. We will review the evidence that exosomes have important physiological (and pathophysiological) functions by mediating communication between kidney cells. Finally, we will consider the potential for exosome-based therapies in kidney disease.

1.2 Exosomes in the kidney and urinary tract: Special considerations

In several respects, the kidneys and urinary tract constitute a "special case" with regards to exosome signaling. First, they have—in the urine—a highly accessible and near-inexhaustible supply of organ-specific EVs, providing an unparalleled bioresource for researchers and clinicians. Second, the urinary tract is lined with distinct spatially-separated cell types that all contribute to a mixed population of EVs in the urine. This constitutes an ideal system in which to study cell-specific EV signaling and furthermore the flow of urine allows for the study of directional EV signaling (from proximal to distal along the urinary tract). Third, the complex micro-anatomy of the glomerular filtration barrier and the tubular-interstitial-vascular interface provides a system in which we can examine the mechanisms through which EVs can travel through the body tissues.

Each of these features presents opportunities and challenges to the researcher and clinician, which we will elaborate on throughout this chapter.

Exosomes. https://doi.org/10.1016/B978-0-12-816053-4.00011-0

1.3 A word on nomenclature

As in other fields, much of the literature on exosomes in nephrology suffers from a confused nomenclature. In particular those studies carried out in the early years of this century, extracellular vesicle sub-populations were either not defined or were defined by a hotchpotch of non-standardized criteria. Therefore the terms "extracellular vesicle", "exosome", "microvesicle", "microparticle", "nanovesicle" (and more besides) were used to refer to populations of vesicles displaying various characteristics. Some terms have even been coined for vesicles peculiar to the urinary tract ("prostasomes").

In this chapter, we will adhere to recent consensus statements regarding nomenclature [1, 2], in which extracellular vesicles are classified according to their biogenesis. We will therefore restrict the term "exosome" to those vesicles displaying properties compatible with origin from a vesiculation event from endosomal pathways (diameter 30–100 nm and enriched markers such as CD63, CD81, flotillin-1 and Tsg101) [3, 4]. We will use the term "microvesicle" to refer to those particles likely to have originated from a vesiculation event at the plasmalemma (diameter 100–1000 nm). When discussing extracellular vesicles that are not so rigorously defined, we will use the broad term "extracellular vesicle" (EV).

2 Urinary exosomes

2.1 Isolation methods

Cells in the kidney and urinary tract release EVs into the urine; these were first characterized in a seminal paper from the Knepper lab [5]. This highly-accessible supply of biomaterial has been enormously helpful in the study of EVs in nephrology. However, the isolation of EVs from urine requires some special methodological considerations.

2.1.1 Interference of urinary proteins with isolation of EVs

Urinary proteins can interact with urinary exosomes and interfere with attempts at their isolation. The most abundant urinary protein, uromodulin (also known as Tamm-Horsfall protein) is structurally related to the zona pellucida proteins and precipitates in the cold (~4°C) and with changes in urine pH and ionic strength [6, 7]. The precipitates form long, rope-like polymeric filaments which are visible on electron microscopy [6, 8]. EVs may become entrapped within a uromodulin mesh and lost in the low-speed pellet during differential centrifugation protocols. This can be circumvented by using reducing agents (e.g. dithiothrietol) to disrupt polymeric uromodulin [6, 7, 9] or by isopycnic centrifugation on a D_2O-sucrose gradient [8].

The confounding effects of urinary protein become more significant in the context of pathological proteinuria, such as in the nephrotic syndrome. Using standard ultracentrifugation protocols, abundant proteins such as albumin and α1-antitrypsin are co-purified with exosomes. To an extent, this problem can be solved by using size-exclusion chromatography to separate the exosomal and free protein fractions [10] or by isopycnic centrifugation on a D_2O-sucrose gradient [11]. Subtle variations in the method used to isolate EVs may have a bearing on the composition of the final sample. For example, fractionation by density gradient ultracentrifugation can yield sub-populations of EVs enriched for aquaporin-2, polycystins-1 and -2 or podocin [8]. These proteins are expressed in different nephron segments and therefore these EV fractions are likely derived from different nephron segments. For example, the podocin-enriched fraction possesses molecular markers compatible with a glomerular origin [11].

2.1.2 Normalization of EVs and EV cargo in urine samples

Another challenge that faces the researcher or clinician when working with urinary EVs is that of normalizing any result. Urine flow rate varies enormously during health and disease and therefore it is probably not particularly meaningful to express results as a concentration (e.g. number of EV particles or micrograms of EV protein per ml or urine). It may be that the flow rate of renal tubular fluid or urine alters the rate at which EVs are released or taken up by the cells lining the urinary tract (to our knowledge this has yet to be rigorously tested). Therefore measuring excretion rates—the number of EV particles excreted in the urine per unit time—may not be sensible either.

Different investigators have adopted different approaches to this problem, often normalizing to urinary creatinine concentration, which is widely used clinically to normalize urinary excretion of protein and other solutes in untimed urine samples. One proposed alternative is to normalize to the concentration of uromodulin because this protein will usually be isolated along with EVs and because the abundance of uromodulin is highly correlated with that of canonical EV markers in the urine of healthy volunteers [6]. However, rates of uromodulin excretion vary significantly between different individuals, so this approach is imperfect when normalization is required to allow comparison of urinary EV products between different subjects [12]. A similar approach is to normalize to total urinary protein—which is a reasonable thing to do, except in subjects with pathological proteinuria.

2.1.3 Methods of isolation of EVs from urine

Other than taking into account the presence of interfering urinary proteins, the only other special consideration for urine—compared to other biofluids—is that sample volume can be large (typically 10s or 100s of milliliters). Differential ultracentrifugation is still considered by many to be the gold-standard method of preparing exosomes from urine; but methods employing filtration, precipitation, immunoaffinity capture or size-exclusion chromatography provide alternatives that can be better suited to large input volumes [13–15]. In a head-to-head comparison of alternative protocols, a modified precipitation protocol gave the highest yield of exosomes and RNA [13].

2.2 Cellular origin and molecular cargo of urinary exosomes

The molecular cargo of urinary EVs has been well-characterized and is similar to that carried by EVs in other biofluids in that it comprises lipids, proteins and nucleic acid and is particularly enriched in non-coding RNA. Urinary EVs exhibit complex surface glycosylation [16].

The concentration of exosomal protein in urine is ~2 μg/mL, accounting for ~3% of total urinary protein [17]. The urinary exosomal proteome (and phosphoproteome) is enriched in apical membrane proteins [9], consistent with their likely biogenesis: that is, the release of EVs directly into the urinary space from the apical surface for cells lining the urinary tract. However cytosolic and nuclear proteins, such as transcription factors, are also detected [18].

The nucleic acid in urinary EV preparations predominantly comprises DNAse-susceptible DNA (presumed to coat the outside of vesicles) and RNAse-resistant RNA (presumed to lie protected within vesicles) [19]. In RNA-sequencing data from healthy human subjects, almost 90% of the reads aligned to ribosomal RNA and of the remainder, ~60% aligned to non-coding RNA sequences [20]. mRNA sequences were enriched in transcripts that are specifically expressed in the nephron, collecting system and lower urinary tract [19, 20].

2.3 Uptake of EVs by kidney cells

The cells lining the urinary tract release EVs into the urine, but they are also able to internalize EVs from the urinary and vascular space. Furthermore, the uptake of EVs (or at least a subset of EVs) by kidney cells appears to be selective and physiologically regulated.

2.3.1 Molecular basis of EV uptake

The cellular uptake of EVs in the kidney has been studied most extensively in the renal tubular epithelial cell. EVs can be taken up across either the apical or basolateral cell membrane [21]. There may be role for apical cilia in facilitating EV uptake: some EVs physically interact with the primary cilium on renal tubular (IMCD) cells and biliary epithelial cells [8].

The cellular uptake of fluorescent EVs (derived from renal tubular cells expressing CD63-GFP) is competitively inhibited by non-fluorescent EVs. Thus EV uptake occurs through saturable pathways [22].

Trypsinization of EVs abolishes their ability to be taken up by kidney cells in vitro and in vivo [23], suggesting that their uptake is dependent on surface protein expression. Some investigators have defined specific surface proteins that are important for EV uptake (for specific cells and specific EVs). For example, blockade of CD29 or CD44 inhibited the uptake of MSC-derived EVs by renal tubular cells [23, 24]. The likely endogenous ligand for CD44, hyaluronic acid, is expressed in the renal interstitium and along the tubular basement membrane [25, 26]. The interaction of podocyte-derived EVs with renal tubular epithelial cells can be inhibited by CD36 blockade [27].

To an extent, the cellular uptake of EVs is selective with respect to the origin of the EVs. Mouse distal renal tubular epithelial cells internalized EVs derived from human proximal renal tubular epithelial cells but not mouse juxtaglomerular cells [21].

These data are all consistent with a model in which the uptake of EVs is dependent on specific interactions between surface proteins on cell and EV (Fig. 1). Such a model is gaining acceptance in the broader EV field: it seems likely that EVs dock selectively with recipient cells through specific receptor-ligand interactions and are then internalized through an active endocytic process [28].

2.3.2 *Physiological regulation of EV uptake and release*

The uptake of EVs by collecting duct cells is known to be regulated by vasopressin and endothelin-1: both established physiological regulators of collecting duct cell function [21, 29]. As all urinary EVs must have passed through the collecting duct, this suggests that the exosomal composition of the urine might

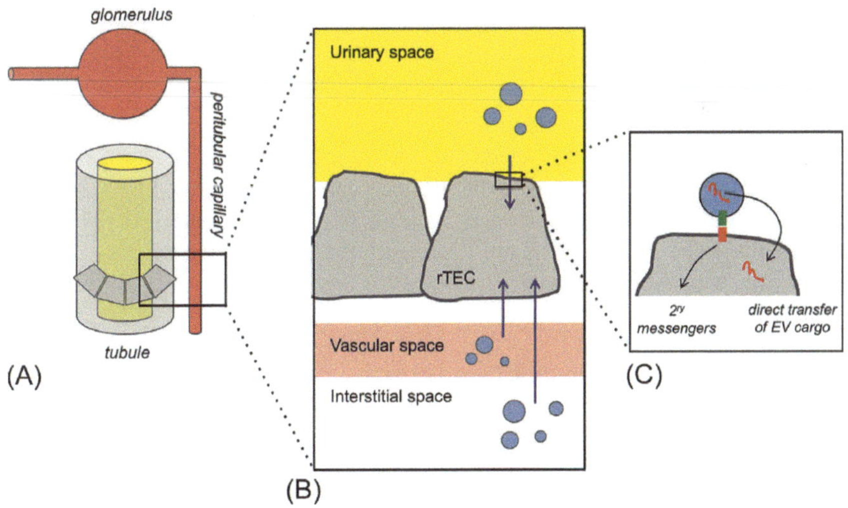

FIG. 1 Model of exosome signalling in the renal tubule. (A and B) In theory, EVs could enter the urinary space through a number of pathways. They could be released by the cells lining the urinary tract or enter from the vascular or interstitial space by crossing the glomerular filtration barrier or the renal tubule. There is some experimental evidence to support all of these routes (see text). (C) In common with other body systems, EVs enter kidney cells by interacting with specific cell-surface molecules. They can influence target cell function through the transfer of a bioactive cargo (nucleic acid/protein/lipid) or by stimulating cell-surface receptors that activate second-messenger cascades.

vary according to vasopressin signaling activity (e.g. with hydration status). In support of this hypothesis, the urinary excretion of EVs expressing podocyte and proximal tubule markers fell after vasopressin treatment in a patient with central diabetes insipidus [21]. The physiological regulators of EV uptake in other (non-vasopressin-responsive) kidney cells has not been extensively investigated.

In proximal tubular cells, EV release was stimulated (at least in vitro) by dopamine and angiotensin II [22]; the authors speculated that both agents might stimulate EV release by raising intracellular calcium levels, but this was not tested directly.

2.3.3 *In vivo delivery of EVs to kidney cells*

One controversial question that has yet to be definitively resolved is whether or not EVs arising from distant organs can enter the urine. On the one hand, the size and charge-profile of EVs should mean that they cannot traverse the glomerular and tubular basement membranes [30, 31]. In keeping with this, the vast majority of RNA and protein molecules detected in urinary EVs from healthy subjects are derived from the kidney and urinary tract [9, 19].

On the other hand, there is evidence that circulating EVs (or EV cargo) are taken up by kidney cells, at least in the context of kidney injury. After fluorescently-labeled EVs are injected into the systemic circulation of mice, the fluorescent label can be detected in renal tubular cells [21, 23, 32]. Similarly, markers of blood cell-derived EVs are detected in the kidney cells of humans and mice with *E. coli*-associated hemolytic uremic syndrome [33]. However, in some systems this phenomenon is only observed in the context of acute kidney injury and not in healthy controls [23, 32].

Remarkably, there is also evidence that intact EVs can pass from the vascular space into kidney cells and the urine. Fluorescently-labeled EVs injected into the circulation were detected in urine by nanotracking analysis [21]. This urinary fluorescent signal was not detected after the injection of free dye and was greatly reduced by detergent treatment of the urine.

3 Urinary exosomes as biomarkers

3.1 Two broad approaches to biomarker research

EVs have the potential to act as biomarkers because their molecular cargo could be used to infer the molecular status of the cell of origin. Proteomic, transcriptomic and lipidomic approaches have been used to identify molecular biomarkers in urinary EVs.

The potential for EVs to act as biomarkers has been explored for two broad applications: as *mechanistic* biomarkers, to provide insight into the function of specific cell types in human disease or animal models and as *clinical* biomarkers, to help diagnose renal or urological diseases.

3.2 Mechanistic biomarkers reporting on renal tubular function

3.2.1 Water transport

The abundance of tubular transport proteins in urinary EVs has been shown to correlate with in vivo transport activity in a variety of settings. In fact the first demonstration of this phenomenon predated the widespread appreciation of urinary EVs: vasopressin was shown to increase urinary aquaporin 2 (AQP2) excretion in patients with central diabetes insipidus [34]. We have since learnt that most urinary AQP2 is excreted in exosomes [35] and the correlation between AQP2 expression in urinary EVs and in vivo vasopressin signalling activity has been replicated several times [29, 36].

3.2.2 Sodium transport

The transport activity of tubular sodium transporters correlates with their abundance in the apical cell membrane and is regulated by post-transcriptional modifications such as phosphorylation. It is therefore plausible that the abundance and/or phosphorylation status of tubular sodium transporters in urinary EVs should correlate with their in vivo transport activity. As proof-of-concept, this was verified in rare Mendelian disorders. Gitelman and Gordon syndrome respectively cause under- and over-activity of the thiazide-sensitive NaCl co-transporter, NCC. In Gitelman syndrome, the abundance of NCC protein in urinary EVs is markedly reduced [37, 38], whereas in Gordon syndrome it is increased [39]. An analogous phenomenon is observed in type I Bartter syndrome, caused by mutations in the furosemide-sensitive Na-K-Cl co-transporter NKCC2 [9, 37].

But does the abundance of sodium transporters in urinary EVs correlate with expression in the kidney during physiological conditions? This has been tested in animal models and humans. Esteva-Font et al. found that this correlation was indeed observed for NCC and NKCC2 in rats during perturbations in dietary sodium content [40]. However, they did not observe changes in urinary exosomal NCC and NKCC2 expression in hypertensive human subjects in response to manipulations in dietary sodium that were sufficient to change the fractional excretion of sodium. Other groups have found increased in the expression of NCC and phosphorylated NCC in urinary exosomes in human subjects with primary aldosteronism: a disease that would be expected to increase renal tubular expression [41]. Similarly, cyclosporin therapy (known to increase NCC expression through its effects on WNK-SPAK signalling) in renal transplant recipients increased the abundance of NCC and NKCC2 in urinary exosomes [42]. Treatment with thiazide diuretics (inhibitors of NCC) increased both the abundance and phosphorylation of NCC in urinary exosomes and this effect was greater in patients who mounted a blood pressure response to the thiazide [43]. This is in keeping with rodent studies, in which chronic thiazide exposure causes increased NCC expression. Similarly, the abundance of NCC and phosphorylated NCC in urinary EVs from kidney transplant recipients was influenced by calcineurin inhibitor exposure and predicted blood pressure response to thiazides [44].

Urinary exosomal NCC expression exhibits diurnal variation in humans and rodents [45, 46]. This is in keeping with the known diurnal variation in kidney expression in rodents, but interestingly diurnal variation was also observed for many other exosomal proteins [45].

3.2.3 Acid-base transport

Similar to the effects observed for water and sodium transporters, the abundance of acid-base transporters in urinary EVs changes in response to perturbations in acid-base homeostasis. In acidosis, the excretion of the B1 subunit of V-ATPase (expressed on acid-excreting intercalated cells) increases whereas that of pendrin (expressed on base-excreting cells) decreases [47, 48]. The opposite is observed in alkalosis.

3.2.4 Tubular profiling

Some groups have attempted to quantify a panel of transporter proteins in urinary EVs in order to obtain a "profile" of transport activity in different tubular segments. This approach was used in an attempt to investigate the mechanism responsible for urinary concentrating and acidification defects in patients with American cutaneous Leishmaniasis. Expression of exosomal transporter proteins changed in a direction that would—if representative of kidney expression—provide a mechanistic explanation. Patients with a urinary concentrating defect had lower AQP2 and (possibly compensatory) higher NKCC2 expression; patients with a urinary acidification defect had higher NHE3, H+-ATPase and pendrin expression [49].

Healthy human subjects treated with an endothelin agonist exhibited increased free water clearance. Analysis of urinary exosomal protein demonstrated that this occurred in the context of increased NKCC2 and AQP2 excretion. This suggests that the aquauresis was caused by increased sodium transport on the loop of Henle (rather than by reduced AQP2 activity in the collecting ducts) and in keeping with this, renal clearance parameters also suggested increased sodium absorption in the loop [50].

In patients with thiazide-induced hyponatremia, urinary exosomal abundance of NCC and AQP2 were reduced and increased respectively relative to normonatremic controls [51]. These changes—if representative of transporter/channel activity in the kidney—could account for the hyponatremia as they would favor sodium loss and water retention. Furthermore these differences persisted several months after thiazides were withdrawn, suggesting that they may have pre-dated thiazide exposure and could thus be potentially used as biomarkers to predict the risk of thiazide-induced hyponatremia.

3.3 Biomarkers of renal disease

Urinary EVs have a clear potential to aid in the diagnosis of rare monogenic diseases (such as those discussed above). However, given the diverse array of nucleic acids, proteins, lipids and other macromolecules carried by urinary EVs, many investigators have used systems biology in an attempt to discover clinically useful signals encoded within the complex urinary EV cargo.

EVs offer several potential advantages as a biomarker reservoir over whole urine or the soluble fraction. First, certain molecular classes are enriched in EVs (e.g. membrane-bound proteins, non-coding RNA). RNA may even be protected within the vesicle core from nuclease degradation. Moreover, by looking for co-expression of markers that are associated with the cell of origin, it might be possible to estimate the abundance of a biomarker of interest in a cell type of interest—thus improving the signal-to-noise ratio compared to whole urine.

3.3.1 EV abundance as a biomarker of cell damage

One hypothesis is that the *number* or *physio-chemical* properties of EVs could provide information about the status of the cells from which they originated. To take a simple example, if the rate of EV release was increased in cell injury then the number of EVs detected in the urine could correlate with cell damage. There is some support for this approach in diabetic kidney disease and pre-eclampsia. In both cases the abundance of podocyte-derived urinary EVs is increased, in keeping with the glomerular damage observed in these conditions [52–54]. In this approach, a single molecular marker is used to define the cell of origin: for example, podocin or nephrin for the podocyte.

More sophisticated approaches have used the co-expression of more than one molecular marker to define EVs derived from subpopulations of cells. For example, CD133+ urinary EVs (i.e. derived from progenitor cells) that co-express glomerular or proximal tubular markers increase in abundance in the days after kidney transplantation. This may suggest that progenitor cells are contributing to glomerular and tubular repair after transplant [55, 56].

3.3.2 EV cargo as biomarkers in disease

Alternatively, the molecular *cargo* of the EVs might provide information about disease states. Transcriptomic, proteomic and lipidomic approaches have discovered potential biomarkers of various renal diseases such as acute tubular injury, glomerular disease and transplant rejection (Table 1); this has been reviewed extensively [72, 56]. To take one example, urinary EVs from a cohort of 15 patients with CKD (of mixed etiology) displayed differential expression of 30 non-coding RNAs, compared to EVs from 10 health controls [66]. This pilot work is encouraging; it remains to be seen whether larger studies in more focused disease groups will discover RNA 'signatures' that can be used to make reliable, clinically useful decisions.

TABLE 1 Biomarkers in urinary EVs that have been tested in human disease subjects.

Cell of origin	Biomarker	Disease	Reference
Glomerular diseases and diabetic kidney disease			
Podocyte	EV number	DKD (in T1DM)	[54]
?	miRNAs	DKD (in T2DM)	[57]
?	EV size, density & proteome	DKD (in T2DM)	[58]
?	Regucalcin	DKD	[59]
Podocyte	WT-1	Podocyte injury/FSGS	[60]
?	Proteome	IgAN and TBMN	[61]
Podocytes	ADAM10	Glomerular diseases	[62]
?	miRNA	SLE and lupus nephritis	[63]
Podocytes	EV number	Pre-eclampsia	[53]
Tubular disorders, AKI, CKD, hypertension			
?	Fetuin A	AKI	[64]
rTECs	Osteoprotegerin	CKD	[65]
rTECs	ncRNA	CKD	[66]
Podocyte		Renovascular hypertension	[67]
?	Proteome	Essential hypertension	[68]
rTECs	NCC, AQP2, PGT	Thiazide-induced hypoNa	[51]
Kidney transplant			
Progenitor	CD133	Graft function	[55]
?	NGAL	Delayed-graft function	[69]
?	Proteome	Rejection/tubular injury	[70]
?	Proteome	Acute rejection	[71]

AKI, acute kidney injury; AQP2, aquaporin-2; CKD, chronic kidney disease; DKD, diabetic kidney disease; FSGS, focal segmental glomerulosclerosis; hypoNa, hyponatraemia; IgAN, IgA nephropathy; NCC, sodium-chloride co-transporter; ncRNA, non-coding RNA; PGT, PGE_2-specific prostaglandin transporter; rTEC, renal tubular epithelial cell; SLE, systemic lupus erythematosus; TBMN, thin basement membrane nephropathy; T2DM, type 2 diabetes mellitus.
? = cell type not known.

As with the work on "mechanistic biomarkers", animal studies have been essential in exploring the correlation between the abundance of a given "injury" biomarker in urinary exosomes and its abundance in the kidney. For example, transcriptomic analysis of urinary EVs from a podocyte injury model showed cystatin C was differentially regulated in injury. This differential expression was mirrored by mRNA and protein expression in podocytes and tubular cells in vivo [73].

4 Functional role of exosomes in the kidney and urinary tract

As in other body systems, accumulating data support the hypothesis that exosomes play a *functional* role in the kidneys and urinary tract: they are not merely cellular waste products. Broadly, they may exert function by interacting with surface receptors on recipient cells or by transferring a functional protein, lipid or nucleic acid cargo to recipient cells. There are numerous reports of EVs modifying the function of kidney cells (Table 2; Fig. 2). The majority describe effects on the renal tubular epithelial cell and we have therefore focused our discussion on these cells.

4.1 Exosome signalling regulating tubulointerstitial cells

4.1.1 Regulation of tubular cell transport function by EV signalling

One of the first reports described how EVs released by vasopressin-stimulated collecting duct cells stimulated water transport in recipient (vasopressin-naïve) collecting duct cells [29]. The abundance of APQ2 in the EVs was increased after vasopressin treatment whereas the abundance of "exosome markers" flotillin-1 and Tsg101 remained constant and we have since learnt that urinary EVs contain functional AQP2 water channels [98]. Therefore it is possible that this effect is mediated by the transfer of functional AQP2 channels but this has yet to be tested rigorously.

Another early report examined signalling from proximal to distal renal tubular cells in vitro. EVs from proximal tubule cells that had been stimulated with a dopamine agonist reduced reactive oxygen species production in recipient distal tubular cells [22]. The mechanism responsible for this effect was not determined but the investigators were able to demonstrate, using fluorescent CD63 and CD9 tags, that EVs were internalised by recipient cells and in some cells the fluorescent label subsequently localized to multivesicular bodies. The potential for EVs to carry signals from the proximal to distal nephron was also examined by Jella et al. They found that exosomes from proximal renal tubular cells decreased the activity of the epithelial sodium channel, ENaC in distal tubular cells, with a greater effect exerted by exosomes released from the apical rather than the basolateral cell surface [91]. They found strong evidence to support a mechanism

TABLE 2 Functional communication by EVs in the kidney.

Donor cell	Recipient cell	Transfer material	Functional effect	Experiment	Reference
EVs from stem cells inducing protection from injury					
MSCs	rTECs	RNA	Enhanced recovery from AKI; resistance to apoptosis	In vitro and in vivo; RNAse as negative control	[23]
MSCs	?	?	Protection against renal injury (5/6 nephrectomy model)	In vivo; effect comparable to infusion of MSCs	[74]
MSCs	?	?	Protection against kidney injury (oxidative stress pathways)	in vitro and in vivo (IRI model)	[75]
MSCs	rTECs	miRNA	Protection from AKI	In vitro; Actinomycin D used to infer transfer of miRNA	[24]
MSCs	rTECs	Hepatocyte growth factor (HGF) mRNA	Protection against tubular injury	In vitro and in vivo; RNAse used to demonstrate RNA-dependent effect	[76]
MSCs	rTECs (HK-2)	miR	Protection against TGFβ-induced EMT; protection against UUO-induced kidney injury	In vitro and in vivo	[77]
MSCs	rTECs	miRNA	Protection from AKI	In vivo; Drosha knock-down negative control	[78]
MSCs	rTECs (NRK-52E)	miRNA (let7c)	Protection from fibrosis	In vitro and in vivo; Cy3-labelled pre-miRNA used to visualize transfer	[79]
MSCs	rTECs	miRNA/mRNA	Recovery from AKI and rTEC proliferation	In vitro and in vivo	[80]
MSCs	rTECs and macrophages	IL-10	Inhibition of renal inflammation	In vivo (porcine model of metabolic syndrome and renal artery stenosis)	[81]

MSCs and HLSCs	Mesangial cells	miR-222	Protection from high-glucose injury	In vitro	[82]
BMDCs	Fibroblasts	miR	Protection against tubulointerstitial fibrosis (tPA signalling pathway)	In vivo and in vitro; demonstrate miR-144 in EVs targets 3′UTR of tPA in renal fibroblasts	[83]
EPCs	rTECs	miRNA	Protection from AKI	In vivo; RNase, Dicer knock-down and miR-antagomir negative controls	[84]
EPCs	Mesangial cells	mRNA	Protection against mesangial injury (Thy1.1 nephritis)	In vitro and in vivo; RNAse used to demonstrate RNA-dependent effect	[85]
Signalling from renal tubular epithelial cells to renal tubular epithelial cells					
? (whole urine)	rTECs	miRNA	Change in expression of solute transport pathways	In vitro	[86]
rTECs	rTECs	Protein (AQP2)	Transfer of water transport capacity	In vitro	[29]
rTECs	rTECs	?	Enhanced recovery from AKI	In vivo	[87]
rTECs	rTECs	miRNA	Down-regulation of miRNA target genes	In vitro and in vivo; exogenous miRNAs as positive control	[21]
rTECs	rTECs	?	Transfer of dopamine signalling pathway activation	In vitro; fluorescently-tagged CD9 and CD63 to demonstrate transfer	[22]
rTECs	rTECs	HIF-1α	Protection from apoptosis	In vitro	[88]
Whole kidney/HK-2	?	?	Hypoxia-induced EVs protect kidneys from IRI	In vitro and in vivo	[89]
rTECs	rTECs	?	Inhibition of wound healing	In vitro; effects of exosomes on EGFR signalling tested	[90]

Continued

TABLE 2 Functional communication by EVs in the kidney.—cont'd

Donor cell	Recipient cell	Transfer material	Functional effect	Experiment	Reference
rTECs (proximal)	rTECs (distal)	GAPDH	ENaC activity	In vitro	[91]
rTECs	rTECs	ATF3 (transcription factor)	Protection against kidney injury (MCP-1 pathway)	In vitro and in vivo	[92]
Signalling between other kidney compartments					
HUVECs/ monocytes	Podocytes	?	Transfer of TNFα-induced pro-inflammatory phenotype	In vitro	[93]
Podocytes	rTECs	None (surface interaction)	Pro-fibrotic signalling	In vitro; blockade of CD36 used to demonstrate surface interaction	[27]
Mesangial	Podocytes	TGFβ1	Transfer of high-glucose induced injury	In vitro	[94]
Leukocytes	Glomerular endothelial & HEK	Kinin B1-receptor	Kinin signalling in vasculitis	In vitro and in vivo in human disease subjects	[95]
rTECs	MSCs	miRNA	Induction of MET	In vitro; miR-mimics as positive control	[96]
rTECs	Macrophages	CCL2 mRNA	Induction of inflammation in response to albuminuria	In vitro and in vivo; GFP-CCL2 fusion construct used to demonstrate transfer of mRNA to macrophages	[97]

?, not determined; AKI, acute kidney injury; AQP2, aquaporin 2; BMDC, bone-marrow-derived cell; EPC, endothelial progenitor cell; HLSC, human liver stem cell; HUVEC, human umbilical vein endothelial cell; IRI, ischemia-reperfusion injury; MET, mesenchymal-to-epithelial transition; miRNA, microRNA; MSC, mesenchymal stem (or stromal) cell; rTEC, renal tubular epithelial cell; UUO, unilateral ureteric obstruction.

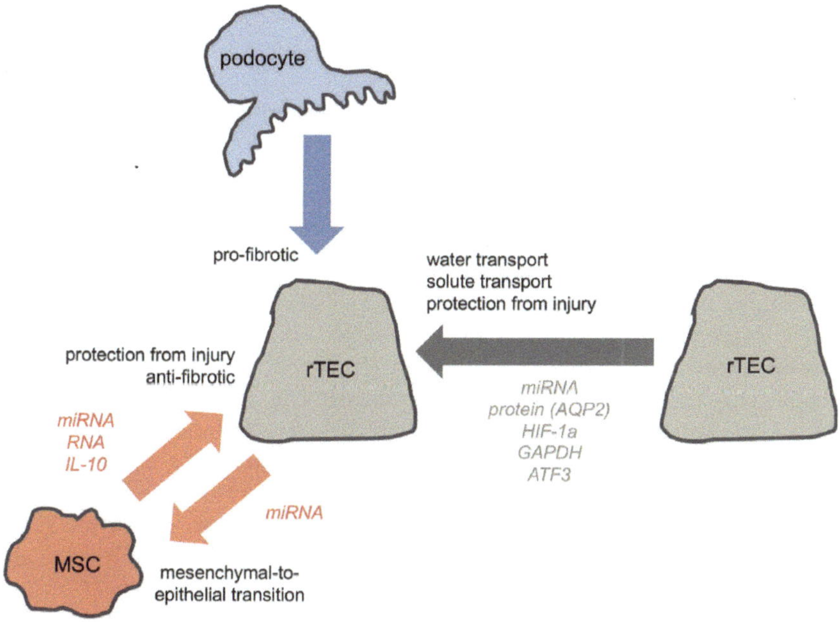

FIG. 2 Functional role of EVs signalling to renal tubular epithelial cells. Several studies have shown that EVs can exert functional effects on renal tubular epithelial cells. These are summarized here; see Table 2 for details.

involving the transfer of GAPDH protein: the exosomes expressed catalytically-active GAPDH, pharmacological inhibition of GAPDH in exosomes abolished their stimulatory effects on ENaC and co-immunoprecipitation experiments suggested a physical interaction between ENaC subunits and GAPDH in distal tubular cells.

Fiona Karet's group took a broader approach, testing the general hypothesis that urinary exosomes might transfer regulatory microRNAs to renal tubular cells. Using small RNA-sequencing, they found that urinary exosomes in healthy volunteers were enriched in miRNAs predicted to target membrane transporters and their regulators. In culture, urinary exosomes were able to regulate the expression of solute transporters in distal and proximal tubular cells [86]. This suggests—but does not directly prove—that miRNAs might form a critical element of the EV cargo responsible for exerting functional effects in target renal tubular cells.

4.1.2 Stem-cell derived EVs confer resistance to kidney injury in renal tubular cells

In the early 2000s, many groups investigated the potential for stem cell therapy to ameliorate organ injury. They found that stem cells could protect against organ injury—including kidney injury—but that this effect was not mediated by engraftment and proliferation of the infused stem cells. Rather, it seemed likely that stem cells migrated to the site of injury where they secreted paracrine factors that re-programmed resident cells (reviewed by [99]). Extracellular vesicles derived from MSCs were found to be sufficient to account for most, if not all, of the protective effect conferred by MSCs in kidney injury models [23, 74].

Camussi's group have published a large series of studies examining the capacity for stem-cell derived EVs to promote renal repair [23, 24, 78, 84, 96]. EVs derived from either mesenchymal stem cells (MSCs) or endothelial progenitor cells (EPCs) are able to confer protection against kidney injury when injected into the systemic circulation of the mouse [23, 78, 84]. Comparing EV preparations enriched in exosomes (100,000 g pellet) or microvesicles (10,000 g pellet), it seems likely that only exosomes exert this therapeutic effect [80]. This therapeutic effect depends on the transfer of critical RNA species to renal tubular cells. The evidence is strongest for microRNAs. The EVs are enriched in miRNA and when this is depleted by general (Dicer or Drosha knockdown) or specific (antagomirs) means, this protective effect is abolished [78, 84].

Other groups have published similar findings. Wang et al. found that MSCs over-expressing miRNA-let7c conferred protection against kidney injury in vivo. In vitro, they showed that MSCs transferred fluorescently-labeled miRNA to renal tubular cells via EVs and that these EVs mitigated the pro-fibrotic effects of TGFβ signalling in renal tubular cells [79]. Dominguez et al. showed that EVs derived from renal tubular epithelial cells were able to accelerate recovery from kidney injury in a rat model [87]. The effect was greater when the EVs were obtained from hypoxic tubular cells: a form of ischemic pre-conditioning.

Taken together, this body of literature supports two key conclusions: that stem-cell-derived exosomes can re-program renal tubular cells in order to protect them from injury and that EV-bound RNA is critical in mediating this effect. We might therefore ask whether the RNA exerts its effects directly or indirectly in the target cell. In order words: are the mRNA molecules delivered to target cells by EVs translated into proteins? Do delivered miRNA molecules enter the RISC (RNA-induced silencing complex) and repress target mRNA expression? This has yet to be answered conclusively but the available evidence supports a direct effect. In vitro, MSC-derived EVs were internalised by renal tubular cells and mitigated the injurious effects of ATP depletion [24]. EV treatment altered miRNA expression within target cells. Experiments in which actinomycin D was used to inhibit transcription suggested that some effects were mediated by direct transfer of miRNA from EVs and others by EV-mediated changes in miRNA transcription in the target cell. There is also evidence that EV-bound miRNA released by *endogenous* stem cells can directly silence target gene expression. Erythropoetin therapy induced protection against kidney injury in a murine ureteric obstruction model. The mechanism was intriguing: erythropoietin stimulated the release of miR-144-containing EVs from bone marrow cells. A miR-144 antagomir blocked this effect and miR-144 targeted the 3'UTR of tissue plasminogen activator in fibroblast cells [83].

miRNA target prediction and gene ontology analyses have been used to predict the key cellular processes that might be regulated by an MSC-derived EV

RNA cargo. These include pathways regulating cell proliferation, apoptosis, fatty acid metabolism, inflammation, matrix-receptor interaction, cell adhesion and cytoskeleton organization [24, 78, 80].

4.1.3 EV signalling from renal tubular cells to other compartments

In addition to being a target of EV-signalling, renal tubular cells are also a *source* of functional EVs. Such signalling might be important in disease states. Renal tubular cells exposed to albumin release exosomes able to induce a pro-inflammatory state in macrophages. These exosome are enriched in mRNA encoding the chemokine CCL2 and experiments using a GFP-CCL2 fusion mRNA construct suggest that EV-bound mRNA is delivered directly to macrophage cells [97]. This pathway could contribute to the tubulointerstitial inflammation seen in albuminuria. Tubular cell-derived EVs have also been shown to induce partial mesenchymal-to-epithelial transition in MSCs, probably by transferring microRNA [96].

4.1.4 EVs signalling to renal tubular cells via cell-surface receptors

RNA transfer mediates many of the functional effects of EVs but this does not preclude contributions from other mechanisms. Podocyte-derived microvesicles induced pro-fibrotic changes in intracellular signalling pathways in cultured renal tubular cells but this effect was completely inhibited by blockade of the surface receptor CD36 [27]. This suggests that either the surface receptor is critical for EV internalization or that the EVs exert their effects by binding the surface receptor and activating intracellular second messenger cascades.

4.2 Exosome-mediated interactions with urinary pathogens

EVs may interact with recipient cells within the same organism but there is also evidence of trans-species communication in the context of urinary tract infections.

Host-derived urinary exosomes are enriched in proteins of the innate immune system including antibacterial peptides. They demonstrate an antibacterial function in vitro, being able to lyse the common urinary pathogen *E. coli* [100]. Conversely, pathogen-derived exosomes can be internalised by host urothelial cells, with implications for host cell function. Exosomes released by the sexually-transmitted parasite *Trichomonas vaginalis* were internalised by urothelial cells, increasing the binding of parasites to the recipient cells [101]. Exosome-mediated interactions between pathogen and host are well-described in other body systems [102].

5 Potential for exosome-based therapies

Given the functional effect that EVs can exert in animal disease models, there is understandable enthusiasm to explore the potential for novel EV-based therapies in human kidney disease. EVs offer a number of theoretical advantages over both conventional small-molecule drugs and cellular therapies. They could be used to re-program multiple pathways in target cells by delivering complex, pleiotropic signals (such as a mix of microRNAs) but they have a stable phenotype, in contrast to cell therapies which might retain the potential for malignancy transformation [103]. Outside of the kidney and urinary tract, proof-of-principle studies have demonstrated that engineered exosomes carrying a small non-coding RNA cargo are able to target cancer cells and improve survival in a mouse cancer model [104].

However, there are several hurdles to the development of safe and effective EV-based therapeutics, not least the need to develop and adequate understanding of basic EV biology. We need to better understand the mechanisms through which EVs exert functional effects and how they are targeted selectively to certain cell types.

The renal tubule is an attractive target for exosome-based therapies. As discussed above, systemically-delivered exosomes enter renal tubular cells and are perhaps even trafficked preferentially to sites of tubular injury. Proof-of-concept studies in rodent models have demonstrated that exosomes delivered in this way can protect against acute kidney injury. Renal tubular cell pathology underpins many clinically important diseases, including most forms of acute kidney injury and chronic kidney disease [105–107]. However intervening to improve outcomes is likely to involve complex reprogramming of co-ordinated cellular processes, such as those driving epithelial-to-mesenchymal transition [108]. With their ability to deliver a pleiotropic cargo, EVs could be ideally placed to effect this.

EVs could be exploited in innovative ways to provide clinical benefits. For example, it may be possible to manipulate the release of endogenous EVs—following on from exploratory studies such as that from Zhou et al. (discussed above) in which erythropoietin increased the release of a population of EVs from bone marrow cells [83]. Or EVs might be used as adjuvant therapies to enhance response to conventional drugs. Microvesicles released from bladder cancer cells were able to potentiate the effects of chemotherapy in a mouse model of bladder cancer by increasing drug delivery to the cell nucleus [109].

6 Conclusions

Exosomes in the kidney and urinary tract have a lot of common with other systems. We are beginning to understand the mechanisms governing exosome release and uptake and the determinants of exosome cargo (Fig. 1). Exosomes can re-program target cell function, probably by transferring a mixed cargo of RNA, protein and lipid; the evidence is strongest for microRNA. Pilot studies show potential for exosomes to become useful biomarkers and therapeutic agents but this has not yet translated into clinical practice.

However, exosomes in the kidneys and urinary tract also have some distinct features. The urine as a source of EVs presents unique challenges and opportunities. The flow of urinary fluid allows for the study of directional EV signalling from proximal to distal along the nephron. The discovery that EVs can pass from the systemic circulation into the urinary space suggests that the kidney provides a model in which we can explore the mechanisms of EV transport in detail.

Renal tubular cells take up EVs from the urinary and vascular space and this process is, at least in part, physiologically regulated (Fig. 1). EV uptake is saturable, dependent on cell-surface receptors and controlled by known hormonal regulators of renal tubular cells. Internalization of exosomes by tubular cells induces changes in cell function. EV-mediated signalling *between* tubular cells (in the same or distant nephron segments) can influence transport function in the target cell; signalling from stem cells to tubular cells influences cell survival, proliferation and susceptibility to injury (Fig. 2).

Excitingly, research into exosomes is in its infancy and there many more open questions than there are answers. In closing, we list some of the pressing unresolved questions regarding exosomes in the kidney in Box 1.

BOX 1 Unresolved research questions regarding exosomes in the kidney and urinary tract.

Fundamental mechanisms of EV release, uptake and transport

- Is EV uptake specific (with respect to the type of EV and type of target cell)? If so, how is this specificity determined?
- What are the physiological regulators of EV release and do these differ in different kidney cell types?
- What are the physiological regulators of EV uptake in different kidney cell types?
- How do EVs travel from the vascular to the urinary space? Is this different in disease states in which the structural integrity of the glomerular or tubular basement membrane is compromised?

Mechanisms of EV signaling

- To what extent are the direct effects of protein and RNA transfer sufficient to account for the functional effect of EV signalling in the kidney?
- To what extent do cell-surface EV interactions account for the functional effects of EV signalling?

Clinical applications

- What is the best method of harvesting urinary EVs for use as biomarkers (optimizing cost, time and stability of the EV cargo)?
- Will the potential utility of urinary EV-based biomarkers be confirmed in large clinical studies?
- How could exogenous EVs be used to deliver therapies in kidney disease?
- How can we modulate endogenous EV signalling to therapeutic effect in kidney disease?

References

[1] Lötvall J, Hill AF, Hochberg F, Buzás EI, Di Vizio D, Gardiner C, Gho YS, Kurochkin IV, Mathivanan S, Quesenberry P, Sahoo S, Tahara H, Wauben MH, Witwer KW, Théry C. Minimal experimental requirements for definition of extracellular vesicles and their functions: a position statement from the International Society for Extracellular Vesicles. J Extracell Vesicles 2014;3:26913.

[2] Witwer KW, Buzás EI, Bemis LT, Bora A, Lässer C, Lötvall J, Nolte-'t Hoen EN, Piper MG, Sivaraman S, Skog J, Théry C, Wauben MH, Hochberg F. Standardization of sample collection, isolation and analysis methods in extracellular vesicle research. J Extracell Vesicles 2013;2. https://doi.org/10.3402/jev.v2i0.20360.

[3] Erdbrugger U, Le TH. Extracellular vesicles in renal diseases: more than novel biomarkers? J Am Soc Nephrol 2015. https://doi.org/10.1681/ASN.2015010074.

[4] van Niel G, D'Angelo G, Raposo G. Shedding light on the cell biology of extracellular vesicles. Nat Rev Mol Cell Biol 2018;19:213–28. https://doi.org/10.1038/nrm.2017.125.

[5] Pisitkun T, Shen R-F, Knepper MA. Identification and proteomic profiling of exosomes in human urine. Proc Natl Acad Sci U S A 2004;101:13368–73. https://doi.org/10.1073/pnas.0403453101.

[6] Fernández-Llama P, Khositseth S, Gonzales PA, Star RA, Pisitkun T, Knepper MA. Tamm-Horsfall protein and urinary exosome isolation. Kidney Int 2010;77:736–42.

[7] Gonzales PA, Zhou H, Pisitkun T, Wang NS, Star RA, Knepper MA, Yuen PST. Isolation and purification of exosomes in urine. In: Rai AJ, editor. The urinary proteome. Totowa, NJ: Humana Press; 2010. p. 89–99.

[8] Hogan MC, Manganelli L, Woollard JR, Masyuk AI, Masyuk TV, Tammachote R, Huang BQ, Leontovich AA, Beito TG, Madden BJ, Charlesworth MC, Torres VE, LaRusso NF, Harris PC, Ward CJ. Characterization of PKD protein-positive exosome-like vesicles. J Am Soc Nephrol 2009;20:278–88. https://doi.org/10.1681/ASN.2008060564.

[9] Gonzales PA, Pisitkun T, Hoffert JD, Tchapyjnikov D, Star RA, Kleta R, Wang NS, Knepper MA. Large-scale proteomics and phosphoproteomics of urinary exosomes. J Am Soc Nephrol 2009;20:363–79. https://doi.org/10.1681/ASN.2008040406.

[10] Rood IM, Deegens JKJ, Merchant ML, Tamboer WPM, Wilkey DW, Wetzels JFM, Klein JB. Comparison of three methods for isolation of urinary microvesicles to identify biomarkers of nephrotic syndrome. Kidney Int 2010;78:810–6. https://doi.org/10.1038/ki.2010.262.

[11] Hogan MC, Johnson KL, Zenka RM, Charlesworth MC, Madden BJ, Mahoney DW, Oberg AL, Huang BQ, Leontovich AA, Nesbitt LL, et al. Subfractionation, characterization, and in-depth proteomic analysis of glomerular membrane vesicles in human urine. Kidney Int 2014;85:1225–37.

[12] Troyanov S, Delmas-Frenette C, Bollée G, Youhanna S, Bruat V, Awadalla P, Devuyst O, Madore F. Clinical, genetic, and urinary factors associated with uromodulin excretion. Clin J Am Soc Nephrol 2016;11:62–9. https://doi.org/10.2215/CJN.04770415.

[13] Alvarez ML, Khosroheidari M, Kanchi Ravi R, DiStefano JK. Comparison of protein, microRNA, and mRNA yields using different methods of urinary exosome isolation for the discovery of kidney disease biomarkers. Kidney Int 2012;82:1024–32. https://doi.org/10.1038/ki.2012.256.

[14] Prunotto M, Farina A, Lane L, Pernin A, Schifferli J, Hochstrasser DF, Lescuyer P, Moll S. Proteomic analysis of podocyte exosome-enriched fraction from normal human urine. J Proteomics 2013;82:193–229. https://doi.org/10.1016/j.jprot.2013.01.012.

[15] Vergauwen G, Dhondt B, Van Deun J, De Smedt E, Berx G, Timmerman E, Gevaert K, Miinalainen I, Cocquyt V, Braems G, Van den Broecke R, Denys H, De Wever O, Hendrix A. Confounding factors of ultrafiltration and protein analysis in extracellular vesicle research. Sci Rep 2017;7:2704. https://doi.org/10.1038/s41598-017-02599-y.

[16] Gerlach JQ, Krüger A, Gallogly S, Hanley SA, Hogan MC, Ward CJ, Joshi L, Griffin MD. Surface glycosylation profiles of urine extracellular vesicles. PLoS ONE 2013;8:e74801. https://doi.org/10.1371/journal.pone.0074801.

[17] Zhou H, Yuen PS, Pisitkun T, Gonzales PA, Yasuda H, Dear JW, Gross P, Knepper MA, Star RA. Collection, storage, preservation, and normalization of human urinary exosomes for biomarker discovery. Kidney Int 2006;69:1471–6.

[18] Zhou H, Cheruvanky A, Hu X, Matsumoto T, Hiramatsu N, Cho ME, Berger A, Leelahavanichkul A, Doi K, Chawla LS, et al. Urinary exosomal transcription factors, a new class of biomarkers for renal disease. Kidney Int 2008;74:613–21.

[19] Miranda KC, Bond DT, McKee M, Skog J, Păunescu TG, Da Silva N, Brown D, Russo LM. Nucleic acids within urinary exosomes/microvesicles are potential biomarkers for renal disease. Kidney Int 2010;78:191–9.

[20] Miranda KC, Bond DT, Levin JZ, Adiconis X, Sivachenko A, Russ C, Brown D, Nusbaum C, Russo LM. Massively parallel sequencing of human urinary exosome/microvesicle RNA reveals a predominance of non-coding RNA. PLoS ONE 2014;9:e96094. https://doi.org/10.1371/journal.pone.0096094.

[21] Oosthuyzen W, Scullion KM, Ivy JR, Morrison EE, Hunter RW, Starkey Lewis PJ, O'Duibhir E, Street JM, Caporali A, Gregory CD, Forbes SJ, Webb DJ, Bailey MA, Dear JW. Vasopressin regulates extracellular vesicle uptake by kidney collecting duct cells. J Am Soc Nephrol 2016. https://doi.org/10.1681/ASN.2015050568.

[22] Gildea JJ, Seaton JE, Victor KG, Reyes CM, Bigler Wang D, Pettigrew AC, Courtner CE, Shah N, Tran HT, Van Sciver RE, Carlson JM, Felder RA. Exosomal transfer from human renal proximal tubule cells to distal tubule and collecting duct cells. Clin Biochem 2014;47:89–94. https://doi.org/10.1016/j.clinbiochem.2014.06.018.

[23] Bruno S, Grange C, Deregibus MC, Calogero RA, Saviozzi S, Collino F, Morando L, Busca A, Falda M, Bussolati B, Tetta C, Camussi G. Mesenchymal stem cell-derived microvesicles protect against acute tubular injury. J Am Soc Nephrol 2009;20:1053–67. https://doi.org/10.1681/ASN.2008070798.

[24] Lindoso RS, Collino F, Bruno S, Araujo DS, Sant'Anna JF, Tetta C, Provero P, Quesenberry PJ, Vieyra A, Einicker-Lamas M, Camussi G. Extracellular vesicles released from mesenchymal stromal cells modulate miRNA in renal tubular cells and inhibit ATP depletion injury. Stem Cells Dev 2014;23:1809–19. https://doi.org/10.1089/scd.2013.0618.

[25] Hansell P, Göransson V, Odlind C, Gerdin B, Hällgren R. Hyaluronan content in the kidney in different states of body hydration. Kidney Int 2000;58:2061–8. https://doi.org/10.1111/j.1523-1755.2000.00378.x.

[26] Herrera MB, Bussolati B, Bruno S, Morando L, Mauriello-Romanazzi G, Sanavio F, Stamenkovic I, Biancone L, Camussi G. Exogenous mesenchymal stem cells localize to the kidney by means of CD44 following acute tubular injury. Kidney Int 2007;72:430–41. https://doi.org/10.1038/sj.ki.5002334.

[27] Munkonda MN, Akbari S, Landry C, Sun S, Xiao F, Turner M, Holterman CE, Nasrallah R, Hébert RL, Kennedy CRJ, Burger D. Podocyte-derived microparticles promote proximal tubule fibrotic signaling via p38 MAPK and CD36. J Extracell Vesicles 2018;7:1432206. https://doi.org/10.1080/20013078.2018.1432206.

[28] French KC, Antonyak MA, Cerione RA. Extracellular vesicle docking at the cellular port: extracellular vesicle binding and uptake. Semin Cell Dev Biol 2017;67:48–55. https://doi.org/10.1016/j.semcdb.2017.01.002.

[29] Street JM, Birkhoff W, Menzies RI, Webb DJ, Bailey MA, Dear JW. Exosomal transmission of functional aquaporin 2 in kidney cortical collecting duct cells: exosome signalling in collecting duct. J Physiol 2011;589:6119–27. https://doi.org/10.1113/jphysiol.2011.220277.

[30] Deregibus MC, Figliolini F, D'Antico S, Manzini PM, Pasquino C, De Lena M, Tetta C, Brizzi MF, Camussi G. Charge-based precipitation of extracellular vesicles. Int J Mol Med 2016;38:1359–66. https://doi.org/10.3892/ijmm.2016.2759.

[31] Lawrence MG, Altenburg MK, Sanford R, Willett JD, Bleasdale B, Ballou B, Wilder J, Li F, Miner JH, Berg UB, Smithies O. Permeation of macromolecules into the renal glomerular basement membrane and capture by the tubules. Proc Natl Acad Sci U S A 2017;114:2958–63. https://doi.org/10.1073/pnas.1616457114.

[32] Grange C, Tapparo M, Bruno S, Chatterjee D, Quesenberry PJ, Tetta C, Camussi G. Biodistribution of mesenchymal stem cell-derived extracellular vesicles in a model of acute kidney injury monitored by optical imaging. Int J Mol Med 2014;33:1055–63. https://doi.org/10.3892/ijmm.2014.1663.

[33] Ståhl A, Arvidsson I, Johansson KE, Chromek M, Rebetz J, Loos S, Kristoffersson A-C, Békássy ZD, Mörgelin M, Karpman D. A novel mechanism of bacterial toxin transfer within host blood cell-derived microvesicles. PLoS Pathog 2015;11:e1004619. https://doi.org/10.1371/journal.ppat.1004619.

[34] Kanno K, Sasaki S, Hirata Y, Ishikawa S, Fushimi K, Nakanishi S, Bichet DG, Marumo F. Urinary excretion of aquaporin-2 in patients with diabetes insipidus. N Engl J Med 1995;332:1540–5.

[35] Wen H, Frøkiær J, Kwon T-H, Nielsen S. Urinary excretion of aquaporin-2 in rat is mediated by a vasopressin-dependent apical pathway. J Am Soc Nephrol 1999;10:1416–29.

[36] Higashijima Y, Sonoda H, Takahashi S, Kondo H, Shigemura K, Ikeda M. Excretion of urinary exosomal AQP2 in rats is regulated by vasopressin and urinary pH. Am J Physiol Renal Physiol 2013;305:F1412–21. https://doi.org/10.1152/ajprenal.00249.2013.

[37] Corbetta S, Raimondo F, Tedeschi S, Syrèn M-L, Rebora P, Savoia A, Baldi L, Bettinelli A, Pitto M. Urinary exosomes in the diagnosis of Gitelman and Bartter syndromes. Nephrol Dial Transplant Off Publ Eur Dial Transpl Assoc—Eur Ren Assoc 2015;30:621–30. https://doi.org/10.1093/ndt/gfu362.

[38] Joo KW, Lee JW, Jang HR, Heo NJ, Jeon US, Oh YK, Lim CS, Na KY, Kim J, Cheong HI, Han JS. Reduced urinary excretion of thiazide-sensitive Na-Cl cotransporter in Gitelman syndrome: preliminary data. Am J Kidney Dis 2007;50:765–73. https://doi.org/10.1053/j.ajkd.2007.07.022.

[39] Mayan H, Attar-Herzberg D, Shaharabany M, Holtzman EJ, Farfel Z. Increased urinary Na-Cl cotransporter protein in familial hyperkalaemia and hypertension. Nephrol Dial Transplant 2008;23:492–6. https://doi.org/10.1093/ndt/gfm641.

[40] Esteva-Font C, Wang X, Ars E, Guillén-Gómez E, Sans L, González Saavedra I, Torres F, Torra R, Masilamani S, Ballarín JA, Fernández-Llama P. Are sodium transporters in urinary exosomes reliable markers of tubular sodium reabsorption in hypertensive patients? Nephron Physiol 2010;114. https://doi.org/10.1159/000274468. p 25–p 34.

[41] van der Lubbe N, Jansen PM, Salih M, Fenton RA, van den Meiracker AH, Danser AJ, Zietse R, Hoorn EJ. The phosphorylated sodium chloride cotransporter in urinary exosomes is superior to prostasin as a marker for aldosteronism. Hypertension 2012;60:741–8.

[42] Esteva-Font C, Guillén-Gómez E, Diaz JM, Guirado L, Facundo C, Ars E, Ballarin JA, Fernández-Llama P. Renal sodium transporters are increased in urinary exosomes of cyclosporine-treated kidney transplant patients. Am J Nephrol 2014;39:528–35. https://doi.org/10.1159/000362905.

[43] Pathare G, Tutakhel OAZ, van der Wel MC, Shelton LM, Deinum J, Lenders JWM, Hoenderop JGJ, Bindels RJM. Hydrochlorothiazide treatment increases the abundance of the NaCl cotransporter in urinary extracellular vesicles of essential hypertensive patients. Am J Physiol Renal Physiol 2017;312:F1063–72. https://doi.org/10.1152/ajprenal.00644.2016.

[44] Tutakhel OAZ, Moes AD, Valdez-Flores MA, Kortenoeven MLA, Vrie MvD, Jeleń S, Fenton RA, Zietse R, Hoenderop JGJ, Hoorn EJ, Hilbrands L, Bindels RJM. NaCl cotransporter abundance in urinary vesicles is increased by calcineurin inhibitors and predicts thiazide sensitivity. PLoS ONE 2017;12:e0176220. https://doi.org/10.1371/journal.pone.0176220.

[45] Castagna A, Pizzolo F, Chiecchi L, Morandini F, Channavajjhala SK, Guarini P, Salvagno G, Olivieri O. Circadian exosomal expression of renal thiazide-sensitive NaCl cotransporter (NCC) and prostasin in healthy individuals. Proteomics Clin Appl 2015;9:623–9. https://doi.org/10.1002/prca.201400198.

[46] Ivy JR, Oosthuyzen W, Peltz TS, Howarth AR, Hunter RW, Dhaun N, Al-Dujaili EAS, Webb DJ, Dear JW, Flatman PW, Bailey MA. Glucocorticoids induce nondipping blood pressure by activating the thiazide-sensitive cotransporter. Hypertension 2016;67:1029–37. https://doi.org/10.1161/HYPERTENSIONAHA.115.06977.

[47] Pathare G, Dhayat N, Mohebbi N, Wagner CA, Cheval L, Neuhaus TJ, Fuster DG. Acute regulated expression of pendrin in human urinary exosomes. Pflugers Arch 2018;470:427–38. https://doi.org/10.1007/s00424-017-2049-0.

[48] Pathare G, Dhayat NA, Mohebbi N, Wagner CA, Bobulescu IA, Moe OW, Fuster DG. Changes in V-ATPase subunits of human urinary exosomes reflect the renal response to acute acid/alkali loading and the defects in distal renal tubular acidosis. Kidney Int 2018;93:871–80. https://doi.org/10.1016/j.kint.2017.10.018.

[49] Oliveira RA, Diniz LF, Teotônio LO, Lima CG, Mota RM, Martins A, Sanches TR, Seguro AC, Andrade L, Silva GB, et al. Renal tubular dysfunction in patients with American cutaneous leishmaniasis. Kidney Int 2011;80:1099–106.

[50] Hunter RW, Moorhouse R, Farrah TE, MacIntyre IM, Asai T, Gallacher PJ, Kerr D, Melville V, Czopek A, Morrison EE, Ivy JR, Dear JW, Bailey MA, Goddard J, Webb DJ, Dhaun N. First-in-man demonstration of direct endothelin-mediated natriuresis and diuresis. Hypertension 2017;70(1):192–200. https://doi.org/10.1161/HYPERTENSIONAHA.116.08832.

[51] Channavajjhala SK, Bramley R, Peltz T, Oosthuyzen W, Jia W, Kinnear S, Sampson B, Martin N, Hall IP, Bailey MA, Dear JW, Glover M. Urinary extracellular vesicle protein profiling and endogenous lithium clearance support excessive renal sodium wasting and water reabsorption in thiazide-induced hyponatremia. Kidney Int Rep 2018;4(1):139–47. https://doi.org/10.1016/j.ekir.2018.09.011.

[52] Burger D, Thibodeau J-F, Holterman CE, Burns KD, Touyz RM, Kennedy CRJ. Urinary podocyte microparticles identify prealbuminuric diabetic glomerular injury. J Am Soc Nephrol 2014;25:1401–7. https://doi.org/10.1681/ASN.2013070763.

[53] Gilani SI, Anderson UD, Jayachandran M, Weissgerber TL, Zand L, White WM, Milic N, Suarez MLG, Vallapureddy RR, Nääv Å, Erlandsson L, Lieske JC, Grande JP, Nath KA, Hansson SR, Garovic VD. Urinary extracellular vesicles of podocyte origin and renal injury in preeclampsia. J Am Soc Nephrol 2017;28:3363–72. https://doi.org/10.1681/ASN.2016111202.

[54] Lytvyn Y, Xiao F, Kennedy CRJ, Perkins BA, Reich HN, Scholey JW, Cherney DZ, Burger D. Assessment of urinary microparticles in normotensive patients with type 1 diabetes. Diabetologia 2017;60:581–4. https://doi.org/10.1007/s00125-016-4190-2.

[55] Dimuccio V, Ranghino A, Praticò Barbato L, Fop F, Biancone L, Camussi G, Bussolati B. Urinary CD133+ extracellular vesicles are decreased in kidney transplanted patients with slow graft function and vascular damage. PLoS ONE 2014;9:e104490. https://doi.org/10.1371/journal.pone.0104490.

[56] Ranghino A, Dimuccio V, Papadimitriou E, Bussolati B. Extracellular vesicles in the urine: markers and mediators of tissue damage and regeneration. Clin Kidney J 2015;8:23–30. https://doi.org/10.1093/ckj/sfu136.

[57] Xie Y, Jia Y, Cuihua X, Hu F, Xue M, Xue Y. Urinary exosomal microRNA profiling in incipient type 2 diabetic kidney disease. J Diabetes Res 2017;2017:6978984. https://doi.org/10.1155/2017/6978984.

[58] Kamińska A, Platt M, Kasprzyk J, Kuśnierz-Cabala B, Gala-Błądzińska A, Woźnicka O, Jany BR, Krok F, Piekoszewski W, Kuźniewski M, Stępień EŁ. Urinary extracellular vesicles: potential biomarkers of renal function in diabetic patients. J Diabetes Res 2016;2016:5741518. https://doi.org/10.1155/2016/5741518.

[59] Zubiri I, Posada-Ayala M, Benito-Martin A, Maroto AS, Martin-Lorenzo M, Cannata-Ortiz P, de la Cuesta F, Gonzalez-Calero L, Barderas MG, Fernandez-Fernandez B, Ortiz A, Vivanco F, Alvarez-Llamas G. Kidney tissue proteomics reveals regucalcin downregulation in response to diabetic nephropathy with reflection in urinary exosomes. Transl Res 2015;166:474–484.e4. https://doi.org/10.1016/j.trsl.2015.05.007.

[60] Zhou H, Kajiyama H, Tsuji T, Hu X, Leelahavanichkul A, Vento S, Frank R, Kopp JB, Trachtman H, Star RA, Yuen PST. Urinary exosomal Wilms' tumor-1 as a potential biomarker for podocyte injury. Am J Physiol Renal Physiol 2013;305:F553–9. https://doi.org/10.1152/ajprenal.00056.2013.

[61] Moon P-G, Lee J-E, You S, Kim T-K, Cho J-H, Kim I-S, Kwon T-H, Kim C-D, Park S-H, Hwang D, Kim Y-L, Baek M-C. Proteomic analysis of urinary exosomes from patients of early IgA nephropathy and thin basement membrane nephropathy. Proteomics 2011;11:2459–75. https://doi.org/10.1002/pmic.201000443.

[62] Gutwein P, Schramme A, Abdel-Bakky MS, Doberstein K, Hauser IA, Ludwig A, Altevogt P, Gauer S, Hillmann A, Weide T, Jespersen C, Eberhardt W, Pfeilschifter J. ADAM10 is expressed in human podocytes and found in urinary vesicles of patients with glomerular kidney diseases. J Biomed Sci 2010;17:3. https://doi.org/10.1186/1423-0127-17-3.

[63] Perez-Hernandez J, Forner MJ, Pinto C, Chaves FJ, Cortes R, Redon J. Increased urinary exosomal microRNAs in patients with systemic lupus erythematosus. PLoS ONE 2015;10:e0138618. https://doi.org/10.1371/journal.pone.0138618.

[64] Zhou H, Pisitkun T, Aponte A, Yuen PST, Hoffert JD, Yasuda H, Hu X, Chawla L, Shen R-F, Knepper MA, Star RA. Exosomal Fetuin-A identified by proteomics: a novel urinary biomarker for detecting acute kidney injury. Kidney Int 2006;70:1847–57. https://doi.org/10.1038/sj.ki.5001874.

[65] Benito-Martin A, Ucero AC, Zubiri I, Posada-Ayala M, Fernandez-Fernandez B, Cannata-Ortiz P, Sanchez-Nino MD, Ruiz-Ortega M, Egido J, Alvarez-Llamas G, Ortiz A. Osteoprotegerin in exosome-like vesicles from human cultured tubular cells and urine. PLoS ONE 2013;8:e72387. https://doi.org/10.1371/journal.pone.0072387.

[66] Khurana R, Ranches G, Schafferer S, Lukasser M, Rudnicki M, Mayer G, Hüttenhofer A. Identification of urinary exosomal noncoding RNAs as novel biomarkers in chronic kidney disease. RNA 2017;23:142–52. https://doi.org/10.1261/rna.058834.116.

[67] Kwon SH, Woollard JR, Saad A, Garovic VD, Zand L, Jordan KL, Textor SC, Lerman LO. Elevated urinary podocyte-derived extracellular microvesicles in renovascular hypertensive patients. Nephrol Dial Transplant 2017;32:800–7. https://doi.org/10.1093/ndt/gfw077.

[68] Damkjaer M, Jensen PH, Schwämmle V, Sprenger RR, Jacobsen IA, Jensen ON, Bie P. Selective renal vasoconstriction, exaggerated natriuresis and excretion rates of exosomic proteins in essential hypertension. Acta Physiol (Oxf) 2014;212:106–18. https://doi.org/10.1111/apha.12345.

[69] Alvarez S, Suazo C, Boltansky A, Ursu M, Carvajal D, Innocenti G, Vukusich A, Hurtado M, Villanueva S, Carreño JE, Rogelio A, Irarrazabal CE. Urinary exosomes as a source of kidney dysfunction biomarker in renal transplantation. Transplant Proc 2013;45:3719–23. https://doi.org/10.1016/j.transproceed.2013.08.079.

[70] Pisitkun T, Gandolfo MT, Das S, Knepper MA, Bagnasco SM. Application of systems biology principles to protein biomarker discovery: urinary exosomal proteome in renal transplantation. Proteomics Clin Appl 2012;6:268–78. https://doi.org/10.1002/prca.201100108.

[71] Sigdel TK, Ng YW, Lee S, Nicora CD, Qian W-J, Smith RD, Camp DG, Sarwal MM. Perturbations in the urinary exosome in transplant rejection. Front Med 2015;1. https://doi.org/10.3389/fmed.2014.00057.

[72] Merchant ML, Rood IM, Deegens JKJ, Klein JB. Isolation and characterization of urinary extracellular vesicles: implications for biomarker discovery. Nat Rev Nephrol 2017;13:731–49. https://doi.org/10.1038/nrneph.2017.148.

[73] Spanu S, van Roeyen CRC, Denecke B, Floege J, Mühlfeld AS. Urinary exosomes: a novel means to non-invasively assess changes in renal gene and protein expression. PLoS ONE 2014;9:e109631. https://doi.org/10.1371/journal.pone.0109631.

[74] He J, Wang Y, Sun S, Yu M, Wang C, Pei X, Zhu B, Wu J, Zhao W. Bone marrow stem cells-derived microvesicles protect against renal injury in the mouse remnant kidney model. Nephrology (Carlton) 2012;17:493–500. https://doi.org/10.1111/j.1440-1797.2012.01589.x.

[75] Zhang G, Zou X, Miao S, Chen J, Du T, Zhong L, Ju G, Liu G, Zhu Y. The anti-oxidative role of micro-vesicles derived from human Wharton-Jelly mesenchymal stromal cells through NOX2/gp91(phox) suppression in alleviating renal ischemia-reperfusion injury in rats. PLoS ONE 2014;9:e92129. https://doi.org/10.1371/journal.pone.0092129.

[76] Ju G, Cheng J, Zhong L, Wu S, Zou X, Zhang G, Gu D, Miao S, Zhu Y, Sun J, Du T. Microvesicles derived from human umbilical cord mesenchymal stem cells facilitate tubular epithelial cell dedifferentiation and growth via hepatocyte growth factor induction. PLoS One 2015;10:e0121534. https://doi.org/10.1371/journal.pone.0121534.

[77] He J, Wang Y, Lu X, Zhu B, Pei X, Wu J, Zhao W. Micro-vesicles derived from bone marrow stem cells protect the kidney both in vivo and in vitro by microRNA-dependent repairing. Nephrology (Carlton) 2015;20:591–600. https://doi.org/10.1111/nep.12490.

[78] Collino F, Bruno S, Incarnato D, Dettori D, Neri F, Provero P, Pomatto M, Oliviero S, Tetta C, Quesenberry PJ, Camussi G. AKI recovery induced by mesenchymal stromal cell-derived extracellular vesicles carrying microRNAs. J Am Soc Nephrol 2015;26:2349–60. https://doi.org/10.1681/ASN.2014070710.

[79] Wang B, Yao K, Huuskes BM, Shen H-H, Zhuang J, Godson C, Brennan EP, Wilkinson-Berka JL, Wise AF, Ricardo SD. Mesenchymal stem cells deliver exogenous microRNA-let7c via exosomes to attenuate renal fibrosis. Mol Ther 2016;24:1290–301. https://doi.org/10.1038/mt.2016.90.

[80] Bruno S, Tapparo M, Collino F, Chiabotto G, Deregibus MC, Soares Lindoso R, Neri F, Kholia S, Giunti S, Wen S, Quesenberry P, Camussi G. Renal regenerative potential of different extracellular vesicle populations derived from bone marrow mesenchymal stromal cells. Tissue Eng Part A 2017;23:1262–73. https://doi.org/10.1089/ten.TEA.2017.0069.

[81] Eirin A, Zhu X-Y, Puranik AS, Tang H, McGurren KA, van Wijnen AJ, Lerman A, Lerman LO. Mesenchymal stem cell-derived extracellular vesicles attenuate kidney inflammation. Kidney Int 2017;92:114–24. https://doi.org/10.1016/j.kint.2016.12.023.

[82] Gallo S, Gili M, Lombardo G, Rossetti A, Rosso A, Dentelli P, Togliatto G, Deregibus MC, Taverna D, Camussi G, Brizzi MF. Stem cell-derived, microRNA-carrying extracellular vesicles: a novel approach to interfering with mesangial cell collagen production in a hyperglycaemic setting. PLoS One 2016;11:e0162417. https://doi.org/10.1371/journal.pone.0162417.

[83] Zhou Y, Fang L, Yu Y, Niu J, Jiang L, Cao H, Sun Q, Zen K, Dai C, Yang J. Erythropoietin protects the tubular basement membrane by promoting the bone marrow to release extracellular vesicles containing tPA-targeting miR-144. Am J Physiol Renal Physiol 2016;310:F27–40. https://doi.org/10.1152/ajprenal.00303.2015.

[84] Cantaluppi V, Gatti S, Medica D, Figliolini F, Bruno S, Deregibus MC, Sordi A, Biancone L, Tetta C, Camussi G. Microvesicles derived from endothelial progenitor cells protect the kidney from ischemia-reperfusion injury by microRNA-dependent reprogramming of resident renal cells. Kidney Int 2012;82:412–27. https://doi.org/10.1038/ki.2012.105.

[85] Cantaluppi V, Medica D, Mannari C, Stiaccini G, Figliolini F, Dellepiane S, Quercia AD, Migliori M, Panichi V, Giovannini L, Bruno S, Tetta C, Biancone L, Camussi G. Endothelial progenitor cell-derived extracellular vesicles protect from complement-mediated mesangial injury in experimental anti-Thy1.1 glomerulonephritis. Nephrol Dial Transplant Off Publ Eur Dial Transpl Assoc—Eur Ren Assoc 2015;30:410–22. https://doi.org/10.1093/ndt/gfu364.

[86] Gracia T, Wang X, Su Y, Norgett EE, Williams TL, Moreno P, Micklem G, Karet Frankl FE. Urinary exosomes contain microRNAs capable of paracrine modulation of tubular transporters in kidney. Sci Rep 2017;7:40601. https://doi.org/10.1038/srep40601.

[87] Dominguez JH, Liu Y, Gao H, Dominguez JM, Xie D, Kelly KJ. Renal tubular cell-derived extracellular vesicles accelerate the recovery of established renal ischemia reperfusion injury. J Am Soc Nephrol 2017;28:3533–44. https://doi.org/10.1681/ASN.2016121278.

[88] Zhang W, Zhou X, Yao Q, Liu Y, Zhang H, Dong Z. HIF-1-mediated production of exosomes during hypoxia is protective in renal tubular cells. Am J Physiol Renal Physiol 2017;313:F906–13. https://doi.org/10.1152/ajprenal.00178.2017.

[89] Zhang G, Yang Y, Huang Y, Zhang L, Ling Z, Zhu Y, Wang F, Zou X, Chen M. Hypoxia-induced extracellular vesicles mediate protection of remote ischemic preconditioning for renal ischemia-reperfusion injury. Biomed Pharmacother 2017;90:473–8. https://doi.org/10.1016/j.biopha.2017.03.096.

[90] Zhou X, Zhang W, Yao Q, Zhang H, Dong G, Zhang M, Liu Y, Chen J-K, Dong Z. Exosome production and its regulation of EGFR during wound healing in renal tubular cells. Am J Physiol Renal Physiol 2017;312:F963–70. https://doi.org/10.1152/ajprenal.00078.2017.

[91] Jella KK, Yu L, Yue Q, Friedman D, Duke BJ, Alli AA. Exosomal GAPDH from proximal tubule cells regulate ENaC activity. PLoS One 2016;11:e0165763. https://doi.org/10.1371/journal.pone.0165763.

[92] Chen H-H, Lai P-F, Lan Y-F, Cheng C-F, Zhong W-B, Lin Y-F, Chen T-W, Lin H. Exosomal ATF3 RNA attenuates pro-inflammatory gene MCP-1 transcription in renal ischemia-reperfusion. J Cell Physiol 2014;229:1202–11. https://doi.org/10.1002/jcp.24554.

[93] Eyre J, Burton JO, Saleem MA, Mathieson PW, Topham PS, Brunskill NJ. Monocyte- and endothelial-derived microparticles induce an inflammatory phenotype in human podocytes. Nephron Exp Nephrol 2011;119:e58–66. https://doi.org/10.1159/000329575.

[94] Wang Y-Y, Tang L-Q, Wei W. Berberine attenuates podocytes injury caused by exosomes derived from high glucose-induced mesangial cells through TGFβ1-PI3K/AKT pathway. Eur J Pharmacol 2018;824:185–92. https://doi.org/10.1016/j.ejphar.2018.01.034.

[95] Kahn R, Mossberg M, Ståhl A-L, Johansson K, Lopatko Lindman I, Heijl C, Segelmark M, Mörgelin M, Leeb-Lundberg LMF, Karpman D. Microvesicle transfer of kinin B1-receptors is a novel inflammatory mechanism in vasculitis. Kidney Int 2017;91:96–105. https://doi.org/10.1016/j.kint.2016.09.023.

[96] Chiabotto G, Bruno S, Collino F, Camussi G. Mesenchymal stromal cells epithelial transition induced by renal tubular cells-derived extracellular vesicles. PLoS ONE 2016;11:e0159163. https://doi.org/10.1371/journal.pone.0159163.

[97] Lv L-L, Feng Y, Wen Y, Wu W-J, Ni H-F, Li Z-L, Zhou L-T, Wang B, Zhang J-D, Crowley SD, Liu B-C. Exosomal CCL2 from tubular epithelial cells is critical for albumin-induced tubulointerstitial inflammation. J Am Soc Nephrol 2018;29:919–35. https://doi.org/10.1681/ASN.2017050523.

[98] Miyazawa Y, Mikami S, Yamamoto K, Sakai M, Saito T, Yamamoto T, Ishibashi K, Sasaki S. AQP2 in human urine is predominantly localized to exosomes with preserved water channel activities. Clin Exp Nephrol 2018;22:782–8. https://doi.org/10.1007/s10157-018-1538-6.

[99] Torres Crigna A, Daniele C, Gamez C, Medina Balbuena S, Pastene DO, Nardozi D, Brenna C, Yard B, Gretz N, Bieback K. Stem/stromal cells for treatment of kidney injuries with focus on preclinical models. Front Med 2018;5. https://doi.org/10.3389/fmed.2018.00179.

[100] Hiemstra TF, Charles PD, Gracia T, Hester SS, Gatto L, Al-Lamki R, Floto RA, Su Y, Skepper JN, Lilley KS, Karet Frankl FE. Human urinary exosomes as innate immune effectors. J Am Soc Nephrol 2014;25:2017–27. https://doi.org/10.1681/ASN.2013101066.

[101] Twu O, de Miguel N, Lustig G, Stevens GC, Vashisht AA, Wohlschlegel JA, Johnson PJ. Trichomonas vaginalis exosomes deliver cargo to host cells and mediate host:parasite interactions. PLoS Pathog 2013;9:e1003482. https://doi.org/10.1371/journal.ppat.1003482.

[102] Coakley G, Maizels RM, Buck AH. Exosomes and other extracellular vesicles: the new communicators in parasite infections. Trends Parasitol 2015;31:477–89. https://doi.org/10.1016/j.pt.2015.06.009.

[103] Robbins PD, Dorronsoro A, Booker CN. Regulation of chronic inflammatory and immune processes by extracellular vesicles. J Clin Invest 2016;126:1173–80. https://doi.org/10.1172/JCI81131.

[104] Kamerkar S, LeBleu VS, Sugimoto H, Yang S, Ruivo CF, Melo SA, Lee JJ, Kalluri R. Exosomes facilitate therapeutic targeting of oncogenic KRAS in pancreatic cancer. Nature 2017;546:498–503. https://doi.org/10.1038/nature22341.

[105] Chevalier RL. The proximal tubule is the primary target of injury and progression of kidney disease: role of the glomerulotubular junction. Am J Physiol Renal Physiol 2016;311:F145–61. https://doi.org/10.1152/ajprenal.00164.2016.

[106] Ferenbach DA, Bonventre JV. Mechanisms of maladaptive repair after AKI leading to accelerated kidney ageing and CKD. Nat Rev Nephrol 2015;11:264–76. https://doi.org/10.1038/nrneph.2015.3.

[107] Kramann R, Kusaba T, Humphreys BD. Who regenerates the kidney tubule? Nephrol Dial Transplant 2015;30:903–10. https://doi.org/10.1093/ndt/gfu281.

[108] Lovisa S, LeBleu VS, Tampe B, Sugimoto H, Vadnagara K, Carstens JL, Wu C-C, Hagos Y, Burckhardt BC, Pentcheva-Hoang T, Nischal H, Allison JP, Zeisberg M, Kalluri R. Epithelial-to-mesenchymal transition induces cell cycle arrest and parenchymal damage in renal fibrosis. Nat Med 2015;21:998–1009. https://doi.org/10.1038/nm.3902.

[109] Jin X, Ma J, Liang X, Tang K, Liu Y, Yin X, Zhang Y, Zhang H, Xu P, Chen D, Zhang T, Lu J, Hu Z, Qin X, Zeng X, Li L, Huang B. Pre-instillation of tumor microparticles enhances intravesical chemotherapy of nonmuscle-invasive bladder cancer through a lysosomal pathway. Biomaterials 2017;113:93–104. https://doi.org/10.1016/j.biomaterials.2016.10.036.

Chapter 12

Extracellular vesicles in neurodegenerative disorders

Imre Mäger[a], Eduard Willms[a], Scott Bonner[a], Andrew F. Hill[b], Matthew J.A. Wood[a]
[a]*University of Oxford, Department of Paediatrics, Oxford, United Kingdom,* [b]*La Trobe University, La Trobe Institute for Molecular Science, Melbourne, VIC, Australia*

1 Introduction

In the central nervous system (CNS), extracellular vesicles (EVs) are an important part of a signaling and communication network which helps to ensure the normal physiological functions of the brain. As compared to other intercellular signal transducers (e.g. neurotransmitters, neurotropic factors, hormones and cytokines), EVs contain a large number of different signaling molecules in the same package. During their biogenesis, EVs become packed with biologically active proteins (both receptors and ligands), lipids, and genetic material (both coding and non-coding RNA), as discussed in detail within other Chapters of this book, and reviewed thoroughly elsewhere [1, 2].

There are multiple factors which define the content and thus the potential biological functions of secreted EVs (Fig. 1). First and foremost, EVs reflect to a certain extent their cells of origin, although many cargoes are specifically enriched within EVs as compared to their parent cells. This, of course, is not surprising because cells can package only those cargoes to EVs which are present in a given cell type itself. However, depending on the cell state and stress conditions, there can be vast changes not only in the gene expression profile but also in the intracellular trafficking of organelles, and in the activation/inactivation of different biological processes (such as the ubiquitin-proteasome system, autophagy, and protein post-translational modifications). The latter, in turn, can affect EV composition by changes in EV secretion pathways (e.g. membrane budding vs secretion via the endo-lysosomal system; secretion via the ESCRT-dependent pathway vs the ceramide-dependent pathway; etc.). Recent studies have indicated very clearly that the level of heterogeneity of EVs secreted by one cell type can be very large [1, 3, 4], and that a number of distinct EV subpopulations can be defined based on the hydrodynamic size of EVs alone [3, 5–7], including small membrane-free particles called exomeres [4], whereas

Exosomes. https://doi.org/10.1016/B978-0-12-816053-4.00012-2

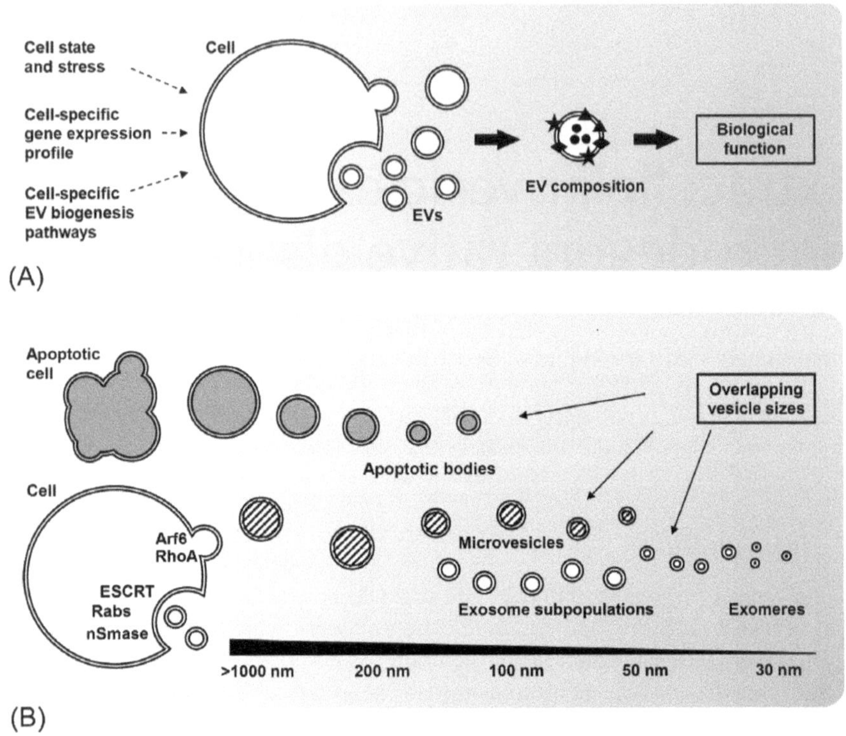

FIG. 1 (A) The biological function of EVs is determined by EV composition, which in turn is affected by the specific gene expression profile of its producing cell type, cell state and stress environment, and cell-specific EV subtype biogenesis pathways which are activated. (B) A number of different EV subpopulations with distinct biogenesis pathways can be secreted by a given cell. While microvesicles are generated by the direct outward budding of the cell membrane, exosome release is related to the trafficking of multivesicular bodies and other components of the endo-lysosomal pathway. Furthermore, several exosome subpopulations exist, most often separated by their hydrodynamic diameter. Separation of EV subpopulations by size only is, however, a difficult task as there can be a considerable overlap between different vesicle types.

functional diversity can exist within these size-defined EV subpopulations [5, 8–10]. This means that by switching between different secretion modes due to stress or other extracellular stimuli, the gross composition of secreted EVs could change greatly.

It should be highlighted that the EV intercellular communication network in the CNS is functioning simultaneously to the network of other signal transduction molecules. Thus, the resulting communication network is likely very complex, especially because it involves many different cell types, both in the context of paracrine and autocrine signaling (Fig. 2). However, studying the entire multifactorial EV-mediated communication network is highly challenging; many current works attempt to address that complexity.

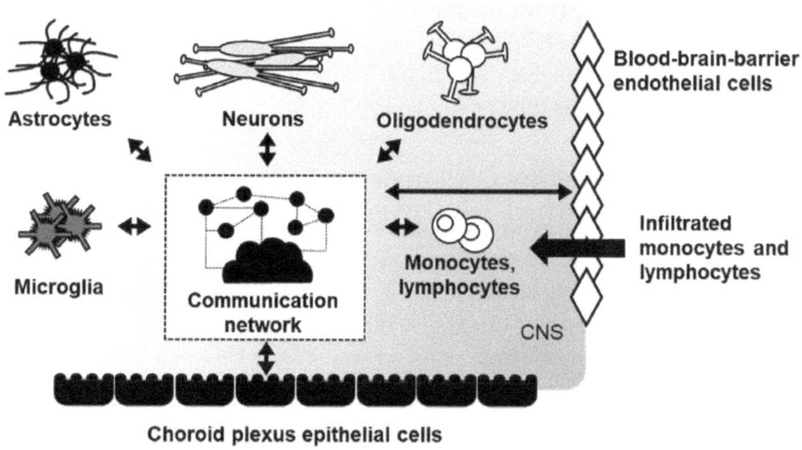

FIG. 2 EVs are secreted by multiple cell types of the CNS, including both resident cells (e.g. astrocytes, neurons, oligodendrocytes and microglia), cells of the blood-brain and blood-CSF barrier (i.e. brain microvascular endothelial cells and choroid plexus epithelial cells), and infiltrated monocytes and lymphocytes in disease. EVs from multiple sources can act both in a paracrine as well as an autocrine manner, forming thereby a complex communication network with multiple interactions between individual cellular components which may be deregulated in numerous pathological conditions.

The majority of evidence about the importance of EV mediated signaling networks in the CNS is perhaps not surprisingly derived from studying different diseases. As discussed below, EVs have been shown or suggested to mediate different aspects of the pathology in a range of neurodegenerative diseases, such as Alzheimer's disease, Parkinson's disease, multiple sclerosis, amyotrophic lateral sclerosis and prion diseases, as focused on in this Chapter as examples (Fig. 3).

The importance of EVs in healthy CNS maintenance is somewhat more difficult to study, partially because EVs have now been shown to contain a number of cargo molecules which are traditionally considered to be secreted as soluble

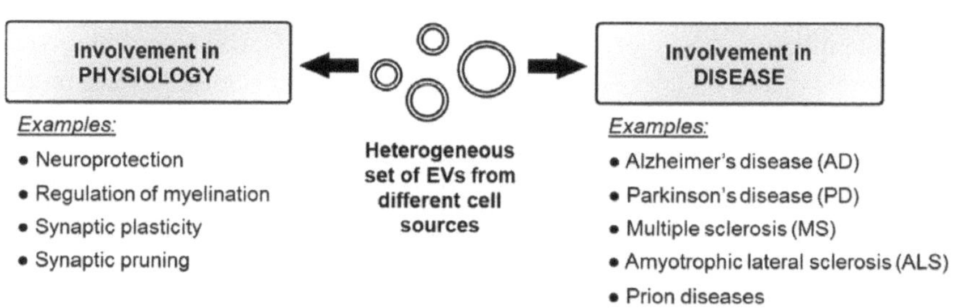

FIG. 3 EVs of virtually all CNS cell types are involved in the pathology of a range of neurodegenerative diseases. On the other hand, EVs are also an important part of the system which ensures correct functioning of the normal CNS physiology. Some of these properties have been attempted to exploit to develop EVs for promoting CNS repair in different disease models.

factors or morphogenes. However, there is a growing body of evidence highlighting the significance of the EV signaling network in the maintenance of many processes in the normal physiology too. In the following sections we will discuss the involvement of EVs in the physiology and pathology of the CNS more in detail.

2 EV-mediated healthy CNS maintenance

A significant amount of evidence of EVs being involved in CNS physiology comes from experiments focusing on neuronal cells. Both neuron-derived EVs themselves as well as EVs secreted by a number of different types of glial cells regulate or support neuronal functions. For example, oligodendrocytes, a type of glial cells involved in myelination of axons, secrete EVs in a Ca^{2+}-dependent manner, which is triggered by the neurotransmitter glutamate [11]. Neuronal activity thereby leads to increased oligodendrocyte EV secretion which in turn provides protection to neurons in oxidative stress and starvation condition by increasing the metabolic activity of neurons [11] potentially in part by the transfer of superoxide dismutase and catalase [12] or other enzymes. In addition to neuroprotection, oligodendrocyte EVs also increase neuronal firing rate, and regulate cellular signaling pathways and gene expression with effects on normal neuronal physiology [12].

However, it should be highlighted that neurons are not the only cell type that can interact with oligodendrocyte EVs. For example, a select group of microglial cells can internalize and degrade oligodendrocyte EVs [13] which may indicate that the amount of oligodendrocyte EVs available to regulate neuronal functions may depend on the activity of microglia. Furthermore, to complicate the EV-mediated signaling network further, other neuronal secreted factors can instead inhibit oligodendrocyte EV production [14], contrary to the increased EV secretion in response to glutamate [14], as explained above. This inhibitory cell communication pathway seems to be related to the auto-regulatory effects of oligodendrocytes to produce myelin via the secretion of self-inhibitory EVs [14], rather than neuronal protection to stress. It is yet to be determined which are the precise differences between the neuroprotective and myelin production-regulatory oligodendrocyte EVs.

While microglia can clear EVs secreted by other cells, they can at the same time secrete their own EVs as well, which have functional consequences in target cells. For example, a population of endocannabinoid-containing microglia EVs can inhibit presynaptic transmission in GABAergic neurons [14]. On the other hand, EVs secreted by microglia in response to ATP can increase the excitatory activity of neuronal cells, including increased exocytosis of glutamate by regulating sphingolipid metabolism [15, 16]. Of note, as sphingolipid metabolic pathways are also involved in EV secretion, it would be important to elucidate, how EV-mediated microglia-neuron interactions feed into other parts of vesicle-mediated CNS cell communication networks, such as protection against stress conditions, as discussed above.

In addition to the regulation of myelination, resistance to stress, and neuronal activity, CNS EVs take part in the modulation of synaptic plasticity. The capacity to regulate synaptic plasticity is mostly attributed to EVs secreted by neurons. During their biogenesis, neuronal EVs become loaded with multiple cargo molecules that can be directly relevant for plasticity, such as the Arc protein [17, 18], AMPA receptor [19], Wnt-family signaling proteins [20–24], ephrins [25], sonic hedgehog (SHH) [26, 27] and Syt4 [28]. While the latter EV-bound molecules are secreted and act mainly across synapses and neuromuscular junctions to support their functions in neuronal plasticity, other neuronal EV components seem to trigger synapse elimination (i.e. synaptic pruning) instead when interacting with microglia. This process seems to be related to the increased activity of microglia to phagocytose neurites, potentially in a complement component C3 dependent manner [29]. The transfer and/or presentation of such signaling molecules across synapses facilitates plasticity-related molecular and phenotypic effects, however, more work needs to be carried out to better understand the role of EVs in regulating synaptic plasticity on a whole organism level, e.g. in the regulation memory, learning and other top level processes.

The examples above represent only a certain proportion of likely roles of EVs in the CNS in the maintenance of its normal physiology. As mentioned earlier, EV communication is complex as it depends on the state and activation status of involved cell types as a whole (Figs. 1 and 2). Many functions of EVs in the CNS are thus revealed more clearly at pathological conditions (Fig. 3) when EV producing cells are defective, over-activated or misregulated in other ways. These functions are discussed below, in sections focusing on the involvement of EVs in CNS pathologies.

3 EVs in neurodegenerative diseases

3.1 Alzheimer's disease

Alzheimer's disease (AD) is the most common form of dementia which severely affects memory and other mental abilities [30]. The exact pathophysiology and the causes of the disease are not completely understood, but many mechanisms have been thoroughly studied and described [31]. Many aspects of AD are related to the increased production of the amyloid beta (Aβ) peptide from its precursor amyloid precursor protein (APP), which in turn is related to the mutations in several other genes such as *APP, PSEN1* and *PSEN2*. Increased levels of Aβ accumulate in extracellular plaques, the presence of which is correlated with the AD pathology. Another hallmark of AD is hyper-phosphorylation of tau proteins which stabilize cytoskeleton/microtubules, particularly in neuronal cells. Hyper-phosphorylated tau proteins aggregate to neurofibrillary tangles by a largely unknown mechanism. Both those events are related to increased neuronal loss, neuro-inflammation, oxidative stress, and activation of microglia and astrocytes.

The understanding that EVs are involved in the pathogenicity of AD has matured via stepwise discoveries [30, 31]. The first indications about the potential role of EVs in the disease were related to the discovery that the Aβ42 peptide, i.e. the isoform of the Aβ peptide that is associated with AD, accumulates in multivesicular bodies (MVBs) in disease [32, 33]. As MVBs are the site of biogenesis of exosomes, one type of EVs, this implies that at least one subtype of EVs can be associated with the plaque-forming Aβ42 peptide. The latter was confirmed in subsequent studies where EV markers Flotillin-1 [34] and Alix [35] were found to co-localize with Aβ in the brains of AD mice and patients, respectively. In support of latter observations, inhibition of an enzyme that is important for exosome biogenesis, neutral sphingomyelinase 2 (nSmase2), has been reported to lead to reduce disease pathology in mouse models [36]. Even though nSmase2 is involved in many cellular processes, i.e. by far not exclusively in the exosome secretion pathway, this evidence may be considered indirect. Nevertheless, reports that astrocyte-derived EVs can stimulate Aβ42 peptide aggregation and concurrently inhibit Aβ42 clearance by microglia cells [36] strongly supports the involvement of the EV network in AD pathology. The latter effect is dependent of ceramide [37], however, it needs to be further clarified if this it is directly related to the presence of ceramide in EVs or to some other EV component in ceramide-positive vesicles.

Interestingly, it seems that EVs themselves can be involved in the pathogenic processing of the APP protein [38]. There are several APP-processing enzymes present in EV preparations, such as BACE1 and other secretases, PSEN1, PSEN2, and ADAM10. Upon APP cleavage, the level of APP C-terminal fragments as well as the cleavage products (incl. Aβ peptides) can be found in EVs [39].

Thus, there are many mechanisms that can lead to increased Aβ42 production and aggregation in the extracellular environment, including EV-mediated secretion, processing and aggregation induction of Aβ. Astrocyte-derived EVs can accelerate Aβ aggregation, leading to the impairment of microglial clearance and degradation of Aβ [36]. At the same time, the high level of extracellular Aβ is related to increased apoptosis of astrocytes themselves, which also leads to the secretion of apoptosis-promoting EVs from these cells [40]. The presence of excessive Aβ also triggers microglia to secrete neurotoxic EVs, however, this could be related to pre-fibrillary Aβ and not the pre-formed aggregates [41]. These microglia-derived EVs can solubilize Aβ aggregates to a certain extent, potentially via interaction with microglia EV lipids [41], contrary to the astrocytic EV-induced Aβ aggregation. Of note, the secretion of neurotoxic EVs by microglia in AD is related not only to the spread of Aβ, but also other proteins, such as the phosphorylated Tau protein [38, 42–45]. Instead of degrading the phosphorylated Tau, microglia have been reported to package this protein into EVs, expel from cells, and induce cytotoxic effects in recipient cells [42].

Above mentioned observations further underline the complicated nature of EV interaction network in disease conditions where certain type of EVs may be protective whereas other pathogenic, depending on their composition. This is illustrated, for example, by the complex nature of nSmase2-dependent secretion of ceramide-containing EVs. Whereas the evidence above suggests that inhibition of nSmase2 activity to inhibit EV secretion might be an interesting therapeutic intervention strategy [36, 41], other evidence points to the opposite direction. For example, the promotion of nSmase2-dependent EV secretion from neurons could be protective at least under some conditions, as it can help to eliminate neurotoxic properties of Aβ and to engage microglia to clear up the toxic peptide [46, 47], and to neutralize Aβ-induced defects in synaptic plasticity [48]. Why neuronal EVs might behave different in that respect compared to astrocytic and microglial EVs remains to be addressed, however one of the hypothesis is that it could be related to the presence of specific glycosphingolipids or the cellular prion protein (PrP^C) on neuronal EVs [30, 49–51] or specific enzymes such as insulin-degrading enzyme (IDE) [52–54].

To summarize the somewhat contradictory findings about the involvement of EVs in AD above, the cumulative effects of EVs in AD pathology may be dependent not only the subtype-specific secretion of EVs of a given cell type but also on the balance and stoichiometry of EV secreted by different cell types in CNS.

3.2 Parkinson's disease

Parkinson's disease (PD) is another common neurodegenerative disease. The cells affected in PD are mainly the dopaminergic neurons in the substantia nigra area of the brain, degeneration of which gives rise to the typical motor symptoms of the PD such as resting tremor, bradykinesia and postural instability [55]. The main contributor to this pathology is α-synuclein (α-syn) which is expressed in the presynaptic region of neurons, but other genes are linked to PD as well (e.g. DJ-1, PINK1, LRRK2) [56]. Even though not all the functions of α-syn are known, it has been established that in PD (as well as in other synucleinopathies) α-syn misfolding, propagation from cell-to-cell, and aggregation to structures called Lewy bodies are key features of the underlying pathology. Emerging evidence suggests that EVs are involved in all of these processes [2, 31].

The role of EVs in the propagation of aggregated α-syn is thought to be somewhat similar to that of the spread of prion protein [57–60], as discussed later in this chapter. While α-syn is clearly detectable in EVs of CNS origin, it is not necessarily the main secretion route of α-syn [57]. A big proportion of α-syn is secreted independently of vesicles, but nevertheless the EV-related smaller fraction of the protein seems to be more important in the pathology [51, 61–63]. It has been argued that the secretion of α-syn may be related to the requirement to adjust the cytosolic concentration of that protein; the failure to do so might be linked to increased probability of misfolding/aggregation of genetic variants and mutated forms of α-syn [56, 64]. Certain genes such as the ATP13A2 seem to be important for the regulation of cytosolic level of α-syn, as suggested by the evidence that high expression of this gene in neurons is related to their higher survival rate and increased secretion of α-syn EVs [65–68].

However, when the toxic form of α-syn is expelled from cells, EVs containing this protein can become neurotoxic when taken up by other neuronal cells [51, 61–63, 69]. Importantly, the EV-associated α-syn is reportedly more neurotoxic than soluble α-syn oligomers [59, 62]. This neurotoxicity can be counteracted by other cells, such as astrocytes and microglia, which can clear extracellular α-syn and reduce the extracellular concentration of neurotoxic α-syn aggregates [70]. However, even though the clearance of α-syn from the extracellular environment might seem a good strategy to reduce neurotoxicity, it actually can lead to the formation of α-syn inclusions in glial cells, microglia over-activation, induction of pro-inflammatory signaling pathways, and reduction of redox enzymes catalase, superoxide dismutase and other antioxidants [70–74]. By secreting EVs, these inflammatory cell to cell signaling pathways are amplified, for example by the induction of tumor necrosis factor-containing EVs from microglia, which can initiate apoptotic processes in recipient cells, including neurons [75].

Another factor contributing to α-syn spread is related to lysosomal dysfunction which accompanies PD, whereby defects in lysosomal processing can increase the fraction of α-syn positive EVs [63]. The change in the secretion of α-*syn*-containing EVs in PD is significant enough that it could be detected in elevated levels not only in the cerebrospinal fluid of patients but also in blood plasma [76, 77]. Not only could this observation be important for developing a sensitive biomarker for PD, but it also sheds additional information on additional potential disease pathology mechanisms, particularly in the spread of the pathology.

The likelihood of uptake of α-syn via EVs is greater than that of soluble α-syn oligomers, as mentioned above, and moreover, the aggregation of soluble α-syn oligomers is enhanced in the presence of increased concentration of EVs, including EVs isolated from the cerebrospinal fluid of PD patients [59, 60, 76]. This seems to suggest that similarly what is observed in AD, as described above, EVs seem to play a role in PD by transferring toxic proteins between neurons, astrocytes and microglia as a signaling network in a positive feedback regulated fashion. Nevertheless, similarly to what was seen I the case of AD, the composition of EVs of PD patients seems to indicate that at least a population of EVs may indeed be neuroprotective [78].

3.3 Multiple sclerosis

Multiple sclerosis (MS) is the most common acquired demyelinating disease that affects the CNS. The disease is characterized by focal plaques of primary demyelination and diffuse neurodegeneration in the grey and white matter of the brain and spinal cord [79]. Risk of disease development has been shown to be linked to genetic factors related to immune modulation [80], and environmental factors such as Epstein-Barr virus infections [81]. However, the ultimate pathogenesis of MS is yet to be fully understood. Thus far, it has been established that autoreactive $CD4^+$ T-cells that target myelin proteins are a key component in MS pathogenesis. What's more, peripheral T-helper $CD4^+$ cell activation has been observed, leading to the generation of pro-inflammatory, autoreactive, Th1 and Th17 T-cell subsets [82]. In addition, B-cells and monocytes/macrophages have also been shown to be active in MS. B cells and $CD8^+$ T-cells constitute a large fraction of the brain-infiltrating cells in MS lesions, although neither the effector cells nor target antigens have been well established thus far [83].

As is the case with multiple other neurodegenerative diseases, there is strong evidence which supports the involvement of immunoregulatory EVs in the pathology of MS. For example, the overall number of secreted EVs in the cerebrospinal fluid is significantly higher in MS patients than that of healthy individuals [84, 85], suggesting imbalance of EV homeostasis in the disease. There are several cell types, including CNS-infiltrated pathology-related cells, that can contribute to the increased EV concentration, leading to specific functional effects.

One of the cell types that secretes pro-inflammatory EVs in MS is microglia. Exogenous microglia EVs, when injected into the brain of mice with experimental autoimmune encephalomyelitis (EAE), which models MS in animals, considerably exaggerate the disease symptoms [84]. Concurrently, aSmase-deficient mice whose EV secretion is impaired are more resistant to EAE [84]. However, it must be noted that aSmase deficiency can reduce EV secretion from other cell types than microglia as well, and that aSmase is also involved in cellular processes other than EV secretion.

The involvement of brain endothelium derived EVs seems highly pertinent to MS pathogenesis too. At certain conditions, endothelial EVs can activate both monocytes [86], and CD4+ and CD8+ T lymphocytes [87]. This occurs via multiple mechanisms, such as the transfer of ICAM-1 receptor (in case of monocytes), and induction of CD40, CD275, MHC II, β2-microglobulin and CCL5 expression (in case of T cells). This can lead to the increased adhesion and trans-endothelial migration of both monocytes [86] and T cells [88, 89]. In addition to the increased trans-endothelial migration, endothelial EVs can induce direct blood-brain barrier (BBB) breakdown as well [90, 91]. The latter is not, however, unique to only endothelial EVs. BBB breakdown can be triggered for example by EVs derived from platelets of MS patients [90, 91], and leukocytes, microglia, and astrocytes via triggering metalloproteinase and caspase 1 containing EVs in response to stimulation with pro-inflammatory cytokines [92, 93]. Cumulatively, these data strongly suggest that exosomes are involved in the facilitation of MS pathogenesis and disease progression, via propagation of inflammatory signals and the transmigration of lymphocytes and myeloid cells through the BBB.

However, in MS, similarly to what was observed in case of AD and PD (see above), not all EVs seem to be pathogenic. There is evidence for the protective benefits of EVs that contain a specific set of cargoes. For example, EVs from glutamate-activated oligodendrocytes support myelination via transfer of myelin proteins and RNA [11, 12], and EVs from interferon gamma-stimulated dendritic cells promote oligodendrocyte progenitor cell growth, increasing myelination, decreasing oxidative stress and promoting remyelination [94]. In addition to supporting myelination, oligodendrocyte EVs increase the resistance of neurons to stress, and enhance their survival and viability, which may have a positive effect on myelination [12]. As such, the role of EVs in this capacity may not just be limited to the development of myelin in healthy individuals, but also the regeneration of damaged myelin sheaths in MS and other demyelinating diseases.

Aside from the implications of EVs in both the pathogenesis and therapeutic intervention of MS, EVs and their cargo have presented themselves as a promising source of biomarkers for the disease. The current diagnostic procedure of MS is a complex one, which is based on clinical findings and MRI imaging results. The complex diagnostic procedure has warranted a thorough search for simpler, faster and more reliable diagnostic methods, and both for prognostic and therapeutic biomarkers [95]. Thusly, over the last decade, efforts have been made by numerous research groups to identify biomarkers of MS, with particular focus on light (NF-L) and heavy (NF-H) neurofilament protein, and astrocyte derived chitinase 3-like 1 (CHI3L1) and glial fibrillar acidic protein (GFAP) [96–99].

However, the conventional biomarker research has suffered from limited success [100]. Consequently, focus has shifted to EVs as a source of biomarkers for MS, which has shown some important promise. For example, the level of myeloid cell-derived EVs in the cerebrospinal fluid of MS patients correlates to MRI lesions in MS patients [84]. Similarly, the number of $CD61^+$, $CD45^+$ and $CD14^+$ EVs in plasma seems to correlate to disease severity [85]. Current successes are not limited to EVs derived from myeloid cells. As there is EV efflux from the CNS, a fraction of oligodendrocyte EVs containing myelin proteins are found in patient serum might be a predictive biomarker as well [101]. There seem to be differences in circulating EV miRNA too between MS patients and controls [102], however, which cells contribute to that difference is largely unknown. While most of the studies focus on biomarker differences in patients versus controls, there are a few reports on normalization of some potential biomarkers but not of others in response to MS treatments [103–105].

It should be emphasized that in MS, similarly to what has been observed in AD and PD, the evidence about the involvement of EVs in disease pathology is clear. Moreover, EVs from different cells seem to have specific functions in the pathophysiology and disease progression, which can be related to cell-specific contents of EVs, and likely even to specific EV subtypes secreted by a given cell at pathological conditions. The importance to consider the entire communication network of all EV types and subpopulations is revealed by these studies as well, particularly in the light that certain kind of EVs can be protective and alleviate the disease.

3.4 Amyotrophic lateral sclerosis

Amyotrophic lateral sclerosis (ALS) is a fatal, progressive neurodegenerative disorder that causes muscle weakness and paralysis, and eventually death, due to the degeneration of motor neurons in the cortex, brain stem and spinal cord. ALS has an incidence of 2–3/100,000 and a prevalence of 6–7/100,000, making it the most common motor neuron disease in Europe [106, 107]. Typically, patient mortality is roughly 2 to 5years following symptom emergence, however some patients have a shorter survival due to respiratory impairment, and some survive for >30years due to slow progression [108, 109].

Cytoplasmic inclusions positive for the nuclear DNA binding protein TAR DNA-binding protein 43 (TDP-43), encoded by *TARDBP*, are the most characteristic neuropathological feature of ALS [110, 111]. TDP-43 positive inclusions are seen in nearly all patients with ALS, including familial ALS [112]. The inclusions contain hyper phosphorylated, ubiquitylated aggregates that mostly consist of C-terminal fragments of TDP-43. Mutations in *TARDBP* are found almost exclusively within the C-terminus, however these mutations account for <1% of ALS cases [113]. In cases of ALS in which TDP-43-positive inclusions are not seen (<3%), aggregates consist of protein produced by mutant genes, most commonly superoxide dismutase 1 (SOD1) or, in rare cases, fused in sarcoma (FUS) [114, 115].

When insoluble TDP-43 from ALS brains is introduced to human neuroblastoma cells expressing TDP-43, TDP-43-positive inclusions are formed, indicating that templated aggregation occurs [116]. Templated aggregation of SOD1 has also been observed in a similar manner [117]. SOD1 aggregates are secreted and endocytosed by neighbouring cells, where they seed aggregation of normal, cytoplasmic mutant SOD1 and thus propagate ALS. When the aggregated SOD1 seed is removed, aggregation persists, suggesting that the newly formed aggregates are capable of continual ALS propagation [118].

The involvement of EVs in ALS is unclear, however there is some evidence to suggest that they do play a role in disease propagation through the spread of aggregates, similarly what can be observed in other neurodegenerative diseases such as AD, PD and prion disease. EVs isolated from conditioned media used to culture TDP-43-expressing neuroblastoma cells show TDP-43 enrichment. Furthermore, TDP-43 has also been detected in EVs isolated from the cerebrospinal fluid (CSF) of ALS patients. Together these findings support the hypothesized transport of aggregates via EVs as a means of ALS propagation [116, 119]. What's more, there is also evidence of SOD1 propagation via exosomes from a cell models of ALS [118, 120].

As of yet, no biomarkers are available for the detection of ALS, although EVs are an attractive source of biomarkers. The analysis of 14 ALS patient serum samples revealed that the miRNA signature of circulating EVs might be a possible biomarker [121]. However, more work is needed to identify which subpopulation of circulating EVs is responsible for this observation and what is the function of those EV miRNA biomarker candidates. In addition, one case study has reported an increase of leukocyte-derived EVs in the CSF of one ALS patient compared to healthy controls [122], and TDP-43 has been detected in EVs isolated from the CSF of an ASL patient cohort [119], although these findings have been relatively inconclusive.

As for the therapeutic benefits of EVs with respect to ALS, there has been some previous evidence to suggest that exosomes originated from adipose derived multipotent stem cells (ASCs) exert neuroprotective effects. This has been observed in an ALS in vitro model based on mutated SOD1-transfected NSC-34 motoneuron-like cells, in which ASC EVs increased ALS motoneuron survival to a certain extent in response to H_2O_2 as pathological insult [123]. ASC derived EVs could also reduce increased SOD-1 aggregation in neuronal cells, from a G93A ALS mouse in vitro model, and restore mitochondrial functions in primary neurons [124]. However, as is the case with both of these studies, as of yet, evidence of these findings is yet to be replicated in vivo.

Even though there is some evidence in the involvement of EVs in ALS pathogenesis, underlined by early successful attempts in EV-based ALS biomarker discovery, there is much less information available to date whether naturally-occurring EV signaling network in the CNS could provide neuroprotection as a part of a disease protection mechanism. However, given the similarities with other neurodegenerative diseases, there will likely be significantly more information available in the future about these matters.

3.5 Prion diseases (transmissible spongiform encephalopathies)

Prion diseases, technically known as transmissible spongiform encephalopathies (TSEs), are a group of progressive neurodegenerative conditions that affect the brain and nervous system. Prion diseases cause impairment of neurological functions, leading to changes in memory and personality, problems with movement, balance, and coordination, which invariably leads to death [125]. Misfolded isoforms (PrP^{Sc}) of the host encoded cellular prion protein (PrP^{C}) are believed to be at the heart of the pathogenesis of prion diseases. Misfolding, aggregation, and accumulation of PrP^{Sc} cause degeneration and subsequent death of neurons, ultimately leading to clusters of liquid filled cavities which give the brain tissue a sponge-like appearance (hence the name spongiform).

Prion diseases differ from other neurodegenerative diseases in their ability to spread within and occasionally between species (i.e. prion diseases are transmissible). The dangerous and potentially devastating implications of this characteristic are highlighted by the bovine spongiform encephalopathy (BSE; 'mad-cow' disease) crisis which occurred predominantly in the UK but also in other parts of Europe from the late 1980s to the 1990s [126]. Unfortunately, prions are not destroyed by cooking or heat-treatment, which meant PrP^{Sc} could be transmitted to humans through consumption of contaminated meat products. Cows are believed to have been infected through animal feed which contained processed remains/bone meal of sheep infected with scrapie, a prion disease affecting sheep [127]. Disturbingly, cattle are also believed to be infected through being fed remains of other infected cows, which were processed into bone meal. The crisis has resulted in two hundred cases of variant Creutzfeldt-Jakob (vCJD) in humans. CJD is the most well-known type of human prion disease. It was first described in the 1920s by the German neurologists Hans Gerhard Creutzfeldt and Alfons Maria Jakob, is not related to BSE exposure and has a worldwide incidence of approximately 1 case per million people per year [128, 129].

However, alongside CJD other prion diseases exist, and likely more are to be discovered. For example, the prion-like spread of α-syn protein in Parkinson's disease was discussed in the respective section above, while some other known prion diseases are very rare, such as the kuru and fatal familial insomnia. Kuru is a very rare type of prion disease that was common among the Fore people of Papua New Guinea. The disease was transmitted among members of the Fore tribe through funerary or ritualistic endocannibalism [130]. Fatal familial insomnia is a rare type of hereditary sleep disorder, characterized by progressive insomnia leading to dementia and death. It is a type of inherited prion disease, associated with mutations in the prion protein gene, which in extremely rare cases can develop sporadically [131].

Prion diseases can occur sporadically, genetically through inheritance of mutations in PRNP gene (e.g. familial CJD and Gerstmann-Sträussler-Scheinker syndrome), or they can be acquired through transmission of pre-existing prions (e.g. variant CJD and kuru). The term prion is coined from 'proteinaceous infectious particle', as indicated by this name, a protein is at the heart of the disease pathology of prion diseases. The physiologically occurring PrP^C(C for cellular) is involved in many different processes, such as neuronal differentiation, synaptic function, and copper homeostasis [132]. As discussed above in the Alzheimer's disease section, PrP^C can even be neuroprotective in neurodegenerative diseases. However, it is the misfolded variant of PrP (PrP^{Sc}) that is mostly relevant to prion diseases.

The secondary protein structure of PrP^C mainly consists of alpha helices and it is soluble and digested by protein degrading proteases. Conversion of PrP^C into toxic PrP^{Sc} (Sc for scrapie) lead to changes in the secondary structure of the protein. The secondary structure is dominated by beta sheets which render it insoluble and highly resistant to digestion by proteases. PrP^{Sc} can promote misfolding of PrP^C and induce misfolding of PrP^C in neighbouring cells. This positive feedback loop leads to disease progression and spread.

Prion proteins have been found associated with exosomes, and PrP^{Sc} bearing EVs can propagate prion infection [132]. PrP^{Sc} specifically reside in lipid rafts present in plasma membrane, where they are linked to these rafts through so called glycophosphatidylinositol (GPI) anchors. Lipid rafts containing GPI-anchored proteins are enriched in EVs and GPI-anchoring has been successfully used as an approach to tether molecules to the EV surface [133]. Studies looking at the inhibition and stimulation of EV biogenesis pathways have been shown to affect transmission of PrP^{Sc} between cells [134–136]. These findings support the role of EVs in the propagation of PrP^{Sc}.

Alternative mechanisms such as direct cell-to-cell contact and tunnelling nanotubes have been proposed to play a role in the intercellular transfer of infectious prion proteins a well [137, 138]. However, it seems that misfolding of PrP^{C} into PrP^{Sc} is triggered in the presence of PrP^{Sc} in multivescular bodies [139], the organelle that is integral to the biogenesis of EVs, as discussed in detail in other chapters of this book. EVs secreted by PrP^{Sc} infected cells change significantly in their morphology, supporting further the involvement of multivesicular bodies and EV in the spread of prion disease [140]. As a positive feedback loop, this can lead to secretion of PrP^{Sc} on EVs, reuptake, and a subsequent round of PrP^{Sc} seeding [132, 140–142]. Of note, those observations may also indicate that in addition to the neurotoxic properties of PrP^{Sc} can be modulated by altered EV composition in response to the prion protein misfolding.

It appears that EVs may be involved in the spread of the neurotoxic prion protein not only in the CNS, but also in the whole organism, as PrP^{Sc} has been found on EVs extracted from blood [143, 144]. Thorough studies on EV biology in relation to PrP^{Sc} secretion can help to shed light whether PrP^{Sc}-containing EVs found in the blood circulation are part of prion disease spread pathology or an indicative sign of disease presence, which could be exploited as an early biomarker of a prion disease [132].

4 Summary and conclusions

EVs are multifaceted and complex cell to cell signaling entities. In neurodegenerative diseases this is reflected by their dualistic roles in some cases both to be disease mediators as well as harbouring neuroprotective roles in some cases. This is likely dependent on the cell type of origin of EVs and disease-specific changes in EV composition. Studying these effects is not straightforward because EVs are highly heterogeneous in their composition and properties. Moreover, EVs from different cell sources are a part of a complex cell communication network, which functions alongside other cell signaling molecules such as cytokines, neurotransmitters, neurotropic factors and hormones. This means that the functions of EVs might depend not only on its cargo composition but also on the state of the cell that is receiving the signal. Both of which factors could be significantly altered in neurodegenerative diseases. Nevertheless, exploiting the properties of EVs and EV sub-fractions in neurodegenerative diseases have shown promise in disease biomarker research, and helped to devise novel treatment strategies of these diseases.

References

[1] Andaloussi SEL, Mäger I, Breakefield XO, Wood MJA. Extracellular vesicles: biology and emerging therapeutic opportunities. Nat Rev Drug Discov 2013;12:347–57.

[2] Thompson AG, et al. Extracellular vesicles in neurodegenerative disease—pathogenesis to biomarkers. Nat Rev Neurol 2016;12:346–57.

[3] Mathieu M, Martin-Jaular L, Lavieu G, Théry C. Specificities of secretion and uptake of exosomes and other extracellular vesicles for cell-to-cell communication. Nat Cell Biol 2019;21:9–17.

[4] Zhang H, et al. Identification of distinct nanoparticles and subsets of extracellular vesicles by asymmetric flow field-flow fractionation. Nat Cell Biol 2018;20:332–43.

[5] Tkach M, Kowal J, Théry C. Why the need and how to approach the functional diversity of extracellular vesicles. Philos Trans R Soc Lond B Biol Sci 2018;373:20160479.

[6] Willms E, et al. Cells release subpopulations of exosomes with distinct molecular and biological properties. Sci Rep 2016;6:22519.

[7] Willms E, Cabañas C, Mäger I, Wood MJA, Vader P. Extracellular vesicle heterogeneity: subpopulations, isolation techniques, and diverse functions in cancer progression. Front Immunol 2018;9:738.

[8] Lai RC, Lim SK. Membrane lipids define small extracellular vesicle subtypes secreted by mesenchymal stromal cells. J Lipid Res 2019;60. https://doi.org/10.1194/jlr.R087411.

[9] Timmers L, et al. Human mesenchymal stem cell-conditioned medium improves cardiac function following myocardial infarction. Stem Cell Res 2011;6:206–14.

[10] Lai RC, et al. MSC secretes at least 3 EV types each with a unique permutation of membrane lipid, protein and RNA. J Extracell Vesicles 2016;1:1–12.

[11] Frühbeis C, et al. Neurotransmitter-triggered transfer of exosomes mediates oligodendrocyte–neuron communication. PLoS Biol 2013;11:e1001604.

[12] Frohlich D, et al. Multifaceted effects of oligodendroglial exosomes on neurons: impact on neuronal firing rate, signal transduction and gene regulation. Philos Trans R Soc Lond B Biol Sci 2014;369:20130510.

[13] Fitzner D, et al. Selective transfer of exosomes from oligodendrocytes to microglia by macropinocytosis. J Cell Sci 2011;124:447–58.

[14] Bakhti M, Winter C, Simons M. Inhibition of myelin membrane sheath formation by oligodendrocyte-derived exosome-like vesicles. J Biol Chem 2011;286:787–96.

[15] Riganti L, et al. Sphingosine-1-phosphate (S1P) impacts presynaptic functions by regulating synapsin I localization in the presynaptic compartment. J Neurosci 2016;36:4624–34.

[16] Antonucci F, et al. Microvesicles released from microglia stimulate synaptic activity via enhanced sphingolipid metabolism. EMBO J 2012;31:1231–40.

[17] Pastuzyn ED, et al. The neuronal gene Arc encodes a repurposed retrotransposon Gag protein that mediates intercellular RNA transfer. Cell 2018;172:275–288.e18.

[18] Ashley J, et al. Retrovirus-like Gag protein Arc1 binds RNA and traffics across synaptic boutons. Cell 2018;172:262–274.e11.

[19] Lachenal G, et al. Release of exosomes from differentiated neurons and its regulation by synaptic glutamatergic activity. Mol Cell Neurosci 2011;46:409–18.

[20] Korkut C, et al. Trans-synaptic transmission of vesicular Wnt signals through Evi/Wntless. Cell 2009;139:393–404.

[21] Holm MM, Kaiser J, Schwab ME. Extracellular vesicles: multimodal envoys in neural maintenance and repair. Trends Neurosci 2018;41:360–72.

[22] Koles K, et al. Mechanism of evenness interrupted (Evi)-exosome release at synaptic boutons. J Biol Chem 2012;287:16820–34.

[23] Beckett K, et al. Drosophila S2 cells secrete wingless on exosome-like vesicles but the wingless gradient forms independently of exosomes. Traffic 2013;14:82–96.

[24] Gross JC, Chaudhary V, Bartscherer K, Boutros M. Active Wnt proteins are secreted on exosomes. Nat Cell Biol 2012;14:1036–45.

[25] Gong J, Körner R, Gaitanos L, Klein R. Exosomes mediate cell contact–independent ephrin-Eph signaling during axon guidance. J Cell Biol 2016;214:35–44.

[26] Vyas N, et al. Vertebrate hedgehog is secreted on two types of extracellular vesicles with different signaling properties. Sci Rep 2015;4:7357.
[27] Matusek T, et al. The ESCRT machinery regulates the secretion and long-range activity of hedgehog. Nature 2014;516:99–103.
[28] Korkut C, et al. Regulation of postsynaptic retrograde signaling by presynaptic exosome release. Neuron 2013;77:1039–46.
[29] Bahrini I, Song J, Diez D, Hanayama R. Neuronal exosomes facilitate synaptic pruning by up-regulating complement factors in microglia. Sci Rep 2015;5:7989.
[30] Malm T, Loppi S, Kanninen KM. Exosomes in Alzheimer's disease. Neurochem Int 2016;97:193–9.
[31] Vella L, Hill A, Cheng L. Focus on extracellular vesicles: exosomes and their role in protein trafficking and biomarker potential in Alzheimer's and Parkinson's disease. Int J Mol Sci 2016;17:173.
[32] Takahashi RH, et al. Intraneuronal Alzheimer Aβ42 accumulates in multivesicular bodies and is associated with synaptic pathology. Am J Pathol 2002;161:1869–79.
[33] Verbeek MM, Otte-Höller I, Fransen JAM, de Waal RMW. Accumulation of the amyloid-β precursor protein in multivesicular body-like organelles. J Histochem Cytochem 2002;50:681–90.
[34] Langui D, et al. Subcellular topography of neuronal Aβ peptide in APPxPS1 transgenic mice. Am J Pathol 2004;165:1465–77.
[35] Rajendran L, et al. Alzheimer's disease beta-amyloid peptides are released in association with exosomes. Proc Natl Acad Sci U S A 2006;103:11172–7.
[36] Dinkins MB, Dasgupta S, Wang G, Zhu G, Bieberich E. Exosome reduction in vivo is associated with lower amyloid plaque load in the 5XFAD mouse model of Alzheimer's disease. Neurobiol Aging 2014;35:1792–800.
[37] Dinkins MB, et al. The 5XFAD mouse model of Alzheimer's disease exhibits an age-dependent increase in anti-ceramide IgG and exogenous administration of ceramide further increases anti-ceramide titers and amyloid plaque burden. J Alzheimers Dis 2015;46:55–61.
[38] Perez-Gonzalez R, Gauthier SA, Kumar A, Levy E. The exosome secretory pathway transports amyloid precursor protein carboxyl-terminal fragments from the cell into the brain extracellular space. J Biol Chem 2012;287:43108–15.
[39] Sharples RA, et al. Inhibition of γ-secretase causes increased secretion of amyloid precursor protein C-terminal fragments in association with exosomes. FASEB J 2008;22:1469–78.
[40] Wang G, et al. Astrocytes secrete exosomes enriched with proapoptotic ceramide and prostate apoptosis response 4 (PAR-4): potential mechanism of apoptosis induction in Alzheimer disease (AD). J Biol Chem 2012;287:21384–95.
[41] Joshi P, et al. Microglia convert aggregated amyloid-β into neurotoxic forms through the shedding of microvesicles. Cell Death Differ 2014;21:582–93.
[42] Asai H, et al. Depletion of microglia and inhibition of exosome synthesis halt tau propagation. Nat Neurosci 2015;18:1584–93.
[43] Simon D, Garcia-Garcia E, Royo F, Manuel Falcon-Perez J, Avila J. Proteostasis of tau. Tau overexpression results in its secretion via membrane vesicles. FEBS Lett 2012;586:47–54.
[44] Saman S, et al. Exosome-associated tau is secreted in tauopathy models and is selectively phosphorylated in cerebrospinal fluid in early Alzheimer disease. J Biol Chem 2012;287:3842–9.
[45] Fiandaca MS, et al. Identification of preclinical Alzheimer's disease by a profile of pathogenic proteins in neurally derived blood exosomes: a case-control study. Alzheimers Dement 2015;11:600–607.e1.

[46] Yuyama K, Sun H, Mitsutake S, Igarashi Y. Sphingolipid-modulated exosome secretion promotes clearance of amyloid-β by microglia. J Biol Chem 2012;287:10977–89.
[47] Yuyama K, et al. Decreased amyloid-β pathologies by intracerebral loading of glycosphingolipid-enriched exosomes in Alzheimer model mice. J Biol Chem 2014;289:24488–98.
[48] An K, et al. Exosomes neutralize synaptic-plasticity-disrupting activity of Aβ assemblies in vivo. Mol Brain 2013;6:47.
[49] Falker C, et al. Exosomal cellular prion protein drives fibrillization of amyloid beta and counteracts amyloid beta-mediated neurotoxicity. J Neurochem 2016;137:88–100.
[50] Chiasserini D, et al. Proteomic analysis of cerebrospinal fluid extracellular vesicles: a comprehensive dataset. J Proteomics 2014;106:191–204.
[51] Bellingham SA, Guo BB, Coleman BM, Hill AF. Exosomes: vehicles for the transfer of toxic proteins associated with neurodegenerative diseases? Front Physiol 2012;3. https://doi.org/10.3389/fphys.2012.00124.
[52] Tamboli IY, et al. Statins promote the degradation of extracellular amyloid β-peptide by microglia via stimulation of exosome-associated insulin-degrading enzyme (IDE) secretion. J Biol Chem 2010;285:37405–14.
[53] Bulloj A, Leal MC, Xu H, Castano EM, Morelli L. Insulin-degrading enzyme sorting in exosomes: a secretory pathway for a key brain amyloid-beta degrading protease. J Alzheimers Dis 2010;19:79–95.
[54] Sanderson RD, Bandari SK, Vlodavsky I. Proteases and glycosidases on the surface of exosomes: newly discovered mechanisms for extracellular remodeling. Matrix Biol 2017. https://doi.org/10.1016/j.matbio.2017.10.007.
[55] Olanow CW, Stern MB, Sethi K. The scientific and clinical basis for the treatment of Parkinson disease (2009). Neurology 2009;72:S1–136.
[56] Chistiakov DA, Chistiakov AA. α-Synuclein-carrying extracellular vesicles in Parkinson's disease: deadly transmitters. Acta Neurol Belg 2017;117:43–51.
[57] Lee H-J. Intravesicular localization and exocytosis of -synuclein and its aggregates. J Neurosci 2005;25:6016–24.
[58] Lee H-J, Suk J-E, Bae E-J, Lee S-J. Clearance and deposition of extracellular α-synuclein aggregates in microglia. Biochem Biophys Res Commun 2008;372:423–8.
[59] Danzer KM, et al. Exosomal cell-to-cell transmission of alpha synuclein oligomers. Mol Neurodegener 2012;7:42.
[60] Gray M, et al. Acceleration of α-synuclein aggregation by exosomes. J Biol Chem 2015;290:2969–82.
[61] Desplats P, et al. Inclusion formation and neuronal cell death through neuron-to-neuron transmission of -synuclein. Proc Natl Acad Sci U S A 2009;106:13010–5.
[62] Emmanouilidou E, et al. Cell-produced -synuclein is secreted in a calcium-dependent manner by exosomes and impacts neuronal survival. J Neurosci 2010;30:6838–51.
[63] Alvarez-Erviti L, et al. Lysosomal dysfunction increases exosome-mediated alpha-synuclein release and transmission. Neurobiol Dis 2011;42:360–7.
[64] Ali SF, Binienda ZK, Imam SZ. Molecular aspects of dopaminergic neurodegeneration: gene-environment interaction in parkin dysfunction. Int J Environ Res Public Health 2011;8:4702–13.
[65] Gitler AD, et al. α-Synuclein is part of a diverse and highly conserved interaction network that includes PARK9 and manganese toxicity. Nat Genet 2009;41:308–15.
[66] Kong SMY, et al. Parkinson's disease-linked human PARK9/ATP13A2 maintains zinc homeostasis and promotes α-Synuclein externalization via exosomes. Hum Mol Genet 2014;23:2816–33.

[67] Ramonet D, et al. PARK9-associated ATP13A2 localizes to intracellular acidic vesicles and regulates cation homeostasis and neuronal integrity. Hum Mol Genet 2012;21:1725–43.

[68] Ramirez A, et al. Hereditary parkinsonism with dementia is caused by mutations in ATP13A2, encoding a lysosomal type 5 P-type ATPase. Nat Genet 2006;38:1184–91.

[69] Hansen C, et al. α-Synuclein propagates from mouse brain to grafted dopaminergic neurons and seeds aggregation in cultured human cells. J Clin Invest 2011;121:715–25.

[70] Lee H-J, et al. Direct transfer of α-synuclein from neuron to astroglia causes inflammatory responses in synucleinopathies. J Biol Chem 2010;285:9262–72.

[71] Halliday GM, Stevens CH. Glia: initiators and progressors of pathology in Parkinson's disease. Mov Disord 2011;26:6–17.

[72] Vekrellis K, Xilouri M, Emmanouilidou E, Rideout HJ, Stefanis L. Pathological roles of α-synuclein in neurological disorders. Lancet Neurol 2011;10:1015–25.

[73] Alvarez-Erviti L, Couch Y, Richardson J, Cooper JM, Wood MJA. Alpha-synuclein release by neurons activates the inflammatory response in a microglial cell line. Neurosci Res 2011;69:337–42.

[74] Uttara B, Singh A, Zamboni P, Mahajan R. Oxidative stress and neurodegenerative diseases: a review of upstream and downstream antioxidant therapeutic options. Curr Neuropharmacol 2009;7:65–74.

[75] Chang C, et al. Exosomes of BV-2 cells induced by alpha-synuclein: important mediator of neurodegeneration in PD. Neurosci Lett 2013;548:190–5.

[76] Stuendl A, et al. Induction of α-synuclein aggregate formation by CSF exosomes from patients with Parkinson's disease and dementia with Lewy bodies. Brain 2016;139:481–94.

[77] Shi M, et al. Plasma exosomal α-synuclein is likely CNS-derived and increased in Parkinson's disease. Acta Neuropathol 2014;128:639–50.

[78] Tomlinson PR, et al. Identification of distinct circulating exosomes in Parkinson's disease. Ann Clin Transl Neurol 2015;2:353–61.

[79] Lemus HN, Warrington AE, Rodriguez M. Multiple sclerosis: mechanisms of disease and strategies for myelin and axonal repair. Neurol Clin 2018;36:1–11.

[80] Ascherio A, Munger KL. Environmental risk factors for multiple sclerosis. Part I: the role of infection. Ann Neurol 2007;61:288–99.

[81] Hohlfeld R, Wekerle H. Autoimmune concepts of multiple sclerosis as a basis for selective immunotherapy: from pipe dreams to (therapeutic) pipelines. Proc Natl Acad Sci U S A 2004;101:14599–606.

[82] Sospedra M, Martin R. Immunology of multiple sclerosis. Annu Rev Immunol 2005;23:683–747.

[83] Hohlfeld R, Dornmair K, Meinl E, Wekerle H. The search for the target antigens of multiple sclerosis, part 2: CD8+ T cells, B cells, and antibodies in the focus of reverse-translational research. Lancet Neurol 2016;15:317–31.

[84] Verderio C, et al. Myeloid microvesicles are a marker and therapeutic target for neuroinflammation. Ann Neurol 2012;72:610–24.

[85] Sáenz-Cuesta M, et al. Circulating microparticles reflect treatment effects and clinical status in multiple sclerosis. Biomark Med 2014;8:653–61.

[86] Jy W, et al. Endothelial microparticles (EMP) bind and activate monocytes: elevated EMP-monocyte conjugates in multiple sclerosis. Front Biosci 2004;9:3137–44.

[87] Wheway J, Latham SL, Combes V, Grau GER. Endothelial microparticles interact with and support the proliferation of T cells. J Immunol 2014;193:3378–87.

[88] Barry OP, Praticò D, Savani RC, FitzGerald GA. Modulation of monocyte-endothelial cell interactions by platelet microparticles. J Clin Invest 1998;102:136–44.

[89] Quandt J, Dorovini-Zis K. The beta chemokines CCL4 and CCL5 enhance adhesion of specific CD4+ T cell subsets to human brain endothelial cells. J Neuropathol Exp Neurol 2004;63:350–62.

[90] Marcos-Ramiro B, et al. Microparticles in multiple sclerosis and clinically isolated syndrome: effect on endothelial barrier function. BMC Neurosci 2014. https://doi.org/10.1186/1471-2202-15-110.

[91] Alexander JS, et al. Blood circulating microparticle species in relapsing-remitting and secondary progressive multiple sclerosis. A case-control, cross sectional study with conventional MRI and advanced iron content imaging outcomes. J Neurol Sci 2015;355:84–9.

[92] Hakulinen J, Sankkila L, Sugiyama N, Lehti K, Keski-Oja J. Secretion of active membrane type 1 matrix metalloproteinase (MMP-14) into extracellular space in microvesicular exosomes. J Cell Biochem 2008;105:1211–8.

[93] Sarkar A, Mitra S, Mehta S, Raices R, Wewers MD. Monocyte derived microvesicles deliver a cell death message via encapsulated caspase-1. PLoS ONE 2009;4:e7140.

[94] Pusic AD, Pusic KM, Clayton BLL, Kraig RP. IFNγ-stimulated dendritic cell exosomes as a potential therapeutic for remyelination. J Neuroimmunol 2014;266:12–23.

[95] Palmer AJ, Colman S, O'Leary B, Taylor BV, Simmons RD. The economic impact of multiple sclerosis in Australia in 2010. Mult Scler 2013;19:1640–6.

[96] Lee JY, Taghian K, Petratos S. Axonal degeneration in multiple sclerosis: can we predict and prevent permanent disability? Acta Neuropathol Commun 2014;2:97.

[97] Comabella M, et al. Cerebrospinal fluid chitinase 3-like 1 levels are associated with conversion to multiple sclerosis. Brain 2010;133:1082–93.

[98] Borràs E, et al. Protein-based classifier to predict conversion from clinically isolated syndrome to multiple sclerosis. Mol Cell Proteomics 2016;15:318–28.

[99] Axelsson M, et al. Glial fibrillary acidic protein: a potential biomarker for progression in multiple sclerosis. J Neurol 2011;258:882–8.

[100] Trentini A, et al. N-acetylaspartate and neurofilaments as biomarkers of axonal damage in patients with progressive forms of multiple sclerosis. J Neurol 2014;261:2338–43.

[101] Galazka G, Mycko MP, Selmaj I, Raine CS, Selmaj KW. Multiple sclerosis: serum-derived exosomes express myelin proteins. Mult Scler 2018;24:449–58.

[102] Selmaj I, et al. Global exosome transcriptome profiling reveals biomarkers for multiple sclerosis. Ann Neurol 2017;81:703–17.

[103] Jimenez JJ, et al. Elevated endothelial microparticle—monocyte complexes induced by multiple sclerosis plasma and the inhibitory effects of interferon-β1b on release of endothelial microparticles, formation and transendothelial migration of monocyte-endothelial microparticle. Mult Scler J 2005;11:310–5.

[104] Lowery-Nordberg M, et al. The effects of high dose interferon-β1a on plasma microparticles: correlation with MRI parameters. J Neuroinflammation 2011;8:43.

[105] Dawson G, Qin J. Gilenya (FTY720) inhibits acid sphingomyelinase by a mechanism similar to tricyclic antidepressants. Biochem Biophys Res Commun 2011;404:321–3.

[106] Mandrioli J, Faglioni P, Merelli E, Sola P. The epidemiology of ALS in Modena, Italy. Neurology 2003;60:683–9.

[107] Chio A, et al. Epidemiology of ALS in Italy: a 10-year prospective population-based study. Neurology 2009;72:725–31.

[108] Stambler N, Charatan M, Cedarbaum JM. Prognostic indicators of survival in ALS. Neurology 1998;50:66–72.

[109] de Carvalho M, et al. Motor neuron disease presenting with respiratory failure. J Neurol Sci 1996;139:117–22.

[110] Arai T, et al. TDP-43 is a component of ubiquitin-positive tau-negative inclusions in frontotemporal lobar degeneration and amyotrophic lateral sclerosis. Biochem Biophys Res Commun 2006;351:602–11.

[111] Neumann M, et al. Ubiquitinated TDP-43 in frontotemporal lobar degeneration and amyotrophic lateral sclerosis. Science 2006;314:130–3.

[112] Mackenzie IRA, Frick P, Neumann M. The neuropathology associated with repeat expansions in the C9ORF72 gene. Acta Neuropathol 2014;127:347–57.

[113] Rutherford NJ, et al. Novel mutations in TARDBP(TDP-43) in patients with familial amyotrophic lateral sclerosis. PLoS Genet 2008;4:e1000193.

[114] Shibata N, et al. Intense superoxide dismutase-1 immunoreactivity in intracytoplasmic hyaline inclusions of familial amyotrophic lateral sclerosis with posterior column involvement. J Neuropathol Exp Neurol 1996;55:481–90.

[115] Deng HX, et al. FUS-immunoreactive inclusions are a common feature in sporadic and non-SOD1 familial amyotrophic lateral sclerosis. Ann Neurol 2010;67:739–48.

[116] Nonaka T, et al. Prion-like properties of pathological TDP-43 aggregates from diseased brains. Cell Rep 2013;4:124–34.

[117] Chia R, et al. Superoxide dismutase 1 and tgSOD1G93A mouse spinal cord seed fibrils, suggesting a propagative cell death mechanism in amyotrophic lateral sclerosis. PLoS ONE 2010;5:e10627.

[118] Münch C, O'Brien J, Bertolotti A. Prion-like propagation of mutant superoxide dismutase-1 misfolding in neuronal cells. Proc Natl Acad Sci U S A 2011;108:3548–53.

[119] Feneberg E, et al. Limited role of free TDP-43 as a diagnostic tool in neurodegenerative diseases. Amyotroph Lateral Scler Frontotemporal Degener 2014;15:351–6.

[120] Gomes C, Keller S, Altevogt P, Costa J. Evidence for secretion of Cu,Zn superoxide dismutase via exosomes from a cell model of amyotrophic lateral sclerosis. Neurosci Lett 2007;428:43–6.

[121] Saucier D, et al. Identification of a circulating miRNA signature in extracellular vesicles collected from amyotrophic lateral sclerosis patients. Brain Res 2018;1708:100–8.

[122] Zachau AC, et al. Leukocyte-derived microparticles and scanning electron microscopic structures in two fractions of fresh cerebrospinal fluid in amyotrophic lateral sclerosis: a case report. J Med Case Reports 2012;6:274.

[123] Bonafede R, et al. Exosome derived from murine adipose-derived stromal cells: neuroprotective effect on in vitro model of amyotrophic lateral sclerosis. Exp Cell Res 2016;340:150–8.

[124] Lee M, et al. Adipose-derived stem cell exosomes alleviate pathology of amyotrophic lateral sclerosis in vitro. Biochem Biophys Res Commun 2016;479:434–9.

[125] Scheckel C, Aguzzi A. Prions, prionoids and protein misfolding disorders. Nat Rev Genet 2018;19:405–18.

[126] Ducrot C, Arnold M, de Koeijer A, Heim D, Calavas D. Review on the epidemiology and dynamics of BSE epidemics. Vet Res 2008;39:15.

[127] Greig JR. Scrapie in sheep. J Comp Pathol 1950;60:263–6.

[128] Richardson EP, Masters CL. The nosology of Creutzfeldt-Jakob disease and conditions related to the accumulation of PrPCJD in the nervous system. Brain Pathol 1995;5:33–41.

[129] Creutzfeldt HG. Über eine eigenartige herdförmige erkrankung des zentralnervensystems (Vorläufige mitteilung). Zeitschrift für die gesamte Neurol und Psychiatre 1920;57:1–18.

[130] Gajdusek DC, Zigas V. Degenerative disease of the central nervous system in New Guinea; the endemic occurrence of kuru in the native population. N Engl J Med 1957;257:974–8.

[131] Llorens F, Zarranz J-J, Fischer A, Zerr I, Ferrer I. Fatal familial insomnia: clinical aspects and molecular alterations. Curr Neurol Neurosci Rep 2017;17:30.

[132] Cheng L, Zhao W, Hill AF. Exosomes and their role in the intercellular trafficking of normal and disease associated prion proteins. Mol Aspects Med 2018;60:62–8.
[133] Kooijmans SAA, et al. Display of GPI-anchored anti-EGFR nanobodies on extracellular vesicles promotes tumour cell targeting. J Extracell Vesicles 2016;5:31053.
[134] Guo BB, Bellingham SA, Hill AF. The neutral sphingomyelinase pathway regulates packaging of the prion protein into exosomes. J Biol Chem 2015;290:3455–67.
[135] Guo BB, Bellingham SA, Hill AF. Stimulating the release of exosomes increases the intercellular transfer of prions. J Biol Chem 2016;291:5128–37.
[136] Vilette D, et al. Efficient inhibition of infectious prions multiplication and release by targeting the exosomal pathway. Cell Mol Life Sci 2015;72:4409–27.
[137] Gousset K, et al. Prions hijack tunnelling nanotubes for intercellular spread. Nat Cell Biol 2009;11:328–36.
[138] Kanu N, et al. Transfer of scrapie prion infectivity by cell contact in culture. Curr Biol 2002;12:523–30.
[139] Yim Y-I, et al. The multivesicular body is the major internal site of prion conversion. J Cell Sci 2015;128:1434–43.
[140] Coleman BM, Hanssen E, Lawson VA, Hill AF. Prion-infected cells regulate the release of exosomes with distinct ultrastructural features. FASEB J 2012;26:4160–73.
[141] Vella LJ, et al. Packaging of prions into exosomes is associated with a novel pathway of PrP processing. J Pathol 2007;211:582–90.
[142] Bellingham SA, Coleman BM, Hill AF. Small RNA deep sequencing reveals a distinct miRNA signature released in exosomes from prion-infected neuronal cells. Nucleic Acids Res 2012;40:10937–49. https://doi.org/10.1093/nar/gks832.
[143] Saá P, et al. First demonstration of transmissible spongiform encephalopathy-associated prion protein (PrPTSE) in extracellular vesicles from plasma of mice infected with mouse-adapted variant Creutzfeldt-Jakob disease by in vitro amplification. J Biol Chem 2014;289:29247–60.
[144] Cervenakova L, et al. Are prions transported by plasma exosomes? Transfus Apher Sci 2016;55:70–83.

Chapter 13

Extracellular vesicles in fibrotic diseases: New applications for fibrosis diagnosis and treatment

Tsukasa Kadota[a,b], Nobuyoshi Kosaka[a], Yu Fujita[a,b], Jun Araya[b], Kazuyoshi Kuwano[b], Takahiro Ochiya[a,c]

[a]*Division of Molecular and Cellular Medicine, National Cancer Center Research Institute, Tokyo, Japan,* [b]*Division of Respiratory Diseases, Department of Internal Medicine, The Jikei University School of Medicine, Tokyo, Japan,* [c]*Institute of Medical Science, Tokyo Medical University, Tokyo, Japan*

1 Introduction

Fibrosis is commonly characterized by the formation and accumulation of extra cellular matrix (ECM) components, leading to the development of remodeling in tissues and organs. Such extensive tissue remodeling causes organ failure and can result in death. It is estimated that fibrotic disorders contribute to up to 45% of all deaths in the United States [1]. Although this estimation includes atherosclerosis, fibrotic diseases have a variety of high mortality and less common conditions such as idiopathic pulmonary fibrosis (IPF), liver cirrhosis, cardiac fibrosis, systemic sclerosis and renal fibrosis. Thus, we now recognize that fibrogenesis is an important diseases process caused by a wide array of the cellular and molecular mechanisms. However, current treatment strategies for fibrosis are of limited number and efficacy. These facts highlight the need for greater comprehension of the fibrosis pathogenesis and to identification of potential therapeutic approaches.

Extracellular vesicles (EVs) include a wide variety of small membranous vesicles, ranging from approximately 50nm to a few micrometers in size, which are released into the extracellular environment by almost all cell types [2, 3]. EVs are often categorized as exosomes, microvesicles, and apoptotic bodies, based on their size, biogenesis, and secretion mechanisms [2, 3]. Exosomes are generated by the inward and reverse budding of an endosomal membrane and are released into the extracellular fluid by the fusion of the multivesicular bodies (MVBs) with the plasma membrane [2, 3]. Microvesicles, which are larger than exosomes, are generated from the plasma membrane via shedding or budding

Exosomes. https://doi.org/10.1016/B978-0-12-816053-4.00013-4

in normal circumstances or in response to stimuli [2, 3]. Apoptotic bodies are a few micrometers in diameter and are released from the plasma membrane during cell apoptosis via indiscriminate blebbing. Although the origins of these vesicles have been defined, current methodologies cannot distinguish among those different types of EVs [4]. Thus, in this chapter, we use the term EV as a general term for all types of vesicles in the extracellular fluid [5]. Recently, EVs have identified as regulatory mediators of intercellular communication through the transfer of their contents such as proteins, messenger RNAs (mRNAs) and microRNAs (miRNAs) [6, 7], in diverse biological and pathological processes including fibrogenesis. In addition, there is overwhelming evidence that EVs have potential to be used as biomarkers of pathological states or as therapeutic agents. Therefore, investigating the roles of EVs in fibrotic diseases could not only advance our understanding of the pathogenesis of this condition but also provide new clinical applications for the diagnosis and treatment of patients with fibrotic diseases. In this chapter, we outline our understanding of how tissue injury and repair lead to fibrosis. We also summarize the current knowledge of the involvement of EVs in fibrosis diseases including liver, lung and heart. Furthermore, we propose EV-based biomarkers and treatments for fibrotic diseases.

2 Mechanisms of fibrosis

Common pathological mechanisms for fibrosis development can be existed in many chronic diseases or injuries. Indeed, involved signaling pathways and mediators related to fibrogenesis in fibrosis are similar across various tissues and organs. For instance, Makarev et al. revealed a number of common pathways between lung and liver fibrosis, such as transforming growth factor (TGF-β), interleukin-6 (IL-6), and integrin-linked kinase signaling [8]. Some antifibrotic drugs such as pirfenidone were also effective against fibrosis across different tissues and organs [9, 10]. Moreover, fibrotic diseases in different tissues and organs can interact with each other. For instance, heart and kidney simultaneously develop fibrosis along with aging through imbalance between the natriuretic peptide system (NPS), renin-angiotensin-aldosterone system (RAAS) and TGF-β1 pathways [11]. Therefore, revealing core molecular mechanisms shared by fibrosis in multiple organs is crucial to understanding the pathogenesis of fibrotic diseases.

Fibrogenesis has been recognized as a part of wound-healing process in response to tissue injury. Fibrogenesis is required to rapidly repair and fix damage through myofibroblast proliferation and extracellular matrix (ECM) deposition in all organs [12]. When an injury naturally resolves, the fibrogenic response is usually limited and resorption of extracellular matrix proteins occurs, which promotes organ repair. On the other hand, when chronic or iterative injury persists, the unremitting activation of effector cells causes a state of overactive wound healing and loss of the normal regenerative process. As a result, the exaggerated and disorganized deposition of ECM ultimately impairs the architecture and functioning of the organ [13].

The wound healing process is orchestrated by a complex layer of regulatory mechanisms, including interplay of a multitude of cell types and molecular systems. In most chronic fibrotic disorders, inflammation can be initial step for fibrogenesis [14] (Fig. 1A). Various stress-induced injury of resident epithelial cells and often endothelial cells leads to production of a variety of inflammatory mediators, such as cytokines, chemokines, and other factors, resulting in stimulation and recruitment of a wide range of inflammatory cells, such as macrophages and neutrophils. These inflammatory cells elicit the activation of mesenchymal cells, such as fibroblasts, through the profibrotic mediators, including platelet-derived growth factor (PDGF), TGF-β1 and IL-13 [15, 16]. In particular, TGF-β1 plays a prominent role in inducing the differentiation of precursor cells into myofibroblasts. The other way to develop fibrosis is the epithelium-driven fibrotic response without inflammatory cell infiltration, represented by IPF[17] (Fig. 1B). In IPF pathogenesis, injured and aberrant activated lung epithelial cells induce the migration of mesenchymal cells and their differentiation to myofibroblasts that can occur without a primary immunopathogenic component [17]. In both inflammatory and non-inflammatory mechanisms, myofibroblasts rapidly produce a variety of ECM to maintain the integrity of the injured tissue during repair and further increase production of cytokines, chemokines, growth factors and other factors in an autocrine and paracrine fashion, which further trigger and promote fibrosis [18].

3 The involvement of EVs in pathogenesis of organ fibrosis

Bioactive soluble mediators of fibrogenesis include chemokines (CCL2, MCP-1, MIP-1β), cytokines (TGF-β1, PDGF, IL-1β, IL-6, IL-13, IL-2, IL-33, and TNF-α), angiogenic factors (VEGF), growth factors (PDGF), peroxisome proliferator-activated receptors (PPARs), caspases, and components of the renin–angiotensin–aldosterone system (ANG II) [13, 19]. Furthermore, non-peptide mediators such as reactive oxygen species (ROS) and lipid mediators have been identified as crucial factors relevant in fibrogenesis [19]. Recently, EVs have emerged as new regulators of fibrogenesis. In this section, we reviewed the growing body of evidence that suggests that EVs are involved in the process of fibrosis in several organs, including liver, lung and heart (Table 1).

3.1 EVs and hepatic fibrosis

Liver fibrosis is a common outcome in chronic liver diseases such as viral hepatitis, alcohol abuse and non-alcoholic fatty liver disease, which can lead to liver cirrhosis and hepatic failure. The activation of sinusoidal hepatic stellate cells (HSC) is a critical event of liver fibrosis, as these cells become the primary source of extracellular matrix in liver upon injury [29]. Following liver injury, HSCs lose their lipid droplets, migrate to injured sites and acquire myofibroblast differentiation, which secrete excessive extracellular matrix proteins.

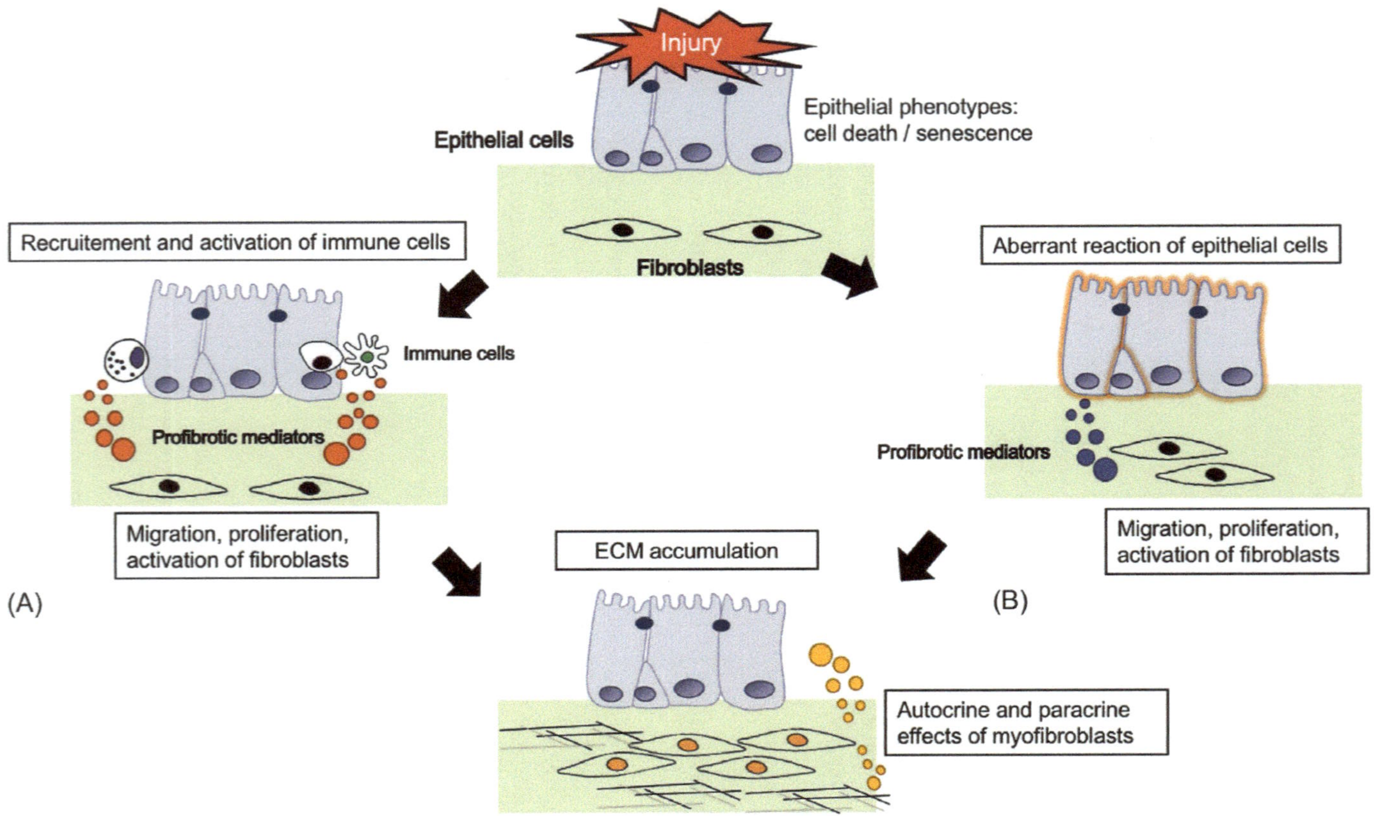

FIG. 1 Schematic representation of fibrosis progression. A wide range of injurious stimuli lead to epithelial cell injury, which induces cell death including apoptosis and/or cellular senescence. In the most cases (A), injured epithelial cells recruit and activate various immune cells, including macrophages and lymphocytes, which in turn, release profibrotic mediators. This response leads to expansion, recruitment, and/or activation of tissue myofibroblasts, the principal source of the pathologic ECM that characterizes organ fibrosis. In contrast, noninflammatory epithelial-driven fibrotic response is the other way to develop fibrosis, which is represented by idiopathic pulmonary fibrosis (B). Injured epithelial cells produce various mediators, leading to fibroblast activation and ECM accumulation that can occur without a primary immunopathogenic component. Once the fibrotic response is established, activated fibroblasts have autocrine and paracrine effects on its microenvironment cells such as epithelial cells and immune cells.

TABLE 1 Extracellular vesicles in organ fibrosis pathogeneis

Organ	Source	Recipient cell	Cargo	Effect	References
Liver	HSC	HSC, hepatocyte	miR-214	CCN2 suppression	[20]
	HSC	HSC	Twistl	CCN2 suppression	[21]
	HSC	HSC	CCN, CCN mRNA	|-	[22]
	Endothelial cell	HSC	Sphingosine kinase 1	HSC migration	[23]
Lung	–	Lung fibroblast	Wint5	Fibroblast proliferation	[24]
	Endothelial cell	Lung fibroblast	–	Fibroblast migration capacity	[25]
Heart	Fibroblast	Cardiomyocyte	miR-23-3p	Development of cardiac hypertrophy	[26]
	Cardiomyocyte	Fibroblast	Hsp90	Development of cardiac fibrosis	[27]
	Cardiac fibroblast	Cardiomyocyte	miRNAs	Nrf2 dysregulation	[28]

HSC, hepatic stellate cell CCN, ysteine-rich-61/connective tissue growth factor/nephroblastoma. Nrf2, Nuclear factor erythroid 2-related factor 2.

To date, there are several reports regarding the function of EVs in liver fibrosis development [20, 23]. CCN2 is a member of the cysteine-rich-61/connective tissue growth factor/nephroblastoma-overexpressed (CCN) family of proteins, which plays a pivotal role in various fibrotic diseases. In liver fibrosis, CCN2 is increasingly expressed in activated HSCs, which directly promote mitogenesis, chemotaxis, and fibrogenesis [30]. Charrier et al. described that HSC-derived EVs contained increased CCN2 or CCN2 mRNA during HSC activation and transferred to other quiescent or activated HSCs [22]. Chen et al. demonstrated that miR-214 directly regulates CCN2 expression in experimental liver fibrosis or during HSC activation and was transferred from HSCs to neighboring HSCs or hepatocytes via EVs [20]. The same group also reported a unique function for twistrelated protein 1 (Twist1), a member of the basic helix-loop-helix family of DNA-binding transcription factors in CCN2-dependent fibrogenesis [21]. HSC-derived EVs contain high levels of Twist1, which drives miR-214 expression and results in CCN2 expression in recipient HSC cells via EV transfer during HSC activation. In addition to the regulation of HSC activation, EVs are also associated with HSC migration. Wang et al. demonstrated that sphingosine kinase 1 in endothelial cell-derived EVs regulates HSC signaling and migration through fibronectin-integrin-dependent EV adherence and dynamin-dependent EV internalization [23]. In liver cirrhosis, angiogenesis contributes to vascular remodeling during cirrhosis. Witek et al. found that activated HSCs and reactive cholangiocytes transfer Hedgehog ligand-containing EVs to sinusoidal endothelial cells (SECs), resulting in increased hepatic SEC activation markers [31]. Therefore, these EVs may promote vascular remodeling during cirrhosis.

3.2 EVs and pulmonary fibrosis

Idiopathic pulmonary fibrosis (IPF) is the most common form of fibrotic pulmonary diseases and is a chronic, progressive, irreversible and fatal lung disease characterized by diffuse alveolar epithelial cell injury and structural remodeling [17]. IPF is typically refractory to antiinflammatory therapies such as glucocorticosteroids [32]. Although the etiology is still not clearly delineated, one of the most well-accepted theories in IPF pathogenesis is the aberrant activation of alveolar epithelial cells and fibroblasts in an aging lung [33].

Currently, there are a few studies on the roles of EVs in IPF pathogenesis. Martin-Medina et al. showed that increased WNT5A is secreted on EVs in bronchoalveolar lavage fluid from IPF patients, which promotes proliferation of lung fibroblasts [24]. Bacha et al. found that the levels of circulating endothelial microparticles isolated from plasma are significantly higher in severe IPF patients, which stimulated the migration capacity of lung fibroblasts [25]. In addition, we revealed that cigarette smoke extract-induced human bronchial epithelial cell-derived EVs promote myofibroblast differentiation in lung fibroblasts through autophagy regulation by the EV miR-210 [34]. Because cigarette smoke is an important risk factor for IPF, the EV-induced myofibroblast differentiation could be involved in the pathogenesis of IPF. Furthermore, our recent study showed that lung fibroblast-derived EVs from IPF patients increase mitochondrial reactive oxygen species (mtROS) and associated mitochondrial damage in lung epithelial cells, leading to mtROS-mediated activation of the DNA damage response and subsequent epithelial cell senescence. Mechanistically, significantly upregulated miR-23b-3p and miR-494-3p in the EVs are responsible for suppressing SIRT3, resulting in the EV-induced phenotypes in lung epithelial cells. Senescent cells not only impair regeneration but also the secretion of bioactive molecules, which is crucial for the initiation and progression of fibrosis in IPF [33, 35]. These results suggest that EVs act as important mediators in IPF pathogenesis (T. Kadota et al., unpublished data).

3.3 EVs and cardiac fibrosis and hypertrophy

Cardiac fibrosis results from pathological myocardial remodeling triggered by most myocardial diseases. In the heart, cardiac fibroblasts are the most abundant cell type and are central regulators of the production of extracellular matrix. Cardiac fibroblasts also communicate with cardiomyocytes via paracrine factors such as TGF-β and fibroblast growth factor (FGF)-2 [36, 37]. In addition, unlike other organs, the heart has very limited regenerative capacity. Consequently, extensive scarring is necessary to prevent rupture following myocardial infarction and other injuries [26]. This compensation mechanism provokes pathological changes such as chamber dilation and cardiomyocyte hypertrophy, ultimately leading to the development of heart failure.

Considering the roles of EVs in cardiac fibrosis and hypertrophy pathogenesis, Bang et al. showed that miR-21-3p in cardiac fibroblast-derived EVs transfers to cardiomyocytes, inducing cardiac hypertrophy by downregulating Sorbin and SH3 Domain Containing 2 (SORBS2) or PDZ and LIM Domain 5 (PDLIM5), both known as regulators of cardiac muscle structure and function [27]. On the other hand, Datta et al. revealed that the mechanism of Hsp90-mediated modulation of IL-6 from cardiomyocytes either in secreted form or in EV form, in which IL-6 alters the fibrotic responses via biphasic activation of STAT-3 signaling in fibroblasts during the development of cardiac fibrosis [28]. In addition, Tian et al. identified EV-associated miRNA-mediated communication between cardiac fibroblasts and cardiomyocytes, contributing to the dysregulation of the Nuclear factor erythroid 2-related factor 2 (Nrf2) /antioxidant response element (AREs) signaling pathway and potentially leading to cardiac fibrosis [38]. Nrf2 is a major regulator of these AREs. Additionally, many Nrf2-regulated enzymes have been implicated in the pathogenesis of heart diseases and are associated with oxidative stress-mediated cardiac remodeling and heart failure [39]. These findings indicate that EV-mediated interactions between cardiomyocytes and fibroblasts contribute to cardiac fibrosis.

The involvement of cardiomyocyte-derived EVs in cardiac repair by exercise was also investigated. In a mouse model of type 2 diabetes, an exercising group had increased levels of cardiomyocyte-derived EVs in the extracellular space, vessel walls and regions lining cardiomyocytes and had upregulated miR-29b and miR-455. Interestingly, these miRNAs in the EVs attenuated fibrosis and cardiomyocyte uncoupling through suppression of MMP-9, a determinant of ECM degradation, which suggests the involvement of miRNAs in cardiomyocyte-derived EVs in cardiac fibrosis pathogenesis [40].

4 The involvement of EVs in fibrotic disease diagnosis

The risk of mortality increases exponentially with increase in fibrosis stage, as evidenced for non-alcoholic fatty liver [41], pulmonary fibrosis [42] and chronic kidney diseases [43]. Thus, the early and accurate detection of the presence and extent of organ fibrosis has a potential benefit. The gold standard for diagnosis and staging of organ fibrosis is histology, which is usually performed on fine-needle biopsies or surgically obtained samples. However, assessment of organ fibrosis by histology has several limitations: invasiveness for obtaining histology, difficulty of repeat biopsy, and lack of representative for the distribution of fibrosis. Therefore, many additional methods have been proposed and already used to noninvasively assess the stage of organ fibrosis such as imaging, mechanical and functional tests, and liquid biopsies.

Recently, EVs have been identified as novel disease biomarkers for several reasons. First, EVs reflect the physiological state and microenvironment of their parental cells, and most cells secrete EVs that contain specific proteins, nucleic acids and lipids [44, 45]. Second, EVs are found in blood and other bodily fluids. Third, EVs are very stable in the extracellular environment because of their phospholipid bilayers. To date, numerous EV-derived proteins and miRNAs have been investigated as potentially useful biomarkers for various diseases [46, 47]. In this section, we review the EV biomarker of fibrosis in liver and lung.

4.1 EV-based biomarker in liver fibrosis

Several studies documented the potential utility of EVbased biomarkers in the degree and existence of liver fibrosis (Table 2). Morattini et al. found that full-length soluble receptortype tyrosineprotein phosphatase γ (sPTPRG) isoforms in EVs have been identified as a biomarker for liver injury in human and mouse plasma [48]. PTPRG belongs to a family of enzymes that remove phosphate groups from phosphotyrosine residues of specific intracellular targets and behave as tumor suppressor genes subjected to loss of heterozygosity deletion, hypermethylation, and point mutation. They showed a positive correlation between high plasma levels of sPTPRG and liver damage in human plasma with low or high ALT level. In another study, soluble CD81 was increased in the serum EV fraction in patients with chronic hepatitis C compared with healthy

TABLE 2 Extracellular vesicles in organ fibrosis pathogeneis

Diseases	Body fluid	Cargo	Control	References
Liver injury	Serum, plasma	sPTPRG	People without liver	[48]
Hepatitis C		soluble CD81	Healthy individuals	[49]
HBV and HCV		miRNAs	Healthy individuals	[50]
CHC	Serum, plasma	CD4+ microparticle	Healthy individuals	[51]
NAFL/NASH	Serum, plasma	CD14+ microparticle	Healthy individuals	[51]
IPF	Plasma	miR-21-5p	Healthy individuals	[52]
IPF	Sputum	miR-142-3p	Healthy individuals	[53]

sPTPRG, soluble receptor-type tyrosine-protein phosphatase γ.

individuals and recovered patients. EVassociated soluble CD81 levels were also associated with plasma ALT levels and severe forms of liver fibrosis [49]. Moreover, Lambrecht et al. investigated the expression profile of circulating EV-associated miRNAs of fibrotic patients with early stage HBV and HCV, which suggest the potential use of these EV-associated miRNAs as markers for early stages of liver fibrosis [50]. Furthermore, Kornek et al. revealed that microparticles from immune cells such as CD4+ and CD8+T cells and invariant natural killer T cells and macrophages/monocytes (CD14+) in serum have potential to assess the histological grade, severity and biopsy stage in patients with CHC and nonalcoholic fatty liver (NAFL) or nonalcoholic steatohepatitis (NASH) [51].

4.2 EV-based biomarkers in pulmonary fibrosis

There are several studies about biomarkers of IPF using EVs (Table 2). Makiguchi et al. showed that the level of miR-21 in EVs is higher in serum from patients with IPF compared to healthy controls [52]. Njock et al. also found that miR-142-3p levels in EVs derived from sputum in IPF patients are inversely correlated with the IPF severity, as evaluated by diffusing capacity of the lungs for carbon monoxide/alveolar volume [53]. These results indicated that circulating EVs have potential to be promising biomarkers for diagnosis and disease severity.

4.3 Current limitations of EVs analysis in disease diagnosis

Although there are a great number of advantages to using EVs as biomarkers, there are several challenges that currently limit their clinical applicability. First, novel methods to isolate and enrich EVs are required to analyze thousands of clinical samples. Currently, the most commonly used method for isolating EVs is ultracentrifugation. However, the ultracentrifugation method for EV isolation needs expensive equipment, is time consuming, and requires high volume of sample. Although alternative methods that depend on EV surface markers are being studied such as microfluidic chips, magnetic beads, and flow cytometry [54], the diagnostic performance of these methods in clinical practice remains unknown. Second, current isolation technologies make it difficult to distinguish different EV subpopulations. EVs are mainly separated by size via ultracentrifugation. However, studies show that the EV family is heterogenous when separated by size [55]. Third, the mechanisms regarding EV biogenesis are required to be elucidated. Therefore, further research is needed to develop technologies at a reasonable price to isolate highly pure EVs for disease biomarker analysis.

5 EVs as new applications for fibrosis treatment

5.1 Reversibility of organ fibrosis

Traditionally, fibrosis has been considered an irreversible process, the end stage of disordered wound healing, and thus refractory to therapy. However, accumulating evidence now suggests that fibrosis is to a large extent reversible, even at later stages [56]. In animal models, fibrosis is reversible with complete or near complete restitution of the organ structure in most cases after the inciting fibrogenic agent is removed. For instance, in a well-established bleomycin-induced murine pulmonary fibrosis model, spontaneous fibrosis resolution with restitution of tissue architecture can occur 3–4weeks after the intratracheal administration [57]. In contrast to animal models where the inducer of fibrosis is well defined, human fibrotic diseases are often multifactorial, and the degree of fibrosis resolution appears to depend on the organ resolution capacity and the mechanism of the fibrosis diseases. The most compelling evidence for the resolution of organ fibrosis in humans is observed in the liver. It is known that liver cirrhosis caused by hepatitis infection demonstrates marked resolution after antiviral treatment [58]. In contrast, heart and lung display very limited regenerative capacity [56]. Nonetheless, recovery of the organ functions such as left ventricular ejection function in cardiac fibrosis and forced vital capacity in IPF has been documented with presumably antifibrotic interventions [59].

With respect to optimal strategies for fibrotic diseases, eradication of the underlying cause of injury may be the most efficacious approach to fibrosis resolution, as exemplified by antiviral treatment of liver cirrhosis. When etiology is unclear or effective treatments for the underlying cause of injury are unavailable, therapeutics targeting fibrosis should be considered. However, although many interventions have been investigated with the aim of fibrosis resolution, few have shown efficacy in patients with fibrosis [56]. Therefore, new therapeutic approaches are needed to overcome fibrotic diseases.

5.2 The roles of EVs as new applications for fibrosis treatment

EVs have received much attention for their potential applications as a source of regenerating or immunomodulatory agents or as a drug delivery vehicle. There are several advantages to using EV-based therapy. First, EVs show low immunogenicity and toxicity and are stable in tissues and in circulation. EVs, especially exosomes, have already been used for antitumor vaccines in several clinical trials. In lung cancer, phase I and II clinical trials using dendritic cell-derived EVe (Dex) have been performed [60, 61]. Although grade-three hepatotoxicity occurred in one of 22 NSCLC patients in a phase II study, the Dex therapy was feasible and well-tolerated. Second, EVs can be selectively and quickly taken up by target cells, which may target specific cell types or tissues [62]. Third, they have unique characteristics such as a lipid bilayer architecture, nanoscale size and stability in the blood circulation and tissue [63]. In this section, we summarized the utility of EVs in therapy applications for reversing fibrosis.

5.3 Mesenchymal stem cell-derived EVs

EVs secreted from stem cells are naturally loaded with various molecules derived from the stem cells. There are various types of stem cell-derived EVs studied for fibrosis in experimental models and in clinical trials, such as embryonic stem cells [64], induced pluripotent stem cells [65], and multipotent/unipotent adult stem cell lineages such as mesenchymal stem cells (MSCs) [66]. In particular, MSCs have been intensively studied and appear to be a promising treatment, especially in the fields of fibrotic diseases.

MSCs are multipotent stem cells that have the capacity to self-renew and differentiate into mesenchymal lineages, such as bone, cartilage, muscle and adipose, under specific conditions of culture [67]. Homing and migration are other distinctive functions of MSCs. Moreover, they can orchestrate tissue repair, immunomodulation, and antiinflammation and/or immunosuppression due to the secretion of various cytokines and soluble factors [68]. Currently, preclinical and clinical trials have shown the ability of MSCs to improve outcomes in various diseases, such as cardiovascular diseases, stroke, spinal cord injury, kidney injury, lung injury and graft-versus-host disease (GvHD) [66]. Importantly, the meta-analysis of 36 clinical trials did not reflect any serious complications related to MSC injections [69].

Numerous studies have suggested that MSC-derived EVs (MSC-EVs) may be an alternative to MSCs. In addition, use of MSC-derived EVs will allow for the delivery of biologically active molecules to injured organs without administration of heterologous divergent cells [66]. Thus, to date, many studies have investigated the use of MSC-EVs for fibrosis therapy.

In liver fibrosis, Li et al. found that EVs derived from human umbilical cord MSCs reduced CCL4-induced liver injury in vivo through inhibiting the TGF-β1/Smad signaling pathway and inhibiting the epithelial-mesenchymal transition of hepatocytes [70]. Another study showed that EV-associated miR-125b from chorionic plate-derived MSCs suppressed the activation of Hedgehog signaling, which promoted reduced liver fibrosis [71]. Moreover, it has been demonstrated that EVs released from adipose tissue-derived MSCs inhibited the proliferation and activation of the human primary HSCs. These results suggested that EVs have antifibrotic effects via shuttling effective EV cargos from MSCs.

In lung fibrosis, Shentu et al. have shown that human bone marrow (BM)-MSC-derived EVs can block TGFβ1-induced myofibroblastic differentiation [72]. Human BM-MSC-derived EVs are taken up by fibroblasts and may utilize a Thy-1-integrin interaction-dependent pathway to facilitate cell-cell communication by EVs. The cell surface protein Thy-1, a GPI-anchored glycoprotein, is highly expressed on the MSC cell surface and is known to interact with integrin. The BM-MSC-derived EVs are enriched for several miRNAs, including miR-630, which targets the profibrotic genes that are upregulated in IPF fibroblasts [72].

In the heart, MSC-derived EVs mediate cardiac tissue repair through various mechanisms such as modulating the injured tissue environment, inducing angiogenesis, promoting proliferation, and preventing apoptosis [73, 74]. Moreover, administration of MSC-derived EVs after myocardial infarction enhances myocardial repair [75]. In addition, Ma et al. showed that EVs from Akt-overexpressing human umbilical cord MSCs improve cardiac regeneration and promote angiogenesis through PDGFD activation [76].

5.4 Body fluid-derived EVs

EVs isolated from body fluid have also been shown to have a therapeutic effect in fibrosis models. For instance, serum EVs from healthy individuals are therapeutic for liver fibrosis due to their ability to attenuate activation of hepatic stellate cells, hepatocyte injury and inflammation. EV treatment improved liver function, reduced apoptosis of hepatocytes, suppressed inflammatory response in the injured liver, reduced release of hepatic or circulating proinflammatory cytokines, reduced inflammatory infiltration, and reduced circulating aspartate aminotransferase/alanine aminotransferase levels in CCL4-injured or thioacetic acid-injured fibrosis mouse models [77]. Another study showed that EVs isolated from rat plasma have the ability to protect the heart from ischemia reperfusion injury [78]. The effect is mediated by HSP70 present on EVs, stimulating TLR4 signaling and leading to the activation of ERK1/2, p38 MAPK, and subsequent HSP27 phosphorylation in cardiomyocytes. In addition, Beltrami et al. found that EVs derived from human pericardial fluid have shown the capacity to promote therapeutic angiogenesis in vivo in a mouse limb ischemia model, suggesting their importance in the context of cardiovascular protection and repair. They mechanistically implicated that the EVs transfer a functional let-7b-5p to endothelial cells, which reduce their TGFBR1 expression and induce their proangiogenic ability [79]. Overall, although the underlying mechanism is not fully understood, body fluid-derived EVs from healthy individuals can be beneficial for patients with fibrotic diseases.

5.5 Current limitations of EV-based therapy

EV-based therapy for fibrotic diseases has several remaining challenges. First, standardized and quality-controlled protocols for EV isolation should be developed. The appropriate isolation method would deliver high-quality and uniform EVs with reproducible purity and potency. Second, effective large-scale production of EVs is required for clinical applicability of the therapy because of low productivity of EVs. Third, a stricter characterization of EVs is urgently needed to better explore vesicle biogenesis and applied EVs in clinical settings. Fourth, it is important to develop appropriate preclinical in vivo models to determine EV dosage, biodistribution, toxicity and immunogenicity. In particular, in vivo imaging of EVs is needed to visualize the kinetics of EVs in the body and quantify the amount of EVs delivered to target damaged tissue [80].

6 Conclusions and perspectives

Recent evidence indicates that fibrosis is reversible in fibrotic diseases under some circumstances. Eradication of the underlying deleterious stimuli should be the most effective strategy toward fibrosis resolution. Where treating the cause of fibrosis is not possible, it appears that inhibiting the deleterious mediators involved in fibrosis development is a promising approach. EVs are important regulators of cell-cell communication in both physiologic and pathologic processes. The findings described herein confirm that EVs play a pivotal role in the pathogenesis of fibrotic diseases. Thus, inhibiting deleterious EVs could be the potential therapeutic options for fibrotic diseases. In addition, accumulating evidence indicates that EVs derived from stem cells or healthy body fluid have a therapeutic effect in fibrotic diseases. Although significant challenges still remain in the clinical application of EVs, further research and development of EVs will likely help patients with fibrosis management and treatment in the future.

Acknowledgements

This work was supported by the Practical Research for Innovative Cancer Control (17ck0106366h001) from Japan Agency for Medical Research and Development, AMED.

References

[1] Wynn TA. Fibrotic disease and the T(H)1/T(H)2 paradigm. Nat Rev Immunol 2004;4:583–94.

[2] Robbins PD, Morelli AE. Regulation of immune responses by extracellular vesicles. Nat Rev Immunol 2014;14:195–208.

[3] Colombo M, Raposo G, Théry C. Biogenesis, secretion, and intercellular interactions of exosomes and other extracellular vesicles. Annu Rev Cell Dev Biol 2014;30:255–89.
[4] Witwer KW, et al. Standardization of sample collection, isolation and analysis methods in extracellular vesicle research. J Extracell Vesicles 2013;2:20360.
[5] Théry C, et al. Minimal information for studies of extracellular vesicles 2018 (MISEV2018): a position statement of the international society for extracellular vesicles and update of the MISEV2014 guidelines. J Extracell Vesicles 2018;7:1535750.
[6] Raposo G, Stoorvogel W. Extracellular vesicles: exosomes, microvesicles, and friends. J Cell Biol 2013;200:373–83.
[7] Kosaka N, et al. Secretory mechanisms and intercellular transfer of microRNAs in living cells. J Biol Chem 2010;285:17442–52.
[8] Makarev E, et al. Common pathway signature in lung and liver fibrosis. Cell Cycle 2016;15:1667–73.
[9] Azuma A, et al. Double-blind, placebo-controlled trial of pirfenidone in patients with idiopathic pulmonary fibrosis. Am J Respir Crit Care Med 2005;171:1040–7.
[10] Khanna D, et al. An open-label, phase II study of the safety and tolerability of pirfenidone in patients with scleroderma-associated interstitial lung disease: the LOTUSS trial. J Rheumatol 2016;43:1672–9.
[11] Sangaralingham SJ, et al. Cardiorenal fibrosis and dysfunction in aging: Imbalance in mediators and regulators of collagen. Peptides 2016;76:108–14.
[12] Gurtner GC, Werner S, Barrandon Y, Longaker MT. Wound repair and regeneration. Nature 2008;453:314–21.
[13] Wynn TA. Cellular and molecular mechanisms of fibrosis. J Pathol 2008;214:199–210.
[14] Rockey DC, Bell PD, Hill JA. Fibrosis—a common pathway to organ injury and failure. N Engl J Med 2015;372:1138–49.
[15] Kisseleva T, Brenner DA. Fibrogenesis of parenchymal organs. Proc Am Thorac Soc 2008;5:338–42.
[16] Wynn TA, Vannella KM. Macrophages in tissue repair, regeneration, and fibrosis. Immunity 2016;44:450–62.
[17] King TE, Pardo A, Selman M. Idiopathic pulmonary fibrosis. The Lancet 2011;378:1949–61.
[18] Kendall RT, Feghali-Bostwick CA. Fibroblasts in fibrosis: novel roles and mediators. Front Pharmacol 2014;5:123.
[19] Weiskirchen R, Weiskirchen S, Tacke F. Organ and tissue fibrosis: molecular signals, cellular mechanisms and translational implications. Mol Aspects Med 2018;*65:2–15*. https://doi.org/10.1016/j.mam.2018.06.003.
[20] Chen L, et al. Epigenetic regulation of connective tissue growth factor by MicroRNA-214 delivery in exosomes from mouse or human hepatic stellate cells. Hepatology 2014;59:1118–29.
[21] Chen L, Chen R, Kemper S, Charrier A, Brigstock DR. Suppression of fibrogenic signaling in hepatic stellate cells by Twist1-dependent microRNA-214 expression: role of exosomes in horizontal transfer of Twist1. Am J Physiol Gastrointest Liver Physiol 2015;309:G491–9.
[22] Charrier A, et al. Exosomes mediate intercellular transfer of pro-fibrogenic connective tissue growth factor (CCN2) between hepatic stellate cells, the principal fibrotic cells in the liver. Surgery 2014;156:548–55.
[23] Wang R, et al. Exosome adherence and internalization by hepatic stellate cells triggers sphingosine 1-phosphate-dependent migration. J Biol Chem 2015;290:30684–96.
[24] Martin-Medina A, et al. Increased extracellular vesicles mediate WNT-5A signaling in idiopathic pulmonary fibrosis. Am J Respir Crit Care Med 2018;198. https://doi.org/10.1164/rccm.201708-1580OC.

[25] Bacha NC, et al. Endothelial microparticles are associated to pathogenesis of idiopathic pulmonary fibrosis. Stem Cell Rev 2018;14:223–35.
[26] Travers JG, Kamal FA, Robbins J, Yutzey KE, Blaxall BC. Cardiac fibrosis: the fibroblast awakens. Circ Res 2016;118:1021–40.
[27] Bang C, et al. Cardiac fibroblast-derived microRNA passenger strand-enriched exosomes mediate cardiomyocyte hypertrophy. J Clin Investig 2014;124:2136–46.
[28] Datta R, et al. Myocyte-derived Hsp90 modulates collagen upregulation via biphasic activation of STAT-3 in fibroblasts during cardiac hypertrophy. Mol Cell Biol 2017;37:552–625.
[29] Hernandez-Gea V, Friedman SL. Pathogenesis of liver fibrosis. Annu Rev Pathol Mech Dis 2011;6:425–56.
[30] Gressner OA, Gressner AM. Connective tissue growth factor: a fibrogenic master switch in fibrotic liver diseases. Liver Int 2008;28:1065–79.
[31] Witek RP, et al. Liver cell–derived microparticles activate hedgehog signaling and alter gene expression in hepatic endothelial cells. Gastroenterology 2009;**136**:320–330.e2.
[32] Gross TJ, Hunninghake GW. Idiopathic pulmonary fibrosis. N Engl J Med 2001; 345:517–25.
[33] Selman M, Pardo A. Revealing the pathogenic and aging-related mechanisms of the enigmatic idiopathic pulmonary fibrosis. An integral model. Am J Respir Crit Care Med 2014;189:1161–72.
[34] Fujita Y, et al. Suppression of autophagy by extracellular vesicles promotes myofibroblast differentiation in COPD pathogenesis. J Extracell Vesicles 2015;4:28388.
[35] Kadota T, et al. Emerging role of extracellular vesicles as a senescence-associated secretory phenotype: Insights into the pathophysiology of lung diseases. Mol Aspects Med 2018;60:92–103.
[36] Kakkar R, Lee RT. Intramyocardial fibroblast myocyte communication. Circ Res 2010;106:47–57.
[37] Takeda N, et al. Cardiac fibroblasts are essential for the adaptive response of the murine heart to pressure overload. J Clin Investig 2010;120:254–65.
[38] Tian C, Gao L, Zimmerman MC, Zucker IH. Myocardial infarction-induced microRNA-enriched exosomes contribute to cardiac Nrf2 dysregulation in chronic heart failure. Am J Physiol Heart Circ Physiol 2018;314:H928–39.
[39] Ashrafian H, et al. Fumarate is cardioprotective via activation of the Nrf2 antioxidant pathway. Cell Metab 2012;15:361–71.
[40] Chaturvedi P, Kalani A, Medina I, Familtseva A, Tyagi SC. Cardiosome mediated regulation of MMP9 in diabetic heart: role of mir29b and mir455 in exercise. J Cell Mol Med 2015;19:2153–61.
[41] Dulai PS, et al. Increased risk of mortality by fibrosis stage in nonalcoholic fatty liver disease: systematic review and meta-analysis. Hepatology 2017;65:1557–65.
[42] Ley B, Collard HR, King TE. Clinical course and prediction of survival in idiopathic pulmonary fibrosis. Am J Respir Crit Care Med 2011;183:431–40.
[43] Tonelli M, et al. Chronic kidney disease and mortality risk: a systematic review. J Am Soc Nephrol 2006;17:2034–47.
[44] de Jong OG, et al. Cellular stress conditions are reflected in the protein and RNA content of endothelial cell-derived exosomes. J Extracell Vesicles 2012;1:18396.
[45] Beninson LA, Fleshner M. Exosomes: an emerging factor in stress-induced immunomodulation. Semin Immunol 2014;26:394–401.
[46] Kadota T, Yoshioka Y, Fujita Y, Kuwano K, Ochiya T. Extracellular vesicles in lung cancer-from bench to bedside. Semin Cell Dev Biol 2017;67:39–47.

[47] Fujita Y, Yoshioka Y, Ochiya T. Extracellular vesicle transfer of cancer pathogenic components. Cancer Sci 2016;107:385–90.

[48] Moratti E, Vezzalini M, Tomasello L, Giavarina D, Sorio C. Identification of protein tyrosine phosphatase receptor gamma extracellular domain (sPTPRG) as a natural soluble protein in plasma. PLoS ONE 2015;10:e0119110.

[49] Welker M-W, et al. Soluble serum CD81 is elevated in patients with chronic hepatitis C and correlates with alanine aminotransferase serum activity. PLoS ONE 2012;7:e30796.

[50] Lambrecht J, et al. Circulating ECV-associated miRNAs as potential clinical biomarkers in early stage HBV and HCV induced liver fibrosis. Front Pharmacol 2017;8:56.

[51] Kornek M, et al. Circulating microparticles as disease-specific biomarkers of severity of inflammation in patients with hepatitis C or nonalcoholic steatohepatitis. Gastroenterology 2012;143:448–58.

[52] Makiguchi T, et al. Serum extracellular vesicular miR-21-5p is a predictor of the prognosis in idiopathic pulmonary fibrosis. Respir Res 2016;17:110.

[53] Njock M-S, et al. Sputum exosomes: promising biomarkers for idiopathic pulmonary fibrosis. Thorax 2018;*74:309–312*. https://doi.org/10.1136/thoraxjnl-2018-211897.

[54] Xu R, Greening DW, Zhu H-J, Takahashi N, Simpson RJ. Extracellular vesicle isolation and characterization: toward clinical application. J Clin Investig 2016;126:1152–62.

[55] Bobrie A, Colombo M, Krumeich S, Raposo G, Théry C. Diverse subpopulations of vesicles secreted by different intracellular mechanisms are present in exosome preparations obtained by differential ultracentrifugation. J Extracell Vesicles 2012;1:18397.

[56] Jun J-I, Lau LF. Resolution of organ fibrosis. J Clin Investig 2018;128:97–107.

[57] Hecker L, et al. Reversal of persistent fibrosis in aging by targeting Nox4-Nrf2 redox imbalance. Sci Transl Med 2014;**6**:231ra47.

[58] Marcellin P, et al. Regression of cirrhosis during treatment with tenofovir disoproxil fumarate for chronic hepatitis B: a 5-year open-label follow-up study. Lancet 2013;381:468–75.

[59] Karimi-Shah BA, Chowdhury BA. Forced vital capacity in idiopathic pulmonary fibrosis—FDA review of pirfenidone and nintedanib. N Engl J Med 2015;372:1189–91.

[60] Morse MA, et al. A phase I study of dexosome immunotherapy in patients with advanced non-small cell lung cancer. J Transl Med 2005;3:9.

[61] Besse B, et al. Dendritic cell-derived exosomes as maintenance immunotherapy after first line chemotherapy in NSCLC. OncoImmunology 2016;5:e1071008.

[62] Hoshino A, et al. Tumour exosome integrins determine organotropic metastasis. Nature 2015;527:329–35.

[63] Lener T, et al. Applying extracellular vesicles based therapeutics in clinical trials—an ISEV position paper. J Extracell Vesicles 2015;4:30087.

[64] Khan M, et al. Embryonic stem cell-derived exosomes promote endogenous repair mechanisms and enhance cardiac function following myocardial infarction. Circ Res 2015;117:52–64.

[65] Hu G-W, et al. Exosomes secreted by human-induced pluripotent stem cell-derived mesenchymal stem cells attenuate limb ischemia by promoting angiogenesis in mice. Stem Cell Res Ther 2015;6:10.

[66] Fujita Y, Kadota T, Araya J, Ochiya T, Kuwano K. Clinical application of mesenchymal stem cell-derived extracellular vesicle-based therapeutics for inflammatory lung diseases. J Clin Med 2018;7:355.

[67] Nardi NB, da Silva Meirelles L. Mesenchymal stem cells: isolation, in vitro expansion and characterization. In: Handbook of experimental pharmacology. 2008. p. 249–82. https://doi.org/10.1007/978-3-540-77855-4_11.

[68] Salem HK, Thiemermann C. Mesenchymal stromal cells: current understanding and clinical status. Stem Cells 2010;28:585–96.
[69] Lalu MM, et al. Safety of cell therapy with mesenchymal stromal cells (SafeCell): a systematic review and meta-analysis of clinical trials. PLoS ONE 2012;7:e47559.
[70] Li T, et al. Exosomes derived from human umbilical cord mesenchymal stem cells alleviate liver fibrosis. Stem Cells Dev 2013;22:845–54.
[71] Hyun J, Wang S, Kim J, Kim GJ, Jung Y. MicroRNA125b-mediated Hedgehog signaling influences liver regeneration by chorionic plate-derived mesenchymal stem cells. Sci Rep 2015;5:14135.
[72] Shentu T-P, et al. Thy-1 dependent uptake of mesenchymal stem cell-derived extracellular vesicles blocks myofibroblastic differentiation. Sci Rep 2017;7:18052.
[73] Lai RC, et al. Exosome secreted by MSC reduces myocardial ischemia/reperfusion injury. Stem Cell Res 2010;4:214–22.
[74] Barile L, et al. Extracellular vesicles from human cardiac progenitor cells inhibit cardiomyocyte apoptosis and improve cardiac function after myocardial infarction. Cardiovasc Res 2014;103:530–41.
[75] Zhang Z, et al. Pretreatment of cardiac stem cells with exosomes derived from mesenchymal stem Cells enhances myocardial repair. J Am Heart Assoc 2016;5:e002856.
[76] Ma J, et al. Exosomes derived from Akt-modified human umbilical cord mesenchymal stem cells improve cardiac regeneration and promote angiogenesis via activating platelet-derived growth factor D. Stem Cells Transl Med 2017;6:51–9.
[77] Chen L, et al. Therapeutic effects of serum extracellular vesicles in liver fibrosis. J Extracell Vesicles 2018;7:1461505.
[78] Vicencio JM, et al. Plasma exosomes protect the myocardium from ischemia-reperfusion injury. J Am Coll Cardiol 2015;65:1525–36.
[79] Beltrami C, et al. Human pericardial fluid contains exosomes enriched with cardiovascular-expressed MicroRNAs and promotes therapeutic angiogenesis. Mol Ther 2017;25:679–93.
[80] Adamiak M, Sahoo S. Exosomes in myocardial repair: advances and challenges in the development of next-generation therapeutics. Mol Ther 2018;26:1635–43.

Chapter 14

Mechanisms of exosome-mediated immune cell crosstalk in inflammation and disease

Todd W. Costantini, Raul Coimbra, Brian P. Eliceiri
Division of Trauma, Surgical Critical Care, Burns and Acute Care Surgery, Department of Surgery, University of California San Diego, San Diego, CA, United States

1 Transfer of exosomes during immune cell-interactions participate in the inflammatory response

Tissue repair and regeneration in healthy tissues is a coordinated response of multiple cell types that has until recently been considered to be mediated by direct inter-cellular interactions and by the release of secreted factors [1–5]. The discovery of exosomes in the 1980s as products of multi-vesicular endosomes (MVEs) whose intraluminal vesicles (now called exosomes) were released from the surface of maturing blood reticulocytes [6–8] has led to an understanding of exosomes as ubiquitous mechanisms for cell-cell communication that is critically important in inflammation [9–12]. Recent studies of exosomes have focused on mechanisms of their biogenesis, their biological activity in mediating cellular signaling at a distance, characterizations of their payload, exosomes as biomarkers, and exosomes as therapeutics. In addition, advances in mouse and human genetics and exosome analysis technology have transformed a class of extracellular vesicles from being a constitutive class of secreted vesicles, to being recognized as regulated bioactive mediators of cell biology, immunology and inflammation. For example, exosomes as mediators of cell-cell communication has emerged as an important mechanism for the regulation of immune responses by professional antigen presenting cells (APCs) [13–19]. APCs include dendritic cells (DCs), and present MHC I and MHC II complexes with peptides through canonical direct cellular interactions as well as by exosomes present on the APC surface through two different models. First, co-stimulation of regulatory molecules present on the APC surface along with exosomes also on the APC surface contain MHCI/II, B7-1, B7-2 and other molecules that regulate the docking of DCs to

Exosomes. https://doi.org/10.1016/B978-0-12-816053-4.00014-6

acceptor cells [9, 20–22], a mechanism termed cross-dressing. Second, APCs can uptake exosomes that lead to changes in the expression of cytokines that affects expression of co-stimulatory molecules and release of APC-derived exosomes for presentation to T cells. While APCs are more efficient than exosomes in activating naïve T cells, antigens present on isolated exosomes can mediate immune-stimulation or immunosuppression based on studies in models of cancer [23–28], infection [29], tolerance [30], immune privilege [31] and arthritis [32, 33]. Immunomodulatory cytokines can also reprogram the biological activity of the exosomes released, demonstrating the plasticity of exosomes and their potential to regulate inflammation signaling at a distance, especially in primary tumors and injury models where local tissue damage mediates immune-surveillance responses in sentinel lymph nodes [9]. The capacity for exosomes to amplify immune responses [4, 5, 20] has been demonstrated by transfer of macrophage-derived exosomes to T cells with subsequent activation of lymphocyte expansion and differentiation, termed an "immune synapse" [29, 34, 35]. Immune responses regulated by tumor-derived exosomes [10, 36, 37] include control of tolerance and immunogenicity [32, 38–41] by direct presentation of tumor derived antigens such as carcinoembryonic antigen (CEA), melan-A, mesothelin, and immunomodulatory molecules such as FasL, RAIL, CD154/CD40L and programmed death ligand-1 (PD-L1) [31, 42–44].

Since exosomes can mediate both immune-stimulatory and immunosuppressive activities, characterization of the cellular source, mechanism of release, payload and effector cell is an important consideration for distinguishing the biological relevance of exosomes in healthy baseline vs. disease models. For example, following bacterial or viral infections, exosomes isolated from infected macrophages confer protection in the presence of APCs and T cells from mice showing that exosomes regulate immune responses in acute models [45–48]. In contrast, tumor-derived exosomes mediate immunosuppression by blocking differentiation and cytokine release in effector cells, and may present PD-L1 and CD40, both known to suppress T cell responses. In contrast, tumor-derived exosomes mediate immunosuppression upon APCs in the tumor microenvironment, as well as secondary responses in the circulation. The systemic effects of tumor-derived exosomes include reprogramming of bone marrow-derived progenitor cells [26], while the presence of specific integrins on the surface of exosomes defines the tropism of tumor-derived exosomes that can direct their homing and biological activity in the local microenvironment and in secondary sites [28]. While beyond the scope to comprehensively summarize in this review, the study of exosomes in infection and cancer has established a central role for exosomes in immunomodulation and provided a foundation for the development of diagnostics and therapeutics, by exploiting the capacity for tumor-derived exosomes to affect specific immune responses mediating immunosurveillance.

2 Relevance of exosome biogenesis and secretion pathways in health and disease

Exosomes mediate intercellular cross-talk signaling in numerous biological systems, signaling that is perturbed in disease, and leads to distinct exosome profiles. Although exosomes are often assumed by investigators in the field to be a "snapshot" of the proteins on the plasma membrane surface of the cells from which the exosomes are released, they represent a more complex cargo from their cell of origin. In fact, proteins present in, and on, the surface of exosomes released from fusion of the multivesicular endosomes (MVEs) with the plasma membrane may also have an intracellular source.

To define the biogenesis of exosomes, and the effects of heath and disease on this machinery, an overview of the biochemistry of endosomal vesicle trafficking machinery that sorts intracellular vesicles to the lysosome, or to the MVE, shows that there are specific cellular mediators of processes that "edit" the payload of vesicles prior to fusion and release from the plasma membrane. The regulation of the biogenesis of exosomes has identified proteins that are associated with secretory and endocytic pathways such as Endosomal Sorting Complexes Responsible for Transport (ESCRTs), lipids and tetraspanins. Various in vitro screening and functional studies have classified ESCRTs into ESCRT-0 (e.g., Hrs, STAM), ESCRT-I (e.g., TSG101), ESCRT-II, and ESCRT-III (e.g., CHMP4C). Sphingomyelinases (nSMase2) and ceramide production are also essential for exosome biogenesis, while Rabs and related activating proteins (i.e., Rab GTPase activating proteins, RabGAPs) [49–53] are implicated in the secretion step of exosome release. As Rabs are part of the superfamily of GTPases, Rab utilization may be cell type-specific. In vitro screening has identified Rab 27 as mediating up to 50% of exosome secretion, consistent with double knockout mouse of the Rab27a and Rab27b, with the activity of these Rabs mediated regulatory RabGAPs. These studies suggest that parallel mechanisms are involved in exosome biogenesis and secretion, with additional pathways required for the intracellular sorting of specific MHCI/II, tetraspanins, and integrins into secreted exosomes.

The loading of microRNAs can be regulated by specific RNA-binding proteins, which when loaded along with proteins into exosomes by the protein folding and packaging machinery, begins to provide an insight into the complexity of exosome payload. The packaging of specific miRNAs, although at a low absolute concentration per exosome, involves specific RNA binding proteins that likely facilitate packaging in the source cell, stabilization in the exosome, and perhaps incorporation into the RNA processing pathways in recipient cells thus allowing for communication of miRNA message between cells [54]. Protein payload as defined primarily by immunoblotting, immuno-gold electron microscopy antibody staining, and proteomics, demonstrates that exosome proteins are diverse and include those mediating antigen presentation (MHC I, MHC II), adhesion (tetraspanins and integrins), membrane trafficking

(Annexins, Rabs, Arfs, clathrin), ESCRT proteins (Alix, Tsg101), heat shock response (Hsp70, Hsp90), cytoskeleton (actin, cofilin, tubulin), enzymes (kinases, enolases), signal transduction (14-3-3, syntenin), and lipid rafts (flotilins) [55]. The International Society for Extracellular Vesicles (ISEV) establishes standards for reporting, with Exocarta (www.exocarta.org) being a database of exosome payload providing comprehensive databases detailing the protein, nucleic acid and lipid composition of exosomes. While isolation methods and quality control vary for exosome isolation and the characterization of exosome payloads, it is clear that the population and content of exosomes can be regulated by various factors and disease processes. Therefore, based on the intracellular origin for exosomes, the contribution of constitutive release of exosomes, and presence of small vesicles resulting from the pinching-off of the plasma membrane into the extracellular space (i.e., microvesicles), it will be to focus future studies of exosomes in a cell-type specific manner, under disease-specific conditions.

Single cell transcriptomics of complex organ systems such as the brain has identified distinct gene expression profiles that are cell type-specific. This approach has greatly advanced an understanding of gene superfamilies in which there is substantially greater cell-type specific expression of kinases, transcription factors and transporters. By extension, we propose that the mechanisms regulating exosome release may be cell type-specific, and regulated by disease state. Thus, the challenge for in vivo mechanistic studies will be the definition of specific pathways and for biomarker studies, the detection of more complex exosome signatures in easily accessed tissues and fluids such as plasma, urine, saliva and cerebrospinal fluid.

3 Mechanisms of exosome release in vivo in disease

Advances in the understanding of mechanisms of exosome release that carry biological payloads have provided the molecular basis for the study of exosomes in disease models with particular attention to their potential for therapeutics and diagnostics. With a focus on exosomes with biological activity upon the immune system we review recent discoveries in the area of (1) exosomes and oncology, and (2) exosomes in non-CNS injury and inflammation. The field of oncology has been revolutionized by recent advances in immune checkpoint inhibitor therapy that has established the central importance of targeting the immune system to activate/re-activate host adaptive immune cell populations to recognize and reduce tumor burden. Paget's "Seed and Soil" hypothesis proposed that the tumor microenvironment would provide important factors that would support the growth and metastasis of tumors that would be dependent on localized accumulation of aberrant ECM proteins, growth factors/chemokines, immune cells, and more recently exosomes. The negative consequences of exosomes accumulating in the tumor microenvironment that carry payloads that could support tumor malignancy while suppressing immune responses, would also provide the potential for the detection of circulating exosomes as biomarkers in plasma. Under a generalized overview of cancer as a wound that doesn't heal, the mechanisms of action of tumor-associated exosomes can provide testable hypotheses for models of acute and chronic injury and inflammation.

4 Exosomes and immune checkpoint inhibition in oncology

Exosomes are known to mediate specific tumor-induced immune response [11, 56, 57] that may be associated with how an individual will respond to immune checkpoint inhibitors and potentially guide the development of individualized treatment plans. Recent advances that we detail in subsequent sections utilize multiplex analysis of exosome profiles [58, 59] and single exosome particle analysis to define specific surface targets (i.e., VFC) [60, 61]. Targets relevant to the regulation of APCs and the activation of T cells can be selected on the basis of their known function and abundance in immune cell signaling and crosstalk [11, 62, 63]. Analyses can be performed with only a small sample of peripheral blood for exosome profiling and quantification of target epitopes from plasma with commercially available instrumentation and appropriate quality assurance (QA) and quality control (QC) standards [64–66]. It has been established in experimental tumor models that exosomes are one of the important "atypical" mechanisms of antigen presentation that activate T cells [67–69]. While several studies have correlated plasma exosomes with tumor progression [70–73], thus establishing their clinical relevance, there are no published studies that analyze exosomes as a measure of the efficacy of immune checkpoint inhibitor therapy. Since quantitative approaches now exist for the rapid assessment of exosome size and number, followed by surface profiling to assess exosome source and complexity, there exists a strategy for EV analysis of specific epitopes that is well-aligned with the clinical relevance of those epitopes with adaptive immune cell signaling. While, the transformative potential of the widespread implementation of these profiling strategies in stratified patient populations, while cross-compared with existing molecular and clinical standards, the further study of the biological activity of exosomes in experimental models supports the goals of analyzing exosomes in human cancer. In the case of exosome analyses where freezing and storage of biologically active specimens is more practical than for cells, plasma is an ideal for sampling from healthy volunteers and cancer patients alike, further enhancing translation to clinic.

5 Exosomes in trauma and ischemia/reperfusion injury

Traumatic injury causing hemorrhagic shock (T/HS) leads to tissue injury where ischemia/reperfusion triggers a cascade of inflammatory reactions leading to a systemic inflammatory response syndrome (SIRS) [74]. Although this response is required as part of the initial inate response to injury, and resolves in most patients as they recover, uncontrolled SIRS can lead to distant organ injury, such as acute lung injury, and multiple organ failure [75]. While the pathophysiology of post-injury multiple organ failure in T/HS remains poorly understood, the biological activity of exosomes for the delivery of biologically active immunomodulatory factors from injured gut epithelium, immune cells, enteric nervous system and vasculature that drain into the mesenteric lymph (ML) can regulate a significant component of SIRS. We have postulated that ML exosomes regulate immune cell activation in both ML and mesenteric lymph nodes (MLN), and when they disseminate into the systemic circulation, they can induce secondary organ injury. The inflammation-relevant functional payload of these exosomes includes miRNA, protein and lipids that can be used to define immune signaling, as a biomarker, or deployed as therapeutic exosomes. We have previously shown that APCs in the ML activate T cells in mesenteric lymph nodes (MLNs), and more recently profiled exosomes to show that immune-relevant epitopes on the surface of exosomes such as MHCII can be used as markers that distinguish sham vs. injury in ML. Nucleic acid and proteomic analysis of exosomes in the ML shows that exosomes bear distinct miRNA and protein payloads following injury, an opportunity to further study exosomes as therapeutic targets and diagnostics (i.e., "theranostics"). For example, exosomes can be engineered by loading with specific regulatory miRNAs to attenuate inflammation by re-introduction into the mesenteric lymphatics or peripheral circulation of injured animals. With the substantial neural-immune regulatory network mediating gut homeostasis, we have explored the possibility of neural regulation of inflammation in the gut wall that can be detected by changes in the exosome profile. In this example, upon electrical stimulation of the vagus nerve, a well-defined mediator of anti-inflammatory parasympathetic inputs on the gut and other organs, we have observed protective effects following trauma and ischemic injury such as improved gut epithelial integrity, reduced immune cell activation, and defined changes in the protein payload of exosomes in the ML. These studies provided the scientific premise for the study of systemic effects of ML exosomes in secondary organs such as the lung. Exosomes traffic though the ML into the thoracic duct and enter the systemic circulation where fluorescently labeled exosomes can be identified in the lung and associated with an increase in markers of lung injury [76]. These observations are consistent with studies in the fields of oncology and immunology where dissemination of tumor/infection/inflammation-associated exosomes via draining lymph nodes has been associated with disease progression and the capacity to propagate intercellular signaling at a distance [77, 78]. For example, it is known that ML contains inflammatory mediators [79] and several studies have shown that (a) ligation of the ML duct attenuates lung inflammation and (b) injection of post-shock ML into the circulation of naïve animals recapitulates a lung-injury phenotype [80, 81]. Our own studies of exosomes in animal models have demonstrated a similar role for exosomes as inflammatory mediators in the ML using a well-established experimental approach that is consistent with the pathophysiology of the shock and resuscitation phases of trauma and ischemia in humans.

6 Exosomes and inflammatory bowel disease

Inflammatory bowel disease (IBD), including Crohn's Disease and Ulcerative Colitis, is a chronic immune-mediated disease of the gastrointestinal tract that afflicts over 1.5 million Americans and is characterized by episodes of remission and relapse [82, 83]. As a rule, IBD is also associated with an activation of immune cells, recruitment of inflammatory factors, and an expansion of mesenteric lymphatic vasculature [84, 85]. Yet a role for ML in the progression of IBD remains unknown. Our group [13, 76, 86–90] has identified pro-inflammatory biological activity in ML that is mediated by exosomes, and proposed that the neuro-immune axis regulates gut inflammation via ML. Exosomes released into the mesenteric lymphatic system travel through the thoracic duct into the subclavian vein and into the systemic circulation [91]. We have shown that pro-inflammatory biological activity in ML is mediated by exosomes. Therefore, ML exosomes released after gut injury are pro-inflammatory to macrophages and suppress dendritic cell activation. Because exosomes released from tumor, epithelial and immune cells mediate intercellular signaling, we propose that gut wall-derived exosomes modulate inflammation in IBD. The capacity for activated immune cells to stimulate neuronal circuits that regulate innate and adaptive immunity [92] has led to several gut injury studies in which stimulation of the vagus nerve provides protection to the gut wall by enhancing barrier integrity and reducing inflammation [93–95]. As detailed above, our previous studies have shown that vagal nerve stimulation attenuates gut inflammation, decreases the toxicity of ML, and reduces peripheral tissue inflammation and subsequent distant organ injury [13, 76, 87–90]. With a focus on ML exosomes, it will be important to determine the functional relevance of specific exosome payloads, such as miRNAs that mediate intercellular communication by regulating signal transduction pathways in target cells. Whether neural regulation of the enteric immune cells alters the phenotype of ML exosomes in IBD and other chronic diseases is an open question, but we suggest that neural control of gut immunity and systemic inflammatory responses can be monitored by analysis of exosome profile and biological activity in the draining ML.

7 Exosomes and wound healing

Clinical observations, experimental findings and dogma in the field of wound healing and tissue repair hold that the resolution of injury comprises four main phases of hemostasis, inflammation, vascularization, and epithelialization [96]. We suggest that exosomes carry payloads of protein and nucleic acids that reflect the phases of injury, affecting the profile, payload and biological activity of exosomes [13, 76, 97]. The coordinated phases of the tissue repair response contrast with the dysregulation of the response that define chronic wounds associated with obesity, diabetes, aging, infection, and scarring following burn injury [98, 99]. We suggest that the role of exosomes in mediating the immune response in a chronic wound may have parallels with the biology of exosomes in the tumor microenvironment. In contrast, the exosome-mediated inflammation observed in healthy tissue may provide a template for the exosome proteins and nucleic acids that promote the biology of pro-reparative exosomes and the resolution of tissue injury. We have linked the biochemical activity of specific mediators of exosome biogenesis with the release of exosome populations with a payload and biological activity that regulates the balance of pro-inflammatory vs. pro-reparative macrophages by focusing on the biochemical activity of RabGAPs. We have established a model based on subcutaneous implants of synthetic polyvinylalcohol (PVA) sponges for the in vivo production of exosomes where the cells releasing the exosomes (i.e., the donor cells) can be targeted by local gene delivery to test the biochemical activity of specific mediators of exosome biogenesis. While most genetic studies defining exosome biogenesis to date have used cultured cell lines, the PVA implants when targeted with lentiviral gene delivery enables the collection of recruited leukocytes and the exosomes released into the conditioned media, referred to as wound fluid since the PVA model is a model of sterile inflammation and wound macrophage recruitment. Therefore, the PVA model can be used as an in vivo 3D scaffold that generates a higher density of $CD9^+$, $CD63^+$, and $CD81^+$ exosomes, without the limitations of cell culture plastic and calf serum generally used to grow cells in vitro [100]. PVA can also be coated with ECM proteins to modify the microenvironment of leukocytes recruited and therefore, exosomes released. Protein epitope analysis for ECM-relevant integrins can then be used to predict the tropism for exosome subpopulations in a wound bed and provides many unexplored opportunities to engineer the tropism of exosomes and validate their physiological relevance in vivo [28].

8 Exosomes and diabetes

Chronic diseases such as diabetes are significant risk factors for complications of wound healing, trauma, and infection. The pro-inflammatory state of the diabetic wound bed, liver, and adipose tissue is characterized by pro-inflammatory cytokines such as TNFα, and insulin resistance. Recent studies show that adipose tissue macrophage-derived exosomes are released into the circulation and contain miRNAs that mediate insulin resistance. Uptake of miRNA-containing exosomes can mediate cellular insulin action and systemic insulin sensitivity that has led to the identification of PPARγ as a therapeutic target. Circulating cell free nucleic analyses, including miRNAs have been the subject of extensive studies in human patient samples and animal models, however, a confounding factor for these studies is the often non-overlapping population of miRNAs. The use of mice that have a tissue-specific knockout for Dicer, the gene regulating the miRNA processing enzyme, can be used to define the functional requirement for specific miRNAs- a genetic tool to refine the subset of miRNAs that are functionally relevant in a disease process [101].

9 Potential for species-specific mechanisms of exosome release

Gene duplications and rearrangements in the evolution of the species have emerged as important mediators of fundamental developmental processes with such genes termed human-specific or uniquely human genes (UHGs). In the neurosciences, specific UHGs have evolved from gene families such as Notch (i.e., Notch2NL) that mediate the formation of neurons, and others from the Ras GTPase superfamily (i.e., ARHGAP11b) that regulate the formation of cortical folds. In the area of exosome biogenesis, members of the RabGTPase activating protein family (i.e., TBC1D3) regulate growth factor signaling, macropinocytosis, a process known to enhance antigen uptake and presentation, and exosome biogenesis. Interestingly, TBC1D3 has also been implicated in the formation of cortical folds, an activity that highlights the importance of further studies of UHGs in the neurosciences. In the context of exosomes and the regulation of the formation and release, there have been extensive studies of Rabs and RabGAPs, in particular, based on the size of their respective superfamilies. The over-arching hurdle in the study of genetic mediators of exosome formation is determining their functional requirement. While most studies to date have used yeast and vertebrate cell lines, the use of knockout mice to target specific exosome biogenesis has been slow. Furthermore, the relevance of most of the members of gene families such as the Rabs or RabGAPs have been determined to date in vitro. In combination with the possibility that some members of a family that are sequence-related may in fact regulate un-related pathways, there is also the likelihood that family members are expressed in different cell types or in different times in development. We have considered the possibility that species-specific expression of genes regulating exosome biogenesis may also need to be interpreted in the context of evolution and the selection pressures on the immune system. We have recently proposed that the immune system, and the adaptive immune system in particular, are likely hotspots for the emergence of UHGs. Like in neural development, the inflammatory response is a primary defense response to infection and injury, and it is clear that the fundamental differences in the composition of immune cells in humans and rodent animal models may confound the translation of experimental studies to the clinical setting. We suggest that as increasingly comprehensive screens identify novel and known regulators of exosome biogenesis and secretion, the relevance of in vitro and in vivo models will need to be considered, especially species- and cell-type specific functions of exosome regulatory genes.

10 Exosomes and technology-driven advances

Exosomes are isolated from conditioned media or biological fluids using various techniques that include serial centrifugation followed by filtration and ultracentrifugation or bead enrichment. These considerations are critical in the deployment of analytical assays of clinical samples in animal models or humans. For example, while variations in isolation methods and quality control limit absolute comparisons, light scattering methods such as Nanoparticle Tracking Analysis (NTA), provide a standardized quantification technique for exosomes stratified based on size. However, NTA does not address the payload or diversity of exosomes, beyond differences in size. Furthermore, debate exists in the field regarding the source of exosomes as extracellular vesicles that arise from intracellular MVEs, vs. microvesicles, which pinch off of the plasma membrane. Although microvesicles have been defined as being released from a different source, the plasma membrane, small microvesicles in the 30–150nm range could confound interpretations. Bead-based enrichment of specific subpopulations of exosomes has advanced the understanding of the complexity of exosomes that could be deployed for the analysis of exosomes in health and disease, however, these techniques require preparative quantities for miRNA or proteomics.

A technology that addresses the complexity of exosomes in biological fluids is vesicle flow cytometry (VFC) that uses an optimized flow cytometry analysis procedure based on the use of fluorescent membrane dyes. VFC is particularly important for the analysis of exosomes of clinical samples since it addresses three hurdles in the field of exosome clinical research that are essential for analyses of patient plasma samples. A hurdle to the limited adoption of VFC technology is the heterogeneity of surface epitopes on individual exosomes that are generally not defined for the analysis of exosomes in patient plasma samples. The VFC workflow minimizes the bias of specific antibodies based on abundance and affinity for first step analysis by instead incorporating a sensitive lipophilic fluorescent dye, (i.e., di-8-ANEPPS), that fluoresces in the lipid bilayer [102], allowing one to reproducibly detect exosomes on the basis of fluorescent-triggering on the membrane dye in a sample independent of size and antibody specificity. To address the limitation of available validated antibodies for VFC, multiplex assays can be used to screen for levels on the surface of exosomes. An important caveat is that there a great range of antibody affinities of different epitopes, thus limiting cross-interpretation unless appropriate calibration techniques are performed. The combination of quantitative approaches now exist for the rapid assessment of exosome size, number, profile, payload, and biological activity can provide a liquid biopsy based on a blood draw from patient cohorts and matched volunteers to define the pathophysiological role of exosomes in health and disease.

11 Conclusion

Exosomes are increasingly being recognized for their importance in mediating cell-cell communication and the immune response in normal health and across a spectrum of human disease. Understanding the regulation of exosome processing and release, as well as the cargo carried both on the surface and within the exosome will be critical to better define exosome biology. Combined with improved technology to better characterize subsets of exosomes with molecular studies defining their processing and release, exosomes may be an important diagnostic and therapeutic target in the future.

References

[1] Lopez-Verrilli MA, Court FA. Exosomes: mediators of communication in eukaryotes. Biol Res 2013;46(1):5–11. https://doi.org/10.4067/S0716-97602013000100001. 23760408.

[2] Raposo G, Stoorvogel W. Extracellular vesicles: exosomes, microvesicles, and friends. J Cell Biol 2013;200(4):373–83. Epub 2013/02/20, https://doi.org/10.1083/jcb.20121113823420871. PMCID 3575529.

[3] Yanez-Mo M, Siljander PR, Andreu Z, Zavec AB, Borras FE, Buzas EI, Buzas K, Casal E, Cappello F, Carvalho J, Colas E, Cordeiro da Silva A, Fais S, Falcon-Perez JM, Ghobrial IM, Giebel B, Gimona M, Graner M, Gursel I, Gursel M, Heegaard NH, Hendrix A, Kierulf P, Kokubun K, Kosanovic M, Kralj-Iglic V, Kramer-Albers EM, Laitinen S, Lasser C, Lener T, Ligeti E, Line A, Lipps G, Llorente A, Lotvall J, Mancek-Keber M, Marcilla A, Mittelbrunn M, Nazarenko I, Nolte-t Hoen EN, Nyman TA, O'Driscoll L, Olivan M, Oliveira C, Pallinger E, Del Portillo HA, Reventos J, Rigau M, Rohde E, Sammar M, Sanchez-Madrid F, Santarem N, Schallmoser K, Ostenfeld MS, Stoorvogel W, Stukelj R, Van der Grein SG, Vasconcelos MH, Wauben MH, De Wever O. Biological properties of extracellular vesicles and their physiological functions. J Extracell Vesicles 2015;4:27066. https://doi.org/10.3402/jev.v4.27066. 25979354. PMCID: PMC4433489.

[4] Gutierrez-Vazquez C, Villarroya-Beltri C, Mittelbrunn M, Sanchez-Madrid F. Transfer of extracellular vesicles during immune cell-cell interactions. Immunol Rev 2013;251(1):125–42. https://doi.org/10.1111/imr.12013. 23278745. PMCID: PMC3740495.

[5] Mittelbrunn M, Gutierrez-Vazquez C, Villarroya-Beltri C, Gonzalez S, Sanchez-Cabo F, Gonzalez MA, Bernad A, Sanchez-Madrid F. Unidirectional transfer of microRNA-loaded exosomes from T cells to antigen-presenting cells. Nat Commun 2011;2:282. https://doi.org/10.1038/ncomms1285. 21505438. PMCID: PMC3104548.

[6] Harding CV, Heuser JE, Stahl PD. Exosomes: looking back three decades and into the future. J Cell Biol 2013;200(4):367–71. https://doi.org/10.1083/jcb.201212113. 23420870. PMCID 3575527.

[7] Harding C, Heuser J, Stahl P. Receptor-mediated endocytosis of transferrin and recycling of the transferrin receptor in rat reticulocytes. J Cell Biol 1983;97(2):329–39. 6309857. PMCID: PMC2112509.

[8] Pan BT, Johnstone RM. Fate of the transferrin receptor during maturation of sheep reticulocytes in vitro: selective externalization of the receptor. Cell 1983;33(3):967–78, 6307529.

[9] Bianco NR, Kim SH, Morelli AE, Robbins PD. Modulation of the immune response using dendritic cell-derived exosomes. Methods Mol Biol 2007;380:443–55. https://doi.org/10.1007/978-1-59745-395-0_28. 17876111.

[10] Hao S, Bai O, Yuan J, Qureshi M, Xiang J. Dendritic Cell-Derived Exosomes Stimulate Stronger CD8(+) CTL Responses and Antitumor Immunity than Tumor Cell-Derived Exosomes. Cell Mol Immunol 2006;3(3):205–11. PubMed PMID: WOS:000208926200006.

[11] Pitt JM, Charrier M, Viaud S, Andre F, Besse B, Chaput N, Zitvogel L. Dendritic cell-derived exosomes as immunotherapies in the fight against cancer. J Immunol 2014;193(3):1006–11. https://doi.org/10.4049/jimmunol.1400703. 25049431.

[12] Quah BJ, O'Neill HC. The immunogenicity of dendritic cell-derived exosomes. Blood Cells Mol Dis 2005;35(2):94–110. Epub 2005/06/25 https://doi.org/10.1016/j.bcmd.2005.05.00215975838.

[13] Kojima M, Gimenes-Junior JA, Langness S, Morishita K, Lavoie-Gagne O, Eliceiri B, Costantini TW, Coimbra R. Exosomes, not protein or lipids, in mesenteric lymph activate inflammation: unlocking the mystery of post-shock multiple organ failure. J Trauma Acute Care Surg 2017;82(1):42–50. https://doi.org/10.1097/TA.0000000000001296. 27779585.

[14] Lee H, Zhang D, Zhu Z, Dela Cruz CS, Jin Y. Epithelial cell-derived microvesicles activate macrophages and promote inflammation via microvesicle-containing microRNAs. Sci Rep 2016;6:35250. https://doi.org/10.1038/srep35250. 27731391. PMCID: PMC5059671.

[15] Li D, Liu J, Guo B, Liang C, Dang L, Lu C, He X, Cheung HY, Xu L, Lu C, He B, Liu B, Shaikh AB, Li F, Wang L, Yang Z, Au DW, Peng S, Zhang Z, Zhang BT, Pan X, Qian A, Shang P, Xiao L, Jiang B, Wong CK, Xu J, Bian Z, Liang Z, Guo DA, Zhu H, Tan W, Lu A, Zhang G. Osteoclast-derived exosomal miR-214-3p inhibits osteoblastic bone formation. Nat Commun 2016;7:10872. https://doi.org/10.1038/ncomms10872. 26947250. PMCID: PMC4786676.

[16] Shenoda BB, Ajit SK. Modulation of Immune Responses by Exosomes Derived from Antigen-Presenting Cells. Clin Med Insights Pathol 2016;9(Suppl 1):1–8. https://doi.org/10.4137/CPath.S39925. 27660518. PMCID: PMC5024790.

[17] Garzetti L, Menon R, Finardi A, Bergami A, Sica A, Martino G, Comi G, Verderio C, Farina C, Furlan R. Activated macrophages release microvesicles containing polarized M1 or M2 mRNAs. J Leukoc Biol 2014;95(5):817–25. https://doi.org/10.1189/jlb.0913485. 24379213.

[18] Truman J-P, Al Gadban MM, Smith KJ, Jenkins RW, Mayroo N, Virella G, Lopes-Virella MF, Bielawska A, Hannun YA, Hammad SM. Differential regulation of acid sphingomyelinase in macrophages stimulated with oxidized low-density lipoprotein (LDL) and oxidized LDL immune complexes: role in phagocytosis and cytokine release. Immunology 2012;136(1):30–45. https://doi.org/10.1111/j.1365-2567.2012.03552.x. PubMed PMID: WOS:000302399200005.

[19] Zhang Y, Liu D, Chen X, Li J, Li L, Bian Z, Sun F, Lu J, Yin Y, Cai X, Sun Q, Wang K, Ba Y, Wang Q, Wang D, Yang J, Liu P, Xu T, Yan Q, Zhang J, Zen K, Zhang C-Y. Secreted monocytic miR-150 enhances targeted endothelial cell migration. Mol Cell 2010;39(1):133–44. https://doi.org/10.1016/j.molcel.2010.06.010. PubMed PMID: WOS:000280139200014.

[20] Mittelbrunn M, Vicente-Manzanares M, Sanchez-Madrid F. Organizing polarized delivery of exosomes at synapses. Traffic 2015;16(4):327–37. https://doi.org/10.1111/tra.12258. PubMed PMID 25614958.

[21] Thery C, Ostrowski M, Segura E. Membrane vesicles as conveyors of immune responses. Nat Rev Immunol 2009;9(8):581–93. https://doi.org/10.1038/nri2567. 19498381.

[22] Valenti R, Huber V, Iero M, Filipazzi P, Parmiani G, Rivoltini L. Tumor-released microvesicles as vehicles of immunosuppression. Cancer Res 2007;67(7):2912–5. https://doi.org/10.1158/0008-5472.CAN-07-0520. 17409393.

[23] Viaud S, Thery C, Ploix S, Tursz T, Lapierre V, Lantz O, Zitvogel L, Chaput N. Dendritic cell-derived exosomes for cancer immunotherapy: what's next? Cancer Res 2010;70(4):1281–5. https://doi.org/10.1158/0008-5472.can-09-3276. PubMed PMID: WOS:000278485700002.

[24] Rak J. Extracellular vesicles—biomarkers and effectors of the cellular interactome in cancer. Front Pharmacol 2013;4:21. https://doi.org/10.3389/fphar.2013.00021. 23508692. PMCID: 3589665.

[25] Logozzi M, De Milito A, Lugini L, Borghi M, Calabro L, Spada M, Perdicchio M, Marino ML, Federici C, Iessi E, Brambilla D, Venturi G, Lozupone F, Santinami M, Huber V, Maio M, Rivoltini L, Fais S. High levels of exosomes expressing CD63 and caveolin-1 in plasma of melanoma patients. PLoS One 2009;4(4):e5219 https://doi.org/10.1371/journal.pone.0005219. 19381331. PMCID: 2667632.

[26] Peinado H, Aleckovic M, Lavotshkin S, Matei I, Costa-Silva B, Moreno-Bueno G, Hergueta-Redondo M, Williams C, Garcia-Santos G, Ghajar C, Nitadori-Hoshino A, Hoffman C, Badal K, Garcia BA, Callahan MK, Yuan J, Martins VR, Skog J, Kaplan RN, Brady MS, Wolchok JD, Chapman PB, Kang Y, Bromberg J, Lyden D. Melanoma exosomes educate bone marrow progenitor cells toward a pro-metastatic phenotype through MET. Nat Med 2012;18(6):883–91. https://doi.org/10.1038/nm.2753. 22635005. PMCID: 3645291.

[27] Kharaziha P, Ceder S, Li Q, Panaretakis T. Tumor cell-derived exosomes: a message in a bottle. Biochim Biophys Acta 2012;1826(1):103–11. Epub 2012/04/17 https://doi.org/10.1016/j.bbcan.2012.03.00622503823.

[28] Hoshino A, Costa-Silva B, Shen TL, Rodrigues G, Hashimoto A, Tesic Mark M, Molina H, Kohsaka S, Di Giannatale A, Ceder S, Singh S, Williams C, Soplop N, Uryu K, Pharmer L, King T, Bojmar L, Davies AE, Ararso Y, Zhang T, Zhang H, Hernandez J, Weiss JM, Dumont-Cole VD, Kramer K, Wexler LH, Narendran A, Schwartz GK, Healey JH, Sandstrom P, Labori KJ, Kure EH, Grandgenett PM, Hollingsworth MA, de Sousa M, Kaur S, Jain M, Mallya K, Batra SK, Jarnagin WR, Brady MS, Fodstad O, Muller V, Pantel K, Minn AJ, Bissell MJ, Garcia BA, Kang Y, Rajasekhar VK, Ghajar CM, Matei I, Peinado H, Bromberg J, Lyden D. Tumour exosome integrins determine organotropic metastasis. Nature 2015;527(7578):329–35. https://doi.org/10.1038/nature15756. 26524530.

[29] Aline F, Bout D, Amigorena S, Roingeard P, Dimier-Poisson I. Toxoplasma gondii antigen-pulsed-dendritic cell-derived exosomes induce a protective immune response against T. gondii infection. Infect Immun 2004;72(7):4127–37. https://doi.org/10.1128/IAI.72.7.4127-4137.2004. 15213158. PMCID: PMC427397.

[30] Karlsson M, Lundin S, Dahlgren U, Kahu H, Pettersson I, Telemo E. "Tolerosomes" are produced by intestinal epithelial cells. Eur J Immunol 2001;31(10):2892–900. https://doi.org/10.1002/1521-4141(2001010)31:10<2892::AID-IMMU2892>3.0.CO;2-I. 11592064.

[31] Frangsmyr L, Baranov V, Nagaeva O, Stendahl U, Kjellberg L, Mincheva-Nilsson L. Cytoplasmic microvesicular form of Fas ligand in human early placenta: switching the tissue immune privilege hypothesis from cellular to vesicular level. Mol Hum Reprod 2005;11(1):35–41. https://doi.org/10.1093/molehr/gah129. 15579659.

[32] Kim SH, Bianco N, Menon R, Lechman ER, Shufesky WJ, Morelli AE, Robbins PD. Exosomes derived from genetically modified DC expressing FasL are anti-inflammatory and immunosuppressive. Mol Ther 2006;13(2):289–300. https://doi.org/10.1016/j.ymthe.2005.09.015. 16275099.

[33] Kim SH, Lechman ER, Bianco N, Menon R, Keravala A, Nash J, Mi Z, Watkins SC, Gambotto A, Robbins PD. Exosomes derived from IL-10-treated dendritic cells can suppress inflammation and collagen-induced arthritis. J Immunol 2005;174(10):6440–8. 15879146.

[34] Colino J, Snapper CM. Exosomes from bone marrow dendritic cells pulsed with diphtheria toxoid preferentially induce type 1 antigen-specific IgG responses absence of free antigen. J Immunol 2006;177(6):3757–62. PubMed PMID: WOS:000240475300029.

[35] Colino J, Snapper CM. Dendritic cell-derived exosomes express a *Streptococcus pneumoniae* capsular polysaccharide type 14 cross-reactive antigen that induces protective immunoglobulin responses against pneumococcal infection in mice. Infect Immun 2007;75(1):220–30. https://doi.org/10.1128/iai.01217-06. PubMed PMID: WOS:000243230500023.

[36] Viaud S, Terme M, Flament C, Taieb J, Andre F, Novault S, Escudier B, Robert C, Caillat-Zucman S, Tursz T, Zitvogel L, Chaput N. Dendritic cell-derived exosomes promote natural killer cell activation and proliferation: a role for NKG2D ligands and IL-15R alpha. PLoS One 2009;4(3). https://doi.org/10.1371/journal.pone.0004942. PubMed PMID: WOS:000265499200006.

[37] Hao S, Bai O, Li F, Yuan J, Laferte S, Xiang J. Mature dendritic cells pulsed with exosomes stimulate efficient cytotoxic T-lymphocyte responses and antitumour immunity. Immunology 2007;120(1):90–102. https://doi.org/10.1111/j.1365-2567.2006.02483.x. PubMed PMID: WOS:000242662500011.

[38] Yang X, Meng S, Jiang H, Zhu C, Wu W. Exosomes Derived from Immature Bone Marrow Dendritic Cells Induce Tolerogenicity of Intestinal Transplantation in Rats. J Surg Res 2011;171(2):826–32. https://doi.org/10.1016/j.jss.2010.05.021. PubMed PMID: WOS:000296957900080.

[39] Peche H, Renaudin K, Beriou G, Merieau E, Amigorena S, Cuturi MC. Induction of tolerance by exosomes and short-term immunosuppression in a fully MHC-mismatched rat cardiac allograft model. Am J Transplant 2006;6(7):1541–50. https://doi.org/10.1111/j.1600-6143.2006.01344.x. PubMed PMID: WOS:000238465400010.

[40] Peche H, Heslan M, Usal C, Amigorena S, Cuturi MC. Presentation of donor major histocompatibility complex antigens by bone marrow dendritic cell-derived exosomes modulates allograft rejection. Transplantation 2003;76(10):1503–10. https://doi.org/10.1097/01.tp.0000092494.75313.38. PubMed PMID: WOS:000186833400022.

[41] Bianco NR, Kim SH, Ruffner MA, Robbins PD. Therapeutic effect of exosomes from indoleamine 2,3-dioxygenase-positive dendritic cells in collagen-induced arthritis and delayed-type hypersensitivity disease models. Arthritis Rheum 2009;60(2):380–9. https://doi.org/10.1002/art.24229. PubMed PMID: WOS:000263276400011.

[42] Monleon I, Martinez-Lorenzo MJ, Monteagudo L, Lasierra P, Taules M, Iturralde M, Pineiro A, Larrad L, Alava MA, Naval J, Anel A. Differential secretion of Fas ligand- or APO2 ligand/TNF-related apoptosis-inducing ligand-carrying microvesicles during activation-induced death of human T cells. J Immunol 2001;167(12):6736–44. https://doi.org/10.4049/jimmunol.167.12.6736. PubMed PMID: WOS:000172613400007.

[43] Munich S, Sobo-Vujanovic A, Buchser WJ, Beer-Stolz D, Vujanovic NL. Dendritic cell exosomes directly kill tumor cells and activate natural killer cells via TNF superfamily ligands. Oncoimmunology 2012;1(7):1074–83. https://doi.org/10.4161/onci.20897. PubMed PMID: WOS:000316279900008.

[44] Zuccato E, Blott EJ, Holt O, Sigismund S, Shaw M, Bossi G, Griffiths GM. Sorting of Fas ligand to secretory lysosomes is regulated by mono-ubiquitylation and phosphorylation. J Cell Sci 2007;120(1):191–9. https://doi.org/10.1242/jcs.03315. PubMed PMID: WOS:000243723800019.

[45] Beauvillain C, Ruiz S, Guiton R, Bout D, Dimier-Poisson I. A vaccine based on exosomes secreted by a dendritic cell line confers protection against T. gondii infection in syngeneic and allogeneic mice. Microbes Infect 2007;9(14–15):1614–22. https://doi.org/10.1016/j.micinf.2007.07.002. PubMed PMID: WOS:000252333100014.

[46] Giri PK, Schorey JS. Exosomes derived from M-bovis BCG infected macrophages activate antigen-specific CD4(+) and CD8(+) T cells in vitro and in vivo. PLoS One 2008;3(6). https://doi.org/10.1371/journal.pone.0002461. PubMed PMID: WOS:000263280700042.

[47] Keryer-Bibens C, Pioche-Durieu C, Villemant C, Souquere S, Nishi N, Hirashima M, Middeldorp J, Busson P. Exosomes released by EBV-infected nasopharyngeal carcinoma cells convey the viral Latent Membrane Protein 1 and the immunomodulatory protein galectin 9. BMC Cancer 2006;6. https://doi.org/10.1186/1471-2407-6-283. PubMed PMID: WOS:000243656700001.

[48] Testa JS, Apcher GS, Comber JD, Eisenlohr LC. Exosome-driven antigen transfer for MHC class II presentation facilitated by the receptor binding activity of influenza hemagglutinin. J Immunol 2010;185(11):6608–16. https://doi.org/10.4049/jimmunol.1001768. PubMed PMID: WOS:000284311500026.

[49] Barr F, Lambright DG. Rab GEFs and GAPs. Curr Opin Cell Biol 2010;22(4):461–70. https://doi.org/10.1016/j.ceb.2010.04.007. 20466531. PMCID: PMC2929657.
[50] Somsel Rodman J, Wandinger-Ness A. Rab GTPases coordinate endocytosis. J Cell Sci 2000;113(Pt 2):183–92, 10633070.
[51] Frasa MA, Koessmeier KT, Ahmadian MR, Braga VM. Illuminating the functional and structural repertoire of human TBC/RABGAPs. Nat Rev Mol Cell Biol 2012;13(2):67–73. https://doi.org/10.1038/nrm3267. 22251903.
[52] Hsu C, Morohashi Y, Yoshimura S, Manrique-Hoyos N, Jung S, Lauterbach MA, Bakhti M, Gronborg M, Mobius W, Rhee J, Barr FA, Simons M. Regulation of exosome secretion by Rab35 and its GTPase-activating proteins TBC1D10A-C. J Cell Biol 2010;189(2):223–32. https://doi.org/10.1083/jcb.200911018. 20404108. PMCID: 2856897.
[53] Pan X, Eathiraj S, Munson M, Lambright DG. TBC-domain GAPs for Rab GTPases accelerate GTP hydrolysis by a dual-finger mechanism. Nature 2006;442(7100):303–6. https://doi.org/10.1038/nature04847. 16855591.
[54] Shurtleff MJ, Temoche-Diaz MM, Karfilis KV, Ri S, Schekman R. Y-box protein 1 is required to sort microRNAs into exosomes in cells and in a cell-free reaction. elife 2016;5:e19276https://doi.org/10.7554/eLife.19276. 27559612. PMCID: PMC5047747.
[55] Lo Cicero A, Stahl PD, Raposo G. Extracellular vesicles shuffling intercellular messages: for good or for bad. Curr Opin Cell Biol 2015;35:69–77. https://doi.org/10.1016/j.ceb.2015.04.013. 26001269.
[56] Pardoll DM. The blockade of immune checkpoints in cancer immunotherapy. Nat Rev Cancer 2012;12(4):252–64. https://doi.org/10.1038/nrc3239. 22437870. PMCID: PMC4856023.
[57] Soung YH, Ford S, Zhang V, Chung J. Exosomes in cancer diagnostics. Cancers (Basel) 2017;9(1):8. https://doi.org/10.3390/cancers9010008. 28085080. PMCID: PMC5295779.
[58] Koliha N, Heider U, Ozimkowski T, Wiemann M, Bosio A, Wild S. Melanoma affects the composition of blood cell-derived extracellular vesicles. Front Immunol 2016;7:282. https://doi.org/10.3389/fimmu.2016.00282. 27507971. PMCID: PMC4960424.
[59] Koliha N, Wiencek Y, Heider U, Jungst C, Kladt N, Krauthauser S, Johnston IC, Bosio A, Schauss A, Wild S. A novel multiplex bead-based platform highlights the diversity of extracellular vesicles. J Extracell Vesicles 2016;5:29975. https://doi.org/10.3402/jev.v5.29975. 26901056. PMCID: PMC4762227.
[60] Groot Kormelink T, Arkesteijn GJ, Nauwelaers FA, van den Engh G, Nolte-'t Hoen EN, Wauben MH. Prerequisites for the analysis and sorting of extracellular vesicle subpopulations by high-resolution flow cytometry. Cytometry A 2016;89(2):135–47. https://doi.org/10.1002/cyto.a.22644. 25688721.
[61] van der Vlist EJ, Nolte-'t Hoen EN, Stoorvogel W, Arkesteijn GJ, Wauben MH. Fluorescent labeling of nano-sized vesicles released by cells and subsequent quantitative and qualitative analysis by high-resolution flow cytometry. Nat Protoc 2012;7(7):1311–26. https://doi.org/10.1038/nprot.2012.065. 22722367.
[62] Nolte-'t Hoen EN, Buschow SI, Anderton SM, Stoorvogel W, Wauben MH. Activated T cells recruit exosomes secreted by dendritic cells via LFA-1. Blood 2009;113(9):1977–81. https://doi.org/10.1182/blood-2008-08-174094. 19064723.
[63] Nolte-'t Hoen EN, van der Vlist EJ, de Boer-Brouwer M, Arkesteijn GJ, Stoorvogel W, Wauben MH. Dynamics of dendritic cell-derived vesicles: high-resolution flow cytometric analysis of extracellular vesicle quantity and quality. J Leukoc Biol 2013;93(3):395–402. Epub 2012/12/19, https://doi.org/10.1189/jlb.0911480. 23248328.
[64] Arraud N, Linares R, Tan S, Gounou C, Pasquet JM, Mornet S, Brisson AR. Extracellular vesicles from blood plasma: determination of their morphology, size, phenotype and concentration. J Thromb Haemost 2014;12(5):614–27. https://doi.org/10.1111/jth.12554. 24618123.

[65] Chandler WL. Measurement of microvesicle levels in human blood using flow cytometry. Cytometry B Clin Cytom 2016;90(4):326–36. https://doi.org/10.1002/cyto.b.21343. 26606416.

[66] Inal JM, Kosgodage U, Azam S, Stratton D, Antwi-Baffour S, Lange S. Blood/plasma secretome and microvesicles. Biochim Biophys Acta 2013;1834(11):2317–25. https://doi.org/10.1016/j.bbapap.2013.04.005. 23590876.

[67] Kambayashi T, Laufer TM. Atypical MHC class II-expressing antigen-presenting cells: can anything replace a dendritic cell? Nat Rev Immunol 2014;14(11):719–30. https://doi.org/10.1038/nri3754. 25324123.

[68] Turpin D, Truchetet ME, Faustin B, Augusto JF, Contin-Bordes C, Brisson A, Blanco P, Duffau P. Role of extracellular vesicles in autoimmune diseases. Autoimmun Rev 2016;15(2):174–83. https://doi.org/10.1016/j.autrev.2015.11.004. 26554931.

[69] Willms E, Johansson HJ, Mager I, Lee Y, Blomberg KE, Sadik M, Alaarg A, Smith CI, Lehtio J, El Andaloussi S, Wood MJ, Vader P. Cells release subpopulations of exosomes with distinct molecular and biological properties. Sci Rep 2016;6:22519. https://doi.org/10.1038/srep22519. 26931825. PMCID: PMC4773763.

[70] Akagi T, Kato K, Kobayashi M, Kosaka N, Ochiya T, Ichiki T. On-chip immunoelectrophoresis of extracellular vesicles released from human breast cancer cells. PLoS One 2015;10(4). https://doi.org/10.1371/journal.pone.0123603. 25928805. PMCID: PMC4415775.

[71] Caivano A, Laurenzana I, De Luca L, La Rocca F, Simeon V, Trino S, D'Auria F, Traficante A, Maietti M, Izzo T, D'Arena G, Mansueto G, Pietrantuono G, Laurenti L, Musto P, Del Vecchio L. High serum levels of extracellular vesicles expressing malignancy-related markers are released in patients with various types of hematological neoplastic disorders. Tumour Biol 2015;36(12):9739–52. https://doi.org/10.1007/s13277-015-3741-3. 26156801.

[72] van Eijndhoven MA, Zijlstra JM, Groenewegen NJ, Drees EE, van Niele S, Baglio SR, Koppers-Lalic D, van der Voorn H, Libregts SF, Wauben MH, de Menezes RX, van Weering JR, Nieuwland R, Visser L, van den Berg A, de Jong D, Pegtel DM. Plasma vesicle miRNAs for therapy response monitoring in Hodgkin lymphoma patients. JCI Insight 2016;1(19):e89631. https://doi.org/10.1172/jci.insight.89631. 27882350. PMCID: PMC5111516.

[73] Ostenfeld MS, Jensen SG, Jeppesen DK, Christensen LL, Thorsen SB, Stenvang J, Hvam ML, Thomsen A, Mouritzen P, Rasmussen MH, Nielsen HJ, Orntoft TF, Andersen CL. miRNA profiling of circulating EpCAM(+) extracellular vesicles: promising biomarkers of colorectal cancer. J Extracell Vesicles 2016;5:31488. https://doi.org/10.3402/jev.v5.31488. 27576678. PMCID: PMC5005366.

[74] Xiao W, Mindrinos MN, Seok J, Cuschieri J, Cuenca AG, Gao H, Hayden DL, Hennessy L, Moore EE, Minei JP, Bankey PE, Johnson JL, Sperry J, Nathens AB, Billiar TR, West MA, Brownstein BH, Mason PH, Baker HV, Finnerty CC, Jeschke MG, Lopez MC, Klein MB, Gamelli RL, Gibran NS, Arnoldo B, Xu W, Zhang Y, Calvano SE, McDonald-Smith GP, Schoenfeld DA, Storey JD, Cobb JP, Warren HS, Moldawer LL, Herndon DN, Lowry SF, Maier RV, Davis RW, Tompkins RG. A genomic storm in critically injured humans. J Exp Med 2011;208(13):2581–90. https://doi.org/10.1084/jem.20111354. 22110166. PMCID: PMC3244029.

[75] Moore EE. Claude H. Organ, Jr. memorial lecture: splanchnic hypoperfusion provokes acute lung injury via a 5-lipoxygenase-dependent mechanism. Am J Surg 2010;200(6):681–9. https://doi.org/10.1016/j.amjsurg.2010.05.010. 21146002. PMCID: PMC3031087.

[76] Kojima M, Gimenes-Junior JA, Chan TW, Eliceiri BP, Baird A, Costantini TW, Coimbra R. Exosomes in postshock mesenteric lymph are key mediators of acute lung injury triggering the macrophage activation via Toll-like receptor 4. FASEB J 2018;32(1):97–110. https://doi.org/10.1096/fj.201700488R. 28855278.

[77] Ying W, Riopel M, Bandyopadhyay G, Dong Y, Birmingham A, Seo JB, Ofrecio JM, Wollam J, Hernandez-Carretero A, Fu W, Li P, Olefsky JM. Adipose tissue macrophage-derived exosomal miRNAs can modulate in vivo and in vitro insulin sensitivity. Cell 2017;171(2):372–84. e12, https://doi.org/10.1016/j.cell.2017.08.03528942920.

[78] Kamerkar S, LeBleu VS, Sugimoto H, Yang S, Ruivo CF, Melo SA, Lee JJ, Kalluri R. Exosomes facilitate therapeutic targeting of oncogenic KRAS in pancreatic cancer. Nature 2017;546(7659):498–503. https://doi.org/10.1038/nature22341. 28607485. PMCID: PMC5538883.

[79] Deitch EA. Gut lymph and lymphatics: a source of factors leading to organ injury and dysfunction. Ann N Y Acad Sci 2010;1207(Suppl 1):E103–11. https://doi.org/10.1111/j.1749-6632.2010.05713.x. 20961300.

[80] Deitch EA, Adams C, Lu Q, Xu DZ. A time course study of the protective effect of mesenteric lymph duct ligation on hemorrhagic shock-induced pulmonary injury and the toxic effects of lymph from shocked rats on endothelial cell monolayer permeability. Surgery 2001;129(1):39–47. Epub 2001/01/10. 11150032.

[81] Senthil M, Watkins A, Barlos D, Xu DZ, Lu Q, Abungu B, Caputo F, Feinman R, Deitch EA. Intravenous injection of trauma-hemorrhagic shock mesenteric lymph causes lung injury that is dependent upon activation of the inducible nitric oxide synthase pathway. Ann Surg 2007;246(5):822–30. Epub 2007/10/31 https://doi.org/10.1097/SLA.0b013e3180caa3af. 17968175.

[82] Betteridge JD, Armbruster SP, Maydonovitch C, Veerappan GR. Inflammatory bowel disease prevalence by age, gender, race, and geographic location in the U.S. military health care population. Inflamm Bowel Dis 2013;19(7):1421–7. https://doi.org/10.1097/MIB.0b013e318281334d. 23518811.

[83] Wagnerova A, Babickova J, Liptak R, Vlkova B, Celec P, Gardlik R. Sex differences in the effect of resveratrol on DSS-induced colitis in mice. Gastroenterol Res Pract 2017;2017. https://doi.org/10.1155/2017/8051870. 28465680. PMCID: PMC5390549.

[84] Kim KW, Song JH. Emerging roles of lymphatic vasculature in immunity. Immune Netw 2017;17(1):68–76. https://doi.org/10.4110/in.2017.17.1.68. 28261022. PMCID: PMC5334124.

[85] Wang Y, Oliver G. Current views on the function of the lymphatic vasculature in health and disease. Genes Dev 2010;24(19):2115–26. https://doi.org/10.1101/gad.1955910. 20889712. PMCID: PMC2947764.

[86] AlSharari SD, Bagdas D, Akbarali HI, Lichtman PA, Raborn ES, Cabral GA, Carroll FI, McGee EA, Damaj MI. Sex differences and drug dose influence the role of the alpha7 nicotinic acetylcholine receptor in the mouse dextran sodium sulfate-induced colitis model. Nicotine Tob Res 2017;19(4):460–8. https://doi.org/10.1093/ntr/ntw245. 27639096.

[87] Kojima M, Costantini TW, Eliceiri BP, Chan TW, Baird A, Coimbra R. Gut epithelial cell-derived exosomes trigger post-trauma immune dysfunction. J Trauma Acute Care Surg 2017;84(2):257–264. Epub 2017/12/02, https://doi.org/10.1097/ta.0000000000001748, 29194317.

[88] Morishita K, Coimbra R, Langness S, Eliceiri BP, Costantini TW. Neuroenteric axis modulates the balance of regulatory T cells and T-helper 17 cells in the mesenteric lymph node following trauma/hemorrhagic shock. Am J Physiol Gastrointest Liver Physiol 2015;309(3):G202–8. Epub 2015/06/06, https://doi.org/10.1152/ajpgi.00097.2015, 26045612.

[89] Morishita K, Costantini TW, Eliceiri B, Bansal V, Coimbra R. Vagal nerve stimulation modulates the dendritic cell profile in posthemorrhagic shock mesenteric lymph. J Trauma Acute Care Surg 2014;76(3):610–7. discussion 7-8, https://doi.org/10.1097/ta.0000000000000137. 24553526.

[90] Morishita K, Costantini TW, Ueno A, Bansal V, Eliceiri B, Coimbra R. A pharmacologic approach to vagal nerve stimulation prevents mesenteric lymph toxicity after hemorrhagic shock. J Trauma Acute Care Surg 2015;78(1):52–8. discussion 8-9, https://doi.org/10.1097/ta.0000000000000489. 25539203.

[91] Ionac M. One technique, two approaches, and results: thoracic duct cannulation in small laboratory animals. Microsurgery 2003;23(3):239–45. https://doi.org/10.1002/micr.10136. 12833325.

[92] Chavan SS, Pavlov VA, Tracey KJ. Mechanisms and therapeutic relevance of neuro-immune communication. Immunity 2017;46(6):927–42. https://doi.org/10.1016/j.immuni.2017.06.008. 28636960. PMCID: PMC5578398.

[93] Costantini TW, Bansal V, Krzyzaniak M, Putnam JG, Peterson CY, Loomis WH, Wolf P, Baird A, Eliceiri BP, Coimbra R. Vagal nerve stimulation protects against burn-induced intestinal injury through activation of enteric glia cells. Am J Physiol Gastrointest Liver Physiol 2010;299(6):G1308–18. https://doi.org/10.1152/ajpgi.00156.2010. 20705905. PMCID: PMC3774266.

[94] Lowry DM, Morishita K, Eliceiri BP, Bansal V, Coimbra R, Costantini TW. The vagus nerve alters the pulmonary dendritic cell response to injury. J Surg Res 2014;192(1):12–8. https://doi.org/10.1016/j.jss.2014.06.012. 25005822.

[95] Costantini TW, Putnam JG, Sawada R, Baird A, Loomis WH, Eliceiri BP, Bansal V, Coimbra R. Targeting the gut barrier: identification of a homing peptide sequence for delivery into the injured intestinal epithelial cell. Surgery 2009;146(2):206–12. https://doi.org/10.1016/j.surg.2009.05.007. 19628075. PMCID: PMC4251594.

[96] Koh TJ, DiPietro LA. Inflammation and wound healing: the role of the macrophage. Expert Rev Mol Med 2011;13. https://doi.org/10.1017/S1462399411001943. 21740602. PMCID: PMC3596046.

[97] Silva AM, Teixeira JH, Almeida MI, Goncalves RM, Barbosa MA, Santos SG. Extracellular vesicles: immunomodulatory messengers in the context of tissue repair/regeneration. Eur J Pharm Sci 2017;98:86–95. https://doi.org/10.1016/j.ejps.2016.09.017. 27644894.

[98] Larouche J, Sheoran S, Maruyama K, Martino MM. Immune regulation of skin wound healing: mechanisms and novel therapeutic targets. Adv Wound Care 2018;7(7):209–31. https://doi.org/10.1089/wound.2017.0761. 29984112. PMCID: PMC6032665.

[99] Novak ML, Koh TJ. Phenotypic transitions of macrophages orchestrate tissue repair. Am J Pathol 2013;183(5):1352–63. https://doi.org/10.1016/j.ajpath.2013.06.034. 24091222. PMCID: PMC3969506.

[100] Patel DB, Santoro M, Born LJ, Fisher JP, Jay SM. Towards rationally designed biomanufacturing of therapeutic extracellular vesicles: impact of the bioproduction microenvironment. Biotechnol Adv 2018;36:2051–9. https://doi.org/10.1016/j.biotechadv.2018.09.001. 30218694.

[101] Thomou T, Mori MA, Dreyfuss JM, Konishi M, Sakaguchi M, Wolfrum C, Rao TN, Winnay JN, Garcia-Martin R, Grinspoon SK, Gorden P, Kahn CR. Adipose-derived circulating miRNAs regulate gene expression in other tissues. Nature 2017;542(7642):450–5. https://doi.org/10.1038/nature21365. 28199304. PMCID: PMC5330251.

[102] Stoner SA, Duggan E, Condello D, Guerrero A, Turk JR, Narayanan PK, Nolan JP. High sensitivity flow cytometry of membrane vesicles. Cytometry A 2016;89(2):196–206. Epub 2015/10/21, https://doi.org/10.1002/cyto.a.2278726484737.

Chapter 15

Exosomes in metabolic syndrome

Soazig Le Lay[a], Ramaroson Andriantsitohaina[a,b], M. Carmen Martinez[a,b]
[a]SOPAM, U1063, INSERM, UNIV ANGERS, SFR ICAT, Bat IRIS-IBS, Angers, France, [b]Angers University Hospital, Angers, France

1 Introduction

As indicated by the last report of the World Health Organization, lifestyle changes such as physical inactivity due to the increasingly sedentary and increased intake of energy-dense foods rich in fat and glucose, lead to an increase of non-communicable diseases including cancers, chronic respiratory diseases, cardiovascular diseases, and diabetes [1]. Genetics and many environmental factors, such as insufficient sleep, tobacco and/or alcohol consumption or endocrine disruptors seem to be also involved in the development of these chronic diseases. More in detail, the presence of several troubles associated with overweight, hypertension, and biochemical alterations, chiefly dyslipidemia and hyperglycemia, accounts for an approximate twofold increased risk of atherosclerotic cardiovascular diseases leading to myocardial infarction, cerebrovascular stroke and peripheral arterial diseases. The conjunction of these alterations defines the metabolic syndrome (MetS), a cluster of abdominal obesity, insulin resistance (IR), elevated blood pressure (BP), and obesity-related dyslipidemia consisting in low levels of HDL-cholesterol and high levels of triglycerides.

Despite the efforts made by various health organizations to define the MetS, some differences still persist with regard to the values of blood pressure, waist circumference illustrating central obesity or plasma HDL levels. Nevertheless, central obesity and the insulin resistance seem to be the main and preliminary screening tools to describe MetS patients (Fig. 1). Indeed, the prevalence of MetS is increasing in all countries independently of the economic developmental level showing the clinical and public health importance of MetS management. In occidental countries, the increase of the prevalence of MetS is directly correlated with the increase of obesity. For instance, MetS prevalence is estimated in approximately 25% of all adults with increased prevalence in advanced ages; however, recent data highlight the prevalence increase in young adults [2].

Exosomes. https://doi.org/10.1016/B978-0-12-816053-4.00015-8

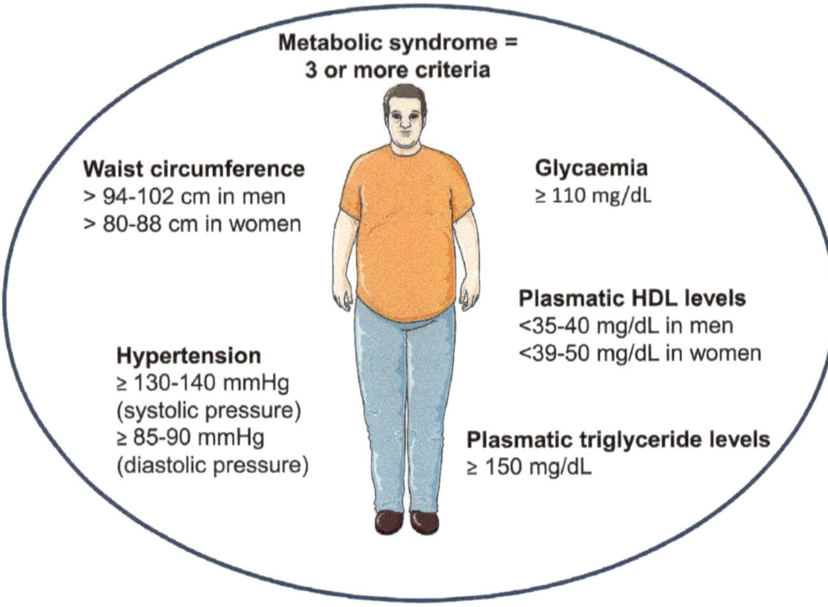

FIG. 1 Criteria for the diagnosis of the metabolic syndrome according to the following associations: WHO, World Health Organization; NCEP-ATPIII, National Cholesterol Education Program-Adult Treatment. Panel III, IDF, International Diabetes Foundation; HDL, high density lipoprotein.

2 Different components of MetS

The interlacing of the five criteria defining the MetS makes difficult to establish the participation of each criterion separately. As indicated above, since patients with MetS are frequently obese and resistant to insulin, it is plausible that both criteria contribute to the development of MetS in greater proportion than hypertension and dyslipidemia [3].

2.1 Obesity

In MetS patients, central or visceral obesity is one of the most important factors in its ability to increase the risk for the development of IR which would facilitate the accumulation of fat in the heart, atherosclerosis and hypertension [4, 5]. In more detail, adipocytes as well as adipose tissue stromal cells including pro-inflammatory macrophages [6] produce pro-atherogenic adipokines such as IL-6, TNFα and MCP-1 implicated in the installation of oxidative stress and chronic inflammation [7]. These effects are reversible as shown by the decrease in the production of pro-inflammatory molecules associated with a reduction in macrophage infiltration after bariatric surgery [8].

2.2 Dyslipidemia

Two MetS criteria concern lipid disorders: on the one hand, an increase in the plasma concentration of triglycerides and, on the other hand, a decrease in that of HDL cholesterol [5, 9]. These lipid disorders directly reflect excessive consumption of saturated fatty acids, partially hydrogenated unsaturated fatty acids and carbohydrates [10]. In vitro, it has been shown that fatty acid or LDL overload induces hepatocytes apoptosis [11]. Also, animals fed a high-fat diet (HFD) have a lipid accumulation in the liver that can lead in some cases to cirrhosis [12]. Finally, the accumulation of LDL in monocytes-macrophages infiltrated into the vascular wall facilitates their transformation into foam cells leading to the formation of atheromatous plaque [13].

2.3 Arterial hypertension

During MetS, the increase in systolic and diastolic blood pressure reflects a complex interaction between different systems. In fact, the secretion of angiotensin II by the adipose tissue can stimulate the production of aldosterone, which plays a crucial role in the control of blood pressure [14]. Moreover, in hyperinsulinemic patients, an increase in the activity of the sympathetic nervous system has been described [15], mainly exacerbated activation of carotid bodies [16] which could lead to the development of hypertension.

2.4 IR and diabetes

The activation of the insulin receptor by its ligand leads to a signaling cascade responsible for increasing glycogen synthesis, glucose uptake, and protein synthesis to increase energy stores. However, in certain pathophysiological situations, this signaling pathway is failing resulting in a reduction of the metabolic effects of insulin, also called IR, in the target tissues (adipose tissue, liver, vascular endothelium and skeletal muscle). IR is characterized by a mobilization of fatty acids in adipocytes that will accumulate in skeletal muscle. In the liver, IR increases glycogenolysis and gluconeogenesis, which contributes to hyperglycemia. With respect to the endothelium, IR reduces nitric oxide (NO) production and increases vasoconstriction, and consequently, hypertension [17]. Finally, IR is the leading cause of type 2 diabetes mellitus (T2DM), characterized by fasting hyperglycaemia. In the long term, the high concentration of glucose in the blood affects certain tissues and organs (eyes, nervous system, heart, vessels, kidneys, etc.) responsible for their failure. All these data show that a vicious circle is set up and that all these factors contribute to the genesis and maintenance of the MetS.

3 EVs: Biomarkers of MetS components

Risk factors defining the MetS are often intrinsically related and, thereby, contribute to MetS genesis and maintenance. Although it is often difficult to distinguish the contribution of each individual MetS component, clinical studies have demonstrated their involvement in exosome production which suggests these small extracellular vesicles (EVs) may be used as prognostic and/or diagnostic biomarkers of metabolic complications (Table 1).

3.1 Obesity

Visceral obesity is considered as one of the most deleterious MetS component, since it highly associates with IR, which favors ectopic lipid accumulation, atherosclerosis and hypertension. Moreover, obesity is characterized by a chronic low-grade inflammation state characterized by pro-inflammatory macrophage infiltration. [8].

TABLE 1 Exosomes as potential biomarkers for diagnostic/prognostic of metabolic complications

	Characteristics	References
Obesity	Increased circulating exosomes in obese patients	[18, 19]
	Positive correlation between circulating exosomes and HOMA IR	[20, 21]
Dyslipidemia	Saturated fatty acids exposure enhanced exosome secretion by adipocytes, hepatocytes and muscle cells	[22–24]
	Positive correlation between circulating exosomes and elevated serum triglycerides	[25]
Hypertension	Angiotensin-II infusion in rats increase serum exosomes	[26]
	Increase urinary exosomes following RAAS activation	[27]
	Correlation between sodium transporters expression levels and hypertension development	[27, 28]
T2DM	Increased circulating exosomes in diabetic patients	[21, 29]
	Specific increase of erythrocyte-derived exosomes	[21]
	IR enhances exosome secretion	[21]

HOMA, homeostasis model assessment; RAAS, renin angiotensin aldosterone system; T2DM, type 2 diabetes mellitus; IR, insulin resistance.

Others and we have shown that both circulating microvesicle and exosome concentrations raise with obesity, and correlate with body mass index (BMI) suggesting that adipose tissue-derived EVs may contribute to such circulating EV concentration increase [18, 19]. Pre-clinical studies have confirmed the ability of adipose cells to secrete exosomes in abundance, secretion moreover enhanced by fat overloading conditions [22, 30]. Despite evidences reporting the presence of adipocyte-derived exosomes in the human circulation [31], contribution of fat-derived exosomes in the overall circulating exosome pool is unknown, mainly due to the lack of a reliable and specific marker to identify specifically fat-derived exosomes. Nonetheless, positive correlation was established between the number of exosomes released by omental adipose tissue and the Homeostasis Model Assessment (HOMA) IR index suggesting the role of fat-derived exosomes in the development of T2DM [20, 21].

3.2 Dyslipidemia

Hypercholesterolemia and hypertriglyceridemia are often reflecting lipid and glucose-enriched diets. In vitro exposure of hepatocytes, muscular cells or adipocytes to high glucose or lipid concentrations enhances exosome secretion [22–24]. Particularly, exosomes produced following palmitic acid-treatment of cells, a saturated fatty acid known to mediate inflammation and insulin-resistant processes, induce fibrotic activation in hepatic cells and IR in muscle cells [24]. Circulating exosome number most strongly correlates with elevated serum triglycerides, by comparison to other metabolic parameters, in MetS patients suggesting that fatty acid overloading will also favor EV secretion in humans [25].

3.3 Hypertension

Hypertension is one of the most important cardiovascular risk factors favoring macrophage infiltration of blood vessels. In fact, arterial pressure elevation occurring in MetS reflects the complex interaction operating between different regulating systems. Secretion of angiotensin-II by adipose tissue participate to aldosterone production, thereby arterial pressure regulation. Sympathetic nervous system activation induced by hyperinsulinemia also contribute to hypertension.

Whereas different publications reported increased circulating concentration of large EVs (namely microvesicles) in hypertensive animal models [32] and demonstrated their potential as predictive/diagnostic marker of atherothrombotic pathologies in humans [33], a limited number of studies focused on circulating exosomes in the context of hypertension. One study reported that hypertensive rats, following angiotensin-II infusion, display an increase of serum exosomes, which are able to upregulate the expression levels of inflammation markers when incubated on endothelial cells in vitro [26]. These results suggest that endothelial damage occurring in hypertension may be partially induced by cargos carried by circulating exosomes. Besides, studies have reported an alteration of urinary exosome content following the renin-angiotensin-aldosterone system activation [27]. Particularly, expression levels of exosomal sodium transporters strongly correlate with hypertension development [27, 28] revealing the promising pronostic application of urinary exosomes to evaluate hypertension.

3.4 IR and type 2 diabetes mellitus (T2DM)

T2DM has been associated with higher levels of both circulating large (microvesicles) and small (exosomes) vesicles [21, 29]. This is particularly true for erythrocyte-derived exosomes shown to be significantly higher with diabetes [21]. These authors moreover demonstrated that IR state enhanced EV secretion identifying a vicious circle where obesity, a major factor risk for T2DM, could initiate EV secretion that could be further entertained by IR and associated chronic inflammation.

3.5 All MetS components gathered

The interlocking of the five definitive criteria characterizing the MetS makes the process difficult to establish the participation of each criterion separately. To date, only circulating microvesicles subclasses have been phenotyped by flow cytometry with regard to the context of MetS. It appears that MetS patients display higher circulating amount of platelets, endothelial, erythrocyte-derived microvesicles by comparison to healthy subjects [34, 35]. Moreover, tissue factor-positive microvesicles are associated with components of the MetS [36] whereas Cystatin C-positive EVs correlate with metabolic complications of obesity development [37].

4 EVs: Bioeffectors of the MetS

4.1 Exosomes and metabolic complications of obesity: A role for fat-derived exosomes?

In recent works, Li et al. [38] and Nie et al. [39] have shown that increased circulating exosomal miR-29a and miR-194 levels in obese patients were strongly associated with impaired human cardiac function, including ejection fraction and N-terminal pro b-type natriuretic peptide levels, a biomarker for the heart failure. Furthermore, in vitro treatment of mouse cardiomyocytes with circulating exosomes from obese patients evoked cardiomyocyte mitochondrial inactivity, and similar results were obtained in mice fed with a HFD. These data highlight that both exosomal miRNA (miR-29a and miR-194) can be an interesting target against obesity-induced cardiac dysfunctions.

Numerous studies have highlighted that exosomes released by adipose tissue can act as a mode of communication between fat and others cells and tissues (Fig. 2). Such network of exosome-mediated exchange has been recently

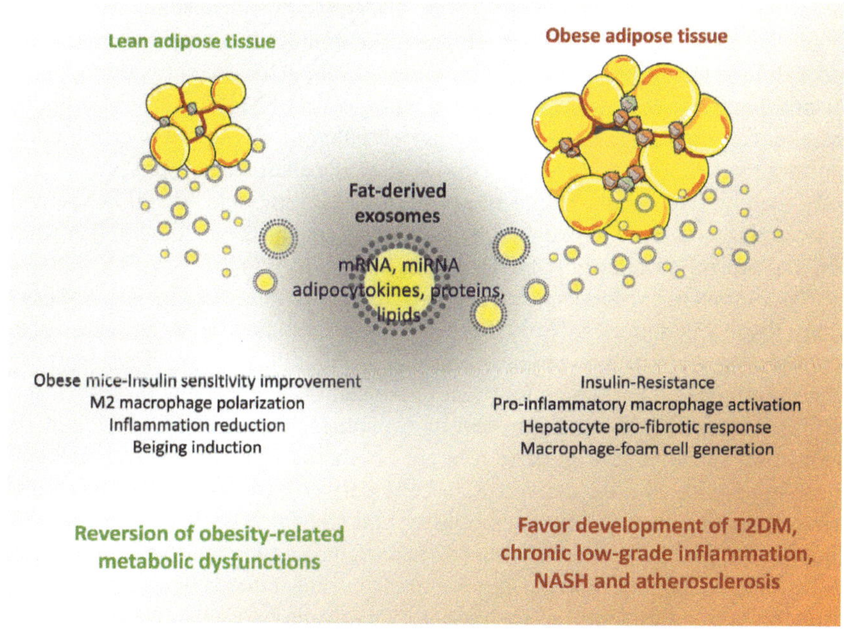

FIG. 2 Metabolic effects of adipose-tissue derived exosomes. Whereas exosomes from lean adipose tissue possess protective effects including the reversion of obesity-related metabolic dysfunctions, exosomes from obese adipose tissue favor development of T2DM, chronic low-grade inflammation, NASH and atherosclerosis.

illustrated in adipose tissue where caveolin-1 protein has been shown to use these exosomes to traffic between endothelial cells and adipocytes thereby replacing protein caveolin-1 levels in adipocytes despite gene disruption [40].

This exosome-based cell material exchange does not restrict to proteins since adipose tissue-derived exosomal miRNAs have been shown to regulate gene expression in liver and muscle to modulate in vivo insulin sensitivity by conveying circulating miRNAs [41, 42]. Previous reports have in fact identified the ability of obese fat-derived exosomes to induce IR state and macrophage activation [43, 44]. This pro-inflammatory response is likely related to the activation of the TLR4 signaling pathway, since mice invalidated for this receptor are unaffected by deleterious effects of obese fat-derived exosomes [43]. Different cargos of fat-derived exosomes have been shown to be involved in this immune response including inflammatory adipocytokines [19, 45] or specific miRNA such as miR-155 which might act by targeting the pivotal regulator of cytokine signaling pathways SOCS1 (suppressor of cytokine signaling 1) [42, 44] or miR-99b which in turn inhibits Fibroblast Growth Factor 21 (FGF21) in the liver [41].

Interestingly, exosomes from visceral adipose tissue, but not from subcutaneous adipose tissue, induced macrophage foam cell generation through the down-regulation of ATP-binding cassette transporter (ABCA1 and ABCG1)-mediated cholesterol efflux [46]. Similarly, only visceral adipose tissue-derived exosomes from HFD-fed mice increased pro-inflammatory macrophage phenotype and cytokine secretion. These effects were accompanied by activation of NF-κB-p65 pathway. Remarkably, intravenous injection of HFD-visceral adipose tissue-derived exosomes intensely enhanced atherosclerosis in hyperlipidemic apolipoprotein E-deficient mice, without affecting the plasma lipid profile and body weight. This study suggests a proatherosclerotic role for HFD-visceral adipose tissue-derived exosomes, by regulating macrophage foam cell formation and pro-inflammatory polarization.

By contrast to exosomes produced in obese conditions, lean fat-derived exosomes display beneficial effects. Hence, exosomes obtained from adipose tissue macrophages from lean mice improved glucose tolerance and insulin sensitivity when administered to obese mice [42]. These authors suggest that the effects are mediated by miR-155 overexpressed by exosomes since miR-155 KO animals are insulin sensitive and glucose tolerant compared to controls. Treatment of obese mice with adipose-derived stem cells-derived exosomes also lead to improved insulin sensitivity, reduced obesity, and alleviated hepatic steatosis [47]. These exosomes are able to activate M2 macrophage polarization, inflammation reduction, and beiging in white adipose tissue of diet-induced obese mice, likely through the transactivation of arginase-1 by exosome-carried active STAT3.

Altogether, these results illustrate the complexity of fat-derived exosomes, which gather proteins, genetic material and lipids reflecting the metabolic state of producing cells. Hence, it is likely that a combination of factors carried by these fat-derived exosomes will be involved in the crosstalk between stroma-vascular fraction (including endothelial cells, macrophages, preadipocytes, etc.) and adipocytes to induce metabolic actions on tissue/cell targets.

4.2 Exosomes as promoters of endothelial dysfunction

As reviewed by Zamani et al. [48], cardiac, endothelial and vascular cells are able to generate exosomes harboring a variety of molecules including miRNAs which may be transferred to recipient cells and modulate their functions. Other types of cells such as adipocytes and hepatocytes can release exosomes and act on cardiovascular cells. Transfer of information between cells through exosomes has been evidenced even in basal (unstimulated) conditions. For instance, exosomes from non-stimulated human aortic smooth muscle cells carrying miRNA-221/222, which is involved in autophagy process regulation, induces downregulation of LC3 II, ATG5, and Beclin-1 expression in human endothelial cells, reflecting an inhibition of endothelial autophagy [49]. By contrast, other authors have shown that endothelial exosomes enriched with miR-143 and miR-145 modulate gene expression in smooth muscle cells and induce an atheroprotective effect via reduction of atherosclerotic lesion formation in an atherosclerotic mouse model [50]. Taken together, these data illustrate the complexity of the dialog between vascular smooth muscle cells and endothelial cells.

Recently, a direct effect of exosomes from the serum of diabetic *db/db* mice on endothelial function has been described. In fact, internalized exosomes into endothelial cells of intact aortas profoundly impair endothelium-dependent relaxations of aortas of nondiabetic *db/+* mice. Interestingly, these exosomes carry arginase 1, an enzyme that reduces the availability of L-arginine, which is the substrate for eNOS-mediated NO production in endothelium resulting to a reduced NO bioavailability. When arginase 1 was silenced, the vaso-harmful effects of *db/db*-derived exosomes was markedly diminished showing the cellular mechanism by which diabetic mouse exosomes induce endothelial dysfunction [47]. Moreover, during diabetes, cardiomyocyte-derived exosomes inhibit endothelial cell proliferation, migration and tube-like formation resulting in an anti-angiogenic effect in a model of T2DM in rats. These effects are mediated by miR-320-enriched exosomes [51].

4.3 Role of exosomes in blood pressure regulation

Circulating exosomes from spontaneously hypertensive rats (SHR) can regulate blood pressure in normotensive rats. Indeed, SHR-derived exosomes significantly increased systolic blood pressure and structural changes of thoracic aorta in Wistar-Kyoto (WKY) rats, while WKY-derived exosomes decreased these parameters in SHR. Moreover, wet weight and perivascular fibrosis of left ventricles in WKY were significantly increased by SHR-derived exosomes, while the fibrosis but not ventricular weight was significantly decreased by WKY-derived exosomes in SHR. This work underlines the crucial role of circulating exosomes in the regulation of blood pressure [52].

Also, exosomes from adventitial fibroblasts isolated from the aorta of SHR were compared with those generated from WKY rats. Interestingly, exosomes from SHR, but not from WKY rats, promoted vascular smooth muscle cell migration. In addition, exosomes from SHR increased angiotensin II and angiotensin-converting enzyme (ACE) contents and ACE activity in vascular smooth muscle cells of both rat strains (WKY and SHR). These effects were prevented by a neutral sphingomyelinase inhibitor GW4869 able to block exosome production, an angiotensin II type 1 receptor antagonist, or an inhibitor of angiotensin-converting enzyme. Altogether, these results indicate the transfer of ACE via the exosomes from adventitial fibroblasts of SHR to vascular smooth muscle cells, resulting in an increase of their migration [53].

The reported differences between the effects of exosomes of SHR or WKY rats might be related, at least in part, with their miRNA content. Accordingly, the analysis of the miRNA expression profile of plasma exosomes in SHR and WKY rats has revealed that percentages of miRNA in the total small RNA isolated from SHRs and WKYs were not significantly different; however, 27 miRNAs were significantly differentially expressed between SHR and WKY exosomes, including 23 up-regulated and 4 down-regulated in SHR exosomes as compared to WKY exosomes [54].

Circulating exosomes from angiotensin II-evoked hypertensive rats evoke similar results than in vitro exosomes generated from angiotensin II-treated macrophages, that is, ICAM1 over-expression in human coronary artery endothelial cells. These effects are linked to a reduced level of miR-17, a negative regulator of ICAM1 expression, in exosomes from hypertensive rats. These data suggest that exosomes from macrophages transfer pro-inflammatory messages to endothelial cells [26].

4.4 Exosomes as vectors promoting lipid disorders

Exosomes secretion from adipocytes, hepatic and muscular cells can be enhanced by saturated fatty acids [22, 24, 55]. This fatty acids class is known to be particularly deleterious, since they induce IR though enhancement of ceramide and diacylglycerol synthesis. Exosomes secreted following palmitic acid exposure are indeed enriched in ceramides and able to transfer these lipotoxic lipids to muscular cells or macrophages [24, 55]. Lipid-enriched exosomes also display fibrotic and inflammatory properties as illustrated by the ability of obese adipocyte-derived exosomes to induce a pro-fibrotic response in hepatocytes [56]. Altogether, these data underline the potential role that exosomes might play in the development of non-alcoholic steatohepatitis and their use for prognostic/diagnostic purposes in order to assess the progression of this disease [55].

Since exosomes are enriched in cholesterol and other lipids, some authors have evaluated whether they can be involved in atherogenic processes. Exosomes generated from activated human $CD4^+$ T lymphocytes induced cholesterol accumulation into THP-1 monocytes subsequent to exosome internalization *via* the phosphatidylserine receptor [57]. Interestingly, since lipid droplets found in exosome-treated monocytes contained free cholesterol and cholesterol esters, and that this can occur within an atheromatous plaque, these data suggest that exosomes might be considered as atherogenic factors under cholesterol excess conditions [57].

5 Exosomes as a new mode of communication between hosts and microbiota

This last decade has highlighted the importance of the gut microbiota in the development of metabolic complications. Dysbiosis of microbiota is associated with obesity and the severity of liver diseases. EVs might represent a new mode of communication since they will relay information between the host and its microbiota acting as metabolic relays. Thus, intestinal epithelial cells produce exosomes enriched in some small RNAs that will be internalized by bacteria and will consequently modulate growth and microbial gene expression pattern [57a]. By contrast, exosomes isolated from the gut microbiota of obese mice induce IR on in vitro and in vivo models [58]. EVs therefore appear as a new player in the complex communication linking the host to its microbiota.

6. Conclusion

Altogether, these data suggest that exosomes could represent predictive biomarkers of cardiometabolic diseases and participate to the development of metabolic dysfunctions through their ability to mediate intercellular communication. A better understanding of the composition of exosomes and their actions could lead to the design of innovative strategies to modulate their biogenesis and actions. These can introduce changes in medical practice, moving from traditional care approaches to personalized medicine based on exosome better knowledge regarding their compositions and biological activities.

Acknowledgments

This work was partially supported by INSERM, Université d'Angers (France), the Société Francophone du Diabète and Région Pays de Loire (Mibiogate).

References

[1] collaborators NCDC. NCD Countdown 2030: worldwide trends in non-communicable disease mortality and progress towards sustainable development goal target 3.4. Lancet 2018;392(10152):1072–88.

[2] Nolan PB, et al. Prevalence of metabolic syndrome and metabolic syndrome components in young adults: a pooled analysis. Prev Med Rep 2017;7:211–5.

[3] Alberti KG, et al. Harmonizing the metabolic syndrome: a joint interim statement of the International Diabetes Federation Task Force on Epidemiology and Prevention; National Heart, Lung, and Blood Institute; American Heart Association; World Heart Federation; International Atherosclerosis Society; and International Association for the Study of Obesity. Circulation 2009;120(16):1640–5.

[4] Brede S, et al. Clinical scenario of the metabolic syndrome. Visc Med 2016;32(5):336–41.

[5] Han TS, Wu FC, Lean ME. Obesity and weight management in the elderly: a focus on men. Best Pract Res Clin Endocrinol Metab 2013;27(4):509–25.

[6] Weisberg SP, et al. Obesity is associated with macrophage accumulation in adipose tissue. J Clin Invest 2003;112(12):1796–808.

[7] Le Lay S, et al. Oxidative stress and metabolic pathologies: from an adipocentric point of view. Oxid Med Cell Longev 2014;2014:908539.

[8] Cancello R, et al. Increased infiltration of macrophages in omental adipose tissue is associated with marked hepatic lesions in morbid human obesity. Diabetes 2006;55(6):1554–61.

[9] Grundy SM. Adipose tissue and metabolic syndrome: too much, too little or neither. Eur J Clin Invest 2015;45(11):1209–17.

[10] Kastorini CM, Panagiotakos DB. Dietary patterns and prevention of type 2 diabetes: from research to clinical practice; a systematic review. Curr Diabetes Rev 2009;5(4):221–7.

[11] Gan LT, et al. Hepatocyte free cholesterol lipotoxicity results from JNK1-mediated mitochondrial injury and is HMGB1 and TLR4-dependent. J Hepatol 2014;61(6):1376–84.

[12] Marengo A, Rosso C, Bugianesi E. Liver cancer: connections with obesity, fatty liver, and cirrhosis. Annu Rev Med 2016;67:103–17.

[13] Expert Dyslipidemia Panel of the International Atherosclerosis Society Panel Memebers. An International Atherosclerosis Society Position Paper: global recommendations for the management of dyslipidemia—full report. J Clin Lipidol 2014;8(1):29–60.

[14] Bomback AS, Klemmer PJ. Interaction of aldosterone and extracellular volume in the pathogenesis of obesity-associated kidney disease: a narrative review. Am J Nephrol 2009;30(2):140–6.

[15] Lambert EA, et al. Should the sympathetic nervous system be a target to improve cardiometabolic risk in obesity? Am J Physiol Heart Circ Physiol 2015;309(2):H244–58.

[16] Ribeiro MJ, et al. Carotid body denervation prevents the development of insulin resistance and hypertension induced by hypercaloric diets. Diabetes 2013;62(8):2905–16.

[17] Lopes HF, et al. Visceral adiposity syndrome. Diabetol Metab Syndr 2016;8:40.

[18] Stepanian A, et al. Microparticle increase in severe obesity: Not related to metabolic syndrome and unchanged after massive weight loss. Obesity (Silver Spring) 2013;21(11):2236–43.

[19] Amosse J, et al. Phenotyping of circulating extracellular vesicles (EVs) in obesity identifies large EVs as functional conveyors of macrophage migration inhibitory factor. Mol Metab 2018;18:134–42.

[20] Kranendonk ME, et al. Human adipocyte extracellular vesicles in reciprocal signaling between adipocytes and macrophages. Obesity (Silver Spring) 2014;22(5):1296–308.

[21] Freeman DW, et al. Altered extracellular vesicle concentration, cargo and function in diabetes mellitus. Diabetes 2018;67(11):2377–88.

[22] Durcin M, et al. Characterisation of adipocyte-derived extracellular vesicle subtypes identifies distinct protein and lipid signatures for large and small extracellular vesicles. J Extracell Vesicles 2017;6(1), 1305677.

[23] Lee YS, et al. Exosomes derived from palmitic acid-treated hepatocytes induce fibrotic activation of hepatic stellate cells. Sci Rep 2017;7(1):3710.

[24] Aswad H, et al. Exosomes participate in the alteration of muscle homeostasis during lipid-induced insulin resistance in mice. Diabetologia 2014;57(10):2155–64.

[25] Kobayashi Y, et al. Circulating extracellular vesicles are associated with lipid and insulin metabolism. Am J Physiol Endocrinol Metab 2018;315(4):E574–82.

[26] Osada-Oka M, et al. Macrophage-derived exosomes induce inflammatory factors in endothelial cells under hypertensive conditions. Hypertens Res 2017;40(4):353–60.

[27] Qi Y, et al. Activation of the endogenous renin-angiotensin-aldosterone system or aldosterone administration increases urinary exosomal sodium channel excretion. J Am Soc Nephrol 2016;27(2):646–56.

[28] van der Lubbe N, et al. The phosphorylated sodium chloride cotransporter in urinary exosomes is superior to prostasin as a marker for aldosteronism. Hypertension 2012;60(3):741–8.

[29] Li S, et al. Cell-derived microparticles in Patients with type 2 diabetes mellitus: a systematic review and meta-analysis. Cell Physiol Biochem 2016;39(6):2439–50.

[30] Lazar I, et al. Adipocyte exosomes promote melanoma aggressiveness through fatty acid oxidation: a novel mechanism linking obesity and cancer. Cancer Res 2016;76(14):4051–7.

[31] Connolly KD, et al. Evidence for adipocyte-derived extracellular vesicles in the human circulation. Endocrinology 2018;159(9):3259–67.

[32] Lopez Andres N, et al. Increased microparticle production and impaired microvascular endothelial function in aldosterone-salt-treated rats: protective effects of polyphenols. PLoS One 2012;7(7), e39235.

[33] Boulanger CM, et al. Extracellular vesicles in coronary artery disease. Nat Rev Cardiol 2017;14(5):259–72.

[34] Agouni A, et al. Endothelial dysfunction caused by circulating microparticles from patients with metabolic syndrome. Am J Pathol 2008;173(4):1210–9.

[35] Helal O, et al. Increased levels of microparticles originating from endothelial cells, platelets and erythrocytes in subjects with metabolic syndrome: relationship with oxidative stress. Nutr Metab Cardiovasc Dis 2011;21(9):665–71.
[36] Diamant M, et al. Elevated numbers of tissue-factor exposing microparticles correlate with components of the metabolic syndrome in uncomplicated type 2 diabetes mellitus. Circulation 2002;106(19):2442–7.
[37] Kranendonk ME, et al. Extracellular vesicle markers in relation to obesity and metabolic complications in patients with manifest cardiovascular disease. Cardiovasc Diabetol 2014;13:37.
[38] Li F, et al. Exosomal microRNA-29a mediates cardiac dysfunction and mitochondrial inactivity in obesity-related cardiomyopathy. Endocrine 2019;63(3):480–8.
[39] Nie H, Pan Y, Zhou Y. Exosomal microRNA-194 causes cardiac injury and mitochondrial dysfunction in obese mice. Biochem Biophys Res Commun 2018;503(4):3174–9.
[40] Crewe C, et al. An endothelial-to-adipocyte extracellular vesicle axis governed by metabolic state. Cell 2018;175(3):695–708, e13.
[41] Thomou T, et al. Adipose-derived circulating miRNAs regulate gene expression in other tissues. Nature 2017;542(7642):450–5.
[42] Ying W, et al. Adipose tissue macrophage-derived exosomal miRNAs can modulate in vivo and in vitro insulin sensitivity. Cell 2017;171(2):372–84, e12.
[43] Deng ZB, et al. Adipose tissue exosome-like vesicles mediate activation of macrophage-induced insulin resistance. Diabetes 2009;58(11):2498–505.
[44] Zhang Y, et al. Adipocyte-derived microvesicles from obese mice induce M1 macrophage phenotype through secreted miR-155. J Mol Cell Biol 2016;8(6):505–17.
[45] Kranendonk ME, et al. Effect of extracellular vesicles of human adipose tissue on insulin signaling in liver and muscle cells. Obesity (Silver Spring) 2014;22(10):2216–23.
[46] Xie Z, et al. Adipose-derived exosomes exert proatherogenic effects by regulating macrophage foam cell formation and polarization. J Am Heart Assoc 2018;7(5).
[47] Zhang H, et al. Serum exosomes mediate delivery of arginase 1 as a novel mechanism for endothelial dysfunction in diabetes. Proc Natl Acad Sci U S A 2018;115(29):E6927–36.
[48] Zamani P, et al. The therapeutic and diagnostic role of exosomes in cardiovascular diseases. Trends Cardiovasc Med 2018;29(6):313–23.
[49] Li L, et al. Human aortic smooth muscle cell-derived exosomal miR-221/222 inhibits autophagy via a PTEN/Akt signaling pathway in human umbilical vein endothelial cells. Biochem Biophys Res Commun 2016;479(2):343–50.
[50] Hergenreider E, et al. Atheroprotective communication between endothelial cells and smooth muscle cells through miRNAs. Nat Cell Biol 2012;14(3):249–56.
[51] Wang X, et al. Cardiomyocytes mediate anti-angiogenesis in type 2 diabetic rats through the exosomal transfer of miR-320 into endothelial cells. J Mol Cell Cardiol 2014;74:139–50.
[52] Otani K, et al. Plasma exosomes regulate systemic blood pressure in rats. Biochem Biophys Res Commun 2018;503(2):776–83.
[53] Tong Y, et al. Exosome-mediated transfer of ACE (angiotensin-converting enzyme) from adventitial fibroblasts of spontaneously hypertensive rats promotes vascular smooth muscle cell migration. Hypertension 2018;72(4):881–8.
[54] Liu X, et al. miRNA profiling of exosomes from spontaneous hypertensive rats using next-generation sequencing. J Cardiovasc Transl Res 2019;12(1):75–83.
[55] Hirsova P, et al. Lipid-induced signaling causes release of inflammatory extracellular vesicles from hepatocytes. Gastroenterology 2016;150(4):956–67.

[56] Koeck ES, et al. Adipocyte exosomes induce transforming growth factor beta pathway dysregulation in hepatocytes: a novel paradigm for obesity-related liver disease. J Surg Res 2014;192(2):268–75.

[57] Zakharova L, Svetlova M, Fomina AF. T cell exosomes induce cholesterol accumulation in human monocytes via phosphatidylserine receptor. J Cell Physiol 2007;212(1):174–81.

[57a] Liu S, et al. The host shapes the gut microbiota via fecal microRNA. Cell Host Microbe 2016;19(1):32–43.

[58] Choi Y, et al. Gut microbe-derived extracellular vesicles induce insulin resistance, thereby impairing glucose metabolism in skeletal muscle. Sci Rep 2015;5, 15878.

Chapter 16

Potential role of exosomes in reproductive medicine and pregnancy

Soumyalekshmi Nair[a], Carlos Salomon[a,b,c]
[a]*Exosome Biology Laboratory, Centre for Clinical Diagnostics, UQ centre for Clinical Research, Royal Brisbane and Women's Hospital, The University of Queensland, St Lucia, QLD, Australia,* [b]*Department of Clinical Biochemistry and Immunology, University of Concepción, Concepción, Chile,* [c]*Department of Obstetrics and Gynecology, Ochsner Baptist Hospital, New Orleans, LA, United States*

1 Introduction

Pregnancy is a unique physiological state characterized by several maternal adaptations for accommodating and maintaining the growing embryo or fetus. The purview of maternal physiological adaptations in pregnancy, involves changes in endocrine, metabolic, cardiovascular, hematologic and immunological systems, to sustain the proper growth and development of the fetus. Most importantly, the temporal and spatial co-ordination of the physiological events is quintessential for successful pregnancy outcomes and molecules and factors released by various organs play key roles in mediating this phenomenon. Placenta, the principal organ in pregnancy, releases various hormones, growth factors, cytokines, miRNAs and proteins which are crucial for maternal and fetal wellbeing. In recent years, our understanding about intercellular communication has been changed, by growing research in the field of extracellular vesicles (EVs), particularly the nanosized vesicles called 'exosomes'. Interestingly, placenta secretes large quantities of exosomes into maternal circulation during normal and pregnancy complications such as preeclampsia, gestational diabetes mellitus, fetal growth restriction and preterm birth, and play important roles in several different aspects of pregnancy including feto-maternal signaling.

Exosomes. https://doi.org/10.1016/B978-0-12-816053-4.00016-X

2 Extracellular vesicles: Their diversity, biogenesis and secretion

Over the past few decades, there has been immense interest in the field of EVs, as they form a novel and efficient method by which cells communicate [1]. In the beginning EVs were thought to be cellular 'debris', but later it was observed that they can interact with specific target cells and modulate their bioactivity [2]. Typically, EVs are vesicles formed of lipid bilayer, released by cells into the extracellular environment and varies in size from 100 to 1000 nm [3]. Notably, they encapsulate a wide variety of messenger molecules including proteins, lipids, RNAs and DNA. EVs are released from almost every cell type, can cross the physiological barriers and enriched in a variety of body fluids including blood, urine, saliva, breast milk and cerebrospinal fluid [4]. EVs, comprise a heterogeneous vesicle population of different origin and varying morphology. Specifically, there are three subclasses, distinguished by their size and biogenesis, namely (1) exosomes (50–150 nm) (2) microvesicles (0.2–1 μm) and (3) apoptotic bodies (large EVs, above 1 μm) [5]. Exosomes are nanovesicles formed from the endocytic pathway. Exosome biogenesis begins with the inward invagination of the plasma membrane leading to formation of structures called multivesicular bodies (MVBs), from which subsequent invagination leads to formation of intraluminal vesicles (ILVs). Interestingly, selective packaging mechanisms that packages molecular cargo including proteins and miRNAs into the ILVs has been reported [6, 7]. The ILVs when released into the extracellular environment via exocytosis are called 'exosomes'. Hence exosomes are enriched in proteins present in the endosomal pathway such as CD63, CD9, CD81, Tumor Susceptibility Gene 101 (TSG101) and Apoptosis-linked gene 2-interacting protein X (ALIX). In addition to these protein markers, the typical cup-shaped morphology and a buoyant density in a sucrose gradient ranging from 1.13–1.19 g/mL distinguish exosomes from other EVs [1, 8, 9]. Other EV populations such as microvesicles originate from the plasma membrane through direct budding or shedding of the membrane and apoptotic bodies are released after apoptosis of cells [10, 11]. Table 1 represents the biogenesis, characteristics and functions of different sub-populations of EVs.

2.1 Exosomes: Intercellular interactions and functions

Exosomes carry a wide repertoire of molecules which includes proteins (both cytoplasmic and membrane-bound), mRNAs, non-coding RNAs such as miRNAs and long non-noncoding RNAs or lncRNAs, fragments of DNA and lipids and this molecular profile forms a representation of their cells of origin. In this way, they carry the genetic information or potent biological signals, from their cells of origin to a neighboring or distally located recipient cell, mediating intercellular communication. Moreover, exosomes are enriched in various biological fluids [28, 29] owing to their ability to cross physiological barriers such as the endothelial [30] and blood-brain barrier [31] and endowing them with the

TABLE 1 Biogenesis, characteristics and functions of different sub-populations of EVs

Types of EV	Mechanism of biogenesis	Size	Proteins present[a]	Cargo	Method of isolation[b]	Reported functions
Exosomes	Correspond to the ILVs formed inside the late endosomes. Released by fusion of MVBs with the plasma membrane. Proteins belonging to the ESCRT and ESCRT-independent mechanisms (lipids and tetraspanins), Rab GTPases and SNARE are involved in the biogenesis [12]	Present a diameter of 50–150 nm, similar to the diameter of ILVs on electron microscopy [13]	Tetraspanins (CD81, CD63, CD9), ESCRT components, Alix, TSG101, syntenins, flotillins, integrins, disintegrin and metalloproteinase domain-containing protein 10/ADAM10/ADA10, HSC70/HSPA8/HSPC7 [13–15]	Proteins, miRNAs, mRNAs, lncRNAs, small fragments of DNA Devoid of nuclear, mitochondrial, endoplasmic reticulum and golgi complex proteins [13, 14]	Differential centrifugation and ultracentrifugation at 100,000 g for 90 m, flotation in density gradient (sucrose or iodixanol), ultrafiltration, size exclusion chromatography, immunoisolation using antibodies and commercial kits (using precipitation) [16, 17]	Cell to cell communication by transfer of molecular cargo to recipient cells [18, 19] Immunoregulation [20, 21] Tumor progression and metastasis [22, 23]
Microvesicles	Direct budding from the plasma membrane by various mechanisms such as release of early endosomal EVs, rise in intracellular Ca^{+} and remodeling of cytoskeleton, activation of phospholipase D2 etc. [1]	Present a larger diameter up to few micrometers or as small as exosomes, which is <150 nm [13]	CD63, CD9, HSC70/HSPA8/HSPC7, flotillin-1, KIF23, RACGAP, CSE1L, ARF6, EMMPRIN [13–15]	Proteins, miRNAs, mRNAs, small fragments of DNA [14]	Differential centrifugation. Centrifugation at 10000 g for 40 m yields microvesicles [15]	Involved in cellular communication, inflammation and immunity, tumor progression and invasiveness [24, 25]

Continued

TABLE 1 Biogenesis, characteristics and functions of different sub-populations of EVs—cont'd

Types of EV	Mechanism of biogenesis	Size	Proteins present[a]	Cargo	Method of isolation[b]	Reported functions
Apoptotic bodies	Formed from dying cells by apoptosis. Apoptotic cells undergo morphological changes such as membrane blebbing, thin membrane protrusion leading to formation microtubule spikes or apoptopodia which are released as apoptotic bodies [26]	>1 μm	CD63, CD9, HSC70/HSPA8/HSPC7, alpha-actinin-4/ACTN4/ACTN4, GP96 = endoplasmin/HSP90B1/ENPL, Phosphatidylserine, histones, calnexin, cytochrome *c*, CX3CL1 and ICAM-3, ICAM-3, phosphatidylserine, sialylated and glycosylated ligands, MHC II [13–15, 26]	Nuclear proteins, cellular organelles [14]	Differential centrifugation. Generally centrifugation at 2000 g for 20 m yields apoptotic bodies [15]. Also, centrifugation at 400 g to remove cells, and filtration through 1.2 μm filter yield apoptotic bodies [27]	Transport proinflammatory cytokines and damage-associated molecular patterns and promote inflammation Involved in immune cell responses and immunoregulation against bacterial and viral infection [26]

[a] *Proteins present may vary depending upon the cellular origin of the EV and on the proteomic characterization method used.*
[b] *There are variations in the EV isolation method depending on the starting material such as biological fluid or cell culture supernatant and the goal of experiments.*

prospects of ideal therapeutic tools [32]. Of note, presence of specific cell surface molecules on the membrane of exosomes, such as differential expression of exosomal integrins, facilitate their targeting to recipient cells [33, 34].

Exosomes can be taken up by target cells in different ways. In some cases, the classical receptor-ligand interaction between membrane proteins of the exosomes and recipient cells, can activate the downstream signaling that can alter the recipient cell function [35, 36]. In other cases, the internalization of exosomes can be mediated by different mechanisms such as phagocytosis, clathrin-mediated endocytosis, caveolin-dependent endocytosis, micropinocytosis and lipid-raft mediated uptake [37]. Additionally, the direct fusion of exosome membrane with the cell membrane, mediating transfer of their contents within the cell has been reported [38, 39]. However, the mechanisms by which exosomal cargo such as proteins and miRNAs can influence or change the behaviour of the recipient cells requires further studies. Interestingly, the proteins in exosomes are reported to activate or deactivate signaling pathways by altering the phosphorylation events and there by transferring genetic information between cells [18]. Regarding miRNAs in exosomes, they can alter the expression of specific target proteins and change the recipient cell phenotype [40, 41]. Fig. 1 is a schematic representation of extracellular vesicle biogenesis, structure of exosome and their interaction with the target cell.

3 The origin and functions of exosomes in pregnancy

Exosomes has been described to play a crucial role in pregnancy by modulating several processes including maternal immune response and metabolic adaptations [42, 43]. It has been identified that the total circulating exosomes in pregnancy increases across gestation [44] and increases in complications of pregnancy such as Gestational Diabetes Mellitus (GDM) and pre-eclampsia (PE) [45, 46]. Until now, the precise origin, cargo and functions of these exosomes in maternal circulation is not fully understood and warrants further investigation. However, the principal organ in pregnancy, the placenta, secretes exosomes into maternal circulation and placental exosomes have been detected as early as 6 weeks of pregnancy [44]. Remarkably, exosomes derived from placenta form an important link by which the placenta communicates with other maternal tissues in normal and complicated pregnancies.

3.1 Exosomes from placenta

Placenta is a unique feto-maternal organ that secretes hormones and molecules essential for the maintenance of pregnancy and acts as a critical connection between the mother and fetus, by trafficking nutrients, oxygen and metabolic wastes. The basic functional units of placenta, the placental villi are covered by three main types of trophoblast cells namely (1) villous cytotrophoblast (2) extravillous cytotrophoblast (EVT) and (3) syncytiotrophoblast (STB)

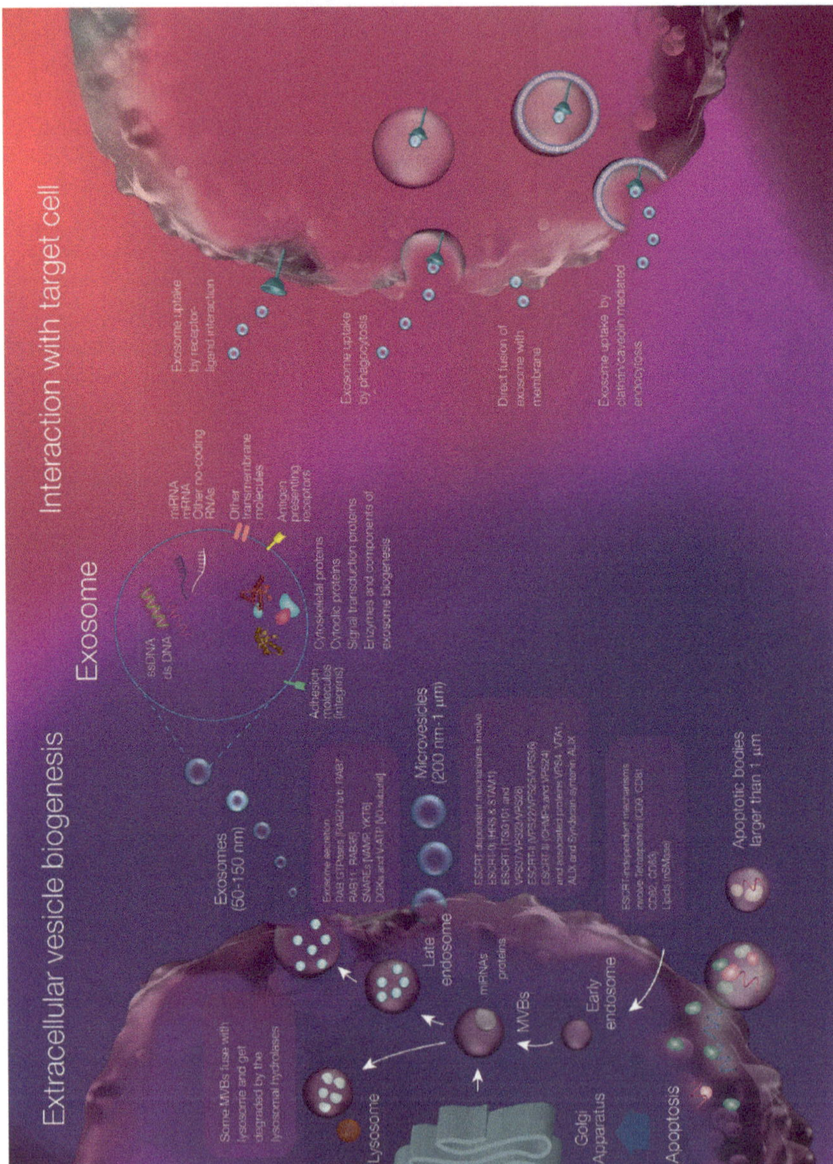

FIG. 1 See figure legend on opposite page.

(which is formed by the fusion of villous cytotrophoblast) [47] and the functions of each set of cells vary. Placenta releases several hormones such as the placental growth hormone, placental lactogen, estrogens and progesterone, and inflammatory mediators such as TNFα and cortisol, which integrate or co-ordinate the maternal physiological changes to the demands of the growing fetus [48, 49]. In addition to hormones and cytokines, placenta releases a rich repertoire of EVs into the maternal circulation throughout gestation [50]. The EVs originate mainly from the STB layer of placenta and four major sub-populations of placental EVs have been identified. They are (1) exosomes (30–100 nm) (2) microvesicles (MVs) (100 nm- 1 μm, also named as STB micro particles) (3) syncytial nuclear aggregates (20–500 μm, cells expel aging nucleus in these vesicles) and (4) apoptotic bodies (1–4 μm) [51]. Exosomes from other sources such as follicular fluid, endometrium, embryo and trophoblast cells may influence the female fertility, implantation and early stages of pregnancy, whereas the placenta STB-derived exosomes are the key players in advanced stages of pregnancy [51, 52]. Also, EVT cell-derived exosomes, has been identified both in maternal circulation and from first trimester placental explants and EVT like cell line Swan71 cells, by the presence of HLA-G antigen [20, 53, 54] (HLA-G is expressed on EVT and not on STB).

Notably, the mechanism of biogenesis, biological functions and effect on maternal physiology varies between each sub-type of placenta-derived EVs. Interestingly, exosomes are unique in their biogenesis as they originate from the endosomal pathway and several mechanisms exist that selectively package the molecular cargo, including proteins and miRNAs into exosomes [6]. Additionally, a plethora of literature demonstrates its potential roles in intercellular communication [1, 55]. The circulating exosome concentration in pregnancy is higher when compared to non-pregnant state and the exosome

---

FIG. 1 Schematic representation of extracellular vesicle biogenesis, the structure of exosome and their interaction with the target cell. There are different subpopulations of extracellular vesicles (1) Exosomes (50–150 nm) originating in the endosomal pathway, (2) Microvesicles (200 nm–1 μm) formed by direct budding of plasma membrane, and (3) Apoptotic bodies (larger than 1 μm) formed by apoptosis in dying cells. Exosome biogenesis begins with the inward invagination of the plasma membrane forming early endosomes. Early endosomes when mature to late endosomes acquire intraluminal vesicles (ILVs) formed by invagination of early endosomal membrane and encapsulate proteins, lipids and nucleic acids which are specifically sorted for packaging. The late endosomes with ILVs are called multivesicular bodies (MVBs), which can either fuse with lysosomes and get degraded or fuse with the plasma membrane and released into the extracellular environment. The important proteins of ESCRT-dependent and ESCRT-independent mechanisms of exosome biogenesis and proteins involved in the exosome secretion has been given. Exosomes present transmembrane receptors and adhesion molecules on their surface and encapsulate variety of proteins including cytosolic, cytoskeletal, signal transduction proteins, proteins of exosome biogenesis machinery and nucleic acids including single and double stranded DNA, mRNAs and non-coding RNAs. At the target cell, exosomes mediate downstream signaling via receptor-ligand interaction or taken up by phagocytosis or clathrin or caveolin mediated endocytosis or release their contents by direct fusion with the cell membrane.

concentration increases in a progressive manner, across gestation, leading to maximum concentration in the 3rd trimester [44, 56].

Microenvironment such as oxygen tension and glucose concentration in the placenta can regulate the biogenesis and release of placenta-derived exosomes. Low oxygen tension or hypoxia increases the release of placenta-derived exosomes and alters the content as well as the bioactivity on cell targets [57]. Similarly, a higher concentration of glucose can increase the release of exosomes from placental cells [58]. The exact mechanisms leading to this is currently not well understood. However, it has been identified that in the presence of hypoxia the Hypoxia Inducible Factor (HIF) can regulate the process of exosome biogenesis [59, 60] as well as in the presence of glucose, activation of intracellular Ca^{2+} channels may play a role in modulating the exosomal pathways [61, 62].

3.2 Detection and isolation of placental exosomes

It has been reported that placental exosomes attributes around 10–25% of the total circulating exosomes during pregnancy [63]. It is important, how the placental exosomes are identified from exosomes originating from other cell types and specifically isolated from the total exosome population. Placental exosomes can be identified specifically by the presence of a protein, Placental Alkaline Phosphatase (PLAP), which forms an integral membrane protein, mainly on the surface of STB [64, 65]. In addition, PLAP expression has been identified in a majority of trophoblast cells in the placenta [66] including the primary cytotrophoblast cells [67] and ED27 trophoblast-like cells isolated from first trimester chorionic villi [68]. Various techniques has been employed for the detection and quantification of $PLAP^+$ exosomes from the total circulating exosomes. Dragovic et al., characterized the $PLAP^+$ exosomes in placental perfusate, using fluorescence NTA, by enriching the $PLAP^+$ exosomes using anti-PLAP antibody coated Dynabeads and labelling them with quantum dots (Qdot) conjugated to anti-PLAP antibody [69]. Similarly, the placental exosomes in maternal circulation were quantified using immunofluorescent NTA with Qdots coupled with $CD63^+$ and $PLAP^+$ antibodies [63, 70]. Additionally, the placental exosome profile in maternal circulation has been characterized by using commercial PLAP ELISA kits [44, 71]. Interestingly, Lai et al., described the isolation of $PLAP^+$ exosomes from maternal circulation using immunoaffinity capture by anti-PLAP coated beads [72]. Most recently, a method for direct isolation of $PLAP^+$ exosomes (without pre-isolation of exosome) using gold-loaded nanoporous ferric oxide nanocubes has been reported [73], opening up new avenues of exosome detection for clinical applications. However, isolation of specific placenta-derived exosomes from maternal circulation is challenging due to the levels of placental exosomes present in maternal circulation (around 15% and depending of the gestational age), relatively lower expression of PLAP on the surface of exosome and highly dependent on the specificity of the PLAP antibody used for procedure.

Several experimental designs have been used to study the role of placental exosomes in pregnancy, including isolation of placental exosomes from maternal plasma/serum, placental cell culture (primary and cell lines), placental perfusion, and placental explants (Table 2 is a representation of the experimental models and isolation methods used for the study of placental exosomes). Maternal plasma has been the most extensively used biological fluid to purify and profile placenta-derived EVs including exosomes. The most commonly used isolation method for obtaining exosomes from maternal plasma is density gradient centrifugation using a sucrose cushion or iodixanol [44, 45, 63]. Subsequently, the placenta derived exosomes were quantified from the total circulating exosomes using any of the previously described methods [44, 67, 70, 71]. Placental exosomes can be isolated from the cell-conditioned media of primary trophoblast cells, trophoblast cell lines such as BeWo, JEG-3 and HTR8 cells and placental explant cultures. The most common method used for the isolation of exosome from cell conditioned media is sequential centrifugation followed by ultracentrifugation or density gradient centrifugation [74, 88, 94]. Another attractive model for the isolation and analysis of placental exosomes is the placental perfusion system based on the dual perfusion of placental cotyledon. Enrichment of EVs and removal of contaminating RBCs in placental perfusate can be achieved by various EV isolation method including ultracentrifugation and filtration [69, 81, 95, 96]. The placental exosomes isolated from biological fluids, in vitro and ex vivo models were quantified and analyzed for their cargo and bioactivity on target cells.

3.3 Bio-distribution of placental exosomes and their effect on target cells

Exosomes released from placenta once enter the circulation can be transported to various target cells and play important roles in mediating the cross talk between placenta and other organs. Tong et al., demonstrated that exosomes derived from first trimester human placenta, when injected in mice, can target specific organs including lungs, liver and kidneys and interact with mesenteric endothelial cells leading to vasodilation via nitric oxide dependent pathway [92]. Furthermore, Miranda et al., identified the presence of human placental exosomes in fetal circulation and demonstrated its relation to fetal growth [70]. In addition, studies in agricultural animals, reported the presence of placental exosomes in umbilical cord blood in sheep and evaluated the transfer of miRNAs via exosomes [97]. This shows that exosomes may be involved in maternal-fetal trafficking, which is the bidirectional passage of maternal cells or molecules to fetal circulation and vice versa. Maternal-fetal trafficking is important for transferring maternal and fetal signals in normal and adverse pregnancy and provides an additional mechanism in protecting the fetus via compensatory mechanisms [98, 99]. However, the mechanism by which placental exosomes cross the specialized barrier of placenta is not clearly known. Concurrently, the establishment of

TABLE 2 Representation of experimental models and isolation methods used for the study of placental exosomes

Experimental model	Exosome isolation method	Key findings	Ref.
Primary human trophoblasts and JEG-3 cells	Ultracentrifugation and density gradient centrifugation	C19MC miRNAs are transferred via exosomes and regulate immunity to viral infections	[74]
BeWo and Jurkat cells	Ultracentrifugation	Exosome miRNA miR-517a-3p targets PRKG1 in natural killer cells and modulate its activity	[75]
Trophoblast cell culture and explant culture	Density gradient centrifugation	Immunomodulatory molecules are secreted from placenta via exosomes	[20]
Sw71 human trophoblast cells	Ultracentrifugation and ultrafiltration	Pro-inflammatory role of trophoblast derived exosomes	[53]
First-trimester primary trophoblast cells	Differential centrifugation and density gradient centrifugation	The exosomal release was significantly increased under high glucose concentration (25 mM) and hypoxia. The exosomes induced the secretion of cytokines from HUVECs compared to control	[76]
Sw71 human trophoblast cells	Ultracentrifugation Ultrafiltration and density gradient centrifugation	Morphologic and proteomic characterization of exosomes released by cultured EVT cells	[77]
Primary trophoblast cell culture	Ultracentrifugation and density gradient centrifugation	C19MC miRNAs are the predominant miRNA species expressed in exosomes released from PHT	[78]
Placenta-derived villous cytotrophoblasts	Differential centrifugation, ExoQuick, Ultrafiltration and Ultracentrifugation	The exosomes are enriched in Syncytin-1 and -2 and synctin-2 was reduced in exosomes in PE	[79]
Cytotrophoblast cells	Density gradient centrifugation	The secretion of exosomes is higher in hypoxia and exosomes from cytotrophoblast increase EVT migration	[57]
Plasma	Density gradient centrifugation	Placental exosomes are identified at 6 weeks of gestations and the concentration of exosome from placenta increased across gestation	[44]

Explant culture	Ultracentrifugation and density gradient centrifugation	Bioactive FasL- and TRAIL-carrying exosomes are secreted by the placenta and have immunomodulatory role in pregnancy	[80]
Dual ex vivo perfusion of placental cotyledon	Ultracentrifugation	Exosomes were characterized using NTA and fluorescence NTA	[81]
Plasma	Density gradient centrifugation	Placental exosome profile in maternal circulation was identified and exosomes can alter endothelial migration	[67]
BeWo cell line	Ultrafiltration and immunoisolation	Placenta-specific miRNAs are extracellularly released via exosomes	[82]
Explant culture, primary trophoblasts and Bewo cell line	Density gradient centrifugation Ultracentrifugation	The presence of syncytin-1 in placental exosomes and its immunosuppressive effects reported	[83]
Plasma and dual ex vivo perfusion of placental cotyledon	Ultracentrifugation	The difficulties and limitations of ex vivo flow cytometric analysis of plasma EVs was analyzed	[84]
Explant culture	Ultracentrifugation Density gradient centrifugation	Exosomes bearing NKG2D ligands are released by human placenta	[85]
Human first trimester EVT cell line and HTR8 cell line	ExoQuick	aPL-induced up-regulation of trophoblast miR-146a-3p is mediated by Toll-like receptor 4 (TLR4), and miR-146a-3p in turn drives the cells to secrete interleukin (IL)-8	[86]
Plasma	Density gradient centrifugation	The number of placental derived exosomes increases with increase in gestation age in normal and GDM pregnancies. The exosomes from GDM pregnancies significantly increased the release of proinflammatory cytokines from endothelial cells	[71]
Plasma	Ultracentrifugation	The differences in the contribution of placental-derived exosomes to total exosomes in maternal circulation and its role in PE was reported	[87]

Continued

TABLE 2 Representation of experimental models and isolation methods used for the study of placental exosomes—cont'd

Experimental model	Exosome isolation method	Key findings	Ref.
JEG-3 and HTR-8/SVneo cell lines	Density gradient centrifugation	The exosomes were characterized for their proteomic cargo and increased the migration of the vascular smooth muscle cells	[88]
Plasma and EVT cells	Density gradient centrifugation	Exosomes isolated from EVT cultured at 8% oxygen increased endothelial cell migration compared to exosomes cultured at 1% oxygen	[89]
Plasma	Size exclusion chroma-tography and ultracentrifugation Size exclusion chromatography Ultracentrifugation	Placental exosomes suppressed T-cell signaling components and identified the role of FasL and PD-L1 in immunoregulation	[90]
Dual ex vivo perfusion of placental cotyledon and plasma	Ultracentrifugation	Functional endothelial nitric oxide (eNOS)synthase is present in the STBEVs of plasma and placenta and lower levels of eNOS is reported in PE	[91]
Plasma	Density gradient centrifugation	Differentially expressed miRNAs (hsa-miR-486-1-5p and hsa-miR-486-2-5p) in exosomes in PE identified	[45]
Plasma	Density gradient centrifugation	Between ~12% and ~25% of circulating exosomes in maternal blood are of placental origin during gestation. Contribution of placental exosomes to the total exosomal population decreases with higher maternal BMI across gestation	[63]
Explant culture	Ultracentrifugation	Nanovesicles from first trimester human placentae localized to the lungs, liver and kidneys 24 h after injection into pregnant mice. Placental nanovesicles rapidly interacted with endothelial cells and mediated vaso-dialation	[92]
Plasma	Density gradient centrifugation	Identified placental exosome profile in fetal growth restriction compared to small for gestational age	[70]
Plasma and placental syncytial lipid rafts	ExoQuick, density gradient centrifugation	The vascular effects of truncated endoglin released via exosomes in the pathophysiology of PE has been studied	[93]

maternal exosomes in fetal system and its role in fetal growth and development awaits further investigation.

Growing evidences demonstrate the potential effect of placental- derived exosomes in their target cells and their significant roles in mediating intercellular communication influencing placentation and maternal-fetal vascular system. Establishment of the fetal-maternal circulation is a key event in placentation which begins with invasion of EVT cells and achieved within 10 weeks of gestation. This invasion by EVTs remodels the uterine spiral arteries (SpA) and establishes the maternal blood flow in the intervillous space. The terminal part of the first trimester occurs is characterized by fusion of cytotrophoblasts to form multinucleated layers of syncytiotrophoblasts, which is bathed in maternal blood, soluble proteins and nutrients and cover a major portion of the surface of the placenta [100]. Hence, the maternal blood in direct contact with the EVTs and syncytiotrophoblasts, receive a major proportion of EVs including exosomes, released from these cells. Exosome signaling has been implicated in the invasion of EVTs, cytotrophoblast fusion and endothelial cell migration and spiral artery remodeling [79, 88, 89, 101, 102].

Specifically, placental cytotrophoblast exosomes under low oxygen tension promote increased migration of extravillous cytotrophoblast (HTR-8) cells indicating placental response to hypoxia [57]. Additionally, exosomes isolated from placental mesenchymal stem cells increase endothelial cell migration and vascular tube formation both in vitro and in vivo indicating vascular adaptation of placenta [57, 103]. Exosomes derived from placental cell lines, JEG-3 (choriocarcinoma cell line) and HTR-8/SVneo (transformed EVT cell line) were characterized for their proteomic cargo and identified a higher proportion of proteins in HRT8/SVneo exosomes to be involved in migration compared to JEG-3 cells [88]. Interestingly, exosomes are involved in trans-cellular metabolism of eicosanoids [104, 105] and reported to carry phospholipases and free fatty acids [104]. These exosome when internalized can influence the metabolic pathways in target cells and contribute to various pathophysiological conditions [105]. Fig. 2 is a schematic representation of exosome biogenesis from the syncytiotrophoblast, the influence of extracellular milieu on the exosome biogenesis machinery and their functional role in their target cells.

The immune-modulatory effects of placental exosomes and its potential role in offering immunotolerance to the developing fetus has been widely studied [42]. Interestingly, a large variety of molecules involved in immunological regulation in pregnancy are closely associated with the exosome biogenesis in placental cells and are secreted in their active forms with exosomes. This includes immunomodulatory molecules such as Fas Ligand (FasL), TNF-related apoptosis-inducing ligand (TRAIL), NKG2D ligands and several members of the B7 family of immune-regulatory ligands, which bind to receptors in lymphocytes and mediate immune response offering immune privilege to fetus [20, 80, 85, 90, 106]. In addition, immune-suppressant molecules such as HLA-G [20, 89] and syncitin-1 [83] are secreted with exosomes, mediating immunotolerance and

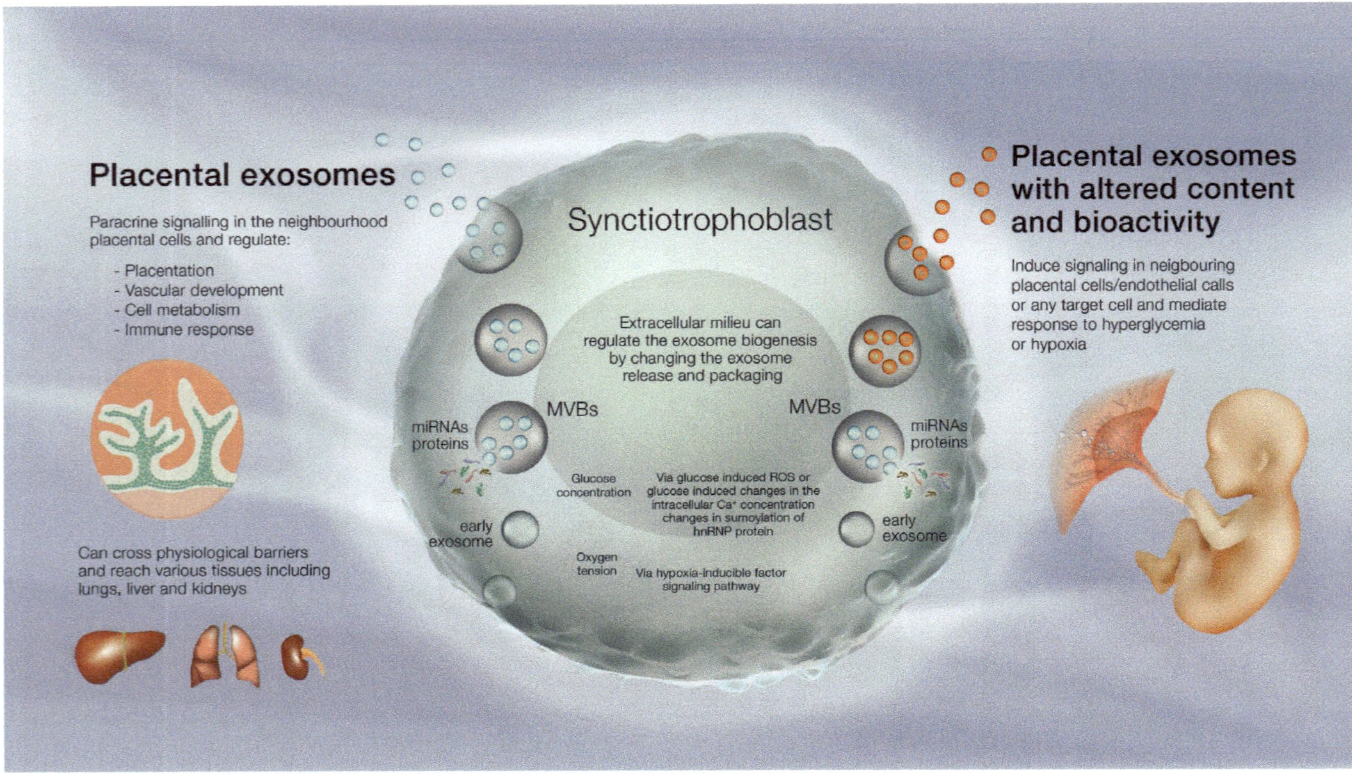

FIG. 2 Schematic representation of exosome biogenesis from the syncytiotrophoblast, the influence of extracellular milieu on the exosome biogenesis machinery and their functional role in their target cells. Placental exosomes released from syncytiotrophoblast can interact with neighboring placental cells, endothelial cells and immune cells mediating placentation, placental vascular development, cell metabolism and immune response. Additionally, placental exosomes can cross physiological barriers and reach various organs including lungs, liver and kidneys. Extracellular environment such as high glucose and low oxygen tension can influence the exosome biogenesis pathway, altering the exosome release and content. High glucose can influence exosome release and bioactivity via inducing release of reactive oxygen species (ROS), increasing the intracellular Ca^{2+} concentration and changing the sumoylation of heterogeneous nuclear ribonucleoprotein (hnRNP). Hypoxia can influence exosome biogenesis via induction of ROS species and Hypoxia Inducible Factor (HIF) signaling.

Th1 to Th2 shift of cytokine profile in pregnancy. Exosomes are also reported to induce pro-inflammatory response and mediate systemic inflammation of pregnancy [53, 54]. For example, placental exosomes increase the recruitment of monocytes from the maternal system into the fetal-maternal interface and increase the release of pro-inflammatory cytokines like IL-1β, IL-6, serpin-E1, granulocyte/monocyte colony stimulating factor and TNF α [54]. Fibronectin present on the exosome membrane binds to α5β1 integrin on the surface of macrophage mediating this effect [53].

Another important method by which exosomes alter the behaviour of their target cells is by exosome-mediated delivery of miRNAs to recipient cells. The chromosome 19 miRNA cluster (C19MC) is a miRNA cluster consisting of 46 miRNAs, expressed solely in human placenta [78]. However, the importance of C19MC for placental heath and pregnancy is not well understood. Delorme-Axford et al., reported C19MC miRNAs namely miR-512-3p and miR-517-3p to transfer viral resistance between trophoblast cells via exosome mediated paracrine delivery and suppress viral replication in placental cells by induction of autophagy [74]. Another C19MC miRNA, miR-517a-3p is carried by placental exosomes and internalized by maternal immune cells regulating their activation and proliferation by targeting the nitric oxide/cGMP signaling pathway [94].

4 Exosomes in complications of pregnancy

4.1 Gestational diabetes mellitus (GDM)

GDM is any degree of glucose intolerance first recognized during pregnancy and is associated with short term and long term adverse outcomes in mother and offspring, leading to significant impact on public health [107]. The metabolic dysfunction in GDM involves decreased insulin secretion from pancreas, decreased hepatic suppression of glucose production and decreased glucose uptake by skeletal muscle [108]. GDM is characterized by poor glycaemic control or hyperglycaemia, hypoxia and changes in the inflammatory or cytokine profile, which contributes to defects in feto-maternal vascularization and placentation [109]. As previously mentioned, hypoxia and a higher level of glucose in the placental cells induce the release of higher concentrations of exosomes, in vitro, when compared to non-diabetic environment [76, 89]. In addition, low oxygen concentrations can alter the bioactivity of exosomes and interestingly, these hypoxic exosomes are reported to increase the release of inflammatory cytokines namely TNF-α and IL-6 and decrease cell migration in their target cells [89]. Exosome released under hyperglycaemia, promotes the release of pro-inflammatory cytokines including IL-6 and TNF-α from endothelial cells [76]. A study performed by Salomon et al., in a longitudinal cohort identified higher concentration of placental exosomes in GDM patients compared to normal pregnant women across gestation and identified that plasma exosomes of GDM women increase the expression of pro-inflammatory cytokines (GM-CSF, IL-4, IL-6, IL-8, IFN-y and TNF-α) from endothelial cells.

[71]. Moreover, exosomes obtained from GDM patients were analyzed for their proteomic cargo, using quantitative, information-independent acquisition (Sequential Windowed Acquisition of All Theoretical Mass Spectra [SWATH]) approach and the major proteins that differ in GDM exosomes compared with exosomes from normal glucose tolerant patients were identified. Among these proteins the most important were the calcium/calmodulin-dependent protein kinase II beta (CAMK2β) and Pappalysin-1 (PAPP-A) which are implicated in pathophysiology of GDM [110].

In addition to proteomic characterization, the exosomal miRNA profile has been identified in GDM by Almohammadi et al., with an upregulation of C19MC miRNAs such as miR-518a-5p, miR-518b, miR-518c, miR-518e, miR-520c-3p, and miR-525-5p in exosomes from GDM patients [111]. Nair et al., analyzed the exosomal miRNA profile in GDM by next generation sequencing of chorionic villi—derived exosomes and identified those miRNAs targeting glucose homeostasis and insulin signaling. Their study also analyzed the effect of GDM chorionic villi-derived exosomes on glucose uptake in skeletal muscles and identified that placental exosomes can alter the glucose uptake in skeletal muscle in normal pregnancy and GDM [112].

Obesity is central to the development of metabolic disorders including GDM [113] and a higher BMI can lead to a higher exosome concentration in maternal circulation with increased release of pro-inflammatory cytokines, IL-6, IL-8 and TNF-α from endothelial cells [63]. Adipocyte-derived exosomes are identified to be key players in mediating the metabolic effects in pregnancy and are implicated in contribution of obesity to the development of GDM [114]. Recently, proteomic profiling of adipocyte- derived exosomes revealed differential protein expression and identified to influence the metabolic pathways in placental cells. Interestingly, adipocyte-exosomes can alter the expression of genes associated with glycolysis and gluconeogenesis in placental cells, thereby indicating their potential role in altering the placental nutrient transfer and excessive fetal growth in GDM [115]. Another important pathophysiological event in GDM is defects in fetoplacental vasculature characterized by endothelial dysfunction induced by hyperglycaemia [109]. Saez et al., reported that exosomes derived from endothelial cells under high glucose altered the endothelial cell activation and migration [116]. Additionally, endothelial cell derived exosomes in GDM modulate the L-arginine/NO signaling pathway and acts in a paracrine manner spreading the pathological phenotype of GDM in the fetoplacental vasculature [117]. Together, these studies establish a potential link between exosome profile and bioactivity with the pathophysiology of GDM. Although the exosome content and bioactivity changes with GDM, its potential role as diagnostic or therapeutic targets in this pregnancy-related complication warrants further studies.

4.2 Pre-eclampsia (PE)

PE is a potentially dangerous placenta-related disorder and a leading cause of maternal and fetal mortality [118]. Improper placentation involving defective spiral artery remodeling leading to placental hypoxia, maternal hypertension and proteinuria as well as abnormal regulation of maternal thrombo-inflammatory systems characterize the pathophysiology of PE [118, 119]. An increased release of circulating microparticles including placenta-derived vesicles eliciting increased inflammatory response is key to the pathophysiology of PE [120, 121]. Additionally, altered expression of proteins and miRNAs in placenta-derived vesicles, characterized by an increased expression of tissue factor, endoglin and fms-like tyrosine kinase (Flt-1) is central to the development of this condition [95]. The concentration of placental exosomes were significantly higher in pre-eclamptic patients compared to their normal controls, with an increased expression of miRNAs has-miR-486-1-5p and has-miR-486-2-5p in PE [45]. Also, it has been reported that, miR-141 carried by STB-derived exosomes, suppressed T cell proliferation and mediated the maternal tolerance to fetus and miR-141 is up-regulated in preeclamptic placenta contributing to the patho-physiology of this condition [122]. Interestingly, Chaparro et al., [123] reported the level of sFlt-1 concentrations in the saliva and gingival crevicular fluid of PE patients to be higher compared to normal ones and the levels of placenta-derived exosomes were found to be elevated in the gingival crevicular fluid of PE patients.

The exact role of exosomes in the development of PE has been studied extensively by analyzing the effect of placenta-derived exosomes in various target cells. The changes in As mentioned before, hypoxia increases the release of exosomes from placenta and hypoxic exosomes induce the release of pro-inflammatory cytokines and decrease cell migration in their target cells [76, 89]. Also, placenta-derived exosomes can influence the endothelial cell migration in vitro and these alterations in endothelial cell migration can contribute to defective spiral artery remodeling and placentation [67, 102]. Changes in exosome release and content can also influence the maternal immunomodulation mediated by these nanovesicles, subsequently contributing to maternal complications such as PE and intrauterine growth restriction (IUGR) [20, 79, 106]. Regarding IUGR, differences in the placental exosome profile in maternal and fetal circulation has been reported, showing a lower contribution of placental-exosomes to total exosomes in fetal growth restriction [70]. In general, exosomes are involved in multiple pathophysiological events such as maternal systemic inflammatory response, thrombo-coagulatory functions, placentation, oxidative response of placenta and vascular remodeling in PE and serves as possible biomarkers for the diagnosis and prognosis of PE [124].

4.3 Preterm birth

Pre-term birth is a pregnancy associated adverse condition characterized by short and long term consequences onchild survival and birth [125]. Not many studies analyzed the exosome profile, content or bioactivity in pre-term birth. Interestingly, in this condition, a 1.8-fold lesser concentration of exosomes than full-term pregnancies (at >36 weeks) is reported [126]. The role of exosomes in inducing premature senescence and sterile inflammation in the fetal membrane and transferring these signals to maternal tissues has been recently under investigation [127]. Exosomes from late gestation carry inflammatory mediators which can act as paracrine mediators of labor and delivery in animal models [128]. Maternal plasma exosomes from patients with pre-term delivery, revealed difference in proteomic cargo compared to term-in-labor patients and these differentially expressed proteins were targeting pathways associated with inflammatory and metabolic signals [129]. Moreover, the circulating exosomal miRNA concentrations differs between mothers delivering term and preterm, targeting pathways associated with TGF-β signaling, p53, and glucocorticoid receptor signaling [130]. Specifically, changes in the concentration and proteomic profile of amniotic fluid-derived exosomes has been reported in pregnancies with pre-term birth [131]. Further, efforts have been made in investigating the predictive and therapeutic role of exosomes in bronchopulmonary dysplasia (BPD), which is a common pathological condition with premature infants [132, 133]. Overall, these studies show that exosomes may influence the molecular mechanism leading to parturition and could be exploited as biological markers of pre-term birth [134]. However, the role of exosomes in influencing the signaling pathways associated with initiation of parturition needs to be investigated further.

5 Conclusions

Exosomes are nanosized extracellular vesicles that originate in the endosomal pathway and play crucial roles in intercellular communication. Exosome signaling has been implicated in various aspects of pregnancy such as maternal metabolic and inflammatory adaptations, maternal-fetal vascular exchange and placentation. Changes in the concentration, cargo and bioactivity of exosomes have been reported in various pathological conditions associated with pregnancy. This ability of exosomes to transfer critical information between cells can be translated into clinical utility by harnessing their potential as attractive biomarker candidates and therapeutic targets in these conditions. Improving the methods to obtain pure populations of placental exosomes and understanding the molecular mechanism by which they regulate key functions related to pregnancy can contribute significantly to the clinical applications of exosomes in reproductive medicine.

Acknowledgments

C.S. is supported by Lions Medical Research Foundation, National Health and Medical Research Council (NHMRC; 1114013), Diabetes Australia, and Fondo Nacional de Desarrollo Científico y Tecnológico (FONDECYT 1170809). S.N. holds a Research Training Program Scholarship from the University of Queensland, funded by the Commonwealth Government of Australia.

References

[1] Colombo M, Raposo G, Théry C. Biogenesis, secretion, and intercellular interactions of exosomes and other extracellular vesicles. Annu Rev Cell Dev Biol 2014;30:255–89.

[2] Lee TH, et al. Microvesicles as mediators of intercellular communication in cancer—the emerging science of cellular 'debris'. Semin Immunopathol 2011;33(5):455–67.

[3] Tkach M, Théry C. Communication by extracellular vesicles: where we are and where we need to go. Cell 2016;164(6):1226–32.

[4] S ELA, et al. Extracellular vesicles: biology and emerging therapeutic opportunities. Nat Rev Drug Discov 2013;12(5):347–57.

[5] Tkach M, Kowal J, Théry C. Why the need and how to approach the functional diversity of extracellular vesicles. PhilosophicalPhilos Trans R Soc B 2017;373(1737):20160479.

[6] Villarroya-Beltri C, et al. sorting it out: regulation of exosome loading. Semin Cancer Biol 2014;28:3–13.

[7] Villarroya-Beltri C, et al. Sumoylated hnRNPA2B1 controls the sorting of miRNAs into exosomes through binding to specific motifs. Nat Commun 2013;4:2980.

[8] Hessvik NP, Llorente A. Current knowledge on exosome biogenesis and release. Cell Mol Life Sci 2018;75(2):193–208.

[9] Colombo M, et al. Analysis of ESCRT functions in exosome biogenesis, composition and secretion highlights the heterogeneity of extracellular vesicles. J Cell Sci 2013;126(24):5553–65.

[10] Muralidharan-Chari V, et al. ARF6-regulated shedding of tumor cell-derived plasma membrane microvesicles. Curr Biol 2009;19(22):1875–85.

[11] Hauser P, Wang S, Didenko VV. Apoptotic bodies: selective detection in extracellular vesicles. Methods Mol Biol 2017;1554:193–200.

[12] Kowal J, Tkach M, Thery C. Biogenesis and secretion of exosomes. Curr Opin Cell Biol 2014;29:116–25.

[13] Tkach M, Kowal J, Thery C. Why the need and how to approach the functional diversity of extracellular vesicles. Philos Trans R Soc Lond B Biol Sci 2018;373(1737):20160479.

[14] Greening DW, et al. Proteomic insights into extracellular vesicle biology-defining exosomes and shed microvesicles. Expert Rev Proteomics 2017;14(1):69–95.

[15] Kowal J, et al. Proteomic comparison defines novel markers to characterize heterogeneous populations of extracellular vesicle subtypes. Proc Natl Acad Sci U S A 2016;113(8):E968–77. https://doi.org/10.1080/14789450.2017.1260450.

[16] Lobb RJ, et al. Optimized exosome isolation protocol for cell culture supernatant and human plasma. J Extracell Vesicles 2015;4:27031.

[17] Li P, et al. Progress in exosome isolation techniques. Theranostics 2017;7(3):789–804.

[18] Al-Nedawi K, et al. Intercellular transfer of the oncogenic receptor EGFRvIII by microvesicles derived from tumour cells. Nat Cell Biol 2008;10(5):619–24.

[19] Valadi H, et al. Exosome-mediated transfer of mRNAs and microRNAs is a novel mechanism of genetic exchange between cells. Nat Cell Biol 2007;9(6):654–9.

[20] Kshirsagar SK, et al. Immunomodulatory molecules are released from the first trimester and term placenta via exosomes. Placenta 2012;33(12):982–90.

[21] Williams JL, et al. Serum exosomes in pregnancy-associated immune modulation and neuroprotection during CNS autoimmunity. Clin Immunol 2013;149(2):236–43.

[22] Zhong H, et al. Induction of a tumour-specific CTL response by exosomes isolated from heat-treated malignant ascites of gastric cancer patients. Int J Hyperthermia 2011;27(6):604–11.

[23] Taylor DD, Gerçel-Taylor C. Tumour-derived exosomes and their role in cancer-associated T-cell signalling defects. Br J Cancer 2005;92(2):305–11.

[24] Warmoes M, Lam SW, van der Groep P, Jaspers JE, et al. Secretome proteomics reveals candidate non-invasive biomarkers of BRCA1 deficiency in breast cancer. Oncotarget 2016;7:63537–48.

[25] Tricarico C, Clancy J, D'Souza-Schorey C. Biology and biogenesis of shed microvesicles. Small GTPases 2016;8(4):220–32.

[26] Caruso S, Poon IKH. Apoptotic cell-derived extracellular vesicles: more than just debris. Front Immunol 2018;9:1486.

[27] Reich III CF, Pisetsky DS. The content of DNA and RNA in microparticles released by Jurkat and HL-60 cells undergoing in vitro apoptosis. Exp Cell Res 2009;315(5):760–8. https://doi.org/10.1016/j.yexcr.2008.12.014.

[28] Andre F, et al. Malignant effusions and immunogenic tumour-derived exosomes. Lancet 2002;360(9329):295–305.

[29] Lässer C, et al. Human saliva, plasma and breast milk exosomes contain RNA: uptake by macrophages. J Transl Med 2011;9:9.

[30] An T, et al. Exosomes serve as tumour markers for personalized diagnostics owing to their important role in cancer metastasis. J Extracell Vesicles 2015;4:27522.

[31] Alvarez-Erviti L, et al. Delivery of siRNA to the mouse brain by systemic injection of targeted exosomes. Nat Biotechnol 2011;29(4):341–5.

[32] Kojima R, et al. Designer exosomes produced by implanted cells intracerebrally deliver therapeutic cargo for Parkinson's disease treatment. Nat Commun 2018;9(1):1305.

[33] Paolillo M, Schinelli S. Integrins and exosomes, a dangerous liaison in cancer progression. Cancer 2017;9(8):95.

[34] Hoshino A, et al. Tumour exosome integrins determine organotropic metastasis. Nature 2015;527(7578):329–35.

[35] Ashiru O, et al. Natural killer cell cytotoxicity is suppressed by exposure to the human NKG2D ligand MICA*008 that is shed by tumor cells in exosomes. Cancer Res 2010;70(2):481–9.

[36] Segura E, et al. CD8+ dendritic cells use LFA-1 to capture MHC-peptide complexes from exosomes in vivo. J Immunol 2007;179(3):1489–96.

[37] Mulcahy LA, Pink RC, Carter DR. Routes and mechanisms of extracellular vesicle uptake. J Extracell Vesicles 2014;3:https://doi.org/10.3402/jev.v3.24641.

[38] Parolini I, et al. Microenvironmental pH Is a key factor for exosome traffic in tumor cells. J Biol Chem 2009;284(49):34211–22.

[39] Montecalvo A, et al. Mechanism of transfer of functional microRNAs between mouse dendritic cells via exosomes. Blood 2012;119(3):756–66.

[40] Ying W, et al. Adipose tissue macrophage-derived exosomal miRNAs can modulate in vivo and in vitro insulin sensitivity. Cell 2017;171(2):372–84. e12.

[41] Yu Y, et al. Adipocyte-derived exosomal MiR-27a induces insulin resistance in skeletal muscle through repression of PPARgamma. Theranostics 2018;8(8):2171–88.

[42] Nair S, Salomon C. Extracellular vesicles and their immunomodulatory functions in pregnancy. Semin Immunopathol 2018;40(5):425–37.

[43] Mitchell MD, et al. Placental exosomes in normal and complicated pregnancy. Am J Obstet Gynecol 2015;213(4, Supplement):S173–81.

[44] Sarker S, et al. Placenta-derived exosomes continuously increase in maternal circulation over the first trimester of pregnancy. J Transl Med 2014;12:204.

[45] Salomon C, Guanzon D, Romero KS, Longo S, Correa P, Illanes SE, Rice GE. Placental exosomes as early biomarker of preeclampsia—potential role of exosomal microRNAs across gestation. J Clin Endocrinol Metab 2017;102(9):3182–94. https://doi.org/10.1210/jc.2017-00672.

[46] Salomon C, et al. The role of placental exosomes in gestational diabetes mellitus. In: Sobrevia L, editor. Gestational diabetes-causes, diagnosis and treatment. IntechOpen; 2013. https://doi.org/10.5772/55298.

[47] Castellucci M, et al. The development of the human placental villous tree. Anat Embryol 1990;181(2):117–28.

[48] Freemark M. Placental hormones and the control of fetal growth. J Clin Endocrinol Metabol 2010;95(5):2054–7.

[49] Freemark M. Regulation of maternal metabolism by pituitary and placental hormones: roles in fetal development and metabolic programming. Horm Res Paediatr 2006;65:41–9. (Suppl. 3).

[50] Adam S, et al. Review: fetal-maternal communication via extracellular vesicles - Implications for complications of pregnancies. Placenta 2017;54:83–8.

[51] Tannetta D, et al. Extracellular vesicles and reproduction-promotion of successful pregnancy. Cell Mol Immunol 2014;11(6):548–63.

[52] Tong M, Chamley LW. Placental extracellular vesicles and feto-maternal communication. Cold Spring Harb Perspect Med 2015;5(3):a023028.

[53] Atay S, Gercel-Taylor C, Taylor DD. Human trophoblast-derived exosomal fibronectin induces pro-inflammatory IL-1beta production by macrophages. Am J Reprod Immunol 2011;66(4):259–69.

[54] Atay S, et al. Trophoblast-derived exosomes mediate monocyte recruitment and differentiation. Am J Reprod Immunol 2011;65(1):65–77.

[55] Thery C, Zitvogel L, Amigorena S. Exosomes: composition, biogenesis and function. Nat Rev Immunol 2002;2(8):569–79.

[56] Nardi Fda S, et al. High levels of circulating extracellular vesicles with altered expression and function during pregnancy. Immunobiology 2016;221(7):753–60.

[57] Salomon C, et al. Hypoxia-induced changes in the bioactivity of cytotrophoblast-derived exosomes. PLoS One 2013;8(11):e79636.

[58] Salomon C, et al. Oxygen tension regulates glucose-induced biogenesis and release of different subpopulations of exosome vesicles from trophoblast cells: a gestational age profile of placental exosomes in maternal plasma with gestational diabetes mellitus. Placenta 2015;36(4):488.

[59] Zhang W, et al. HIF-1-mediated production of exosomes during hypoxia is protective in renal tubular cells. Am J Physiol Renal Physiol 2017;313(4):F906–13.

[60] King HW, Michael MZ, Gleadle JM. Hypoxic enhancement of exosome release by breast cancer cells. BMC Cancer 2012;12:421.

[61] Savina A, et al. Exosome release is regulated by a calcium-dependent mechanism in K562 cells. J Biol Chem 2003;278(22):20083–90.

[62] MacDonald PE, Eliasson L, Rorsman P. Calcium increases endocytotic vesicle size and accelerates membrane fission in insulin-secreting INS-1 cells. J Cell Sci 2005;118(Pt 24):5911–20.

[63] Elfeky O, et al. Influence of maternal BMI on the exosomal profile during gestation and their role on maternal systemic inflammation. Placenta 2017;50:60–9. https://doi.org/10.1016/j.placenta.2016.

[64] Vongthavaravat V, et al. Isolated elevation of serum alkaline phosphatase level in an uncomplicated pregnancy: a case report. Am J Obstet Gynecol 2000;183(2):505–6.

[65] Leitner K, et al. Placental alkaline phosphatase expression at the apical and basal plasma membrane in term villous trophoblasts. J Histochem Cytochem 2001;49(9):1155–64.

[66] Bashiri A, et al. Positive placental staining for alkaline phosphatase corresponding with extreme elevation of serum alkaline phosphatase during pregnancy. Arch Gynecol Obstet 2007;275(3):211–4.

[67] Salomon C, et al. A gestational profile of placental exosomes in maternal plasma and their effects on endothelial cell migration. PLoS One 2014;9(6):e98667.

[68] Kniss DA, et al. ED(27) trophoblast-like cells isolated from first-trimester chorionic villi are genetically identical to HeLa cells yet exhibit a distinct phenotype. Placenta 2002;23(1):32–43.

[69] Dragovic RA, et al. Isolation of syncytiotrophoblast microvesicles and exosomes and their characterisation by multicolour flow cytometry and fluorescence nanoparticle tracking analysis. Methods 2015;87:64–74.

[70] Miranda J, et al. Placental exosomes profile in maternal and fetal circulation in intrauterine growth restriction—Liquid biopsies to monitoring fetal growth. Placenta 2018;64:34–43.

[71] Salomon C, et al. Gestational diabetes mellitus is associated with changes in the concentration and bioactivity of placenta-derived exosomes in maternal circulation across gestation. Diabetes 2016;65(3):598–609.

[72] Lai A, et al. Optimized specific isolation of placenta-derived exosomes from maternal circulation. Methods Mol Biol 2018;1710:131–8.

[73] Boriachek K, et al. Avoiding pre-isolation step in exosome analysis: direct isolation and sensitive detection of exosomes using gold-loaded nanoporous ferric oxide nanozymes. Anal Chem 2019;91:3827–34.

[74] Delorme-Axford E, et al. Human placental trophoblasts confer viral resistance to recipient cells. Proc Natl Acad Sci U S A 2013;110(29):12048–53.

[75] Kambe S, et al. Human exosomal placenta-associated miR-517a-3p modulates the expression of PRKG1 mRNA in Jurkat cells1. Biol Reprod 2014;91(5):129. 1-11.

[76] Rice GE, et al. The effect of glucose on the release and bioactivity of exosomes from first trimester trophoblast cells. J Clin Endocrinol Metab 2015;100(10):E1280–8.

[77] Atay S, et al. Morphologic and proteomic characterization of exosomes released by cultured extravillous trophoblast cells. Exp Cell Res 2011;317(8):1192–202.

[78] Donker RB, et al. The expression profile of C19MC microRNAs in primary human trophoblast cells and exosomes. Mol Hum Reprod 2012;18(8):417–24.

[79] Vargas A, et al. Syncytin proteins incorporated in placenta exosomes are important for cell uptake and show variation in abundance in serum exosomes from patients with preeclampsia. FASEB J 2014;28(8):3703–19.

[80] Stenqvist AC, et al. Exosomes secreted by human placenta carry functional Fas ligand and TRAIL molecules and convey apoptosis in activated immune cells, suggesting exosome-mediated immune privilege of the fetus. J Immunol 2013;191(11):5515–23.

[81] Dragovic RA, et al. Sizing and phenotyping of cellular vesicles using nanoparticle tracking analysis. Nanomedicine 2011;7(6):780–8. https://doi.org/10.1016/j.nano.2011.04.003.

[82] Luo SS, et al. Human villous trophoblasts express and secrete placenta-specific microRNAs into maternal circulation via exosomes. Biol Reprod 2009;81(4):717–29.

[83] Tolosa JM, et al. The endogenous retroviral envelope protein syncytin-1 inhibits LPS/PHA-stimulated cytokine responses in human blood and is sorted into placental exosomes. Placenta 2012;33(11):933–41.

[84] Dragovic RA, et al. Multicolor flow cytometry and nanoparticle tracking analysis of extracellular vesicles in the plasma of normal pregnant and pre-eclamptic women. Biol Reprod 2013;89(6):151.

[85] Hedlund M, et al. Human placenta expresses and secretes NKG2D ligands via exosomes that down-modulate the cognate receptor expression: evidence for immunosuppressive function. J Immunol 2009;183(1):340–51.

[86] Gysler SM, et al. Antiphospholipid antibody-induced miR-146a-3p drives trophoblast interleukin-8 secretion through activation of Toll-like receptor 8. Mol Hum Reprod 2016;22(7):465–74.

[87] Pillay P, et al. Placental exosomes and pre-eclampsia: maternal circulating levels in normal pregnancies and, early and late onset pre-eclamptic pregnancies. Placenta 2016;46:18–25.
[88] Salomon C, et al. Extravillous trophoblast cells-derived exosomes promote vascular smooth muscle cell migration. Front Pharmacol 2014;5:175.
[89] Truong G, et al. Oxygen tension regulates the miRNA profile and bioactivity of exosomes released from extravillous trophoblast cells—liquid biopsies for monitoring complications of pregnancy. PLoS One 2017;12(3):e0174514.
[90] Sabapatha A, Gercel-Taylor C, Taylor DD. Specific isolation of placenta-derived exosomes from the circulation of pregnant women and their immunoregulatory consequences. Am J Reprod Immunol 2006;56(5–6):345–55.
[91] Motta-Mejia C, et al. placental vesicles carry active endothelial nitric oxide synthase and their activity is reduced in preeclampsia. Hypertension 2017;70(2):372–81.
[92] Tong M, et al. Placental nano-vesicles target to specific organs and modulate vascular tone in vivo. Hum Reprod 2017;32(11):2188–98.
[93] Ermini L, et al. A single sphingomyelin species promotes exosomal release of endoglin into the maternal circulation in preeclampsia. Sci Rep 2017;7(1):12172.
[94] Kambe S, et al. Human exosomal placenta-associated miR-517a-3p modulates the expression of PRKG1 mRNA in Jurkat cells. Biol Reprod 2014;91(5):129.
[95] Tannetta DS, et al. Characterisation of syncytiotrophoblast vesicles in normal pregnancy and pre-eclampsia: expression of Flt-1 and endoglin. PLoS One 2013;8(2):e56754.
[96] Gardiner C, et al. Extracellular vesicle sizing and enumeration by nanoparticle tracking analysis. J Extracell Vesicles 2013;2:https://doi.org/10.3402/jev.v2i0.19671.
[97] Cleys ER, et al. Identification of microRNAs in exosomes isolated from serum and umbilical cord blood, as well as placentomes of gestational day 90 pregnant sheep. Mol Reprod Dev 2014;81(11):983–93.
[98] Jeanty C, Derderian SC, MacKenzie TC. Maternal-fetal cellular trafficking: clinical implications and consequences. Curr Opin Pediatr 2014;26(3):377–82.
[99] Kallenbach LR, et al. Maternal background strain influences fetal-maternal trafficking more than maternal immune competence in mice. J Reprod Immunol 2011;90(2):188–94.
[100] Burton GJ, Jauniaux E, Charnock-Jones DS. Human early placental development: potential roles of the endometrial glands. Placenta 2007;28:S64–9. Suppl A.
[101] Salomon C, et al. Hypoxia regulates the response of trophoblast-derived exosomes to hyperglycemia and displays a difference placental exosome profile in plasma from patients with gestational diabetes mellitus. Reprod Sci 2015;22:257a–8a.
[102] Salomon C, et al. Exosomal signaling during hypoxia mediates microvascular endothelial cell migration and vasculogenesis. PLoS One 2013;8(7):e68451.
[103] Zhang HC, et al. Microvesicles derived from human umbilical cord mesenchymal stem cells stimulated by hypoxia promote angiogenesis both in vitro and in vivo. Stem Cells Dev 2012;21:3289–97.
[104] Subra C, et al. Exosomes account for vesicle-mediated transcellular transport of activatable phospholipases and prostaglandins. J Lipid Res 2010;51(8):2105–20.
[105] Record M, et al. Exosomes as new vesicular lipid transporters involved in cell-cell communication and various pathophysiologies. Biochim Biophys Acta 2014;1841(1):108–20.
[106] Frangsmyr L, et al. Cytoplasmic microvesicular form of Fas ligand in human early placenta: switching the tissue immune privilege hypothesis from cellular to vesicular level. Mol Hum Reprod 2005;11(1):35–41.
[107] Metzger BE, Coustan DR. Summary and recommendations of the Fourth International Workshop-Conference on Gestational Diabetes Mellitus. The Organizing Committee. Diabetes Care 1998;21(Suppl 2):B161–7.

[108] Catalano PM. Trying to understand gestational diabetes. Diabet Med 2014;31(3):273–81.

[109] Desoye G, Hauguel-de Mouzon S. The human placenta in gestational diabetes mellitus. The insulin and cytokine network. Diabetes Care 2007;30(Suppl 2):S120–6.

[110] Jayabalan N, et al. Quantitative proteomics by SWATH-MS suggest an association between circulating exosomes and maternal metabolic changes in gestational diabetes mellitus. Proteomics 2018;19:e1800164.

[111] Almohammadi D, et al. C19MC miRNA signatures of placenta-derived exosomes in women diagnosed with gestational diabetes mellitus. In: Diabetes programing-genes and gestation. Endocrine Society; 2016. p. OR27-4.

[112] Nair S, et al. Human placental exosomes in gestational diabetes mellitus carry a specific set of miRNAs associated with skeletal muscle insulin sensitivity. Clin Sci 2018;132:2451–67.

[113] Catalano PM, et al. Longitudinal changes in glucose metabolism during pregnancy in obese women with normal glucose tolerance and gestational diabetes mellitus. Am J Obstet Gynecol 1999;180(4):903–16.

[114] Jayabalan N, et al. Cross talk between adipose tissue and placenta in obese and gestational diabetes mellitus pregnancies via exosomes. Front Endocrinol (Lausanne) 2017;8:239.

[115] Jayabalan N, et al. Adipose tissue exosomal proteomic profile reveals a role on placenta glucose metabolism in gestational diabetes mellitus. J Clin Endocrinol Metabol 2018;104:1735–52. p. jc.2018-01599-jc.2018-01599.

[116] Saez T, et al. Fetoplacental endothelial exosomes modulate high d-glucose-induced endothelial dysfunction. Placenta 2018;66:26–35.

[117] Saez T, et al. Human umbilical vein endothelium-derived exosomes play a role in foetoplacental endothelial dysfunction in gestational diabetes mellitus. Biochim Biophys Acta 2018;1864(2):499–508.

[118] Redman CW, Sargent IL. Latest advances in understanding preeclampsia. Science 2005;308(5728):1592–4.

[119] Laresgoiti-Servitje E. A leading role for the immune system in the pathophysiology of preeclampsia. J Leukoc Biol 2013;94(2):247–57.

[120] Germain SJ, et al. Systemic inflammatory priming in normal pregnancy and preeclampsia: the role of circulating syncytiotrophoblast microparticles. J Immunol 2007;178.

[121] Gohner C, Plosch T, Faas MM. Immune-modulatory effects of syncytiotrophoblast extracellular vesicles in pregnancy and preeclampsia. Placenta 2017;60:S41–51.

[122] Ospina-Prieto S, et al. MicroRNA-141 is upregulated in preeclamptic placentae and regulates trophoblast invasion and intercellular communication. Transl Res 2016;172:61–72.

[123] Chaparro A, et al. Placental biomarkers and angiogenic factors in oral fluids of patients with preeclampsia. Prenat Diagn 2016;36(5):476–82.

[124] Pillay P, et al. Placenta-derived exosomes: potential biomarkers of preeclampsia. Int J Nanomedicine 2017;12:8009–23.

[125] Menon R. Preterm birth: a global burden on maternal and child health. Pathog Glob Health 2012;106(3):139–40.

[126] Taylor DD, Akyol S, Gercel-Taylor C. Pregnancy-associated exosomes and their modulation of T cell signaling. J Immunol 2006;176(3):1534–42.

[127] Menon R, et al. Histological evidence of oxidative stress and premature senescence in preterm premature rupture of the human fetal membranes recapitulated in vitro. Am J Pathol 2014;184(6):1740–51.

[128] Sheller-Miller S, et al. Exosomes cause preterm birth in mice: evidence for paracrine signaling in pregnancy. Sci Rep 2019;9(1):608.

[129] Menon R, et al. Quantitative proteomics by SWATH-MS of maternal plasma exosomes determine pathways associated with term and preterm birth. Endocrinology 2019;160:639–50.
[130] Menon R, et al. Circulating exosomal miRNA profile during term and preterm birth pregnancies: a longitudinal study. Endocrinology 2019;160(2):249–75.
[131] Dixon CL, et al. Amniotic fluid exosome proteomic profile exhibits unique pathways of term and preterm labor. Endocrinology 2018;159(5):2229–40.
[132] Lesage F, Thebaud B. Nanotherapies for micropreemies: stem cells and the secretome in bronchopulmonary dysplasia. Semin Perinatol 2018;42(7):453–8.
[133] Lal CV, et al. Exosomal microRNA predicts and protects against severe bronchopulmonary dysplasia in extremely premature infants. JCI insight 2018;3(5):e93994.
[134] Salomon C, et al. Placental exosomes during gestation: liquid biopsies carrying signals for the regulation of human parturition. Curr Pharm Des 2018;24(9):974–82.

Chapter 17

Exosomes in respiratory disease

Shamila D. Alipoor[a], **Esmaeil Mortaz**[b,c]
[a]*Molecular Medicine Department, Institute of Medical Biotechnology, National Institute of Genetic Engineering and Biotechnology (NIGEB), Tehran, Iran,* [b]*Clinical Tuberculosis and Epidemiology Research Center, National Research Institute of Tuberculosis and Lung Diseases (NRITLD), Shahid Beheshti University of Medical Sciences, Tehran, Iran,* [c]*Department of Immunology, Faculty of Medicine, Shahid Beheshti University of Medical Sciences, Tehran, Iran*

Cell-to-cell communication is the major hallmark of multicellular organisms. Lung is a pair of complicated organs containing a broad range of cells within the parenchyma and airway structures. So, intracellular communication is pivotal in the optimal functioning of lung [1].

Intracellular communication is mediated through multiple mechanisms such as direct contact of cells or via secreted molecules. Recently, a third mechanism has emerged that introduce exosomes for transferring the messages between the live cells [2]. The discovery of exosomes has changed the way the vital functions of body such as physiological and signal orchestration have been understood.

Exosomes are small-sized vesicles (30–100 nm) that embed from almost all cell types and enclose cellular content including functional molecules such as RNA (mRNA and noncoding RNA), DNA (mtDNA, ssDNA, and dsDNA), proteins, lipids, lipidic mediators, etc. (Fig. 1). Exosome secretion allow the cells to release a selective source of molecules for transferring their message and exert the specific modulation in distance cells as well as the removal of unwanted molecules. For example, after being captured, the exosome can trigger various cell signaling processes, impress cell growth in target cells, or promote cell invasion as well as tumor growth in a tumor microenvironment [3].

The exosomes were first described in the mid-1980s referred to small vesicles that were produced by maturating sheep reticulocytes to eliminate some membrane bound proteins and obsolete cellular components [4–6]. These structures were originally considered as the cell's "garbage bin" but further studies demonstrated that B cell-derived exosomes have immune regulatory effect and can stimulate T cell [7, 8]. Additional studies emphasized the involvement of exosomes in many of biological and pathological conditions such as shuttling infectious agents, tumorigenicity, immune modulation as well as neurodegenerative diseases [9, 10].

Exosomes. https://doi.org/10.1016/B978-0-12-816053-4.00017-1

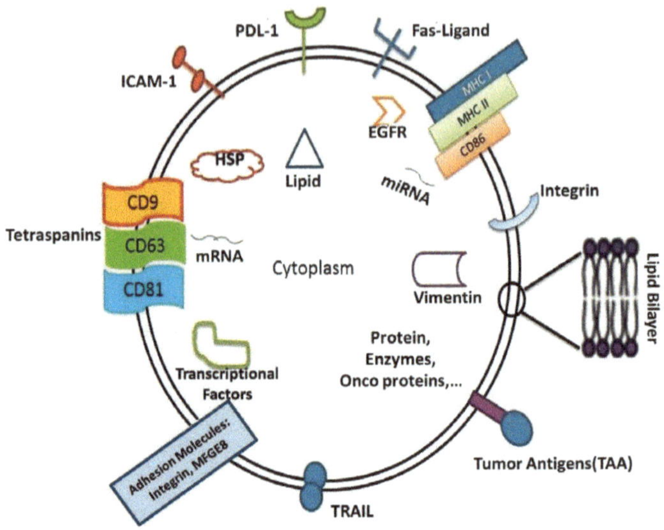

FIG. 1 Schematic diagram of components generally found within exosomes. Exosomes are secreted membrane vesicles released into the extracellular space and shuttle hosts' cellular content such as proteins, lipids, nucleic acids, etc. *HSP*, heat-shock protein; *miRNA*, microRNA; *TAA*, tumor-associated antigen; *TRAIL*, tumor-necrosis-factor-related apoptosis-inducing ligand; *ICAM-1*, intercellular adhesion molecule 1; *PD-L1*, programmed death-ligand 1; *MHC*, major histocompatibility complex; *EGFR*, epidermal growth factor receptor.

The turning point in exosome research was the finding that these microvesicles contribute to genetic exchange between cells by their functional nucleic acid contents such as mitochondrial DNA, messenger RNA as well as small noncoding miRNA which are functional in recipient cells [11]. It was first reported in 2007 that mast cells (MCs)-derived exosomes contain over 1200 mRNAs, which could be translated into proteins after delivery into recipient cells [12]. The presence of miRNAs and mRNAs has been now demonstrated in the exosomes from different origin [11].

The range of biological molecule cargos inside the exosomes may differ based on their cell of origin as well as the physiological condition of the cells when the exosomes are released. Generally, these vesicle contain tetraspanins, integrins, major histocompatibility complex (MHC), adhesion molecules, membrane-trafficking proteins, lipid rafts, membrane transport proteins, and fusion proteins, which are most frequently detected as exosomal content.

Exosomes also carry metabolic enzyme, cytoskeletal proteins, signal transduction proteins, lipid mediators such as cholesterol, diglycerides, phospholipids, chaperones, prostaglandins, and leukotrienes (LTs). Tetraspanins molecules such as CD63, CD81, CD9, and heat-shock protein (Hsp70) are universal exosomal protein content and considered as exosomal markers [13].

ExoCarta is a database of exosomal proteins, RNAs, and lipids [14]. According to this database, 4563 proteins, 194 lipids, 1639 mRNAs, and 764 miRNAs have been identified in the exosomes from different species [14].

Packaging within a lipid bilayers protects exosomal contents such as proteins and miRNAs from degradation, which may be present in body fluids. This results in a relatively long and stable duration providing a great deal of information about the physiological status of the originating cell [15]. On the other hand, packaging of functional molecules inside the exosome gives an advantage of simultaneous delivery of multiple components to recipient cells [16].

Exosomes can be easily found in almost all body fluids such as urine, plasma, breast milk, bronchoalveolar lavage (BAL) fluid, saliva, seminal fluid, amniotic liquid, ascites, synovial fluid, breast milk, and cerebrospinal fluid (CSF), so attract more attention in the field of medical application.

Exosomes properties such as stability in circulation, low immunogenicity, and biocompatibility made them suitable as delivery systems of small RNAs, signaling molecules, or drugs [17, 18]. In addition, since exosomal composition may differ according to the physiological or pathological situations of the cells, this made them attractive as biomarkers for diagnosis or therapy monitoring [19].

As such, exosome can exert autocrine, paracrine, and endocrine effect on the other cells, which can be exploited for therapeutic usage [20]. For example, in cell therapy, exosomes derived from mesenchymal stem cells (MSCs) and other progenitor cells can resemble their parent cell's properties and activity in cytoprotective, angiogenic, and regenerative effects which make them highly interesting for "cell therapy without cells" techniques [21].

In this section, we take a look on exosomes roles in lung microenvironment including their role in lung function and in the pathological condition in respiratory diseases.

1 Exosome in lung microenvironment and pathogenesis

Lung diseases are the most common causes of death worldwide. Air pollutions, smoking, and toxic agents as well as infectious factors have important roles in the increasing incidence of these diseases.

In the lung, bronchial and alveolar epithelial cells (AECs) in the upper and lower airways are the largest surface area exposed to the external environment and provides first line defense against environmental pollutants.

The exposure to toxic smoking agents, microorganisms, or toxic gases may result in airway injury which lead to the development of various lung diseases such as lung cancer, asthma, or chronic obstructive pulmonary disease (COPD).

Lung protection in airway is done by a range of mechanisms including mechanical barriers, mucociliary apparatus accompanied with enzymes, and secretory immunoglobulin A (IgA) in the innate immune system. On the other hand, invading microorganisms can be captured by the dendritic cell (DC) network in the respiratory mucosa which will be transferred to the draining lymph nodes. After the bronchioles, alveolar macrophages (AM), capture the introducing particles in a microenvironment that is rich in IgG, fibronectin, complement elements, surfactant, and other defense elements. Other inflammatory cells, in particular neutrophils, may be recruited depending on the load of pathogens. Excessive inflammatory reactions in airway microenvironment result in epithelial cell damage which leads to further inflammatory immune responses and phenotypic alteration in epithelial cells. This alteration may result in "epithelial-mesenchymal transition"(EMT) which may contribute to airway remodeling resulting in pathological outcome such as COPD, lung cancer, or bronchial asthma.

Based on these facts, paracrine cell-to-cell communication in lung cell network such as epithelial, endothelial, immune, and mesenchymal cells have important roles in the orchestration of physiological homeostasis and pathological condition in respiratory microenvironment.

Exosomes have recently gained more interest because of their roles in intracellular and paracrine communication.

Admyer et al. in 2003 for the first time reported the presence of exosomes containing major histocompatibility and co-stimulatory molecules in BAL obtained from healthy individuals [22].

These exosomes contain miRNAs, functional RNAs, and proteins that can shuttle their functional cargo to other cells that affect cell homeostasis or physiological conditions. For example, AM-derived exosomes promote differentiation of naïve monocytes by delivering mir-223 [23]. The presence of membrane-bound mucins in the epithelial cell-derived exosomes alter surface charge and structural properties of exosomes which can impress local mucociliary defense mechanisms in lung innate immune system [24]. Exosomes have also been shown to be important in the regulation of inflammation pathways within airway by transferring signaling molecules. For example, the delivery of suppressor of cytokine signaling (SOCS)1 by the macrophage-derived exosomes results in the reduced activity of signal transducer and activator of transcription (STAT) in AECs [25]. On the other hand, exosomes have been reported to play pivotal roles in the mechanisms underlying cellular cross talk and final phenotype of disease in asthma [26]. In allergic airway inflammation, epithelial cells produce exosomes with pro-inflammatory effect [27]. Exosomes originated from airway epithelial cells had neutralizing effect on human influenza virus because of their mucins and alpha 2,6-linked sialic acid cargo [28](Fig. 2).

Exosome also may contribute to stress response in the airway. In sarcoidosis, exosomes elevate the production of pro-inflammatory elements such as INF-ɣ, interleukin (IL)-13, and CXCL-8 in respiratory milieu and lead to the development of an inflammatory reaction [29]. Furthermore, infection condition can change the secretion and cargo of exosomes. For example, macrophages infected with mycobacterium secrete exosomes that are enriched in HSP-70 [30]. These

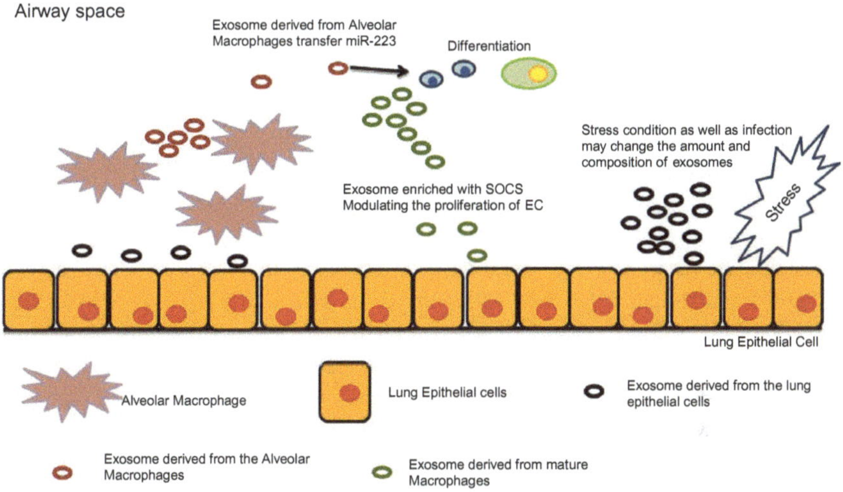

FIG. 2 Exosomes in stressed airway physiology. A broad range of respiratory cell types in the airway can release exosomes. Exposure to various types of stress such as infection or oxidative stress modify exosome composition and enhance their secretion. Injured macrophage-derived exosomes contain a variety of pro-inflammatory cytokines which may affect immunity efficacy. Alveolar macrophage-derived exosomes can control cellular homeostasis and differentiation by shuttling miR-223 and regulate the inflammatory signaling by transferring SOCS (suppressor of cytokine signaling) protein to lung epithelial cells. Lung EC-derived exosomes released into the blood and shuttle modified biomolecules in stress conditions.

exosomes recruit immune cells and active immune system. Pathological condition also alter the miRNA content of exosomes [31–33]. These strategy may be a part of cell stress response.

In hypoxic pulmonary hypertension, mesenchymal stromal cells exert protective activity against lung inflammation via paracrine mechanisms mediated by exosomes [34].

Almost all respiratory cells including lung epithelial cells, fibroblasts, endothelial cells (ECs), and various immune cells release exosomes into the extracellular space and are involved in diverse physiologic processes and maintenance of lung hemostasis. In the following, the roles of these exosomes are discussed.

2 The roles of exosomes derived from the effector immune cells and structural cells in lung microenvironment

Lung structural cells release exosomes that play important roles in lung health or disease [35]. These exosomes have important roles in fine-tuning the intracellular communication in lung microenvironment [1, 32].

These demonstrate that the production of exosomes by lung structural cells increases and their contents change in lung pathological conditions [32].

Among the lung structural cells, airway smooth muscle cells (ASMCs), goblet cells, and epithelial cells are the key cells subjected to change in pathological conditions. There are no studies addressing ASMCs or goblet cell-derived exosomes [36] whereas alveolar or bronchial epithelial cells (BECs) release exosomes with different morphology, that is, size and shape [37].

AECs-derived exosomes enclose important molecules including mucins and sialic acid that contribute to innate immunity mechanisms, so play an important role in the modulation of airway inflammation [37].

Mechanical stress and cytokines such as IL-13 increase exosome release by BECs and induce alteration in the composition of BEC-derived exosomes. BECs stimulated by mechanical stress released exosomes containing a source of tissue factors (TFs), so can trigger TF signals. These exosomes also can promote cell proliferation in monocytes [38]. Mechanical stress during episodes of bronchoconstriction in asthmatics can induce exosome secretion from BECs [38]. In addition, exosomes derived from the bronchial fibroblasts contain transforming growth factor (TGF)-2 that can induce proliferation in BECs upon uptake [39].

Lymphocytes are the key factors of immune function and have important roles in stress responses and inflammatory reactions in lung microenvironment. B cell, as an antigen presetting cell (APC), can trigger asthmatic responses without the involvement of immunoglobulin E (IgE) and T-lymphocytes [40].

IFN-ɣ or IL-4 producing B-lymphocytes also can be involved in the differentiation of naive T0-lymphocytes into T1- or T2-lymphocytes. In addition, B regulatory (Breg) cells downregulate inflammation in hyperresponsive airway [41] and have an important role in providing relief from allergic inflammation in asthma pathologic condition by recruitment of natural Treg (CD4+ CD25+ FoxP3+ cells) to the lung [42].

There are a large number of evidence that demonstrate the production of exosomes by B cells. B cell-derived exosomes resemble their parent phenotype with expression of MHC, integrins (ß1 and ß2), as well as co-stimulatory molecules like CD40, CD80, and CD86 [26]. These exosomes could induce T-cell responses by presenting antigenic peptides [43] and also promote DC maturation by shuttling HSP70 [44]. Exosomes from B cells contain birch peptide (Bet v 1-derived peptides) and can induce proliferation and IL-5 and IL-13 synthesis in T cell. These exosome-associated responses resemble those observed upon direct contact between B and T cells. These findings bypass the need for direct cell-to-cell contact and indicate the influential role of B cell-derived exosomes in the modulation of lung immunity [43].

T-lymphocytes also can release exosomes [45–47] however their biological activity has been not completely understood [48]. Exosome secreted by T cells are saucer-shaped exosomes containing Src-like tyrosin kinases, MHCs, CD3, adhesion molecules CD2, lysosomal-associated membrane protein 1 (LAMP-1), chemokine receptor CXCR4, lymphocyte function-associated antigen-1 (LFA-1), and T-cell receptor subunits [45, 48].

The production of exosomes by T cells is highly induced by TCR activation; however stimulation with mitogenic signals such as phorbol esters and ionomycin does not induce exosome release, so exosome production by T cells is a highly regulated process [45].

Activated T cells also secrete exosomes with Fas ligand that increased the expression of matrix metalloproteinase 9 and promoted tumor invasion in mice lung [49]. T regulatory (Treg; CD8+CD45+ cells) as well as CD4+ T cells release exosomes with antitumor activity that inhibit CD8+ cytotoxic T-lymphocyte in lung microenvironment [46].

Activated CD3+ T cells cross talk with resting autologous T cells via exosomes. In fact, exosomes produced by activated CD3+ T cells together with IL-2 induce cell proliferation and a distinct cytokine profile in the autologous resting CD3+ T cells [50].

Eosinophils are the major granulocytes that produce, store, and release an array of immune mediators in lung microenvironment such as chemokines, lipid mediators, classical eosinophil proteins including eosinophil peroxidase (EPX), eosinophil cationic protein (ECP), and eosinophil-derived neurotoxin (EDN) as well as major basic protein (MBP). So these cells play important roles in the inflammatory pathologic condition such as asthma and allergy [51].

Eosinophils can uptake exosomes via a mechanism of internalization of the whole vesicle. On the other hand, eosinophils release saucer-shaped exosomes with eosinophil characteristic proteins cargo such as MBP and EPO (eosinophil peroxidase) [52].

Eosinophil-derived exosomes also carry proteins that are implicated in different cell function including S100 proteins (S100A8 and S100A9) with antifungal and antibacterial activity [53]; HSP70 involved in protein folding; cell adhesion molecules such as integrins; Filaggrin and periostin as well as metabolic enzymes [52]. These exosomes could resemble their parent cell activity and trigger tissue damage, so can play an important role in the modulation of lung hemostasis as well as response to stress in lung microenvironment.

MCs play an important role in lung health through immune responses to respiratory pathogens and tissue injury. However, MCs may drive disease processes via triggering inflammation and remodeling tissue structure in the case of inappropriate and chronic activation such as asthma, pulmonary fibrosis, pulmonary hypertension, COPD, acute respiratory distress syndrome (ARDS), and lung cancer [54].

Activated MCs release allergic and pro-inflammatory mediators such as histamine, prostaglandins, LTs, tumor necrosis factor (TNF)-α, and IL-13 and promote innate and adaptive immune responses in lung microenvironment [55, 56]. MCs cross talk with lung microenvironment including airway structural cells and other inflammatory cells through different mechanisms including exosomes production. Exosomes are derived from MCs uptake by other immune cells in lung microenvironment and impress their phenotype and biological activities. For example, the uptake of MC-derived exosomes by DCs induces expression of costimulatory MHC class II, CD80, CD86, and CD40 molecules which enable DC to present antigen to T cells [57]. These exosomes can also stimulate the production of cytokines by B- and T-lymphocytes [58].

MC-derived exosomes can trigger cell signaling via cell surface receptors. For example, these exosomes promote IgE production via CD40 surface ligand on B cells in the absence of T cells [59]. In addition, bone marrow-derived mast cells (BMMCs) release exosomes with surface CD63 and OX40L which enable them to induce proliferation and differentiation naive T cells to T2 cells through ligation with their ligand. It has been demonstrated that these exosomes also impress T-cell functions by transferring their cargo [60].

DCs are known as the specialized APCs that process and present antigens to T cells. On the other hand, because of their capacity for phagocytosis activity, they can contribute to innate immunity [61].

DCs release exosomes that possess their parent properties and express MHC as well as co-stimulatory molecules CD86 and CD54 on their surface. These exosomes have ability to stimulate T cells by their surface MHC molecules [36]. Besides, they have potential to induce proliferation and differentiation in T cells via their surface CD86 and interact with LFA-1 on T-lymphocytes by their surface CD54 [62].

DC-derived exosomes may induce T2 response by presenting the allergens to these cells [63]. The enzymatic cargo of these exosomes can convert LTA4 to other LTs such as LTB4 and LTC4 [64]. In addition, the enzymatic cargo of DC-derived exosomes can produce lipid metabolites of arachidonic acid (5-keto eicosatetraenoic acid, KETE, and LTB4) that act as pro-inflammatory metabolites and trigger migration of granulocytes and leukocytes to the site of inflammation [62].

3 Exosomes in lung diseases

The roles and functions of exosomes in the initiation, development as well as treatment of lung disease such as asthma, COPD, sarcoidosis, and tuberculosis are reviewed further in the following sections.

3.1 Asthma

Asthma is a heterogeneous chronic airways disease, characterized by reversible airway obstruction in response to nonspecific stimuli [65]. There is a limited knowledge regarding underlying pathophysiology of asthma which has made the asthma treatment a big challenge. There is a broad phenotypes of asthma with different underlying pathophysiology. Although, two forms of asthma have been defined clinically (T2 and non-T2), asthma is increasingly considered as a syndrome rather than a single disease [66]. Several research programs in asthma issue aims to link asthma classification based on phenotypes with pathophysiological mechanism thereby defining asthma endotypes which will predict drug efficacy [67].

The main aspect of asthma is the pathophysiological effects of activated CD4+ T-cell pro-inflammatory cytokines in response to environmental allergens such as infections or air pollutants. The release of the mediators such as IL-4, 5, and 13 promote activation of MCs, Th2 cells as well as eosinophils which in turn caused reversible obstruction, remodeling, and hyperresponsiveness of airway [26].

In asthma a broad range of cells from the innate and adaptive immune system are involved to trigger an inflammatory reaction on lung epithelial cells. A large amount of lymphocytes, plasma cells, eosinophils, and neutrophils are detectable in the bronchial tissues and mucus of asthmatic airways.

During an asthma attack which is triggered upon airway simulation with allergens, a rapid-phase IgE-mediated decrease in bronchial airflow occur that may be accompanied with a late IgE-mediated reaction and reduced bronchial airflow that may persist up to 4–8 h [56].

In asthma, it is important to have readily accessible biomarkers that define patient subsets to ensure that the correct drug is given to the right patient at the right time.

This is essential from the patients' perspective and for the health-care provider when the current blunt measures such as blood eosinophils do not distinguish differences in underlying pathophysiological process [26].

Exosome context has opened a new window for biomarker discovery in diagnosis and treatment monitoring of asthmatic patients. In addition, regarding the role of exosomes in intracellular communication, they have undeniable effect on pathology of asthma so considering of the role of exosomes in asthma pathological conditions play an important role in the understanding of its underlying mechanism.

3.1.1 Exosomes in asthma

Regarding the biology and structural properties of exosomes, the roles of these microvesicles in the pathobiology of asthma is undeniable. In asthma pathophysiology, inflammation is the main pathogenic deriver and exosomes can promote the chronic inflammatory processes via modulation of the function of immune cells at the levels of recruitment, activation, or differentiation. In asthma, a broad range of immune and structural cells release exosomes such as DCs eosinophils, T cells, epithelial or BECs, etc.

DCs produce and release exosomes that carry co-stimulatory molecules and MHC classes I and II and can trigger allergen-specific T2 cells as their parent cell [68].

Exosomes containing mucine-1 marker are detected in the BAL from the asthmatic patient which indicate their BECs origin [69]. These exosomes can induce the expression of CXCL-8 and LT C4 in target BECs [27]. Acting in a reciprocal manner, IL-13 enhances exosome release by airway epithelial cells and these exosomes in turn promote the proliferation of undifferentiated lung macrophages [70]. Interestingly in an asthma mice model the inhibition of exosome production by GW4869 alleviated the induction of asthmatic features in this model [26].

Eosinophil-derived exosomes containing eosinophilic proteins such as EPO, MBP, and ECP can mimic the function of their parent in driving the progression of inflammatory process in asthmatics via autocrine or paracrine functions [52].

In asthmatic patient the amount of exosome production and release from eosinophils is greater than those released from cells of healthy subjects. These exosomes increase the production of inflammatory mediators such as chemokines, reactive oxygen species (ROS), and nitric oxide (NO) in the recipient eosinophils. In addition, they upregulate the expression of adhesion molecules such as intercellular adhesion molecule 1 (ICAM-1) and integrin 2 [71] in the eosinophils, thus enhance the migration capacity of these cells which is so important in asthma process [72]. Interestingly, eosinophilic exosomes from healthy subject does not have any effect on the function of asthmatic eosinophil [73].

Exosome also can prevent the initiation and development of an allergic response [74]. Exosome from the BAL fluid of a tolerized mice inhibits allergic sensitization upon intranasal delivery. This indicates the potential of exosome-based vaccine in the treatment of asthma and other allergic diseases [74]. CD8+ T-cells produce exosomes that are coated with antigen-specific antibody and contain miR-150. Uptake of these exosomes by T cells has induced antigen-specific tolerance in mice [75].

MC-derived exosomes can bind and decrease serum level of free IgE via their surface FcεRI which in turn restrict the consequence of MC activation [76]. Fibroblast-derived exosomes from asthmatic patient have enhancing effect on the proliferation of epithelial cells due to their TGFβ-2 content. The level of exosomal TGF-β2 in asthmatic fibroblasts were lower than that in healthy subjects [39].

Lipopolysaccharide-stimulated neutrophils release exosomes with proliferative and remodeling effect on the ASMCs [77].

Interaction between MCs and ASMCs in human airway can modulate inflammation and provoke asthma symptoms. MCs-ASMC interaction can lead to enhanced secretion of ASMC cytokines which result in the recruitment of more MCs and promote asthma manifestation [78]. Interaction between MCs and ASMCs can occur without direct contact via MC-derived exosomes and promote asthma process. So the depletion of MC-derived exosomes or other soluble factors may contribute to subside the stimulatory asthma responses [1].

Exosomal miRNA contents in asthmatic exosomes also undergo change. Exosomal dysregulated miRNAs in asthmatic patient were correlated with some clinical features such as eosinophil count or FEV1 [79] and mostly have contributed to pathways related to airway integrity. In addition, the differential exosomal miRNAs profile in SA patients were associated with TGF-ß signaling pathway, the ErbB signaling pathway, and focal adhesion [80] as well as IL-13-mediated events [33].

3.2 Lung cancer

Lung cancer is one of the most common cancer in the world and despite considerable achievements in lung cancer diagnosis and treatment, it still remains the leading cause of cancer-related deaths worldwide. The main challenge of global control of this disease is the lack of early diagnosis methods as well as limited understanding of its underlying pathophysiology mechanisms.

Regarding the physiology and function of exosomes in the intracellular communication, they can influence physiological process of the recipient cell which may result in pathological conditions such as cancer [81].

Tumor cells can produce and release exosomes. The exosomes derived from the tumor cells are known as TEX [82].

TEXs are the main mechanism of cross talk in tumor microenvironment that enable cancerous cells to favor their surrounding environment for their own growth and survival. In fact, the fate of tumor malignancy is determined by the molecular and cellular components in tumor microenvironment [83] and exosomes impress this microenvironment based on their content.

Exosomes derived from turmeric cells transfer a lot of tumorigenic messages from their parent cells by delivering a broad range of immune stimulatory and immune inhibitory factors as well as oncoproteins that induce reprogramming in target cells. For example, TEX contain K-RAS and MET or oncogenic miRNAs, which promote cancer initiation and growth in otherwise healthy cells after delivery [84].

TEXs have the key role in creating pre-metastatic niche and promoting metastasis by directing the disseminated tumor cells to future metastatic sites [85, 86], which is directed by integrin expression on TEX surface [87]. However, TEX may induce an antitumor immune response by modulating natural killer (NK) cell activity and expressing their surface receptors such as killer cell lectin-like receptor K1 (KLRK1 or NKG2D) [88, 89]. In addition; TEX can modulate EMT, cancer-associated fibroblast function, and angiogenesis.

Lung cancer-associated TEXs induce pro-metastatic inflammatory responses. Lung cancer-associated TEXs bind to intracellular toll-like receptors (TLRs) on immune cells via their miR-21 and -29a cargo and in turn trigger activation of NF-κB eventually resulting in metastasis and tumor growth [90] (Fig. 3).

3.2.1 The role of exosomes in immune cell function in lung cancer

In a mouse model of lung cancer, TEXs caused reduction in NK cell percentage in lung tissue [91]. Reduced activity of NK cells in lung tissue is also observed in human lung cancer [92]. NK cells via their C-type lectin-like receptor NKG2D trigger cytotoxic pathways against cancerous cells [93]. TEXs reduce the expression of this receptor by transferring TGF-β1 to the NK cells thereby

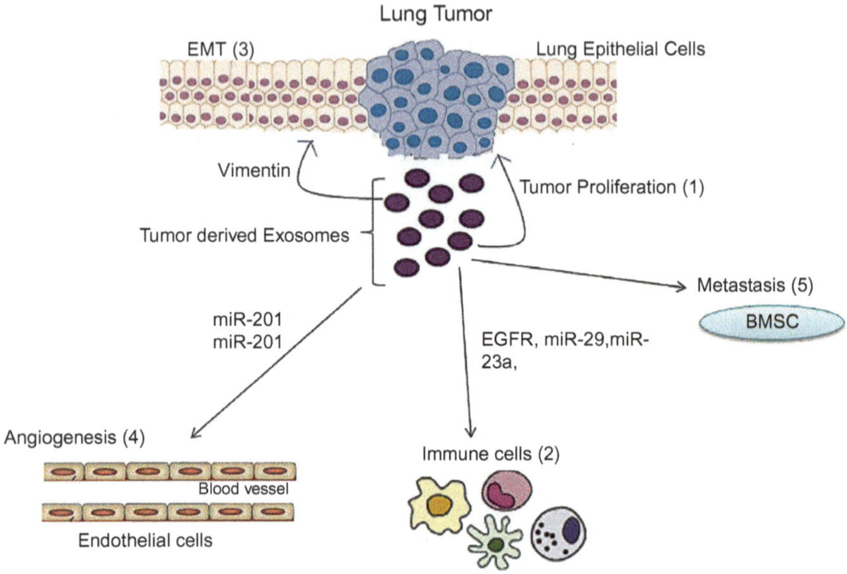

FIG. 3 The function of lung tumor-derived exosome (TEX). TEX impact on the tumor microenvironment by enhancing tumor cell growth and progression (1); modulating immune responses (2); regulating epithelial-mesenchymal transition (EMT) (3); angiogenesis (4); and inducing metastatic behavior in bone marrow progenitors (5).

reducing NK cell cytotoxic activation [94]. TEXs also can downregulate NKG2D expression by shedding NKG2D ligands on the tumor cells, which results in the desensitization and internalization of receptor [95].

In addition, miR-23a cargo from lung cancerous TEXs can directly bind to CD107a molecule which protects NKs from granule-associated damages [96].

TEXs may also use other mechanisms to interfere with NK cell antitumor activity by impressing IL-2-associated pathways [97], perforin, or cyclin D3 production [91] as well as Janus kinase3 (Jak3) activation [91].

Tumor-associated exosomes induce polarization and differentiation of immune cells in tumor microenvironment. DCs represent different phenotypes and activities at the tumor site and may exert either pro-tumorigenic or anti-tumorigenic functions. The function of DC is determined by the nature of tumor microenvironment, which is so intricate and valuable. In fact, although DCs normally act through the induction and maintenance of antitumor immunity, the tumor microenvironment may educate them to promote tumorigenicity, so the antigen-presenting function of DC may be reduced or lost in cancers. DCs also may differentiate to immunosuppressive/tolerogenic regulatory DCs in the tumor microenvironment that subside the function effector T cells and promotes tumorigenic condition. Various mechanisms may be responsible for malfunction of DCs in cancer but tumor-derived exosomes have important role in this concept by transferring a broad range of molecules.

Approximately 80% of exosomes isolated from the lung cancer biopsy have potential to induce tolerogenic DC and regulatory T cell due to their epidermal growth factor receptor (EGFR) cargo and can lead to the inhibition of tumor antigen-specific CD8+ cells activity [98]. miR-203a in TEX decreases TLR4 expression on DCs which in turn lead to a reduced level of down stream cytokines such as TNF-α and IL-12 [98] as well as fault in cellular immunity and DC dysfunction [99]. TEXs also interfere DC maturation via their TGF-β1 cargo and induce myeloid-derived suppressor cells (MDSCs) [100]. MDSCs are a heterogeneous and immature population of cells with myeloid origin that strongly expand in cancer and have remarkable potential to suppress T-cell proliferation and function with frequent secretion of immune inhibitory cytokines such as TGF-β and prostaglandin E2 [100]. TEX also may direct the differentiation of CD14+ monocytes to MDSCs rather than to DC [100]. The presence of heat-shock protein 72 (HSP72) on the surface of TEX enhances suppressive activity of MDSCs via STAT3 pathway and autocrine IL-6 production in MDSCs [101]. Interaction of TEX with monocyte leads to the maintenance of an immature monocyte status with downregulated or loss of costimulatory molecules [102] and human leukocyte antigen-DR along with unchanged CD14 surface expression [103] on monocytes. Injection of mice model of lung cancer with TEX significantly increased recruitment of MDSCs into lung and promote metastasis. TEXs have ability to promote metastasis due to their MyD88 cargo which is involved in integration and transduction of TLR signaling [104].

TEXs contribute to the generation of tumor-associated macrophages (TAMs) from monocyte precursor by delivering functional receptor tyrosine kinase, triggering the mitogen-activated protein kinase (MAPK) pathway and inhibition of apoptosis-related caspases [105]. In lung cancers, TAMs induce STAT3 activation and promote tumor progression [105]. Interaction of TAM and tumor cells is mediated by TEXs. In addition, TAM-derived exosomes have a reciprocal supportive role in providing factors that maintain TAM survival within the inflammatory niche [106]. TAM-derived exosomes contain a specific proteomic signature with higher proteolytic activity [107]. Overall, the existence of different types of cancerous exosome with different functions in tumor microenvironment enforces the role of exosomes as the major player in the promotion of the tumor development and progression.

3.2.2 The role of exosomes in EMT in lung cancer

EMT is an important process in cancer development and metastasis. In this process the epithelial cells lose their cell polarity and adhesion properties, acquire mesenchymal cell properties, and gain a motile trait [108], which enables them to migrate to distant sites allowing metastasis and tumor progression [108].

TEX isolated from the serum of patient with highly metastatic lung cancer show that TEX can induce EMT in recipient human BECs [109]. These TEXs contain high level of vimentin. Vimentin is a member of type III intermediate filament protein family, is widely expressed in mesenchymal cells and is known as marker for EMT [110]. The association of the level of vimentin with metastasis and invasion potency is observed in lung cancer [111]. In addition, it is also detected in prostate [112] and gastric cancers [113].

In lung cancer, vimentin alters tumor cell adhesion properties by modifying the VAV2–Rac1 pathway and focal adhesion kinase activity [113].

TGF-ß-induced EMT in epithelial adenocarcinoma A549 cells alter the exosomal cargo in these cells [114]. The exosomes released from the mesenchymal-like A549 cell contain an enhanced level of β-catenin, vimentin, E-cadherin, and miR-23a in comparison to those from epithelial-like A549 cells.

Mir-23a cargo in these exosomes directly target E-cadherin and regulate TGF-β-induced A549 cell EMT in a smad-dependent manner [115].

3.2.3 The role of exosomes in angiogenesis in lung cancer

Angiogenesis and vascular network formation is a crucial process for cancer progression and metastasis. Different mechanisms contribute to this process but exosomes play an important role in the vascular tube formation by transferring angiogenic factors including VEGF, TGF-β, and fibroblast growth factor.

The presence of pathogenic condition such as hypoxia in tumor microenvironment induce production and release exosomes with specific cargo that gain ability to induce angiogenesis which enable cells to alleviate the stress conditions in the tumor microenvironment [115].

In CL1-5 lung adenocarcinoma cells, hypoxia condition induce angiogenesis by increasing the production of exosomes with enhanced level of miR-23a cargo. Exosomal mir-23a directly target prolyl hydroxylase 1 and 2 (PHD1 and 2), which lead to the accumulation of hypoxia-inducible factor-1 α (HIF-1 α) in ECs. In addition, exosomal miR-23a also increase vascular permeability and cancer trans-endothelial migration by targeting tight junction protein ZO-1 [116].

Tissue inhibitor of metalloproteinases (TIMP)-1 is an important player in lung cancer progression and metastasis [117]. Overexpression of this factor in lung adenocarcinoma cells enhances the expression of tumorigenic miR-210 in these cells and also in their derived exosomes via the modulation of the PI3K/Akt/HIF-1 pathway. These exosomes can promote angiogenesis via downregulation of Ephrin A3 upon uptake by ECs [117].

The expression level of exosomal mir-210 in the serum of lung cancer patients is significantly higher than that in healthy subjects [118]. EGFR cargo in lung cancer-derived exosomes trigger EGFR-dependent responses in ECs which leads to autocrine activation of vascular endothelial growth factor (VEGF) receptor 2 (VEGFR-2) and promotes angiogenesis [119]. Interestingly, injection of exosomes isolated from lung cancer patient into a rat critical limb ischemia model significantly enhanced the expression of VEGFR-2, increased angiogenesis, and improved blood flow in this model [120].

3.2.4 The role of exosomes in metastasis in lung cancer

The role of TEX in the formation of pre-metastatic niches (PMN) and development of metastasis has recently attracted more attention [121]. The importance of exosomes in metastasis in lung cancer was first described by Janowska-Wieczorek et al. in 2005 [122].

In lung cancer, the upregulation of angiopoietin 2 (Angpt2), MMP3, and MMP10 in pre-metastatic stage increase vascular permeability and vascular leaking in the lung. This process is regulated and managed by exosomes.

In addition, melanoma-derived TEXs can increase vascular permeability in the lung by transferring and overexpressing oncogene MET in BM progenitor cells and reprogramming bone marrow progenitors toward a pro-vascular phenotype by increasing the pro-angiogenic c-Kit +Tie2+ cell population in the lung pre-metastatic niche [123]. Vascular leaking promotes extravasation, facilitates attraction of CTCs to the pre-metastatic site [124], and promotes metastasis and tumor progression.

In fact, melanoma TEXs have important role in the cancer metastasis to lung and this effect is associated with the degree of metastatic ability of melanoma. Interestingly, TEX obtained from highly metastatic melanomas in comparison to those from the poorly metastatic showed a greater burden on the lung [123].

The administration of melanoma TEXs into a naïve mice demonstrated lung residency within 24h and increased permeability of lung epithelial cells at the TEX-induced pre-metastatic site. This effect was associated with upregulated expression of pre-metastatic niche effector molecules such as S100A8, s100A9, and TNF-α at the site of TEX injection [125]. TEX administration also caused an upregulation of inflammatory and extracellular matrix (ECM)-related melanoma genes [123].

The platelets play an important role in tumor progression, angiogenesis, and metastasis in lung cancer through the exosomes. Platelet-derived exosomes (PMV) shuttle integrin α2β (CD41) which enhance proliferation and tumor progression in mice [122].

In addition, the exosomal small RNAs in lung cancer trigger TLR3 activation in epithelial cells which in turn increase chemokine production and subsequent recruitment of inflammatory immune cells to the lung, promoting pre-metastatic niche formation [126].

Exosome surface integrins have determinant role in the target cell selection. A specific profiles of surface integrins direct tumor-derived exosomes (TEX) to a specific organ, driving metastatic organotropism [87]. For example, lung fibroblasts with up-regulated s100 gene uptake exosomes with surface α6α4 integrin heterodimer which in turn promote lung PMN formation [87].

Exosomes with surface α6β4 and α6β1 integrins such as those derived from 4175-LuT breast cancer cells promote lung metastasis with preferentially localization in laminin-rich regions of lung [87].

A common metastatic site for NSCLC is bone which causes osteolytic lesions [127]. In NSCLC, cancerous cells with upregulated EGFR [128], releasing exosomes contain amphiregulin (AREG), an EGFR ligand [129], that can trigger EGFR pathway in pre-osteoclasts and enhance the expression of RANKL and proteolytic enzymes which in turn can provoke a vicious cycle driving osteolytic bone metastasis [128]. On the contrary, exosomes derived from highly metastatic bone tumors are preferentially localized to lung and promote metastatic process [130].

The lung is one of the most common metastatic targets and cancer metastatic to the lung is more common than primary lung cancers [130]. The molecular mechanism underlying this tissue-specific metastasis is not completely understood but possibly TEXs play crucial roles in this process. More focused studies is needed to understand the precise mechanism of exosome-specific functions in tumor microenvironment, driving lung metastasis.

3.3 Tuberculosis

Mycobacterium tuberculosis (MTB) has ability to remain passive in human body for years without causing any symptoms which is referred to as "latent tuberculosis." This ability has hampered the disease eradication up to now [131].

Til 2010, some studies have demonstrated the presence of mycobacterial proteins in the exosomes released from the *M. tuberculosis* infected macrophages by western blots.

The comprehensive proteomic analysis was used to identify the protein content of the exosomes derived from the macrophages infected with either live or dead *M. tuberculosis*. The results demonstrated the incorporation of a cluster of 41 mycobacterial proteins in these exosomes [132]. Further analysis of this cluster demonstrated the presence of highly immunogenic *Mycobacterium* proteins including antigen SAT-6 (Rv3875), Ag85 complex (Rv3804c, Rv1886c, and Rv0129c), MPT64 (1980c), and MPT63 (1926c) [133].

Subsequent studies analyzed serum exosomes from tuberculosis (TB) patients and identified 20 mycobacterial proteins in these exosomes with dominant presence of 85b, BfrB, GlcB, and Mpt64 antigens [134]. Interestingly, the pulmonary and extrapulmonary TB were distinguishable based on serum exosomal markers such as MPT64 [134].

The effort to distinguish active and latent TB also showed some evidences of specific exosomal markers [134]. The possibility of latent TB detection is particularly important within an endemic population, which makes it possible to monitor the risk persons and limit the transmission of infection.

The exosomal content from macrophages after treatment with *M. tuberculosis* culture filtrate proteins (CFP) activates both innate and acquired immune responses. The majority of proteins cargo of these exosomes overlapped with those present in the exosomes isolated from *M. tuberculosis*-infected macrophages. These evidences suggested that the exosomes with *M. tuberculosis* antigen cargo may be an appropriate approach for developing tuberculosis CFP-based vaccine [135].

Exosomes may shuttle pathogen-associated molecular patterns (PAMPs) and modulate immune response in this manner [136]. *M. tuberculosis* can induce partial resistance to INF-ɣ in infected macrophages. This effect is performed by Mtb- associated PAMPs such as the 19 kDa lipoprotein and mycolyl-arabino galactan that bind to macrophage receptors such as TLR 2 and result in the suppression of IFN-gamma responsiveness [137]. This behavior can be mimicked by exosomes released from Mtb-infected macrophages [138].

In fact, macrophages that are exposed to the exosomes containing mycobacterial components had limited response to IFN-γ stimulation which indicate that exosomes can act as carriers of PAMPs and lead to the suppression of a host immune response in the target cells [139].

Interestingly, Mtb infection modulate the profile of host miRNA expression [133]. Infected human macrophages with virulent Mtb H37Rv and avirulent *M. bovis* BCG induced patterns of miRNA expression that were mostly similar between the two live mycobacteria, while a substantially different pattern emerged from infection with killed Mtb bacilli [140]. This result reveals the active influence of live intracellular bacteria on host cell miRNA network.

On the other hand, the amount and content of exosomal miRNAs significantly changes after *M. tuberculosis* infection in macrophages. Infection of macrophage-derived monocytes (MDM) with BCG induces release of specific exosomal miRNAs from BCG-infected MDMs that mostly involved in host metabolism and energy production [32]. Additionally, more than 100 mRNAs were unique to the exosome released by *Mtb* infected cells that mostly were involved in the modulation of host immune responses [141]. Overall, these findings emphasize the functional and diagnostic potentials of exosomal mRNAs and miRNAs in tuberculosis.

3.4 COPD

COPD is characterized by slow progressive airflow obstruction, accompanied by infiltration of inflammatory cells, thickening of the alveolar wall and ultimately damage the lung parenchymal and epithelium. The main characteristic of COPD is the presence of inflammation [142, 143] and exosomes have intensive roles in this pathologic condition [143].

In emphysema, injury of the ECs within lung parenchyma play the main role [144]. In patients with stable COPD, the production of endothelial-derived exosomes increases during exacerbation and the level of these particles is significantly correlated with destruction and airflow limitation in the lungs. These microparticles enclose vascular endothelial cadherin, platelet EC adhesion molecules, and *E*-selectin [145]. These findings clearly emphasize the roles of exosomes in lung damage and pathophysiology of COPD. The exposure of airway to irritants such as cigarette smoke (CS) can trigger epithelial cell injury and COPD development by promoting pulmonary capillary vasculature destruction, epithelial cell senescence, acerbation as well as airway remodeling which together results in the loss of lung function [146]. Damaged epithelial cells act as a source of inflammatory mediators such as TNF-α, IL-1α, GM-CSF, TGF-ß, and CXCL-8 that exert their effect through autocrine and paracrine effects.

In this way, TGF-ß promotes myofibroblast differentiation and fibrosis development and induces airway remodeling [147]. The paracrine effect of this mediators, at least in part, is modulated by exosomes [148]. Continues exposure of lung epithelial cells to CS induces the secretion of exosomes that were enriched with CCN1. CCN1 as an extra cellular matrix has important role in remodeling and repair process [149] and through wnt signaling pathways enhances CXCL-8 expression [150]. Hence, CCN1-enriched exosomes could lead to the recruitment of inflammatory cells such as neutrophils to lung mesenchyme or parenchyma via paracrine induction of CXCL-8 secretion which subsequently results in lung tissue fibrosis [150]. CS also increases the level of exosome secretion by lung epithelial cells [151]. Almost 1.5% of COPD patients suffer from Alpha 1 anti-trypsin (AAT1) deficiency [145]. AAT1 is a glycoprotein serine protease inhibitor that protects the lungs from damage due to inflammatory insult [152]. Transportation of AAT1 from lung ECs to the alveolar epithelium and air space by exosomes with rapid uptake by epithelial cells. Efficient trafficking of enriched AAT1 exosomes across the lung endothelial barrier mediate protection against CS [153]. One of the important symptom in COPD is EMT that leads to a small airway remodeling and fibrosis [154]. EMT may be accompanied by angiogenesis in the large airway and leads to development of pro-cancer stromal niche [155]. Regarding the biological roles of exosome in shuttling bioactive molecules and intracellular communication, exosomes impart important effect on EMT process [155]. COPD progression is accompanied by an increase in small airway wall thickness that occur followed by a repair and remodeling process mediated by the fibroblast of epithelial mesenchymal trophic unit (EMTU). The EMTU consists of one epithelial cell layer, the basement membrane zone (BMZ), and mesenchymal cells including fibroblasts [1]. Remodeling process is the result of cross talk between injured and stressed epithelium with the underlying fibroblast layer which result in mesenchymal cell proliferation and remodeling. This process is highly regulated by the exosomes. In addition to being important drivers of COPD pathophysiology, exosomes have potential to serve as biomarkers of COPD progress or improvement [145]. The amount of lung EC damage in COPD may be determined by the level of circulating exosomes. It is suggested that the plasma level of exosomal miRNAs is associated with the degree of damages to the skeletal muscle and can be considered as the potential biomarker of skeletal muscle dysfunction [156]. The different expression of five exosomal miRNAs in BAL fluid of COPD patients were found to be associated with degree of disease. These miRNAs target ribosomal s6 kinase (S6K) which is part of TORC1 signaling pathway and a crucial regulator of skeletal muscle wasting. This finding emphasize the potential of exosomal miRNAs as noninvasive diagnostic biomarkers in COPD progression and treatment [156].

3.5 Sarcoidosis

Pulmonary sarcoidosis is a systematic and inflammatory disease that predominantly affects lungs. Sarcoidosis is characterized by the presence of non-caseating granulomas (NCGs) and interferon producing T cells in affected organ followed by inflammation and tissue damage especially in the lung [157]. Sarcoidosis etiology is not completely understood but based on the molecular characteristic of sarcoidosis pathophysiology, it seems that exosomes play the major role in the initiation and progression of inflammation in this disease. BAL fluid of a sarcoidosis patient contains an increased number of exosomes in comparison to healthy subjects. The exosomes isolated from the BAL fluid from sarcoidosis patients increased the expression of IFN-gamma and IL-13 [158] by peripheral blood mononuclear cells (PBMCs), and CXCL-8 production by epithelial cells [29]. In addition, a different profile of exosomal miRNAs was detected in serum and BAL fluid exosomes in sarcoidosis patients [159]. Overall, the potential roles of exosomes in the initiation and progression of inflammation in sarcoidosis is deniable, however, more sophisticated and practical studies is needed to fulfill the promise of the exosome usage for the treatment of sarcoidosis.

3.6 Other respiratory diseases

In addition to the lung diseases mentioned above, the function of exosomes in some of other lung disorders has been investigated.

In hypoxia, the expression of pro-inflammatory mediators and alternatively activated macrophages increases and cause the progression of hypoxic pulmonary hypertension [34].

Hypoxia triggers endoplasmic reticulum (ER) stress and leads to the secretion of a large number of exosomes from the lung epithelial cells. These exosomes contain high level of caspase-3 that activate AM and promote inflammatory responses resulting in lung injury. These observations emphasize the role of exosomes in cross talk between lung epithelial cells and macrophages toward the development of tissue injury [160].

Idiopathic pulmonary fibrosis (IPF) is a chronic lung disease characterized by a progressive lung fibrosis which leads to irreversible decline in lung function [161].

In exosomes from IPF patient, decreased level of anti-fibrotic miRNAs such as miR-141 and increased level of fibrogenic miRNAs such as miR-7 has been detected compared with healthy subjects. Interestingly, a significant correlation was found between the degree of miR-7 upregulation and the burden of disease and also between miR-125b upregulation and milder disease. These results highlight the potential roles of exosomal miRNAs as biomarker to gain an insight into the pathogenesis of IPF [162].

Cystic fibrosis (CF) is a progressive lung disorder that is characterized by a genetic deficiency in CF transmembrane conductance regulator (CFTR) that affects the chloride channel activity and results in the massive neutrophil granulocyte influx in the airways and mucostasis [163]. Since the exosomes secretion from the BECs is increased in CF patients, so the use of biofluid exosomes as the source of disease biomarkers for noninvasive diagnosis of CF is worth considering [164].

In addition, the high level of prolyl endopeptidase (PE) enzyme is detected in the exosome isolated from the CF patient with persistent bacterial infection.

PE enzyme is involved in the synthesis of chemo-attractant tri-peptide Pro-Gly-Pro (PGP) from collagen by neutrophils [165]. This enzyme is secreted by the airway epithelial cells packaged in exosomes. The production and release of PE are enhanced by lipopolysaccharides (LPS) which act through TLR4.

The presence of enhanced level of PE enzyme in CF patient may be responsible for the hyperinflammatory state observed in these patients [165].

In gut biology, the secretion of exosomes from the gastrointestinal epithelium to the basolateral side is linked to the antigen presentation, whereas luminal release of exosomes from the biliary and intestinal epithelium is highly triggered upon TLR4 signaling and increased upon infection [166]. The differential route of exosome release is worth considering in CF patient. Exosomes produced by airway epithelial cells have particular size and physical properties based on the type and physiological condition of their cell of origin, which may have diagnosis value in aberrant conditions [37]. For example, in CF patient, the amount and type of exosomal surface mucins are altered and characterized by excessive mucus production. Mucin surface determines the size and surface charge of exosomes.

In addition, the alteration in miRNA and protein content of exosomes was detected in CF patient which reflects the cellular change upon an inflammatory challenge. These findings emphasize the role of exosomes in promoting a hyper inflammatory state in the airway microenvironment in CF [167].

A major aim in current CF research is the restoration of CFTR function which has been promising with the emergence of exosome utilization as a target of drug delivery [163]. Usage of exosomes as a vector to transfer of CFTR to human CFTR-deficient nasal epithelial cells was successful and leads to restore CFTR function in these cells in a dose-dependent manner [163]. Exosomes derived from CFTR-positive Calu-3 cells or from A549 cells transduced with an adenoviral vector overexpressing green fluorescent protein (GFP)-tagged CFTR (GFP-CFTR) were able to package and transfer the GFP-CFTR glycoprotein and mRNA (GFP-CFTR) to target cell and correct the genetic defect in human CF cells [163].

4 Clinical applications of exosomes in respiratory diseases: Clinical trials and future perspectives

Exosomes have important roles in the maintenance of the homeostatic state in respiratory microenvironment and supporting lung normal function.

In lung, exosomes secret a broad range of cell types from structural to immune cells. The number and content of these exosomes differ based on the disease status and carried distinct disease-specific constituents. Regarding these evidences, exosomes can provide novel biomarkers in diagnosis and treatment monitoring for a various range of respiratory diseases and may also be used for therapeutic interventions.

Exosomes derived from MSCs is now considered for their potential for wound healing and tissue repair particularly in tissue generation in lung [164].

The MSC-derived exosomes are also of interest because of their antiapoptotic and anti-inflammatory properties [168] that can be effective in lung chronic inflammatory conditions [168]. These exosomes have capacity to accelerate lung tissue regeneration and wound healing which may be effective for soothing and repair of airway remodeling in asthma [168].

In an animal model of ARDS, MSC-derived exosomes induced the expression of keratinocyte growth factor (KGF) in the injured alveolus, improved the protein permeability, and reduced the inflammation in the lung [169]. Clinical trial using exosome-based therapy in ARDS is being now conducted [170].

Using DC-derived exosomes (dexosomes) loaded with tumor antigens with the aim of NSCLC immunotherapy is in phase I clinical trials [171]. The result of this trial showed that these exosomes can induce both innate and adaptive immune responses in the patients and also introduced "dexosome therapy" as a safe and feasible method that may be a good alternative for current therapy methods for cancer or other lung inflammatory diseases.

Further studies are needed to find the effect of this therapy method on disease stability and on long-term survival.

Exosomes can also be harnessed as a trap to subside immune response. MC-derived exosomes can capture serum-free IgE via their surface FCεR1 receptors and in turn limit the subsequent activation of MCs. In addition, CD8+ cell-derived exosomes can induce an antigen-specific tolerance in mice and alleviate allergic contact dermatitis (ACD) in this model [75].

Since the exosomal content reflect the aberrations in their parent cells and because of their stability in the biological fluids, exosomes may serve as a promising "liquid biopsy" biomarkers of lung cancer [172]. Exosome-based biomarkers bypass the need to surgery for tissue biopsy which provide a noninvasive diagnostic approach [172]. Exosomal biomarkers such as miRNAs and proteins have been measured in serum and plasma of lung cancer patients. Microarray-based analysis of serum exosomal miRNAs of lung cancer patient showed that upregulation of miR-21 and miR-4257 is significantly associated with a recurrence of the disease [173]. The analysis of serum exosomal proteins in 276 NSCLC indicated some of them such as NY-ESO-1 have correlation with survival [174].

NSCLC TEXs also show enhanced expression of EGFR [175]. Two tumor suppressor miRNA, miR-51 and miR-373, were reduced in lung cancer patients and this diminution was associated with poor prognosis [175].

In addition, exosomal miRNAs analysis in lung cancer patient showed the variation in the expression level of miR-378a, miR-379, miR-139-5p, miR-200b-5p [70], miR-21 [173], miR-155 [173], miR-23b, miR-10b-5p [173], and miR-4257 [173] in comparison to healthy subjects.

Exosomal miRNAs also have been considered as a marker for therapeutic evaluation in lung cancer. The tumorgenic miR-208a and miR-1246 that, respectively, target p21 and DR5 mRNAs are considered as a marker for resistance to radiotherapy [176].

Despite providing noninvasive methods with higher sensitivity and specificity in diagnosis or therapeutic monitoring of lung cancer; TEX-based biomarkers still face many challenges. The lack of standardized exosome isolation methods and the heterogeneity in cancer-derived exosomes are in these issue [177]. However, there is much interest in combining the information of miRNAs and proteins cargo from lung cancer TEXs with next-generation sequencing and proteomic analysis to drive greater achievement into TEX-based lung cancer diagnosis or therapy.

A new method has been recently suggested the use of surface-enhanced Raman Spectroscopy (SERS) combined with principle component analysis (PCA) to characterize the exosomes surface protein and lipid molecules. Since lipids and proteins have their specific Raman spectra, the exosomes from tumor or normal cells with different pattern of surface lipid and proteins vary in their Raman spectral patterns. This method differentiate lung cancer TEXs from the normal cell exosomes by 95.3% sensitivity and 97.3% specificity [178]. Despite current challenges in exosome congenital methods such as heterogeneity in isolated tumor-derived exosomes as well as requirement of large amounts of highly concentrated sample, this approach may be good choice to be translated in clinical practice [178].

Mass spectrometric quantification of 1369 serum exosomal proteins in 46 patient with advanced NSCLC demonstrated CD91 as a exosomal antigen in lung adenocarcinoma [179].

Another effort in this concept introduced a multivariate extracellular vesicle array (EV Array) as a potential complementary method in the diagnosis of NSCLC with phenotyping the plasma exosomes. Using this approach, a panel of 30 exosomal surface protein markers including CD91, CD317, and EGFR was introduced that could identify patients with 75% specificity [180].

The analysis of urinary exosomes in NSCLC patients with proteomic mass spectrometry showed that the presence of Lucien rich alpha-2-glycoprotein 1 (LRG1) in these exosomes is significantly higher than those from healthy subjects. LRG1 was also highly expressed in NSCLC tissue [181]. Other markers such as CD171, CD151, and tetra-spanin 8 were also reported as potential exosomal biomarkers for NSCLC diagnosis [182].

In addition to diagnosis applications, exosomes has been of particular interest as suitable vehicles for the delivery of drug or nucleic acids to target organ.

Cow milk exosomes loaded with Aferin-A were used for drug delivery in a murine model of lung cancer. The exosome containing Aferin-A apply a tumor inhibitory effect after injection in murine lung cancer models [183].

Exosomes released from the brain endothelial cell line (bEND.3) have the ability to pass through the blood-brain barrier and showed successful effect in the in vivo drug delivery and reduction in VEGF level in a brain tumor [184].

The unique biological properties of exosomes make them valuable targets for developing novel strategies in cancer therapy. Considering that the tumor-derived exosomes shuttle tumor-specific antigens, they are very attractive as anticancer vaccines [185].

Since the target selection of exosomes is organotropism and performed through integrin-mediated signaling [87], thus blocking integrins by decoy peptides can prevent exosome fusion and uptake which in turn inhibit tumor progression [186].

One of the recent advances in immunotherapy of lung cancer is blocking the inflammatory signals and the negative regulators of T-cell activation such as PD-1 and PD-L1 in the tumor microenvironment that is described to be modulated by exosomes [174].

The strategies to block exosomal secretion or inhibit exosome-mediated cellular cross talk may be functional in the prevention of a favorable tumor microenvironment development [185].

However, exosomes may significantly increase the efficacy of immunotherapy in lung cancer by regulating anti-inflammatory signals in tumor microenvironment [187].

Overall, because of their particular properties such as specific targeting, small size, shuttling signaling, and biological molecules, as well as the ability to cross biological barriers, exosomes are considered in wider spectrum of medical applications from cancer diagnosis biomarkers to cancer treatment, drug delivery, and tumor immunotherapy.

Overall, current knowledge highlights the role of exosomes in cellular cross talk within the lung microenvironment and also in pathogenegenicity. Exosomes offer not only noninvasive diagnostic methods but also are highly considered as new potential avenues for lung treatment disorders.

Despite the presence of constraints and inconvenient nature of exosome isolation and purification methods, it is expected that exosomes will be exploited in cancer concepts in the near future. However, further more sophisticated clinical studies are needed to address these limitations in exosome usage before exosome-based technologies can be translated in to clinical application.

References

[1] Alipoor SD, et al. Exosomes and exosomal miRNA in respiratory diseases. Mediators Inflamm 2016;2016, 5628404.

[2] Raposo G, Stoorvogel W. Extracellular vesicles: exosomes, microvesicles, and friends. J Cell Biol 2013;200(4):373–83.

[3] Clark DJ, et al. Triple SILAC quantitative proteomic analysis reveals differential abundance of cell signaling proteins between normal and lung cancer-derived exosomes. J Proteomics 2016;133:161–9.

[4] Johnstone RM, et al. Vesicle formation during reticulocyte maturation. Association of plasma membrane activities with released vesicles (exosomes). J Biol Chem 1987;262(19):9412–20.
[5] Pan BT, Johnstone RM. Fate of the transferrin receptor during maturation of sheep reticulocytes in vitro: selective externalization of the receptor. Cell 1983;33(3):967–78.
[6] Trams EG, et al. Exfoliation of membrane ecto-enzymes in the form of micro-vesicles. Biochim Biophys Acta 1981;645(1):63–70.
[7] Bhatnagar S, Schorey JS. Exosomes released from infected macrophages contain *Mycobacterium avium* glycopeptidolipids and are proinflammatory. J Biol Chem 2007;282(35):25779–89.
[8] Kourembanas S. Exosomes: vehicles of intercellular signaling, biomarkers, and vectors of cell therapy. Annu Rev Physiol 2015;77:13–27.
[9] Jan AT, et al. Perspective insights of exosomes in neurodegenerative diseases: a critical appraisal. Front Aging Neurosci 2017;9:317.
[10] Quek C, Hill AF. The role of extracellular vesicles in neurodegenerative diseases. Biochem Biophys Res Commun 2017;483(4):1178–86.
[11] Eissa NT. The exosome in lung diseases: message in a bottle. J Allergy Clin Immunol 2013;131(3):904–5.
[12] Valadi H, et al. Exosome-mediated transfer of mRNAs and microRNAs is a novel mechanism of genetic exchange between cells. Nat Cell Biol 2007;9(6):654.
[13] Tickner JA, et al. Functions and therapeutic roles of exosomes in cancer. Front Oncol 2014;4:127.
[14] Mathivanan S, Simpson RJ. ExoCarta: a compendium of exosomal proteins and RNA. Proteomics 2009;9(21):4997–5000.
[15] Cheng L, et al. Exosomes provide a protective and enriched source of miRNA for biomarker profiling compared to intracellular and cell-free blood. J Extracell Vesicles 2014;3(1):23743.
[16] Hough KP, et al. Exosomes in immunoregulation of chronic lung diseases. Allergy 2017;72(4):534–44.
[17] Samanta S, et al. Exosomes: new molecular targets of diseases. Acta Pharmacol Sin 2018;39(4):501–13.
[18] Barile L, Vassalli G. Exosomes: therapy delivery tools and biomarkers of diseases. Pharmacol Ther 2017;174:63–78.
[19] Masyuk AI, Masyuk TV, Larusso NF. Exosomes in the pathogenesis, diagnostics and therapeutics of liver diseases. J Hepatol 2013;59(3):621–5.
[20] Andaloussi ELS, et al. Extracellular vesicles: biology and emerging therapeutic opportunities. Nat Rev Drug Discov 2013;12(5):347–57.
[21] Barile L, et al. Extracellular vesicles from human cardiac progenitor cells inhibit cardiomyocyte apoptosis and improve cardiac function after myocardial infarction. Cardiovasc Res 2014;103(4):530–41.
[22] Admyre C, et al. Exosomes with major histocompatibility complex class II and co-stimulatory molecules are present in human BAL fluid. Eur Respir J 2003;22(4):578–83.
[23] Ismail N, et al. Macrophage microvesicles induce macrophage differentiation and miR-223 transfer. Blood 2012;121:984–95. https://doi.org/10.1182/blood-2011-08-374793.
[24] Rose MC, Voynow JA. Respiratory tract mucin genes and mucin glycoproteins in health and disease. Physiol Rev 2006;86(1):245–78.
[25] Fujita Y, et al. Extracellular vesicles in lung microenvironment and pathogenesis. Trends Mol Med 2015;21(9):533–42.
[26] Mortaz E, et al. Exosomes in severe asthma: update in their roles and potential in therapy. Biomed Res Int 2018;2018:2862187.

[27] Kulshreshtha A, et al. Proinflammatory role of epithelial cell-derived exosomes in allergic airway inflammation. J Allergy Clin Immunol 2013;131(4):1194–203, 1203.e1–14.

[28] Kesimer M, et al. Characterization of exosome-like vesicles released from human tracheobronchial ciliated epithelium: a possible role in innate defense. FASEB J 2009;23(6):1858–68.

[29] Qazi KR, et al. Proinflammatory exosomes in bronchoalveolar lavage fluid of patients with sarcoidosis. Thorax 2010;65(11):1016–24.

[30] Cordazzo C, et al. Rapid shedding of proinflammatory microparticles by human mononuclear cells exposed to cigarette smoke is dependent on Ca^{2+} mobilization. Inflamm Res 2014;63(7):539–47.

[31] Alipoor SD, et al. miR-1224 expression is increased in human macrophages after infection with Bacillus Calmette-Guerin (BCG). Iran J Allergy Asthma Immunol 2018;17(3):250–7.

[32] Alipoor SD, et al. Bovis Bacillus Calmette-Guerin (BCG) infection induces exosomal miRNA release by human macrophages. J Transl Med 2017;15(1):105.

[33] Levänen B, et al. Altered microRNA profiles in bronchoalveolar lavage fluid exosomes in asthmatic patients. J Allergy Clin Immunol 2013;131(3):894–903, e8.

[34] Lee C, et al. Exosomes mediate the cytoprotective action of mesenchymal stromal cells on hypoxia-induced pulmonary hypertension. Circulation 2012;126(22):2601–11.

[35] Merluzzi S, et al. Mast cells, basophils and B cell connection network. Mol Immunol 2015;63(1):94–103.

[36] Théry C, et al. Indirect activation of naïve CD4+ T cells by dendritic cell-derived exosomes. Nat Immunol 2002;3(12):1156.

[37] Kesimer M, Gupta R. Physical characterization and profiling of airway epithelial derived exosomes using light scattering. Methods 2015;87:59–63.

[38] Park JA, et al. Tissue factor-bearing exosome secretion from human mechanically stimulated bronchial epithelial cells in vitro and in vivo. J Allergy Clin Immunol 2012;130(6):1375–83.

[39] Haj-Salem I, et al. Fibroblast-derived exosomes promote epithelial cell proliferation through TGF-β2 signalling pathway in severe asthma. Allergy 2018;73(1):178–86.

[40] De Vooght V, et al. B-lymphocytes as key players in chemical-induced asthma. PLoS ONE 2013;8(12):e83228.

[41] Natarajan P, Guernsey LA, Schramm CM. Regulatory B cells in allergic airways disease and asthma. In: Regulatory B cells. Springer; 2014. p. 207–25.

[42] Harris DP, et al. Reciprocal regulation of polarized cytokine production by effector B and T cells. Nat Immunol 2000;1(6):475.

[43] Admyre C, et al. B cell–derived exosomes can present allergen peptides and activate allergen-specific T cells to proliferate and produce TH2-like cytokines. J Allergy Clin Immunol 2007;120(6):1418–24.

[44] Clayton A, et al. Induction of heat shock proteins in B-cell exosomes. J Cell Sci 2005; 118(Pt 16):3631–8.

[45] Blanchard N, et al. TCR activation of human T cells induces the production of exosomes bearing the TCR/CD3/zeta complex. J Immunol 2002;168(7):3235–41.

[46] Zhang H, et al. CD4(+) T cell-released exosomes inhibit CD8(+) cytotoxic T-lymphocyte responses and antitumor immunity. Cell Mol Immunol 2011;8(1):23–30.

[47] Peters PJ, et al. Molecules relevant for T cell-target cell interaction are present in cytolytic granules of human T lymphocytes. Eur J Immunol 1989;19(8):1469–75.

[48] Ventimiglia LN, Alonso MA. Biogenesis and function of T cell-derived exosomes. Front Cell Dev Biol 2016;4:84.

[49] Cai Z, et al. Activated T cell exosomes promote tumor invasion via Fas signaling pathway. J Immunol 2012;188(12):5954–61.

[50] Wahlgren J, et al. Activated human T cells secrete exosomes that participate in IL-2 mediated immune response signaling. PLoS ONE 2012;7(11):e49723.

[51] Stone KD, Prussin C, Metcalfe DD. IgE, mast cells, basophils, and eosinophils. J Allergy Clin Immunol 2010;125(2):S73–80.

[52] Mazzeo C, et al. Exosome secretion by eosinophils: a possible role in asthma pathogenesis. J Allergy Clin Immunol 2015;135(6):1603–13.

[53] Lässer C, et al. Exosomes in the nose induce immune cell trafficking and harbour an altered protein cargo in chronic airway inflammation. J Transl Med 2016;14(1):181.

[54] Virk H, Arthur G, Bradding P. Mast cells and their activation in lung disease. Transl Res 2016;174:60–76.

[55] Brightling CE, et al. Mast-cell infiltration of airway smooth muscle in asthma. N Engl J Med 2002;346(22):1699–705.

[56] Rossios C, et al. Impaired innate immune gene profiling in airway smooth muscle cells from chronic cough patients. Biosci Rep 2017;37(6), BSR20171090.

[57] Skokos D, et al. Mast cell-derived exosomes induce phenotypic and functional maturation of dendritic cells and elicit specific immune responses in vivo. J Immunol 2003;170(6):3037–45.

[58] Tkaczyk C, et al. In vitro and in vivo immunostimulatory potential of bone marrow-derived mast cells on B-and T-lymphocyte activation. J Allergy Clin Immunol 2000;105(1):134–42.

[59] Gauchat J-F, et al. Induction of human IgE synthesis in B cells by mast cells and basophils. Nature 1993;365(6444):340.

[60] Li F, et al. Mast cell-derived exosomes promote Th2 cell differentiation via OX40L-OX40 ligation. J Immunol Res 2016;2016:3623898.

[61] Iwasaki A, Medzhitov R. Control of adaptive immunity by the innate immune system. Nat Immunol 2015;16(4):343.

[62] Esser J, et al. Exosomes from human macrophages and dendritic cells contain enzymes for leukotriene biosynthesis and promote granulocyte migration. J Allergy Clin Immunol 2010;126(5):1032–40, e4.

[63] Vallhov H, et al. Dendritic cell-derived exosomes carry the major cat allergen Fel d 1 and induce an allergic immune response. Allergy 2015;70(12):1651–5.

[64] Liu Y, Cao X. Characteristics and significance of the pre-metastatic niche. Cancer Cell 2016;30(5):668–81.

[65] Gelb AF, Christenson SA, Nadel JA. Understanding the pathophysiology of the asthma-chronic obstructive pulmonary disease overlap syndrome. Curr Opin Pulm Med 2016;22(2):100–5.

[66] Lambrecht BN, Hammad H. The immunology of asthma. Nat Immunol 2015;16(1):45.

[67] Lötvall J, et al. Asthma endotypes: a new approach to classification of disease entities within the asthma syndrome. J Allergy Clin Immunol 2011;127(2):355–60.

[68] Admyre C, et al. Exosomes–nanovesicles with possible roles in allergic inflammation. Allergy 2008;63(4):404–8.

[69] Torregrosa Paredes P, et al. Bronchoalveolar lavage fluid exosomes contribute to cytokine and leukotriene production in allergic asthma. Allergy 2012;67(7):911–9.

[70] Cazzoli R, et al. microRNAs derived from circulating exosomes as noninvasive biomarkers for screening and diagnosing lung cancer. J Thorac Oncol 2013;8(9):1156–62.

[71] Canas JA, et al. Exosomes from eosinophils autoregulate and promote eosinophil functions. J Leukoc Biol 2017;101(5):1191–9.

[72] Sastre B, et al. Novel modulators of asthma and allergy: exosomes and microRNAs. Front Immunol 2017;8:826.

[73] Lässer C. Exosomal RNA as biomarkers and the therapeutic potential of exosome vectors. Expert Opin Biol Ther 2012;12(Suppl. 1):S189–97.
[74] Prado N, et al. Exosomes from bronchoalveolar fluid of tolerized mice prevent allergic reaction. J Immunol 2008;181(2):1519–25.
[75] Rebane A, Akdis CA. MicroRNAs in allergy and asthma. Curr Allergy Asthma Rep 2014;14(4):424.
[76] Xie G, et al. Mast cell exosomes can suppress allergic reactions by binding to IgE. J Allergy Clin Immunol 2018;141(2):788–91.
[77] Dvorak AM. Degranulation and recovery from degranulation of basophils and mast cells. In: Ultrastructure of mast cells and basophils. Karger Publishers; 2005. p. 205–51.
[78] Liu L, Yang J, Huang Y. Human airway smooth muscle cells express eotaxin in response to signaling following mast cell contact. Respiration 2006;73(2):227–35.
[79] Francisco Garcia A, et al. LSC abstract altered small RNA cargo in severe asthma exosomes. Eur Respiratory Soc 2016;48:PP101.
[80] Suzuki M, et al. Altered circulating exosomal RNA profiles detected by next-generation sequencing in patients with severe asthma. Eur Respiratory Soc 2016;48:PA3410.
[81] Andre F, et al. Exosomes for cancer immunotherapy. Ann Oncol 2004;15(Suppl 4):iv141–4.
[82] Muller L, et al. Tumor-derived exosomes regulate expression of immune function-related genes in human T cell subsets. Sci Rep 2016;6:20254.
[83] Haratani K, et al. Tumor immune microenvironment and nivolumab efficacy in EGFR mutation-positive non-small cell lung cancer based on T790M status after disease progression during EGFR-TKI treatment. Ann Oncol 2017;28(7):1532–9.
[84] Sun T, et al. Role of exosomal noncoding RNAs in lung carcinogenesis. Biomed Res Int 2015;2015:125807.
[85] Zhang Y, Wang XF. A niche role for cancer exosomes in metastasis. Nat Cell Biol 2015;17(6):709–11.
[86] Sceneay J, Smyth MJ, Möller A. The pre-metastatic niche: finding common ground. Cancer Metastasis Rev 2013;32(3–4):449–64.
[87] Hoshino A, et al. Tumour exosome integrins determine organotropic metastasis. Nature 2015;527(7578):329.
[88] Théry C, Ostrowski M, Segura E. Membrane vesicles as conveyors of immune responses. Nat Rev Immunol 2009;9(8):581.
[89] Viaud S, et al. Dendritic cell-derived exosomes promote natural killer cell activation and proliferation: a role for NKG2D ligands and IL-15Rα. PLoS ONE 2009;4(3), e4942.
[90] Fabbri M, et al. MicroRNAs bind to Toll-like receptors to induce prometastatic inflammatory response. Proc Natl Acad Sci 2012;109(31):E2110–6.
[91] Whiteside TL. Immune modulation of T-cell and NK (natural killer) cell activities by TEXs (tumour-derived exosomes). Portland Press Limited; 2013.
[92] Szczepanski MJ, et al. Blast-derived microvesicles in sera from patients with acute myeloid leukemia suppress natural killer cell function via membrane-associated transforming growth factor-β1. Haematologica 2011;96(9):1302–9.
[93] Malmberg K-J, et al. Natural killer cell-mediated immunosurveillance of human cancer. Semin Immunol 2017;31:20–9.
[94] Berchem G, et al. Hypoxic tumor-derived microvesicles negatively regulate NK cell function by a mechanism involving TGF-β and miR23a transfer. Oncoimmunology 2016;5(4):e1062968.
[95] Clayton A, et al. Human tumor-derived exosomes down-modulate NKG2D expression. J Immunol 2008;180(11):7249–58.

[96] Cohnen A, et al. Surface CD107a/LAMP-1 protects natural killer cells from degranulation-associated damage. Blood 2013;122(8):1411–8.
[97] Filipazzi P, et al. Recent advances on the role of tumor exosomes in immunosuppression and disease progression. Semin Cancer Biol 2012;22(4):342–9.
[98] Huang S-H, et al. Epidermal growth factor receptor-containing exosomes induce tumor-specific regulatory T cells. Cancer Invest 2013;31(5):330–5.
[99] Zhou M, et al. Pancreatic cancer derived exosomes regulate the expression of TLR4 in dendritic cells via miR-203. Cell Immunol 2014;292(1–2):65–9.
[100] Chen W, et al. Tumor-related exosomes contribute to tumor-promoting microenvironment: an immunological perspective. J Immunol Res 2017;2017:1073947.
[101] Chalmin F, et al. Membrane-associated Hsp72 from tumor-derived exosomes mediates STAT3-dependent immunosuppressive function of mouse and human myeloid-derived suppressor cells. J Clin Invest 2010;120(2):457–71.
[102] Yang C, et al. Tumor-derived exosomes confer antigen-specific immunosuppression in a murine delayed-type hypersensitivity model. PLoS ONE 2011;6(8), e22517.
[103] Valenti R, et al. Tumor-released microvesicles as vehicles of immunosuppression. Cancer Res 2007;67(7):2912–5.
[104] Liu Y, et al. Contribution of MyD88 to the tumor exosome-mediated induction of myeloid derived suppressor cells. Am J Pathol 2010;176(5):2490–9.
[105] Song X, et al. Cancer cell-derived exosomes induce mitogen-activated protein kinase-dependent monocyte survival by transport of functional receptor tyrosine kinases. J Biol Chem 2016;291:8453–64. https://doi.org/10.1074/jbc.M116.716316.
[106] Ades PA. A controversial step forward: a commentary on the 2013 ACC/AHA guideline on the treatment of blood cholesterol to reduce atherosclerotic cardiovascular risk in adults. Coron Artery Dis 2014;25(4):360.
[107] Zhu Y, et al. A comprehensive proteomics analysis reveals a secretory path- and status-dependent signature of exosomes released from tumor-associated macrophages. J Proteome Res 2015;14(10):4319–31.
[108] Lamouille S, Xu J, Derynck R. Molecular mechanisms of epithelial–mesenchymal transition. Nat Rev Mol Cell Biol 2014;15(3):178.
[109] Rahman MA, et al. Lung cancer exosomes as drivers of epithelial mesenchymal transition. Oncotarget 2016;7(34):54852.
[110] Satelli A, Li S. Vimentin in cancer and its potential as a molecular target for cancer therapy. Cell Mol Life Sci 2011;68(18):3033–46.
[111] Havel LS, et al. Vimentin regulates lung cancer cell adhesion through a VAV2–Rac1 pathway to control focal adhesion kinase activity. Oncogene 2015;34(15):1979.
[112] Wei J, et al. Overexpression of vimentin contributes to prostate cancer invasion and metastasis via src regulation. Anticancer Res 2008;28(1A):327–34.
[113] Otsuki S, et al. Vimentin expression is associated with decreased survival in gastric cancer. Oncol Rep 2011;25(5):1235–42.
[114] Kim J, et al. Exosome cargo reflects TGF-β1-mediated epithelial-to-mesenchymal transition (EMT) status in A549 human lung adenocarcinoma cells. Biochem Biophys Res Commun 2016;478(2):643–8.
[115] Cao M, et al. MiR-23a regulates TGF-β-induced epithelial-mesenchymal transition by targeting E-cadherin in lung cancer cells. Int J Oncol 2012;41(3):869–75.
[116] Hsu YL, et al. Hypoxic lung cancer-secreted exosomal miR-23a increased angiogenesis and vascular permeability by targeting prolyl hydroxylase and tight junction protein ZO-1. Oncogene 2017;36(34):4929–42.

[117] Chang Y-H, et al. Down-regulation of TIMP-1 inhibits cell migration, invasion, and metastatic colonization in lung adenocarcinoma. Tumour Biol 2015;36(5):3957–67.

[118] Rabinowits G, et al. Exosomal microRNA: a diagnostic marker for lung cancer. Clin Lung Cancer 2009;10(1):42–6.

[119] Al-Nedawi K, et al. Endothelial expression of autocrine VEGF upon the uptake of tumor-derived microvesicles containing oncogenic EGFR. Proc Natl Acad Sci 2009;106(10):3794–9.

[120] Sheu J-J, et al. Administered circulating microparticles derived from lung cancer patients markedly improved angiogenesis, blood flow and ischemic recovery in rat critical limb ischemia. J Transl Med 2015;13(1):59.

[121] Lobb RJ, Lima LG, Möller A. Exosomes: key mediators of metastasis and pre-metastatic niche formation. Semin Cell Dev Biol 2017;67:3–10.

[122] Janowska-Wieczorek A, et al. Microvesicles derived from activated platelets induce metastasis and angiogenesis in lung cancer. Int J Cancer 2005;113(5):752–60.

[123] Peinado H, et al. Melanoma exosomes educate bone marrow progenitor cells toward a pro-metastatic phenotype through MET. Nat Med 2012;18(6):883.

[124] Peinado H, et al. Pre-metastatic niches: organ-specific homes for metastases. Nat Rev Cancer 2017;17(5):302–17.

[125] Hiratsuka S, et al. The S100A8–serum amyloid A3–TLR4 paracrine cascade establishes a pre-metastatic phase. Nat Cell Biol 2008;10(11):1349.

[126] Liu Y, et al. Tumor exosomal RNAs promote lung pre-metastatic niche formation by activating alveolar epithelial TLR3 to recruit neutrophils. Cancer Cell 2016;30(2):243–56.

[127] Taverna S, et al. Amphiregulin contained in NSCLC-exosomes induces osteoclast differentiation through the activation of EGFR pathway. Sci Rep 2017;7(1):3170.

[128] Dempke WC, Suto T, Reck M. Targeted therapies for non-small cell lung cancer. Lung Cancer 2010;67(3):257–74.

[129] Higginbotham JN, et al. Amphiregulin exosomes increase cancer cell invasion. Curr Biol 2011;21(9):779–86.

[130] Macklin R, et al. Extracellular vesicles secreted by highly metastatic clonal variants of osteosarcoma preferentially localize to the lungs and induce metastatic behaviour in poorly metastatic clones. Oncotarget 2016;7(28):43570.

[131] Velayati AA, et al. Populations of latent *Mycobacterium tuberculosis* lack a cell wall: isolation, visualization, and whole-genome characterization. Int J Mycobacteriol 2016;5(1):66–73.

[132] Giri PK, et al. Proteomic analysis identifies highly antigenic proteins in exosomes from *M. tuberculosis*-infected and culture filtrate protein-treated macrophages. Proteomics 2010;10(17):3190–202.

[133] Kruh-Garcia NA, Wolfe LM, Dobos KM. Deciphering the role of exosomes in tuberculosis. Tuberculosis (Edinb) 2015;95(1):26–30.

[134] Kruh-Garcia NA, et al. Detection of *Mycobacterium tuberculosis* peptides in the exosomes of patients with active and latent *M. tuberculosis* infection using MRM-MS. PLoS ONE 2014;9(7):e103811.

[135] Booton R, Lindsay MA. Emerging role of MicroRNAs and long noncoding RNAs in respiratory disease. Chest J 2014;146(1):193–204.

[136] Schorey JS, et al. Exosomes and other extracellular vesicles in host-pathogen interactions. EMBO Rep 2015;16(1):24–43.

[137] Fortune SM, et al. Mycobacterium tuberculosis inhibits macrophage responses to IFN-gamma through myeloid differentiation factor 88-dependent and -independent mechanisms. J Immunol 2004;172(10):6272–80.

[138] Singh PP, et al. Exosomes released from *M. tuberculosis* infected cells can suppress IFN-γ mediated activation of naïve macrophages. PLoS ONE 2011;6(4):e18564.
[139] Singh PP, et al. Exosomes released from *M. tuberculosis* infected cells can suppress IFN-gamma mediated activation of naive macrophages. PLoS ONE 2011;6(4):e18564.
[140] Furci L, et al. Alteration of human macrophages microRNA expression profile upon infection with *Mycobacterium tuberculosis*. Int J Mycobacteriol 2013;2(3):128–34.
[141] Singh PP, Li L, Schorey JS. Exosomal RNA from *Mycobacterium tuberculosis*-infected cells is functional in recipient macrophages. Traffic 2015;16(6):555–71.
[142] Pearson M. Is the primary mechanism underlying COPD: inflammation or ischaemia? COPD: J Chron Obstruct Pulmon Dis 2013;10(4):536–41.
[143] Barnes PJ, Shapiro S, Pauwels R. Chronic obstructive pulmonary disease: molecular and cellularmechanisms. Eur Respir J 2003;22(4):672–88.
[144] Kratzer A, et al. Endothelial cell adhesion molecule CD146: implications for its role in the pathogenesis of COPD. J Pathol 2013;230(4):388–98.
[145] Takahashi T, Kubo H. The role of microparticles in chronic obstructive pulmonary disease. Int J Chron Obstruct Pulmon Dis 2014;9:303.
[146] Di Stefano A, et al. Cellular and molecular mechanisms in chronic obstructive pulmonary disease: an overview. Clin Exp Allergy 2004;34(8):1156–67.
[147] Ardekani AM, Naeini MM. The role of microRNAs in human diseases. Avicenna J Med Biotechnol 2010;2(4):161.
[148] Li CJ, et al. Novel proteolytic microvesicles released from human macrophages after exposure to tobacco smoke. Am J Pathol 2013;182(5):1552–62.
[149] Moon H-G, et al. CCN1 secretion and cleavage regulate the lung epithelial cell functions after cigarette smoke. Am J Physiol Lung Cell Mol Physiol 2014;307(4):L326–37.
[150] Chew LP, et al. Fast detection of common geometric substructure in proteins. J Comput Biol 1999;6(3–4):313–25.
[151] Letsiou E, et al. Pathologic mechanical stress and endotoxin exposure increases lung endothelial microparticle shedding. Am J Respir Cell Mol Biol 2015;52(2):193–204.
[152] Sorroche PB, et al. Alpha-1 antitrypsin deficiency in COPD patients: a cross-sectional study. Archivos de Bronconeumología (English Edition) 2015;51(11):539–43.
[153] Lockett AD, et al. Active trafficking of alpha 1 antitrypsin across the lung endothelium. PLoS ONE 2014;9(4):e93979.
[154] Sohal SS, Walters EH. Role of epithelial mesenchymal transition (EMT) in chronic obstructive pulmonary disease (COPD). Respir Res 2013;14(1):120.
[155] Vella LJ. The emerging role of exosomes in epithelial–mesenchymal-transition in cancer. Front Oncol 2014;4:361.
[156] Donaldson A, et al. Increased skeletal muscle-specific microRNA in the blood of patients with COPD. Thorax 2013;68:1140–9. https://doi.org/10.1136/thoraxjnl-2012-203129.
[157] Maertzdorf J, et al. Common patterns and disease-related signatures in tuberculosis and sarcoidosis. Proc Natl Acad Sci 2012;109(20):7853–8.
[158] Lin J, et al. Exosomes: novel biomarkers for clinical diagnosis. Scientific World J 2015;2015:657086.
[159] Kishore A, et al. Detection of exosomal miRNA in pulmonary sarcoidosis. Eur Respir J 2014;44(Suppl 58):199.
[160] Moon H, et al. Lung epithelial cell-derived extracellular vesicles activate macrophage-mediated inflammatory responses via ROCK1 pathway. Cell Death Dis 2015;6(12), e2016.

[161] Ley B, Collard HR, King Jr. TE. Clinical course and prediction of survival in idiopathic pulmonary fibrosis. Am J Respir Crit Care Med 2011;183(4):431–40.
[162] Minnis P, et al. Serum exosomes from IPF patients display a fibrotic miRNA profile that correlates to clinical measures of disease severity. Eur Respiratory Soc 2015;46:PA3845.
[163] Vituret C, et al. Transfer of the cystic fibrosis transmembrane conductance regulator to human cystic fibrosis cells mediated by extracellular vesicles. Hum Gene Ther 2016;27(2):166–83.
[164] Porro C, et al. Isolation and characterization of microparticles in sputum from cystic fibrosis patients. Respir Res 2010;11(1):94.
[165] Szul T, et al. Toll-like receptor 4 engagement mediates prolyl endopeptidase release from airway epithelia via exosomes. Am J Respir Cell Mol Biol 2016;54(3):359–69.
[166] Hu G, et al. Release of luminal exosomes contributes to TLR4-mediated epithelial antimicrobial defense. PLoS Pathog 2013;9(4):e1003261.
[167] Batson B, et al. Qualitative and quantitative changes of gel forming mucins and exosomes in response to infection and inflammation in the airways. In: C60. All about cystic fibrosis. American Thoracic Society; 2016. p. A5565.
[168] Huang L, et al. Exosomes in mesenchymal stem cells, a new therapeutic strategy for cardiovascular diseases? Int J Biol Sci 2015;11(2):238.
[169] Zhu YG, et al. Human mesenchymal stem cell microvesicles for treatment of *Escherichia coli* endotoxin-induced acute lung injury in mice. Stem Cells 2014;32(1):116–25.
[170] Wilson JG, et al. Mesenchymal stem (stromal) cells for treatment of ARDS: a phase 1 clinical trial. Lancet Respir Med 2015;3(1):24–32.
[171] Morse MA, et al. A phase I study of dexosome immunotherapy in patients with advanced non-small cell lung cancer. J Transl Med 2005;3(1):9.
[172] Soung YH, et al. Exosomes in cancer diagnostics. Cancer 2017;9(1):8.
[173] Dejima H, et al. Exosomal microRNA in plasma as a non-invasive biomarker for the recurrence of non-small cell lung cancer. Oncol Lett 2017;13(3):1256–63.
[174] Sandfeld-Paulsen B, et al. Exosomal proteins as prognostic biomarkers in non-small cell lung cancer. Mol Oncol 2016;10(10):1595–602.
[175] Alipoor SD, et al. The roles of miRNAs as potential biomarkers in lung diseases. Eur J Pharmacol 2016;791:395–404.
[176] Reclusa P, et al. Exosomes genetic cargo in lung cancer: a truly Pandora's box. Transl Lung Cancer Res 2016;5(5):483.
[177] Zöller M. Exosomes in cancer disease. In: Cancer gene profiling. Springer; 2016. p. 111–49.
[178] Park J, et al. Exosome classification by pattern analysis of surface-enhanced Raman spectroscopy data for lung cancer diagnosis. Anal Chem 2017;89(12):6695–701.
[179] Ueda K, et al. Antibody-coupled monolithic silica microtips for highthroughput molecular profiling of circulating exosomes. Sci Rep 2014;4:6232.
[180] Jakobsen KR, et al. Exosomal proteins as potential diagnostic markers in advanced non-small cell lung carcinoma. J Extracell Vesicles 2015;4(1):26659.
[181] Li Y, et al. Proteomic identification of exosomal LRG1: a potential urinary biomarker for detecting NSCLC. Electrophoresis 2011;32(15):1976–83.
[182] Sandfeld-Paulsen B, et al. Exosomal proteins as diagnostic biomarkers in lung cancer. J Thorac Oncol 2016;11(10):1701–10.
[183] Munagala R, et al. Bovine milk-derived exosomes for drug delivery. Cancer Lett 2016;371(1):48–61.

[184] Yang T, et al. Exosome delivered anticancer drugs across the blood-brain barrier for brain cancer therapy in Danio rerio. Pharm Res 2015;32(6):2003–14.
[185] Inamdar S, Nitiyanandan R, Rege K. Emerging applications of exosomes in cancer therapeutics and diagnostics. Bioeng Transl Med 2017;2(1):70–80.
[186] Wu YJ, et al. Targeting α V-integrins decreased metastasis and increased survival in a nude rat breast cancer brain metastasis model. J Neurooncol 2012;110(1):27–36.
[187] Whiteside TL. Tumor-derived exosomes and their role in cancer progression. In: Advances in clinical chemistry. Elsevier; 2016. p. 103–41.

Chapter 18

Exosomes in retinal diseases

Sarah R. Weber, Mi Zhou, Yuanjun Zhao, Jeffrey M. Sundstrom
Department of Ophthalmology, Penn State College of Medicine, Hershey, PA, United States

1 Introduction

The retina is a specialized sensory tissue that lines the back of the eye (Fig. 1). It functions to sense light and communicates images to the brain in the form of electoral signals. The retina is organized into layers of cells, the major types of which include photoreceptors, horizontal cells, bipolar cells, amacrine cells, and ganglion cells. External to the photoreceptor layer is a monolayer of post-mitotic cells known as the retinal pigment epithelium (RPE) (Fig. 1). These cells carry out numerous specialized functions that are critical in maintaining the health of the overlying neuroretina.

Blood supply to the retina derives from two sources: the central retinal artery, which supplies the inner retina, and the choroid, a rich vascular layer located between the retina and sclera, which supplies the outer retina. The basement membrane of the capillaries in the choroid form the outer layer of Bruch's membrane (Fig. 1), a multi-layered extracellular matrix that is of clinical significance in multiple retinal diseases. The innermost layer of Bruch's is formed by the RPE basement membrane, and additional layers of collagens and elastic fibers form the middle three layers.

Due to the nature of its function, the retina cannot ethically be biopsied from patients. Direct study of retinal tissue or cells thus comes from postmortem eyes, animal models, or cell culture studies, each of which has some form of artifact. An additional method that allows direct analysis of patient-derived tissue is interrogation of ocular fluids, which is frequently accomplished via proteomic approaches. The most proximal biofluid in relation to the retina is the vitreous humor, but aqueous humor from retinal disease patients is frequently studied as well. Both of these fluids can be safely biopsied from patients during surgery or in a clinical setting [1, 2]. Although aqueous and vitreous are both believed to contain proteins relevant to retinal disease and they do show some overlap in their proteomes, in general their protein contents appear to be quite distinct [3]. For anatomical reasons as well as a vast body of evidence demonstrating the

Exosomes. https://doi.org/10.1016/B978-0-12-816053-4.00018-3

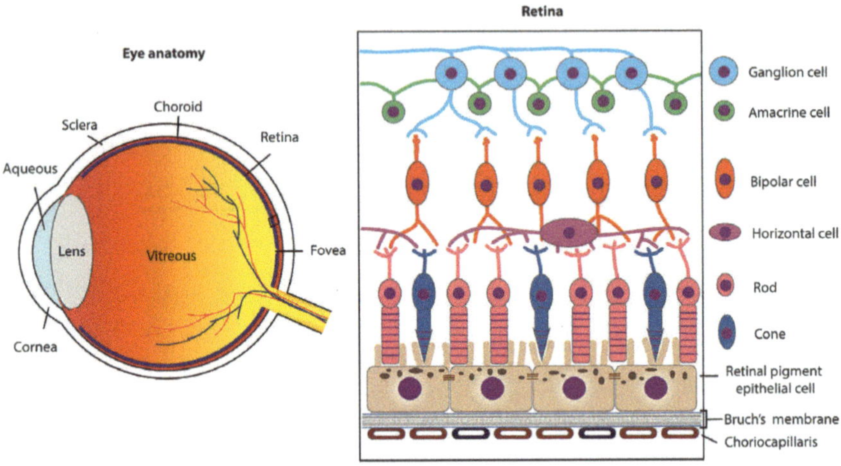

FIG. 1 *Left*: cross-sectional diagram of gross eye anatomy. *Right*: schematic of retinal cell layers, expanded from boxed area on left.

presence of retinal disease proteins in vitreous [4–9], we emphasize interrogation of vitreous rather than aqueous as a proximal biofluid of the retina.

Substantiating the premise that vitreous can provide insight into retinal disease, an abundant population of exosomes, a subclass of extracellular vesicles (EVs), were recently identified in vitreous from patients with no retinal vascular disease. This finding was validated in postmortem donors with no known ocular pathology and in mice [5]. Exosomes have been identified in aqueous humor as well [10].

EVs function in cell-cell communication in both health and disease [11, 12]. As EVs are secreted by most cell types, these functions span a wide array of conditions and systems. Because of their abundance and integral roles in so many pathological processes, they are often interrogated as a biomarker source. EV-based therapeutics are a budding research interest as well. Despite these growing investigational fields, EV research in the eye has not kept pace. Our understanding of the roles of EVs in retinal disease is exceedingly limited, although their extreme abundance in vitreous fluid suggests a potential role in the neighboring retina. Fortunately, studies investigating the role of EVs in general mechanisms of disease that are known to contribute to retinal pathology can provide valuable preliminary findings for application to retina research. These include oxidative stress, endoplasmic reticulum (ER) stress, inflammation and immune responses, angiogenesis, and cancer and are outlined in the following sections. Additional inferences can be drawn from investigations of EVs in the context of disease processes in other organ systems that mirror those that occur in retinal disease. EV studies fulfilling these criteria have been incorporated into the sections below where applicable.

EVs are a heterogenous group vesicles that are commonly divided into subclasses according to size, surface markers, or biogenesis mechanism. Criteria for these classifications have been dynamic throughout the past several years as researchers have expanded the current knowledgebase regarding these vesicles. Two major EV classes are exosomes, which are of endosomal origin and tend to range from 50 to 150 nm, and microvesicles, which are shed from the plasma membrane and generally have diameters between 50 and 500 nm [13]. For the purposes of this chapter, we will refer to EV subtypes according to their classification in the original study and use the general term "EV" for unspecified vesicles as well as to refer to exosomes and microvesicles collectively.

2 General disease mechanisms

2.1 Oxidative stress

Oxidative stress refers to an increase in the amount of biomolecule-damaging reactive oxygen species (ROS) that are present within cells [14]. ROS are generated via normal metabolic processes and are combatted by antioxidant molecules or enzymes. When ROS production exceeds cellular antioxidant capacity and redox balance is disrupted, pathology ensues.

The pathological changes that occur in both inherited and age-related macular degenerations are associated with increased levels of oxidative stress. In these diseases, gradual degeneration of retinal cells leads to blindness. Macular degeneration is marked by the formation of deposits beneath the retina or RPE. In age-related macular degeneration (AMD), these deposits are termed drusen. As deposits increase in number and size, they coalesce, compromising the overlying cells' access to nutrients. Eventually, large patches of retinal tissue become atrophic and die; this lesion is referred to as geographic atrophy (GA) and defines advanced dry AMD. Wet AMD, in contrast, is defined by aberrant formation of new blood vessels (neovascularization; NV) that originate from the choroid and grow into the retina through breaks in Bruch's membrane (Fig. 2).

Occurrence of AMD is linked to oxidative stress-producing environmental factors [15]. Much of the oxidative stress-induced pathology in macular degenerations can be traced back to RPE cell damage. Oxidative stress leads to debris accumulation and cell death, ultimately resulting in drusen and GA in AMD [16]. Additionally, the accumulation of oxidized molecules may disrupt the RPE cells' ability to modulate angiogenesis, leading to NV [16]. Multiple studies have demonstrated a potential role for exosomes in these oxidative stress-induced RPE cell changes. For example, in a study using ARPE-19 cells, Wang et al. examined RPE changes associated with exposure to a toxic element of cigarette smoke, a known AMD risk factor [17], and found an increase in exosome markers CD63, CD81, and LAMP2 [18]. These three proteins were also found in drusen from AMD patients [19].

The role of exosomes in AMD pathology may also extend to GA. Recently, Shah et al. proposed that EVs may be the vector by which RPE damage spreads over long distances from one location to another, resulting in regions of GA.

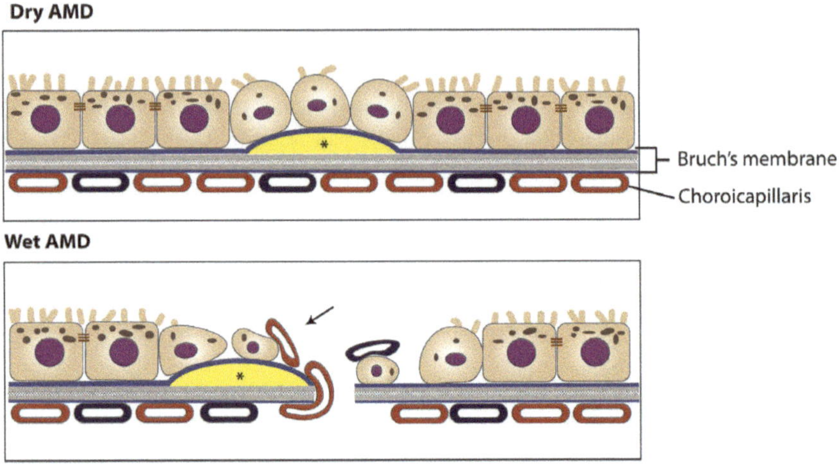

The asterisk (*) represents drusen; the arrow respresents neovascularization.

FIG. 2 Schematic representations of dry and wet AMD pathology.

They showed that EVs derived from cultured primary RPE cells subjected to oxidative stress decreased barrier function in recipient RPE monolayers. This effect was mediated by histone deacetylase 6 [20], which regulates autophagy in response to oxidative stress [21].

Other researchers have demonstrated a potential role of exosomes in angiogenesis under oxidative stress conditions. It is known that ROS stimulate vascular endothelial growth factor (VEGF) expression [22] and that VEGF is a major factor driving angiogenesis in retinal diseases like macular degenerations and diabetic retinopathy (DR) [23]. It has been shown that the number of exosomes secreted by ARPE-19 cells increases upon exposure to oxidative stress and that the content of these vesicles becomes pro-angiogenic, with greater amounts of vascular endothelial growth factor receptor (VEGFR)-1 and VEGFR-2 membrane proteins and encapsulated messenger (m)RNA expression as well as a greater ability to induce angiogenesis in vitro [24]. The phosphorylation states of proteins in ARPE-19 cell-derived exosomes have also been shown to change after exposure to oxidative stress, and these proteins were found to be involved in apoptosis, cell survival, and metabolic pathways [25].

Further investigation into the involvement of EVs in oxidative stress underlying retinal pathology is needed, but studies focusing on RPE cell-derived exosomes suggest the two are interlinked.

2.2 ER stress

The balance between protein biosynthesis and degradation and the ability to fold proteins properly are both critical aspects in the maintenance of cellular homeostasis. The ER is the master regulator of these processes, and when they are out of balance, ER stress occurs. The cellular response to ER stress is the activation of downstream signaling pathways—collectively termed the unfolded protein response (UPR)—that aim to restore balance through proper protein folding, halted translation, increased degradation, or—if the damage is irreparable—apoptosis [26].

ER stress has been implicated in numerous retinal diseases. In multiple inherited retinal degenerations (IRDs), mutant, misfolded proteins may accumulate in the ER, inducing stress. These include IRDs caused by mutations in the genes encoding rhodopsin [27, 28], interphotoreceptor retinoid-binding protein [27], carbonic anhydrase IV [29], and fibulin-3 [30]. Other IRD-causing mutations, such as those in ELOVL fatty acid elongase 4 [31], phosphodiesterase 6B [32], and usherin [33], are associated with upregulation of ER stress and UPR mediators. Both IRDs and AMD share the feature of deposits, which in part are composed of protein aggregates, indicating disrupted proteostasis. Although direct evidence for involvement of exosomes in these conditions is lacking, substantial evidence from studies of analogous processes and diseases in other systems has accumulated.

Recently, it was found that ER stress increases exosome release [34]. Additional studies have suggested that exosomes play a role in prion transmission and aggregation in neurodegenerative diseases in the central nervous system [35]. These diseases are marked by aberrant accumulation of insoluble protein aggregates, such as β-amyloid plaques in Alzheimer's disease, α-synuclein aggregation in Parkinson's disease, and prion plaques in multiple prion diseases. Multiple Aβ peptides—which contribute to β-amyloid plaque formation in Alzheimer's disease—have been shown to be secreted via exosomes [36], and exosomes incubated with Aβ were found to promote protein aggregation in cultured glial cells [37]. Tau, which forms aggregates in both Alzheimer's and Parkinson's diseases, can also be secreted via exosomes [38, 39], and cerebrospinal fluid-derived exosomes from patients with Alzheimer's have been shown to increase tau aggregation in vitro [39]. Similarly, α-synuclein can be secreted via exosomes [40] and is more toxic to recipient cells in its exosome-associated form than in its free form [41]. Prion proteins have also been shown to be associated with exosomes [42], and stimulation of exosome secretion was recently found to increase intercellular prion transmission [43]. In addition to secretion and toxicity, exosomes may play a role in deposit formation in these diseases [44, 45]. Deposit formation is a unifying feature between these diseases and macular degenerations, and deposits in retinal degenerations share characteristics with those of neurodegenerative diseases. In AMD donor eyes, 129 proteins have been identified in sub-retinal deposits, and 20% of components are shared between Alzheimer's plaques and AMD deposits [46]. The deposits that occur in IRDs are phenotypically and compositionally similar to those of AMD, and in at least one IRD, amyloid peptide has been identified in deposits [47]. Thus, it is possible that exosomes play a similar role in deposit formation in macular degenerations.

In addition to this potential role in propagating pathological changes, exosomes may also function in maintenance of proteostasis. Molecular chaperones such as heat shock proteins (HSPs) are major players in the maintenance of proteostasis, as they aid the process of protein folding and regulate degradation of improperly folded proteins [48, 49]. HSPs are thus upregulated in response to stress. Takeuchi et al. showed that the molecular chaperones Hsp40, Hsp70, and Hsp90 are secreted via exosomes. Importantly, exosomes containing these proteins reduced aggregate formation in a neurodegenerative disease model, contributing to maintenance of proteostasis [50]. Increased retinal HSP expression has been observed in AMD [51], although others have shown decreased HSP expression in RPE cells during earlier stages of AMD [52].

Another, smaller chaperone protein, αB-crystallin, is secreted by RPE cells via exosomes and is thought to confer neuroprotection to the abutting photoreceptors [53]. Inhibition of αB-crystallin expression in RPE cell culture resulted in blockage of exosome secretion [54], underlining its integral interaction with exosomes. αB-crystallin has a generalized capacity to reduce protein aggregation and has been associated with numerous neurodegenerative diseases, including Alzheimer's disease, Parkinson's disease, and prion disease [55]. It has also been proposed that αB-crystallin may serve as a biomarker for AMD, as its expression has been shown to be elevated in RPE cells from AMD patients in a manner that appeared to correlate with disease severity [56]. In a proteomic analysis of drusen, αB-crystallin was found to be among the most common proteins identified [57]. Because of its close relationship with exosomes and its association with diseases marked by aberrant protein aggregation, it is feasible that αB-crystallin plays a role in macular degeneration via an exosome-mediated mechanism involving ER stress.

These findings from studies of retinal cells as well as other systems indicate that the ER stress underlying and contributing to retinal pathology may in part be mediated by exosomes. In addition to functioning in the disease state, exosomes may play a role in maintenance of proteostasis in retinal cells and prevention of diseases that are exacerbated by ER stress.

2.3 Inflammation and immune responses

Although the eye is considered immune-privileged, multiple retinal diseases are caused or exacerbated by aberrant inflammation or disrupted immune responses. These include conditions heavily driven by inflammation, such as uveitis, as well as multifactorial conditions in which inflammation is one of multiple integrated components contributing to pathology, such as AMD and DR.

Uveitis is an inflammation of the uvea and can be subclassified according to the particular structures affected as well as the cause of inflammation. Here we focus on autoimmune uveitis affecting the retina.

Uveitis is thought to derive largely from T cell activity and is characterized by perturbances in the levels of diverse cytokines in both ocular fluids and serum [58]. RPE cells are known to play an immunosuppressive role in the eye and specifically function to suppress T cell activity [59, 60], and EVs have been shown to mediate immune activity in other systems [61–63]. Recent evidence suggests that RPE cell-derived EVs may play a role in mediating inflammation in uveitis. Knickelbein et al. demonstrated that RPE cell-derived EVs cultured with peripheral blood mononuclear cells derived from noninfectious uveitis patients inhibited T cell proliferation. When cultured with monocytes, RPE cell-derived EVs increased the amount of monocytes belonging to a subpopulation known to regulate immune responses in patients with noninfectious uveitis [64]. EVs derived from other cell types may be able to play a similar immunosuppressive role. Bai and colleagues recently demonstrated therapeutic effects of mesenchymal stem cell (MSC)-derived exosomes in a rat model of autoimmune uveoretinitis. Specifically, they found that periocular injection of MSC-derived subdued disease intensity by decreasing migration of immune cells, including T cells, into the eye [65].

Although less dramatic than in uveitis, immune activation and inflammation also play a role in AMD. Multiple studies have demonstrated an association between genetic polymorphisms of complement components and AMD [66–68]. In a proteomic analysis of Bruch's membrane and choroidal tissue from maculae of AMD donors, the majority of identified proteins were found to be involved in immunity or defense [69]. The cells responsible for the bulk of immunoregulation in the eye, the RPE, are also the primary cell type affected in AMD. RPE cells secrete regulatory molecules like tumor growth factor beta and express complement-regulating proteins in their membranes [70–72]. The role of inflammation in AMD progression may be partially mediated by exosomes. Among the complement regulatory proteins expressed by the RPE are CD59 and CD46. In a study of maculae from AMD eyes, the levels of these proteins were found to be decreased in areas displaying pathological changes. CD59 and CD46 levels were also decreased in vitro in ARPE-19 cells treated with a complement-triggering compound. As the increased CD59 and CD46 staining in the supernatant displayed significant overlap with that of an exosome marker, the authors of this study proposed that these proteins were released by RPE cells via exosomes [73]. Further substantiating the possibility that exosomes play a role in RPE immunoregulation, a generalized increase in the amount of EVs secreted by RPE cells has been demonstrated in response to stimulation with inflammation-promoting cytokines [64].

In both AMD and DR, senescence—cellular "aging" in which cells lose their ability to divide—has effects that appear to impact inflammation. A major consequence of senescence is adoption of the senescence-associated secretory phenotype (SASP), which is in part characterized by upregulation of a particular body of proinflammatory cytokines and other molecules [74]. Increased expression of SASP markers has been demonstrated in ARPE-19 cells exposed to oxidative stress. Additionally, senescent ARPE-19 cells were shown to increase VEGF expression and decrease complement factor H (CFH) expression [75]. As

RPE-derived VEGF is thought to drive NV in exudative AMD [76], and *CFH* polymorphisms are associated with AMD susceptibility [67], the findings from this study suggest that RPE senescence contributes considerably to AMD pathology. SASP has also been shown to play a pathological role in an ischemic retinopathy mouse model, in which intravitreal delivery of SASP inhibitors decreased retinal NV. In the same study, SASP markers including plasminogen activator inhibitor 1, interleukin (IL)-6, IL-8, and VEGF were identified in vitreous from proliferative diabetic retinopathy (PDR) patients [77]. Vitreous levels of SASP markers including IL-6, IL-8, and C–C motif chemokine 2 have also been shown to correlate strongly with those of VEGF [78], the driver of NV in this disease. Mounting evidence suggests that EVs play a substantial role in senescence and aging (reviewed in [79]), and thus they may play a role in the senescent changes that contribute to AMD and DR. EV secretion has been shown to increase with senescence [80], and may represent one mechanism by which senescent cells remove unwanted DNA [81]. As mentioned above, SASP is associated with upregulation of myriad proinflammatory cytokines. Recently, it was shown that many cytokines, including many that are associated with SASP, can be secreted in association with EVs and that this association is dynamic and system-dependent [82].

Based on the well-established role of exosomes in inflammation in other systems as well as emerging evidence regarding RPE-derived exosomes, senescence, and EV-associated cytokines, it is possible that exosomes contribute to immune activation and inflammation in multiple retinal diseases.

2.4 Retinal and choroidal angiogenesis

Aberrant blood vessel formation, termed NV, occurs in numerous retinal diseases and often results in significant vision loss. NV associated with retinal disease can be classified as retinal or choroidal, named according to the vasculature from which the newly formed, pathological blood vessels extend. Retinal NV occurs in DR, retinopathy of prematurity, retinal vein occlusions, and other retinopathies, whereas choroidal NV is characteristic of exudative AMD, other macular degenerations, myopia, trauma, and others [83]. Although anatomically distinct, both types of NV are driven by shared angiogenic factors and inhibited by common antiangiogenic factors. Major angiogenic factors in the retina include VEGF [84], fibroblast growth factor (FGF)-2 [85], angiopoietins and Tie2 [84], Wnt [86, 87], and matrix metalloproteinases (MMPs) [88], while antiangiogenic factors include pigment epithelium-derived factor (PEDF) [89], angiostatin [90], and endostatin [91]. Few studies have examined the role of exosomes in retinal or choroidal NV specifically, but substantial evidence from other fields has demonstrated an integral role for EVs in angiogenesis in other systems.

RPE-derived exosomes are the focus of the majority of studies examining the role of EVs in retinal or choroidal NV. It was recently shown that exosomes derived from stressed ARPE-19 cells contained increased VEGFR proteins and transcripts and were able to stimulate angiogenesis in vitro [24]. RPE exosomes may also mediate antiangiogenic properties. Naga et al. found that the intravitreally injected anti-VEGF antibody bevacizumab, which is capable of penetrating the retina to be taken up by RPE cells, appears to be released from RPE cells via exosomes [92]. The antiangiogenic factor PEDF, which is apically secreted from RPE cells to prevent aberrant angiogenesis from occurring in the retina, was identified in RPE-derived exosomes and was present in greater proportions in apically secreted vesicles [93]. Another protein with anti-angiogenic properties, peroxisome proliferator-activated receptor gamma, was shown to be elevated in the vitreous and aqueous fluids of PDR patients [94]. As this protein is known to be present in circulating plasma exosomes [95], it is possible that it is exosome-associated in ocular fluids as well.

Additional studies have examined the effects of exosomes derived from other cell types on the retina. Exosomes derived from MSC cells cultured under hypoxic conditions were found to decrease retinal ischemia upon intravitreal injection in a mouse model of oxygen-induced retinopathy [96]. In a laser-induced choroidal NV mouse model, retinal astroglial cell-derived exosomes containing anti-angiogenic molecules were able to suppress leakage of the retinal vasculature and inhibit choroidal NV. In this model, RPE-derived exosomes did not have these effects [97].

Further evidence for the involvement of EVs in NV comes from other systems. Multiple angiogenic factors that drive retinal and choroidal NV have been identified in exosomes or other EVs. Tumor-derived exosomes have been shown to be capable of delivering VEGF and similar growth factors as well as pro-angiogenic mRNAs and miRNAs to other cells [98]. Additionally, exosomes from healthy cells have been shown to stimulate VEGF mRNA expression in vitro [99]. Microvesicle-associated VEGF has been shown to cause more sustained activation of VEGF receptors than free VEGF and to be insensitive to bevacizumab, resulting in an EV-associated VEGF-induced inability to block VEGF signaling [100]. In a glioblastoma model, similar effects were observed with the angiogenic factor FGF-2. In this study, glioblastoma-derived exosomes were enriched with FGF-2 co-receptors and stimulated increased vascularization compared to FGF-2 alone [101]. Tie2 signaling components have also been identified in EVs. Specifically, angiopoietin (Ang) 1 and Ang2 proteins have been shown to be secreted in exosomes [102, 103], while larger EVs have been shown to contain Ang1 mRNA [104]. Numerous studies have reported exosomal secretion of Wnt signaling components [105–109], and it has been shown that a significant portion of active Wnt pathway components are secreted on these vesicles [105, 106]. MMPs are also critical factors for angiogenesis, as they degrade the extracellular matrix (ECM) into which new blood vessels grow. Active MMPs have been previously identified on exosomes [110]. Treatment with exosomes secreted by cancer cells were shown to stimulate secretion of multiple MMPs and VEGFA by recipient cells [111], and cancer cell-derived exosomes overexpressing MMP-13 were found to stimulate angiogenesis [112].

Although studies specific to retinal and choroidal NV are limited, preliminary work suggests that EVs play a role in NV in retinal disease. This concept is substantiated by a much vaster body of evidence demonstrating the pro-angiogenic capacity of EVs in other systems, meriting further investigation into similar mechanisms in the retinal and choroidal vasculature.

2.5 Cancer

Exosomes play a well-established role in cancer, and evidence for their oncologic effects across diverse organ systems continues to mount. They have been shown to mediate immune responses in cancer, influence the tumor microenvironment, and contribute to metastasis. For example, exosomes derived from antigen-presenting cells have been shown to aid immune responses against cancer in mice [113, 114]. In the tumor microenvironment, bidirectional vesicle effects have been observed. In one study, exosomes secreted by prostate cancer cells activated fibroblasts in the tumor microenvironment [115], while in another, exosomes derived from cancer-associated fibroblasts induced a pro-invasion cargo change in breast cancer cell-derived exosomes [108]. Cancer cell-derived exosomes are able to aid in metastasis by contributing to ECM degradation [116, 117], directing organotropism [116], and helping to establish premetastatic niches [118].

In addition to these functional studies, exosomes are increasingly being investigated as biomarkers in cancer, with their micro (mi)RNA commonly the area of focus. Exosomes serve to protect these nucleic acids, which are frequently dysregulated in cancer. Accumulating evidence points to functional roles for miRNAs similar to those described above for exosomes generally [119]. Although limited relative to that of the cancer field in general, some evidence demonstrates exosomal miRNA dysregulation in uveal melanoma. Uveal melanoma is the most common primary intraocular tumor and arises from melanocytes. It has a poor prognosis, with about half of patients succumbing to metastatic disease and the majority of those dying in the first 5 years of disease [120]. Of those who develop metastases, about 90% are to liver [121]. A study by Eldh et al. identified exosomes positive for the melanoma-specific marker Melan-A in the liver perfusates of patients with uveal melanoma, suggesting that tumor-derived exosomes reached the metastatic site in these patients. Profiling of the miRNA content of these vesicles revealed distinct miRNA cargo in comparison to melanoma cell lines [122]. Another study examined miRNAs in native vitreous, vitreous exosomes, and serum in uveal melanoma patients and compared them to non-diseased control patients. They identified 32 miRNAs that were differentially expressed in at least two of the three samples types, with the majority of these changes occurring in native vitreous and vitreous exosomes [123].

Exosomes are widely studied in the context of cancer, and findings from this research can serve to guide further investigation into ocular cancers. Recent studies examining miRNA dysregulation in uveal melanoma lay the groundwork for applying strategies for exosome analysis to the eye.

3 Exosome-based therapeutic approaches

In addition to the above findings suggesting that exosomes contribute to retinal pathology, other data have demonstrated therapeutic effects of exosomes in the retina. The majority of these studies have examined MSC-derived exosomes. These vesicles are frequently favored over stem cells themselves, as the therapeutic effects of MSCs appear to derive mainly from the soluble factors they secrete, and cell-free approaches avoid safety issues intrinsic to cell transplantation [124]. A few studies have examined the effects of MSC-derived exosomes in the retina. Mead et al. observed an increase in retinal ganglion cell survival and axonal regeneration after intravitreal injection of exosomes from bone marrow-derived MSCs in a rat optic nerve crush model. In primary retinal cell cultures, these vesicles demonstrated neuroprotective effects [125]. Similarly, Yu et al. observed a reduction in laser damage, apoptosis, and inflammation after intravitreal injection of MSC-derived exosomes in a laser-induced retinopathy mouse model [126]. As mentioned in the above sections, MSC-derived exosomes also reduced the severity of experimental autoimmune uveitis in rats via inhibition of migration of inflammatory cells into the eye [65]. In an experimental autoimmune uveitis mouse model, MSC-derived EVs prevented disease onset and inhibited immune cell activation [127].

In addition to serving as therapeutics in and of themselves, exosomes may be able to enhance other therapeutic techniques that are already in use. Wassmer et al. demonstrated that exosome-associated adeno-associated virus (AAV) 2 allowed superior gene delivery to the retina after intravitreal injection in mice, relative to conventional AAV2-mediated delivery [128]. As the first gene therapy in the US was recently approved for an IRD, these findings provide a foundational framework for future application to retinal disease in humans.

As the roles of EVs in retinal disease are only beginning to be uncovered, so too are their therapeutic applications in these conditions. Based on the studies outlined here, however, EV-based therapeutics are a promising option for future treatment of retinal disease.

4 Conclusions

It is clear from other fields that EVs play integral roles in both health and disease. Although data demonstrating a role for EVs in retinal health and disease directly is only beginning to accumulate, inferences drawn from other fields substantiate this preliminary evidence and provide a foundational premise for further investigation. Numerous studies have demonstrated roles for EVs in universal pathological mechanisms that underlie retinal diseases and diseases of other tissues. These mechanisms include oxidative stress, ER, stress, inflammation and immune responses, angiogenesis, and cancer. More recent research suggests that EVs can also confer therapeutic benefits to the retina. Further research is needed in order to expand and deepen current understanding of these topics.

References

[1] Van der Lelij A, Rothova A. Diagnostic anterior chamber paracentesis in uveitis: a safe procedure? Br J Ophthalmol 1997;81(11):976–9.
[2] Ghodasra DH, et al. Safety and feasibility of quantitative multiplexed cytokine analysis from office-based vitreous aspiration. Invest Ophthalmol Vis Sci 2016;57(7):3017–23.
[3] Ecker SM, Hines JC, Pfahler SM, Glaser BM. Aqueous cytokine and growth factor levels do not reliably reflect those levels found in the vitreous. Mol Vis 2011;17:2856–63.
[4] Aiello LP, et al. Vascular endothelial growth factor in ocular fluid of patients with diabetic retinopathy and other retinal disorders. N Engl J Med 1994;331(22):1480–7.
[5] Zhao Y, et al. Liquid biopsy of vitreous reveals an abundant vesicle population consistent with the size and morphology of exosomes. Transl Vis Sci Technol 2018;7(3):6.
[6] Schori C, et al. The proteomic landscape in the vitreous of patients with age-related and diabetic retinal disease. Invest Ophthalmol Vis Sci 2018;59(4):AMD31–40.
[7] Klaassen I, et al. Identification of proteins associated with clinical and pathological features of proliferative diabetic retinopathy in vitreous and fibrovascular membranes. PLoS One 2017;12(11):e0187304.
[8] Nobl M, et al. Proteomics of vitreous in neovascular age-related macular degeneration. Exp Eye Res 2016;146:107–17.
[9] Bromberg-White JL, et al. Identification of VEGF-independent cytokines in proliferative diabetic retinopathy vitreous. Invest Ophthalmol Vis Sci 2013;54:6472–80.
[10] Dismuke WM, et al. Human aqueous humor exosomes. Exp Eye Res 2015;132:73–7.
[11] Tetta C, et al. Extracellular vesicles as an emerging mechanism of cell-to-cell communication. Endocrine 2013;44(1):11–9.
[12] Malloci M, et al. Extracellular vesicles: mechanisms in human health and disease. Antioxid Redox Signal 2018;30(6):813–56.
[13] van Niel G, D'Angelo G, Raposo G. Shedding light on the cell biology of extracellular vesicles. Nat Rev Mol Cell Biol 2018;19:213–28.
[14] Schieber M, Chandel NS. ROS function in redox signaling and oxidative stress. Curr Biol 2014;24(10):R453–62.
[15] Wang W, et al. Genetic and environmental factors strongly influence risk, severity and progression of age-related macular degeneration. Signal Transduct Target Ther 2016;1:16016.
[16] Brantley MA, Umfress AC, Sternberg P. Mechanisms of oxidative stress in retinal injury. In: Schachat AP, editor. Ryan's retina. Elsevier Inc; 2018.
[17] Beatty S, et al. The role of oxidative stress in the pathogenesis of age-related macular degeneration. Surv Ophthalmol 2000;45(2):115–34.
[18] Wang AL, et al. Changes in retinal pigment epithelium related to cigarette smoke: possible relevance to smoking as a risk factor for age-related macular degeneration. PLoS One 2009;4(4):e5304.
[19] Wang AL, et al. Autophagy and exosomes in the aged retinal pigment epithelium: possible relevance to drusen formation and age-related macular degeneration. PLoS One 2009;4(1):e4160.

[20] Shah N, et al. Extracellular vesicle-mediated long-range communication in stressed retinal pigment epithelial cell monolayers. Biochim Biophys Acta Mol basis Dis 2018;1864(8):2610–22.

[21] Lee JY, et al. HDAC6 controls autophagosome maturation essential for ubiquitin-selective quality-control autophagy. EMBO J 2010;29(5):969–80.

[22] Fay J, et al. Reactive oxygen species induce expression of vascular endothelial growth factor in chondrocytes and human articular cartilage explants. Arthritis Res Ther 2006;8(6):R189.

[23] Amadio M, Govoni S, Pascale A. Targeting VEGF in eye neovascularization: what's new?: a comprehensive review on current therapies and oligonucleotide-based interventions under development. Pharmacol Res 2016;103:253–69.

[24] Atienzar-Aroca S, et al. Oxidative stress in retinal pigment epithelium cells increases exosome secretion and promotes angiogenesis in endothelial cells. J Cell Mol Med 2016;20(8):1457–66.

[25] Biasutto L, et al. Retinal pigment epithelium (RPE) exosomes contain signaling phosphoproteins affected by oxidative stress. Exp Cell Res 2013;319(13):2113–23.

[26] Hetz C. The unfolded protein response: controlling cell fate decisions under ER stress and beyond. Nat Rev Mol Cell Biol 2012;13(2):89–102.

[27] Li S, et al. Secretory defect and cytotoxicity: the potential disease mechanisms for the retinitis pigmentosa (RP)-associated interphotoreceptor retinoid-binding protein (IRBP). J Biol Chem 2013;288(16):11395–406.

[28] Gorbatyuk MS, et al. Restoration of visual function in P23H rhodopsin transgenic rats by gene delivery of BiP/Grp78. Proc Natl Acad Sci U S A 2010;107(13):5961–6.

[29] Rebello G, et al. Apoptosis-inducing signal sequence mutation in carbonic anhydrase IV identified in patients with the RP17 form of retinitis pigmentosa. Proc Natl Acad Sci U S A 2004;101(17):6617–22.

[30] Marmorstein LY, et al. Aberrant accumulation of EFEMP1 underlies drusen formation in Malattia Leventinese and age-related macular degeneration. Proc Natl Acad Sci U S A 2002;99(20):13067–72.

[31] Karan G, et al. Loss of ER retention and sequestration of the wild-type ELOVL4 by Stargardt disease dominant negative mutants. Mol Vis 2005;11:657–64.

[32] Yang LP, et al. Activation of endoplasmic reticulum stress in degenerating photoreceptors of the rd1 mouse. Invest Ophthalmol Vis Sci 2007;48(11):5191–8.

[33] Tucker BA, et al. Patient-specific iPSC-derived photoreceptor precursor cells as a means to investigate retinitis pigmentosa. elife 2013;2:e00824.

[34] Kanemoto S, et al. Multivesicular body formation enhancement and exosome release during endoplasmic reticulum stress. Biochem Biophys Res Commun 2016;480(2):166–72.

[35] Lim YJ, Lee SJ. Are exosomes the vehicle for protein aggregate propagation in neurodegenerative diseases? Acta Neuropathol Commun 2017;5(1):64.

[36] Rajendran L, et al. Alzheimer's disease beta-amyloid peptides are released in association with exosomes. Proc Natl Acad Sci U S A 2006;103(30):11172–7.

[37] Dinkins MB, et al. Exosome reduction in vivo is associated with lower amyloid plaque load in the 5XFAD mouse model of Alzheimer's disease. Neurobiol Aging 2014;35(8):1792–800.

[38] Saman S, et al. Exosome-associated tau is secreted in tauopathy models and is selectively phosphorylated in cerebrospinal fluid in early Alzheimer disease. J Biol Chem 2012;287(6):3842–9.

[39] Wang Y, et al. The release and trans-synaptic transmission of Tau via exosomes. Mol Neurodegener 2017;12(1):5.

[40] Emmanouilidou E, et al. Cell-produced alpha-synuclein is secreted in a calcium-dependent manner by exosomes and impacts neuronal survival. J Neurosci 2010;30(20):6838–51.

[41] Danzer KM, et al. Exosomal cell-to-cell transmission of alpha synuclein oligomers. Mol Neurodegener 2012;7:42.

[42] Cheng L, Zhao W, Hill AF. Exosomes and their role in the intercellular trafficking of normal and disease associated prion proteins. Mol Asp Med 2018;60:62–8.

[43] Guo BB, Bellingham SA, Hill AF. Stimulating the release of exosomes increases the intercellular transfer of prions. J Biol Chem 2016;291(10):5128–37.

[44] Yuyama K, Yamamoto N, Yanagisawa K. Accelerated release of exosome-associated GM1 ganglioside (GM1) by endocytic pathway abnormality: another putative pathway for GM1-induced amyloid fibril formation. J Neurochem 2008;105(1):217–24.

[45] Grey M, et al. Acceleration of alpha-synuclein aggregation by exosomes. J Biol Chem 2015;290(5):2969–82.

[46] Booij JC, et al. The dynamic nature of Bruch's membrane. Prog Retin Eye Res 2010;29(1):1–18.

[47] Sohn EH, et al. Comparison of drusen and modifying genes in autosomal dominant radial drusen and age-related macular degeneration_Fibulin 3 IHC. Retina 2015;35(1):48–57.

[48] Hartl FU, Bracher A, Hayer-Hartl M. Molecular chaperones in protein folding and proteostasis. Nature 2011;475(7356):324–32.

[49] Takeuchi T. Non-cell autonomous maintenance of proteostasis by molecular chaperones and its molecular mechanism. Biol Pharm Bull 2018;41(6):843–9.

[50] Takeuchi T, et al. Intercellular chaperone transmission via exosomes contributes to maintenance of protein homeostasis at the organismal level. Proc Natl Acad Sci U S A 2015;112(19):E2497–506.

[51] Kaarniranta K, et al. Heat shock proteins as gatekeepers of proteolytic pathways-implications for age-related macular degeneration (AMD). Ageing Res Rev 2009;8(2):128–39.

[52] Nordgaard CL, et al. Proteomics of the retinal pigment epithelium reveals altered protein expression at progressive stages of age-related macular degeneration. Invest Ophthalmol Vis Sci 2006;47(3):815–22.

[53] Sreekumar PG, et al. alphaB crystallin is apically secreted within exosomes by polarized human retinal pigment epithelium and provides neuroprotection to adjacent cells. PLoS One 2010;5(10):e12578.

[54] Gangalum RK, et al. Inhibition of the expression of the small heat shock protein alphaB-crystallin inhibits exosome secretion in human retinal pigment epithelial cells in culture. J Biol Chem 2016;291(25):12930–42.

[55] Horwitz J. The function of alpha-crystallin in vision. Semin Cell Dev Biol 2000;11(1):53–60.

[56] De S, et al. Human retinal pigment epithelium cell changes and expression of alphaB-crystallin: a biomarker for retinal pigment epithelium cell change in age-related macular degeneration. Arch Ophthalmol 2007;125(5):641–5.

[57] Crabb JW, et al. Drusen proteome analysis: an approach to the etiology of age-related macular degeneration. Proc Natl Acad Sci U S A 2002;99(23):14682–7.

[58] Horai R, Caspi RR. Cytokines in autoimmune uveitis. J Interf Cytokine Res 2011;31(10):733–44.

[59] Kaestel CG, et al. The immune privilege of the eye: human retinal pigment epithelial cells selectively modulate T-cell activation in vitro. Curr Eye Res 2005;30(5):375–83.

[60] Jorgensen A, et al. Human retinal pigment epithelial cell-induced apoptosis in activated T cells. Invest Ophthalmol Vis Sci 1998;39(9):1590–9.

[61] Kim SH, et al. Exosomes derived from IL-10-treated dendritic cells can suppress inflammation and collagen-induced arthritis. J Immunol 2005;174(10):6440–8.

[62] Bianco NR, et al. Therapeutic effect of exosomes from indoleamine 2,3-dioxygenase-positive dendritic cells in collagen-induced arthritis and delayed-type hypersensitivity disease models. Arthritis Rheum 2009;60(2):380–9.

[63] Robbins PD, Morelli AE. Regulation of immune responses by extracellular vesicles. Nat Rev Immunol 2014;14(3):195–208.
[64] Knickelbein JE, et al. Modulation of immune responses by extracellular vesicles from retinal pigment epithelium. Invest Ophthalmol Vis Sci 2016;57(10):4101–7.
[65] Bai L, et al. Effects of mesenchymal stem cell-derived exosomes on experimental autoimmune uveitis. Sci Rep 2017;7(1):4323.
[66] Hageman GS, et al. Extended haplotypes in the complement factor H (CFH) and CFH-related (CFHR) family of genes protect against age-related macular degeneration: characterization, ethnic distribution and evolutionary implications. Ann Med 2006;38(8):592–604.
[67] Edwards AO, et al. Complement factor H polymorphism and age-related macular degeneration. Science 2005;308(5720):421–4.
[68] Klein RJ, et al. Complement factor H polymorphism in age-related macular degeneration. Science 2005;308(5720):385–9.
[69] Yuan X, et al. Quantitative proteomics: comparison of the macular Bruch membrane/choroid complex from age-related macular degeneration and normal eyes. Mol Cell Proteomics 2010;9(6):1031–46.
[70] Hirsch L, et al. TGF-beta2 secretion from RPE decreases with polarization and becomes apically oriented. Cytokine 2015;71(2):394–6.
[71] Chen M, Forrester JV, Xu H. Synthesis of complement factor H by retinal pigment epithelial cells is down-regulated by oxidized photoreceptor outer segments. Exp Eye Res 2007;84(4):635–45.
[72] Luo C, et al. Complement expression in retinal pigment epithelial cells is modulated by activated macrophages. Exp Eye Res 2013;112:93–101.
[73] Ebrahimi KB, et al. Oxidized low-density-lipoprotein-induced injury in retinal pigment epithelium alters expression of the membrane complement regulatory factors CD46 and CD59 through exosomal and apoptotic bleb release. Adv Exp Med Biol 2014;801:259–65.
[74] Coppe JP, et al. The senescence-associated secretory phenotype: the dark side of tumor suppression. Annu Rev Pathol 2010;5:99–118.
[75] Marazita MC, et al. Oxidative stress-induced premature senescence dysregulates VEGF and CFH expression in retinal pigment epithelial cells: implications for age-related macular degeneration. Redox Biol 2016;7:78–87.
[76] Ford KM, D'Amore PA. Molecular regulation of vascular endothelial growth factor expression in the retinal pigment epithelium. Mol Vis 2012;18:519–27.
[77] Oubaha M, et al. Senescence-associated secretory phenotype contributes to pathological angiogenesis in retinopathy. Sci Transl Med 2016;8(362):362ra144.
[78] Zhou J, Wang S, Xia X. Role of intravitreal inflammatory cytokines and angiogenic factors in proliferative diabetic retinopathy. Curr Eye Res 2012;37(5):416–20.
[79] Takasugi M. Emerging roles of extracellular vesicles in cellular senescence and aging. Aging Cell 2018;17(2).
[80] Lehmann BD, et al. Senescence-associated exosome release from human prostate cancer cells. Cancer Res 2008;68(19):7864–71.
[81] Takahashi A, et al. Exosomes maintain cellular homeostasis by excreting harmful DNA from cells. Nat Commun 2017;8:15287.
[82] Fitzgerald W, et al. A system of cytokines encapsulated in extracellular vesicles. Sci Rep 2018;8(1):8973.
[83] Vavvas D, et al. In: Schachat AP, editor. Basic mechanisms of pathologic retinal and choroidal angiogenesis, in Ryan's retina. Elsevier Inc; 2018. p. 639–55.

[84] Campochiaro PA. Molecular pathogenesis of retinal and choroidal vascular diseases. Prog Retin Eye Res 2015;49:67–81.
[85] Rousseau B, et al. Involvement of fibroblast growth factors in choroidal angiogenesis and retinal vascularization. Exp Eye Res 2003;77(2):147–56.
[86] Hu Y, et al. Pathogenic role of the Wnt signaling pathway activation in laser-induced choroidal neovascularization. Invest Ophthalmol Vis Sci 2013;54(1):141–54.
[87] Drenser KA. Wnt signaling pathway in retinal vascularization. Eye Brain 2016;8:141–6.
[88] Das A, et al. Retinal neovascularization is suppressed with a matrix metalloproteinase inhibitor. Arch Ophthalmol 1999;117(4):498–503.
[89] Mori K, et al. Pigment epithelium-derived factor inhibits retinal and choroidal neovascularization. J Cell Physiol 2001;188(2):253–63.
[90] Drixler TA, et al. Angiostatin inhibits pathological but not physiological retinal angiogenesis. Invest Ophthalmol Vis Sci 2001;42(13):3325–30.
[91] Bai YJ, et al. Antiangiogenesis effects of endostatin in retinal neovascularization. J Ocul Pharmacol Ther 2013;29(7):619–26.
[92] Aboul Naga SH, et al. Intracellular pathways following uptake of bevacizumab in RPE cells. Exp Eye Res 2015;131:29–41.
[93] Klingeborn M, et al. Directional exosome proteomes reflect polarity-specific functions in retinal pigmented epithelium monolayers. Sci Rep 2017;7(1):4901.
[94] Katome T, et al. Expression of intraocular peroxisome proliferator-activated receptor gamma in patients with proliferative diabetic retinopathy. J Diabetes Complicat 2015;29(2):275–81.
[95] Looze C, et al. Proteomic profiling of human plasma exosomes identifies PPARgamma as an exosome-associated protein. Biochem Biophys Res Commun 2009;378(3):433–8.
[96] Moisseiev E, et al. Protective effect of intravitreal administration of exosomes derived from mesenchymal stem cells on retinal ischemia. Curr Eye Res 2017;42:1–10.
[97] Hajrasouliha AR, et al. Exosomes from retinal astrocytes contain antiangiogenic components that inhibit laser-induced choroidal neovascularization. J Biol Chem 2013;288(39):28058–67.
[98] Konstantinell A, et al. Secretomic analysis of extracellular vesicles originating from polyomavirus-negative and polyomavirus-positive Merkel cell carcinoma cell lines. Proteomics 2016;16(19):2587–91.
[99] Janowska-Wieczorek A, et al. Microvesicles derived from activated platelets induce metastasis and angiogenesis in lung cancer. Int J Cancer 2005;113(5):752–60.
[100] Feng Q, et al. A class of extracellular vesicles from breast cancer cells activates VEGF receptors and tumour angiogenesis. Nat Commun 2017;8:14450.
[101] Monteforte A, et al. (*) Glioblastoma exosomes for therapeutic angiogenesis in peripheral ischemia. Tissue Eng Part A 2017;23(21–22):1251–61.
[102] Ju R, et al. Angiopoietin-2 secretion by endothelial cell exosomes: regulation by the phosphatidylinositol 3-kinase (PI3K)/Akt/endothelial nitric oxide synthase (eNOS) and syndecan-4/syntenin pathways. J Biol Chem 2014;289(1):510–9.
[103] Goetzl EJ, et al. Altered cargo proteins of human plasma endothelial cell-derived exosomes in atherosclerotic cerebrovascular disease. FASEB J 2017;31(8):3689–94.
[104] Tang XD, et al. Mesenchymal stem cell microvesicles attenuate acute lung injury in mice partly mediated by Ang-1 mRNA. Stem Cells 2017;35(7):1849–59.
[105] Beckett K, et al. Drosophila S2 cells secrete wingless on exosome-like vesicles but the wingless gradient forms independently of exosomes. Traffic 2013;14(1):82–96.
[106] Gross JC, et al. Active Wnt proteins are secreted on exosomes. Nat Cell Biol 2012;14(10):1036–45.
[107] Korkut C, et al. Trans-synaptic transmission of vesicular Wnt signals through Evi/Wntless. Cell 2009;139(2):393–404.

[108] Luga V, et al. Exosomes mediate stromal mobilization of autocrine Wnt-PCP signaling in breast cancer cell migration. Cell 2012;151(7):1542–56.
[109] Zhang L, Wrana JL. The emerging role of exosomes in Wnt secretion and transport. Curr Opin Genet Dev 2014;27:14–9.
[110] Shimoda M, Khokha R. Proteolytic factors in exosomes. Proteomics 2013;13(10–11):1624–36.
[111] Chowdhury R, et al. Cancer exosomes trigger mesenchymal stem cell differentiation into pro-angiogenic and pro-invasive myofibroblasts. Oncotarget 2015;6(2):715–31.
[112] You Y, et al. Matrix metalloproteinase 13-containing exosomes promote nasopharyngeal carcinoma metastasis. Cancer Sci 2015;106(12):1669–77.
[113] Ruivo CF, et al. The biology of cancer exosomes: insights and new perspectives. Cancer Res 2017;77(23):6480–8.
[114] Zitvogel L, et al. Eradication of established murine tumors using a novel cell-free vaccine: dendritic cell-derived exosomes. Nat Med 1998;4(5):594–600.
[115] Webber J, et al. Cancer exosomes trigger fibroblast to myofibroblast differentiation. Cancer Res 2010;70(23):9621–30.
[116] Hoshino D, et al. Exosome secretion is enhanced by invadopodia and drives invasive behavior. Cell Rep 2013;5(5):1159–68.
[117] Mu W, Rana S, Zoller M. Host matrix modulation by tumor exosomes promotes motility and invasiveness. Neoplasia 2013;15(8):875–87.
[118] Peinado H, et al. Melanoma exosomes educate bone marrow progenitor cells toward a pro-metastatic phenotype through MET. Nat Med 2012;18(6):883–91.
[119] Salehi M, Sharifi M. Exosomal miRNAs as novel cancer biomarkers: challenges and opportunities. J Cell Physiol 2018;233(9):6370–80.
[120] Kujala E, Makitie T, Kivela T. Very long-term prognosis of patients with malignant uveal melanoma. Invest Ophthalmol Vis Sci 2003;44(11):4651–9.
[121] Diener-West M, et al. Development of metastatic disease after enrollment in the COMS trials for treatment of choroidal melanoma: collaborative ocular melanoma study group report No. 26. Arch Ophthalmol 2005;123(12):1639–43.
[122] Eldh M, et al. MicroRNA in exosomes isolated directly from the liver circulation in patients with metastatic uveal melanoma. BMC Cancer 2014;14:962.
[123] Ragusa M, et al. miRNA profiling in vitreous humor, vitreal exosomes and serum from uveal melanoma patients: pathological and diagnostic implications. Cancer Biol Ther 2015;16(9):1387–96.
[124] Volarevic V, et al. Mesenchymal stem cell-derived factors: immuno-modulatory effects and therapeutic potential. Biofactors 2017;43(5):633–44.
[125] Mead B, Tomarev S. Bone marrow-derived mesenchymal stem cells-derived exosomes promote survival of retinal ganglion cells through miRNA-dependent mechanisms. Stem Cells Transl Med 2017;6(4):1273–85.
[126] Yu B, et al. Exosomes derived from MSCs ameliorate retinal laser injury partially by inhibition of MCP-1. Sci Rep 2016;6:34562.
[127] Shigemoto-Kuroda T, et al. MSC-derived extracellular vesicles attenuate immune responses in two autoimmune murine models: type 1 diabetes and uveoretinitis. Stem Cell Rep 2017;8(5):1214–25.
[128] Wassmer SJ, et al. Exosome-associated AAV2 vector mediates robust gene delivery into the murine retina upon intravitreal injection. Sci Rep 2017;7:45329.

Chapter 19

MSC-exosomes in regenerative medicine

Yueyuan Zhou[a,b], Nobuyoshi Kosaka[a,d], Zhongdang Xiao[b], and Takahiro Ochiya[a,c]

[a]*Division of Molecular and Cellular Medicine, National Cancer Center Research Institute, Tokyo, Japan,* [b]*State Key Laboratory of Bioelectronics, School of Biological Science and Medical Engineering, Southeast University, Nanjing, China,* [c]*Department of Molecular and Cellular Medicine, Institute of Medical Science, Tokyo Medical University, Tokyo, Japan,* [d]*Department of Translational Research for Extracellular Vesicles, Institute of Medical Science, Tokyo Medical University, Tokyo, Japan*

1 Introduction

Regenerative medicine aims at restoring diseased and injured tissues and whole organs through applying engineering and life science principles to promote regeneration. Strategies include the use of materials and de novo generated cells, as well as their various combinations, to contribute to tissue healing and reconstruct missing or damaged tissues and organs [1]. The current trend in the field of regenerative medicine is to use stem cells, including mesenchymal stromal (stem) cells (MSCs). MSCs are multipotent nonhematopoietic adult stem cells that were first identified by Friedenstein et al. in bone marrow in the 1960s [2]. MSCs possess various unique properties, including homing to damaged tissues, multilineage differentiation potential, colony forming and self-renewal abilities [3, 4]. Due to these properties, MSCs have received significant attention as a promising cell-based therapy for treatment of injured, malfunctioning tissues and organs in the field of regenerative medicine (Fig. 1). It was initially thought that MSCs exert their therapeutic effects by migrating to the sites of damage, engrafting and subsequently differentiating into target cells for tissue regeneration [3]. However, several recent studies have indicated that the beneficial actions of MSCs are attributable not only to their differentiation potential but also predominantly to their paracrine activity because only a small fraction of the transplanted MSCs survive and fuse with the existing cells at the site of injury site in host tissues [5–7]. The paracrine function of MSCs is exerted through secretion of soluble factors and release of extracellular vesicles (EVs), such as exosomes. Indeed, these soluble factors, which include growth factors, chemokines, and

Exosomes. https://doi.org/10.1016/B978-0-12-816053-4.00019-5

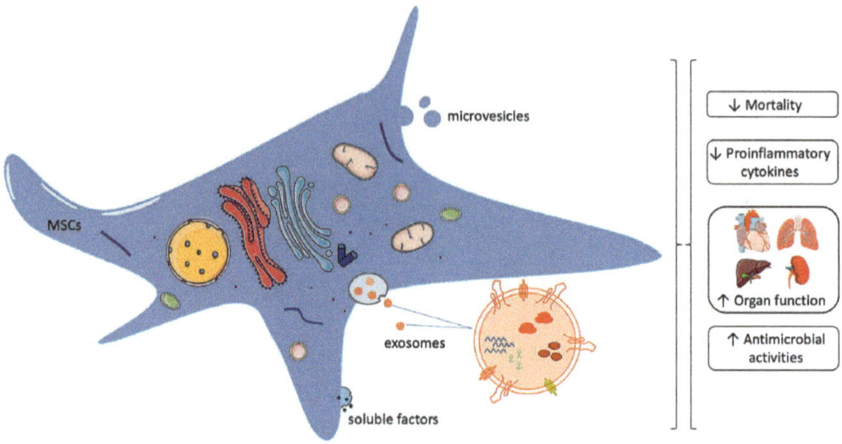

FIG. 1 General summary of MSCs in regenerative medicine.

cytokines, can reduce apoptosis and fibrosis, increase angiogenesis, enhance neuronal survival and differentiation, restrict local inflammation and modulate immune responses [8–10]. Recent studies have revealed that EVs released by MSCs may also play a critical role in the physiological function of these cells and at least in part be responsible for the therapeutic effects of MSCs [11–13].

2 Exosomes derived from mesenchymal stromal cells

Extracellular vesicles (EVs) comprise a heterogeneous group of nanosized lipid bilayer membrane vesicles that are secreted by a variety of cell types, including T cells, B cells, dendritic cells, mast cells, endothelial cells neuronal cells, cancer cells, embryonic cells and MSCs [14, 15]. EVs are categorized into several subpopulations, exosomes, microvesicles and others, based on their origin, size, morphology and cargo content. The classical view of exosome biogenesis holds that they are initially produced by invagination of the endosomal membrane to create multivesicular bodies (MVBs). In fact, exosomes themselves comprise a fairly heterogeneous population in terms of their biochemical composition and phenotype, which is determined by their cell source. As research in the field of exosomes and extracellular vesicles progresses, the nomenclature continues to be defined and refined. It has recently been revealed that exosomes have at least three subpopulations with unique N-glycosylation, protein, lipid, DNA and RNA profiles and biophysical properties [16]. Exosomes, ranging from 40 to 150 nm, with a density of 1.09–1.19 g/mL, were first identified during research on maturation of sheep reticulocytes [17]. Initially, exosomes were considered cellular debris or a means to excrete cellular metabolic waste, but it has been shown that they act as an additional intercellular communication mechanism and are involved in multiple physiological and pathological processes.

Exosomes are enveloped in a lipid bilayer membrane that protects their contents from degradation and enables them to move long distances in tissues and be internalized by recipient cells. The membrane is enriched in annexins; tetraspanins, such as CD9, CD63, and CD81; and heat-shock proteins, including Hsp60, Hsp70, and Hsp90. The internal biological contents of exosomes include a variety of RNA species (e.g., mRNA, miRNA, premiRNA, and long-noncoding RNA), DNA, lipids, and soluble proteins that are transferred to target cells [18–20].

Exosome cargoes are cell-type-specific and are often influenced by the physiological or pathological state of the parental cells, as well as the external stimuli that regulate their secretion, and the molecular mechanisms involved in their biogenesis [21, 22]. Furthermore, it has been shown that exosomes derived from MSCs can inherit their parental cell pleiotropic functions of promoting tissue repair and regeneration [23–25]. It was revealed that some surface receptors (CD29, CD44, CD73, and CD105), signaling molecules (involved in modulation of BMP, MAPK, and PPAR recipient cell signaling pathways), adhesion molecules and MSC-related markers contribute to the therapeutic benefit of MSC-derived EVs [26, 27]. Similarly, MSC-derived exosomes decreased apoptosis and increased cell proliferation in an acute kidney injury model partly through exosome-mediated RNA transfer [28]. Recently, MSC-derived exosomes and microvesicles were distinguished through high-resolution proteomics analysis [29]. In short, exosomes generated from MSCs can act as a targeted delivery system by transporting exogenous biological and chemical molecules for cell-free regenerative medicine. In addition, exosomes derived from MSCs obtained from younger and older hosts showed different miRNA expression patterns, indicating that exosomes are biomarker candidates to assess the state of the parental cells [30, 31]. Therefore, MSC-exosomes hold promise as biomarkers for identification of the specific requested transplant cells for the current stem cell-based therapy in regenerative medicine.

3 Application of MSC-EVs in regenerative medicine

Regenerative medicine utilizes MSCs due to their obvious capabilities, such as homing to injured and inflamed tissues, multilineage differentiation and immunomodulatory effects. However, there are still many challenges in aspects of engrafting, homing and differentiation long-term safety. The clinical limitations of MSC-based regenerative medicine therapy include their immune-associated rejection, genetic instability, low survival rate and function restriction. In addition, tumor formation by MSCs after transplantation is a major risk [32]. Standards for ex-vivo expansion of MSCs, large-scale production, storage and delivery, and quality control should be verified and validated [33].

Recently, MSC-exosomes have shown potential in cell-free regenerative therapy because they offer several superior advantages compared with cells, synthetic nanoparticles, and single molecules. In contrast with whole MSCs, exosomes possess beneficial qualities, such as their lack of a nucleus to avoid neoplastic transformation; enhanced stability allowing long-term preservation and long-transfer of biomolecules in the body; greater ease of surface modification with targeting molecules; and loading capacity for small molecules, proteins and RNAs. In addition, exosomes can also be engineered with distinct receptors or antibodies to transfer therapeutic cargoes to specific cells and tissues. Because exosomes are generated from cells, they show excellent biocompatibility and systemic biodistribution [34, 35]. Furthermore, exosomes host various types of biomolecules, enabling them to contribute to different therapeutic mechanisms simultaneously, which cannot be achieved with traditional small single molecules. Therefore, here, we address the molecular mechanisms of exosomes applied in regenerative medicine.

3.1 Neurological regeneration

With respect to the relationship between MSC-exosomes and neurological disease, stem cell-therapy should be discussed first. Currently, stem cell-therapy is considered a promising therapeutic option for several neurological disorders, including stroke, Parkinson’s disease, amyotrophic lateral sclerosis and Huntington's disease [36]. Below, we introduce studies related to the use of MSC-exosomes as a therapeutic strategy for neurological disease.

Alzheimer's is characterized by the accumulation of **b**-amyloid peptide (A**b**) in the brain due to an imbalance between A**b** production and clearance. Neprilysin (NEP) is the most important A**b**-degrading enzyme in the brain. It was shown that exosomes derived from human adipose tissue mesenchymal stem cells contained a large amount of enzymatically active NEP and could decrease both secreted and intracellular A**b** levels in neuroblastoma cells after coculture, which might be a therapeutic approach for Alzheimer's disease [37]. Similarly, exosomes isolated from stem cells derived from the dental pulp of human exfoliated deciduous teeth (SHEDs) suppressed 6-OHDA-induced apoptosis in dopaminergic neurons in a 3D culture environment, suggesting that exosomes derived from SHEDs are a promising new therapeutic tool in the treatment of Parkinson's disease [38]. Exosomes generated from bone marrow mesenchymal stem cells significantly improved functional recovery by promoting neurite remodeling, neurogenesis, and angiogenesis after systemic administration in a rat stroke model [39]. It was similarly reported that MSC-exosomes led to an improvement in neurological impairment, long-term neuroprotection and enhanced angioneurogenesis. These effects were in parallel with adjusted postischemic immune responses in a mouse stroke model [40]. Ophelders et al. found that MSC-exosomes protected brain function by reducing the neurological sequelae following hypoxic-ischemic injury [41]. In another study, it was demonstrated that treatment with MSC-exosomes significantly ameliorated inflammation-induced neuronal cellular degeneration, reduced microgliosis and prevented reactive astrogliosis, resulting in improved long-lasting cognitive function [42].

miRNAs associated with MSC-exosomes are also critical in neurological regeneration. Xin et al. demonstrated that MSC-exosomes transferred miR-133b to neuron cells and astrocytes, which subsequently benefited neurite remodeling and functional recovery in a rat stroke model [43]. Xin et al. reported that MSC-exosomes enriched with the miR-17-92 cluster enhanced oligodendrogenesis, neurogenesis, neural plasticity and neurological recovery via PTEN targeting to activate the PI3K/Akt/mTOR/GSK-3b signaling pathway [44]. Loss of retinal ganglion cells (RGCs) and their axons is one of the primary causes of blindness, resulting from either traumatic (optic neuropathy) or degenerative eye disease. Exosomes derived from bone marrow MSCs promoted statistically obvious survival of RGCs and regeneration of their axons while partially preventing RGC axon loss and RGC dysfunction. These exosomes transferred their cargo into the inner retinal layers, and the effects were reliant on miRNA [45]. Traumatic brain injury (TBI) involves a set of secondary pathological and/or functional alterations within the brain due to a sudden external force and is a major cause of death and long-term disability worldwide [46]. It has been demonstrated that bone marrow MSC-exosomes induced reproduction of endothelial cells in the lesion boundary zone and dentate gyrus and immature and mature neurons and reduced neuroinflammation in rats after TBI [47]. In another TBI mouse model, isolated MSC-exosomes suppressed neuroinflammation after TBI and rescued pattern separation and spatial learning impairments; therefore, these exosomes interrupted the self-perpetuating cycle of tissue destruction and inflammation [48]. Exosomes derived from MSCs reduced neuroinflammation by suppressing the activation of astrocytes and microglia and promoted neurogenesis, possibly by targeting the neurogenic niche, and these effects contributed to nervous tissue repair and functional recovery after TBI [49].

Although newly reported studies have revealed that MSC-exosomes improve functional recovery, promote neurogenesis and reduce neuroinflammation, the exact cellular and molecular mechanisms involved are unclear. Hence, mechanistic studies should be conducted to take full advantage of MSC-exosomes and support their potential in neurological regeneration.

3.2 Cardiovascular regeneration

Cardiovascular disease is the major cause of death and disability not only in developed countries but worldwide [50]. A major cause of cardiovascular disease is damage to the vasculature resulting in gradual build-up of atherosclerotic plaques that can partially occlude vessels, followed by the distal myocardium, and lead to ischemia under situations of increased cardiac demand. It is crucial to quickly reperfuse the vessel to prevent death of the ischemic myocardium. However, reperfusion may cause additional injury. The application of MSC-exosomes in cardiovascular regeneration was first reported in 2010 by Lai et al. [51]. Exosomes isolated from human embryonic stem cell-derived MSC-conditioned medium (CM) were shown to mediate cardioprotection by reducing infarct size during myocardial ischemia/reperfusion injury in a mouse model. Subsequently, another study by the same group showed that MSC-exosome treatment in a myocardial infarction mouse model resulted in decreased infarct size, enhanced NADH and ATP levels, and reduced oxidative stress. Meanwhile, MSC-exosomes increased phosphorylated-Akt and phosphorylated-GSK-3**b** levels and reduced the phosphorylated-c-JNK level in ischemic/reperfused hearts. Therefore, MSC-exosomes can improve cardiac function by restoring bioenergetics, reducing oxidative stress and activating pro-survival signaling pathways [52]. Other researchers have found that intramyocardial administration of MSC-derived exosomes markedly enhanced blood flow recovery, reduced infarct size and preserved cardiac systolic and diastolic performance; thus, MSC-exosomes protected cardiac tissue through stimulation of neoangiogenesis [53]. In addition, it was reported that MSC-exosomes significantly promoted tube formation of human umbilical vein endothelial cells, impaired T-cell function, reduced infarct size, and preserved cardiac systolic and diastolic performance, hence stimulating neovascularization and restraining the inflammation response [54].

Recently, various studies have focused on the role of miRNAs delivered by MSC-exosomes in cardiovascular regeneration. It was shown that miR-221 delivered by exosomes derived from MSCs downregulated p53 and upregulated modulator of apoptosis (PUMA) expression, hence, significantly cardioprotection [55]. Another study by the same group reported that MSC-exosomes transferred miR-19a in vitro and into ischemic myocardium in vivo, producing a greater cardioprotective effect by reducing the expression of phosphatase and tensin homolog (PTEN) and subsequent activation of the Akt and ERK signaling pathways [56]. Li et al. found that MSC-exosomes inhibited cardiac fibrosis and inflammation and improved cardiac function in a rat myocardial infarction model. In addition, they verified that MSC-exosomes and MSCs have similar miRNA expression profiles, indicating that MSC-exosomes could replace MSCs for cardiac repair. In addition, several specific and critical miRNAs present in MSC-exosomes were significantly different from those in MSCs, suggesting that MSC-exosomes may work even better than MSCs in cardiac regeneration. For instance, increased miR-29 expression and reduced miR-34 expression in both MSC-exosomes and MSCs might be responsible for the therapeutic effects of MSC-exosomes. The significantly lower expression of miR-21 and miR-15 in exosomes compared with MSCs likely helps explain the superior benefits of exosomes over MSCs [57]. However, another study investigating exosomal miR-21 revealed that human endometrium-derived MSCs (EnMSCs) conferred better cardioprotection than human bone marrow MSCs and adipose tissue MSCs because of the enhanced paracrine actions of secreted exosomes enriched with miR-21. Antagonism of miR-21 by anti-miR treatment abolished the antiapoptotic and angiogenic effects of EnMSCs, and these results indicated that exosomal miR-21 is a potential mediator of the cardioprotection effect of EnMSCs via the PTEN/Akt pathway [58]. Cardiac injury and dysfunction are commonly observed in sepsis and result in poor blood perfusion in multiple tissues [59]. It was reported that miR-233 contained in MSC-exosomes attenuated sepsis-trigged myocardial depression downregulated the expression of Sema3A and Stat3, leading to a reduction in the inflammatory response and cell death [60]. Zhang et al. found that pretreatment of cardiac stem cells with MSC-exosomes decreased cardiac fibrosis and increased survival and capillary density, thereby improving cardiac function in a rat myocardial infarction model [61].

Several pathways are involved in the therapeutic effects of exosomes in cardiovascular disease: (1) exosomes produced from intramyocardially injected stem cells exert paracrine effects; (2) resident cardiac stem cells may cause autotrophic stimulation of themselves or other cell types in the heart; (3) exosomes injected intramyocardially can affect different types of cells directly; (4) exosomes administered systemically interact with cells of the cardiovascular system, such as endothelial, blood, and cardiac cells [62].

3.3 Hepatic regeneration

The liver is a vital organ that possesses a powerful regeneration capability and exerts detoxification functions in the body. Liver disease, including viral hepatitis, alcoholic liver disease, nonalcoholic fatty liver disease, and associated end-stage liver disease, is a global health concern. Several studies have been conducted to examine the use of MSC-exosomes for treatment of liver injury. Xu et al. established carbon tetrachloride (CCl_4)-induced liver fibrosis mouse models. Then, the mouse livers were directly injected with human umbilical cord MSC-exosomes. It was revealed that transplantation of human umbilical cord MSC-exosomes reduced the surface fibrous capsules and softened their textures and alleviated hepatic inflammation and collagen deposition by inhibiting epithelial-to-mesenchymal transition [63]. In another CCl_4-induced liver injury mouse model, MSC-exosomes inhibited acetaminophen (APAP)- and hydrogen peroxide (H_2O_2)-induced hepatocyte apoptosis, thereby eliciting hepatoprotective effects against toxicant-induced injury primarily through activation of proliferative and regenerative responses [64]. Additionally, human umbilical cord MSC-exosomes transferred glutathione peroxidase 1 (GPX1) and promoted the recovery of hepatic oxidant injury [65]. In a hepatic ischemia-reperfusion (I/R) injury rat model, after human-induced pluripotent stem cell-derived MSC (hiPSC-MSC)-exosomes were injected systemically via the inferior vena cava, the infiltration of inflammatory cells was inhibited, the release of inflammatory factors was reduced, and liver oxidative stress was alleviated. These results suggest that hiPSC-MSC-exosomes confer liver protection in I/R injury via suppression of inflammatory responses, attenuation of the oxidative stress response and inhibition of apoptosis [66]. Du et al. found that hiPSC-MSC-exosomes suppressed hepatocyte necrosis and sinusoidal congestion in the ischemia/reperfusion injury model. Moreover, hiPSC-MSC-exosomes promoted primary hepatocyte proliferation, increased the activity of sphingosine kinase and synthesis of sphingosine-1-phosphate (S1P) [67]. Human menstrual blood-derived stem cells (MenSCs) are a novel source of MSCs that provide the advantage of being easy to collect and isolate. MenSC-exosomes were found to possess the ability to alleviate fulminant hepatic failure because they expressed various cytokines, such as IL-6 and angiopoietin-1, and reduced the number of liver mononuclear cells and the amount of the active apoptotic protein caspase-1 in fulminant hepatic failure [68]. In response to liver injury, IL-6 mediated the acute phase response and induced both cytoprotective and mitogenic functions. IL-6-induced signaling pathways have been verified to play a critical role in early onset as well as progression and maintenance of the liver regenerative process [69]. In conclusion, MSC-exosomes can inhibit the storm of inflammatory factors, suppress apoptosis, promote angiogenesis, provide energy support, promote hepatocyte proliferation and contribute to biliary restoration to reverse liver failure and may be an alternative to MSC therapies.

3.4 Renal regeneration

Evidence has shown that the kidney is a target organ of several diseases (for example, hypertension, anemia, and dyslipidemia, among others), and when kidney physiology is compromised, it can initiate or exacerbate other pathophysiological conditions, such as cardiovascular disease [70]. In a rat model of established chromic kidney disease (CKD), MSC-derived conditioned medium promoted therapeutic rescue shown by decreasing CKD progression and reducing hypertension and glomerular injury, suggesting that the paracrine factors, including cytokines and exosomes secreted by MSCs, contributed to the reno protective effects [71]. In a cisplatin-induced acute kidney injury (AKI) rat model, MSC-exosome treatment reduced the number of apoptotic cells, oxidative stress, and activation of the p38 mitogen-activated protein kinase (p38 MAPK) pathway. It was revealed that MSC-exosomes promoted cell proliferation through activation of the ERK1/2 pathway, indicating that MC-exosomes could repair cisplatin-induced AKI in rats and NRK-52E cell injury by ameliorating oxidative stress and cell apoptosis, promoting cell proliferation in vivo and in vitro [72]. In addition, Zou et al. reported that exosomes generated from human Wharton-Jelly MSCs (hWJMSCs) ameliorated renal injury at both the acute and chronic stage after ischemic AKI. They observed that after administration of hWJMSC-exosomes renal cell apoptosis was mitigated and proliferation was enhanced, the expression of CX3CL1 was down-regulated and the number of $CD68^{+}$ macrophages in the kidney was decreased. These results indicate that hWJMSC-exosomes exert therapeutic effects through anti-inflammatory activity via suppression of CX3CL1 [73]. In kidney injury following ischemia/reperfusion, the chemokine CCR2 and its receptor CX3CR1 mediate signaling pathways that regulate monocyte/macrophage trafficking and hence contribute to inflammation-related kidney injury [74]. However, exosomes secreted from MSCs enriched in CCR2 were able to reduce the concentration of free CCL2, thereby suppressing its ability to recruit or activate macrophages and consequently improving renal injury repair in ischemia/reperfusion-induced renal injury [75]. Another cytokine, IL-10, was found to be involved in the renoprotective effects of MSC-exosomes. In a porcine model of metabolic syndrome and renal artery stenosis, adipose tissue MSC-derived exosomes attenuated renal inflammation and improved medullary oxygenation and fibrosis, whereas the renoprotective therapeutic effects were blunted in pigs treated with IL10-depleted exosomes [76]. Additionally, miRNAs transferred by MSC-exosomes participate in AKI recovery. Collino et al. decreased global miRNA expression in exosomes via knockdown of Drosha in MSCs and treated glycerol-induced mouse models with these exosomes. They found that global downregulation of miRNAs enriched in MSC-exosomes halted the renal regenerative effects, suggesting that the delivery of miRNAs mediated by exosomes plays a critical role in renoprotective function [77]. Recently, application of MSC-exosomes to patients has been conducted to assess the safety and therapeutic efficacy of exosomes derived from human cord blood in ameliorating the progression of grade III and IV chronic kidney disease. The results showed that administration of cord blood MSC-exosomes was safe and could ameliorate the inflammatory immune response and improve the overall kidney function in grade III and IV CKD patients [78]. Taken together, accumulated evidence suggests that MSC-exosomes possess renoprotective potential through their biological cargoes, including miRNAs and proteins.

3.5 Skeletal, chondral, and muscular regeneration

Pathological destructive bone diseases, including osteoporosis, osteoarthritis and rheumatoid arthritis, are associated with a persistent decrease in patient quality of life and are considered to present a major global health problem [79–81]. It was reported that adipose tissue MSC-derived exosomes preconditioned by tumor necrosis factor-alpha (TNF-**a**) could promote proliferation and differentiation of primary osteoblastic cells by suppressing the Wnt signaling pathway [82]. These results indicated that MSC-exosomes might potentially be used for bone tissue regeneration and replace direct stem cell transplantation for bone repair and regeneration. Zhang et al. reported that exosomes isolated from human-induced pluripotent stem cell-derived MSCs could enhance the osteoinductivity of **b**-TCP by activating the PI3K/Akt signaling pathway in human bone marrow MSCs [83]. Another study demonstrated that MSC-exosomes promoted orderly cartilage regeneration in an immunocompetent rat osteochondral defect model. Weekly intra-articular injections of MSC-exosomes promoted early cell infiltration and proliferation, reduced the number of apoptotic cells in the reparative tissue, and induced polarization of synovial macrophages into a regenerative M2 phenotype, thus facilitating orderly cartilage and subchondral bone generation [84]. In addition, injection of MSC-exosomes rescued the slow fracture healing in $CD9^{-/-}$ femur fracture mouse models. Because the levels of bone repair-related cytokines, including monocyte chemotactic protein-1 (MCP-1), MCP-3, and stromal cell-derived factor-1, in exosomes were lower compared with levels in conditioned medium, it was suggested that bone repair might be partly mediated by other exosome cargoes, such as miRNAs. miR-21, an anti-apoptotic miRNA, was most abundant in MSC-exosomes [85]. Previous studies have also demonstrated that miR-21 promotes osteogenic differentiation of MSCs by targeting small mothers against decapentaplegic 7 (Smad7) [86] and the PI3K/**b**-catenin pathway [87], and local injection of miR-21 overexpressing MSCs promoted fracture healing in a rat model [88]. In an established rat osteochondral defect model, weekly intra-articular injections of human embryonic MSC-derived exosomes led to an enhanced gross appearance and improved histological scores. Furthermore, by 12weeks, exosome-treated defects displayed complete restoration of cartilage and subchondral bone that exhibited characteristic features, including hyaline cartilage with good surface regularity, complete bonding to adjacent cartilage, and extracellular matrix deposition, that closely resembled those of age-matched unoperated controls compared with PBS-treated defects with fibrous reproduction [89]. In addition to enhanced osteogenesis, exosomes derived from human-induced pluripotent stem cell-derived MSCs (hiPSC-MSCs) enhanced angiogenesis in osteoporotic rats. The in vitro results showed that hiPSC-MSC-exosomes promoted cell proliferation and alkaline phosphatase (ALP) activity, and upregulated mRNA and protein expression of osteoblast-related genes in bone marrow MSCs derived from ovariectomized rats [90]. MSC-exosomes promote skeletal muscle regeneration and cartilage and subchondral bone restoration. Nakamura et al. reported that human bone marrow MSC-secreted exosomes enhanced myogenesis and angiogenesis in vitro and skeletal muscle regeneration in an in vivo model of muscle injury. They also verified that miR-494 in the MSC-exosomes might be partially responsible for the promotion of muscle regeneration [91]. Treatment with MSC-exosomes and cardiac-specific overexpression of miR-494 have been reported to protect against ischemia/reperfusion-induced cardiac injury [51, 53, 92]. For restoration of bone defects through the endochondral pathway, the main challenges include the neovascular system to recruit mesenchymal progenitors, chondrogenic differentiation and hypertrophic maturation of these progenitors, and remodeling of hypertrophic cartilage [93]. However, exosome-based therapy may overcome one or more of these bottlenecks. In conclusion, MSC-exosomes show potential advantages in the bone defect microenvironment. They exert therapeutic effects through their abundant contents and work in a multifunctional manner.

3.6 Cutaneous regeneration

Following a skin injury, the damaged tissue is repaired through coordinated biological actions that constitute the cutaneous healing response. Wound healing is a complex process involving a cascade of molecular and cellular events, including cell migration, inflammation, angiogenesis, granulation formation, re-epithelialization, and extracellular matrix (ECM) remodeling [94].

A variety of studies have focused on the beneficial accelerating effects of MSC-exosomes on cutaneous wound healing. It was reported that bone marrow MSC-derived exosomes were internalized by human umbilical vein endothelial cells and induced dose-dependent increases in tube formation by endothelial cells. Moreover, these MSC-exosomes activated several signaling pathways involved in wound healing, such as Akt, ERK, and STAT3 pathways, and induced the expression of a number of growth factors, including hepatocyte growth factor (HGF), insulin-like growth factor-1 (IGF-1), nerve growth factor (NGF), and stromal-derived growth factor-1 (SDF-1). Thereby, MSC-exosomes were shown to enhance proliferation and migration of fibroblasts derived from normal donors and chronic wound patients in a dose-dependent manner [95]. It was also revealed that exosomes generated from human-induced pluripotent stem cell-derived MSCs accelerated re-epithelialization, reduced scar widths and promoted maturity of collagen after injection. Not only was the generation of newly formed vessels enhanced, but their maturation in wound sites was accelerated. In vitro results demonstrated that these exosomes stimulated the proliferation and migration of human dermal fibroblasts and HUVECs and increased the secretion of Type I and III collagen and elastin in fibroblasts and tube formation by HUVECs in a dose-dependent manner [96]. The specific expression pattern of miRNAs enriched in MSC-exosomes was verified to be at least in part responsible for the therapeutic wound healing effects of MSC-exosomes. During wound healing, exosomes isolated from umbilical cord-derived MSCs contained a group of distinct miRNAs (miR-21, -23a, -125b, and -145) that were correlated with negative regulation of myofibroblast formation through suppression of the transforming growth factor-**b**2/SMAD2 signaling pathway [97]. Another study reported that miR-125a transferred by adipose tissue MSC-exosomes repressed the expression of the angiogenesis inhibitor delta-like 4 (DDL4) and modulated endothelial cell angiogenesis by promoting formation of endothelial tip cells [98]. In a severe burn rat model, miR-181c in human umbilical cord MSC-exosomes effectively alleviated inflammation by suppressing the TLR4 signaling pathway [63]. Additionally, human umbilical cord MSC-exosomes promoted beta-catenin nuclear translocation and induced increased expression of proliferating cell nuclear antigen, cyclin D3, N-cadherin, and **b**-catenin and decreased E-cadherin expression. Furthermore, activation of the Wnt4/**b**-catenin signaling pathway participated in the enhanced angiogenesis [99]. These results suggested that human umbilical cord MSC-exosomes display strong promise for wound healing applications. Flap necrosis is the most frequent postoperative complication encountered in reconstructive surgery, and the main reason might be deficient neovascularization and diminished levels of proangiogenic factor production. It has been found that exosomes from adipose tissue MSCs can significantly increase the survival and capillary density of flaps subjected to ischemia-reperfusion injury, and this MSC-exosome activity was possibly mediated through the IL-6/STAT3 pathway [100]. Though the importance of the effects of exosomes derived from MSCs on the wound healing process is well established and is commonly accepted in most studies in this field, many of the detailed mechanisms underlying the functions of the bioactive molecules delivered by exosomes are still unknown. More research is required to clarify how exosomes regulate the wound healing process.

4 Exosomal modification in regenerative medicine

Exosomes have shown great potential for therapeutics and possess various properties allowing them to be an alternative to stem-cell-based therapies for support of endogenous repair and enhancement of existing regenerative medicine approaches. The biocompatibility of exosomes, with suitable modification, could increase the stability and efficacy of therapeutics while enhancing cellular uptake. In general, exosome modification primarily involves modification of their surface with specific ligands and loading with therapeutic agents, such as short-interfering RNA (siRNA), miRNA, DNA, proteins and small molecules.

4.1 Engineering the exosomal surface

Exosomes represent a new type of long-distance transfer of bioactive molecules due to the protection from degradation and deactivation provided by the lipid bilayer membrane. Theoretically, exosomes can be engineered to target specific cell types. Exosomes derived from MSCs inherit characteristics that can influence tissue responses to injury, infection, and disease. Successful treatment for ischemic stroke requires safe and effective delivery of drugs, including a variety of chemicals, peptides and genetic molecules; however, the main obstacle is the blood-brain barrier (BBB) [101, 102]. Tian et al. proposed a simple, rapid and efficient method to conjugate ligands that have high affinity to integrins onto MSC-exosomal surfaces through bio-orthogonal copper-free azide alkyne cycloaddition. First, they established a cerebral ischemia mouse model. The cycloconjugated MSC-exosomes were then injected intravenously and were shown to target the lesion region of the ischemic brain and enter microglia, neurons and astrocytes. In addition, these exosomes preloaded with curcumin promoted the release of pro-inflammatory cytokines and activation of microglia [103].

It has been reported that the targeted dendritic exosomes conjugate with the neuron-specific RVG peptide Lamp2b were able to transfer GAPDH siRNA and BACE1 siRNA across the BBB. Subsequently, mRNA and **b**-amyloid expression in the brain were significantly suppressed [104]. Based on the therapeutic effects of MSC-exosomes in Alzheimer's disease [37], these adipose tissue MSC-derived exosomes could be modified with Lamp2b to target recipient neuroblastoma cells, thereby decreasing secreted and intracellular A**b** levels. In another example, exosomes were modified with GE11 peptide, which specifically bound to EGFR on their surfaces delivered miRNA to xenograft breast cancer cells in mouse models [105].

4.2 Engineering effective exosome contents

The exosome lipid bilayer membrane surrounds and contains a hydrophilic core. Based on a thorough understanding of the exosome structure, another critical aspect of exosome modification is exosomal content. Therapeutic agents, such as small molecules, proteins, and nucleic acids, can be incorporated into exosomes using two major approaches: active or passive encapsulation [106, 107]. These different approaches lead to different loading efficiencies and drug stabilities in the extracellular vesicles. Passive cargo-preloading strategies are relatively simple and are mainly divided into two types: incubation of the drug with exosomes or with parental cells. The hydrophobicity of small chemical molecules accounts for the significantly higher loading because hydrophobic drugs can interact with the lipid layers of the membrane [108]. For instance, curcumin is a hydrophobic polyphenol compound that is soluble in aqueous solution. It was reported that curcumin could be incorporated into exosomes through incubation with exosomes derived from mouse lymphoma cells at 22°C in phosphate-buffered saline (PBS) for 5min before sucrose gradient centrifugation and protected mice against lipopolysaccharide (LPS)-induced septic shock [108]. The in vivo and in vitro results demonstrated that the target specificity was determined by exosomes, and an improvement in curcumin activity was achieved by directing curcumin to inflammatory cells. Moreover, therapeutic but not toxic effects were observed. Similarly, catalase, a potent antioxidant, was loaded into exosomes through incubation at room temperature. Catalase-loaded exosomes protected SNpc neurons against oxidative stress in mice with acute brain inflammation [109]. In addition, hydrophobically modified siRNAs (hsiRNAs) were efficiently loaded into exosomes derived from U87 cells upon co-incubation and delivered to mouse primary cortical neurons and efficiently silenced up to 35% of the target mRNA, Huntington mRNA [110]. Active drug loading refers to electroporation, sonication, extrusion, freeze/thaw cycles, permeabilization, and click chemistry methods.

4.2.1 *Electroporation*

This strategy involves creating small pores in exosome membranes via application of an electrical field to exosomes suspended in a conductive solution. The electrical current disturbs the phospholipid bilayer, resulting in the formation of temporary pores, which allow for free diffusion of the drug into the interior of the vesicle. Then, the integrity of the exosomal membrane is recovered after the drug loading process. Electroporation has been applied to preload several types of cargos, including siRNA and miRNA. GAPDH and BACE1 siRNAs were electroporated at 400V and 125μF into exosomes and exhibited the greatest siRNA retention. Furthermore, transfer of this siRNA by exosomes successfully knocked down GAPDH and BACE1 expression [104]. Similar results were shown in the report by Wahlgren et al. They used exosomes from plasma as a drug delivery system to T cells and monocytes and preloaded them with siRNA against mitogen-activated protein kinase 1 (MAPK1) via electroporation [111]. In another example, two selected siRNAs against RAD51 and RAD52 were electroporated into exosomes, and these exosomes were verified to deliver the siRNAs to recipient cells and strongly suppressed the expression of RAD51 [112]. However, it has been reported that electroporation of exosomes with siRNA is accompanied by extensive siRNA aggregate formation, which might lead to overestimation of the amount of siRNAs actually loaded into exosomes [113].

A miR-155 mimic was introduced into exosomes via electroporation and significantly increased miR-155 levels in primary mouse hepatocytes and the liver of miR-155 knockout mice [114]. In addition to siRNA and miRNA, small molecule drugs have been loaded into exosomes via electroporation. Doxorubicin was electroporated into exosomes derived from immature dendritic cells with an encapsulation efficiency of up to 20% [115]. Johnsen et al. reported that electroporation did not change the endogenous stimulatory capacity of adipose tissue MSC-exosomes to induce glioblastoma multiforme cell proliferation but induced adverse morphological changes, including aggregation of the exosomes [116].

4.2.2 Sonication

After being mixed with drugs and proteins, exosomes can be sonicated with a homogenizer probe. The mechanical shear force from the sonicator probe compromises the exosome membrane integrity and allows the drug to diffuse into the exosomes during membrane deformation. Paclitaxel was loaded into exosomes derived from RAW 264.7 macrophages through sonication, and the loading capacity was nearly 30%, higher than that achieved via incubation (~2%) and electroporation (~5.5%) [117]. In another study, catalase was loaded into exosomes using several different methods, including sonication, saponin, freeze-thaw cycles and extrusion. Additionally, it has been verified that sonication of exosomes might alter the surface protein content as well as the lipid bilayer organization and lead to increased exosomal interactions with cellular membranes of target cells. Furthermore, extensive reformation and reshaping of exosomes upon sonication and extrusion enabled catalase diffusion across relatively tight and highly structured lipid bilayers and resulted in a high loading efficiency [109].

4.2.3 Freeze/thaw cycles

After incubation of exosomes with drugs at room temperature, the mixture can be rapidly frozen at −80°C or in liquid nitrogen and then thawed at room temperature. Repeating this freeze/thaw cycle at least 3 times ensures the effectiveness and efficiency of encapsulation. This method can optimize the properties of the exosome surface to decrease immunogenicity and increase colloidal stability and improve the half-life of exosomes in blood [118]. Based on a nanoparticle tracking analysis, this strategy has been shown to induce aggregation of exosomes and lead to a broad size distribution of the drug-loaded exosomes. Additionally, the exosomes preloaded with catalase using freeze/thaw cycles displayed intermediate loading efficiency values compared with those loaded with the enzyme via incubation at room temperature, sonication, extrusion and permeabilization. However, these particles were considerably larger in size, likely due to aggregation [109].

4.2.4 Permeabilization

Typically, permeabilization is applied to cells after fixation in order to detect intracellular antigens [119]. Because the membrane of exosomes is similar to that of cells in composition and structure, exosomes can be permeabilized to make them permeable to antibodies. Two general types of reagents are commonly used for permeabilization: organic solvents, such as methanol; and detergents, such as saponin and Triton X-100 [119]. It was reported that saponin-assisted permeabilization increased drug loading into exosomes by 11-fold compared with passive loading without saponin [106].

Other classifications of exosome-based drug formulations include (A) loading of naïve exosomes isolated from parental cells ex vivo; (B) loading of parental cells with a drug, which is then released from exosomes; and (C) transfection/infection of parental cells with DNA encoding therapeutically active compounds, specifically, miRNA and siRNA, which are then released from exosomes [120]. The previously described active or passive encapsulation strategies involved incorporation of drug into isolated exosomes. As a second approach, parental cells can be preloaded with exogenous compounds. The mouse MSC cell line SR4987 was incubated with paclitaxel and then released a significant amount of exosomes enriched with paclitaxel [121]. Similarly, treatment of resistant HepG2 cells with anticancer drugs (paclitaxel, carboplatin, etoposide and irinotecan hydrochloride) induced the cells to secrete exosomes with upregulated expression of heat shock proteins (HSPs) on the surface, subsequently leading to superior immunogenicity in inducing HSP-specific NK cell responses [122]. In addition, it was eported that chemotherapeutic-loaded nanovesicles were produced via breakdown of monocytes or macrophages using serial extrusion through filters with diminishing pore sizes. These nanoparticles possessed natural cell targeting ability by maintaining the topology of plasma membrane proteins and induced TNF-α-stimulated endothelial cell death in vitro and reduced tumor growth in vivo [123]. It was also shown that exogenous compounds could be packaged in extracellular vesicles (EVs) through engineering the parental cells via liposomes, and the EVs mediated autonomous intercellular migration of the compounds through multiple cancer cell layers [124]. Finally, isolation and purification of drug-preloaded exosomes from genetically modified parental cells has been commonly used for manufacturing exosome-based formulations. Generally, DNA encoding miRNA, siRNA, DNA or mRNA sequences is transfected or electroporated into source cells, and then, the genes are encapsulated into exosomes derived from these parental cells (Table 1).

GAPDH-siRNA and BACE1-siRNA loading into exosomes via electroporation was demonstrated strong mRNA (60%) and protein (62%) knockdown of BACE1, and these siRNA-loaded exosomes displayed therapeutic effects in Alzheimer's disease in mouse models [104]. miRNAs can suppress protein levels by degrading mRNA or inhibiting translation of target genes, and exosomes can deliver miRNAs to target recipient cells. Exosomes derived from MSCs transfected with a miR-146b expression plasmid decreased EGFR and NF-**k**B protein levels in glioma cells in vitro. The in vivo results showed that the volume of glioma tumors were reduced after treatment with the miR-146-loaded exosomes [128]. In another neuronal disease, exosomes generated from MSCs transfected with anti-miR-9 were internalized by glioblastoma multiforme (GBM) cells and decreased the expression of the drug transporter gene MDR1 in chemoresistant GBM cells and increased the chemosensitivity of GBM cells [129]. It was reported that exosomes isolated from MSCs and loaded with miR-124 subsequently delivered miR-124 to infarct sites, promoting cortical neural progenitors to obtain a neuronal identity and protecting against ischemic injury via robust cortical neurogenesis [132]. miR-124 is also involved in Huntington's disease; decreased miR-124 expression increases RE1-silecing transcription factor (REST) levels, resulting in repression of key target genes, such as brain-derived neurotrophic factor [144]. It was reported that exosomes derived from human Wharton Jelly MSCs transfected with miR-30 can be delivered to injured renal tubular epithelial cells and restore the expression of miR-30 in injured rat kidney [135]. In addition, DNA is another type bioactive molecule that can be carried to target cells and tissues. The BCR/ABL hybrid gene, which is involved in the pathogenesis of chronic myeloid leukemia (CML), was transferred from exosomes derived from donor K562 cells to neutrophils and decreased their phagocytic activity in vitro [145]. Lamichhane et al. reported that exogenous linear DNA could be loaded into exosomes via electroporation in quantities sufficient to yield an average of hundreds of DNA molecules per vesicle. They suggested that the DNA loading efficiency and capacity in exosomes was dependent on DNA size, with linear DNA molecules less than 1000bp in length being more efficiently loaded compared with larger linear DNA and plasmid DNA molecules. It was further shown that exosomes delivered the exogenous DNA to recipient cells [139]. In addition to DNA, microvesicles, including exosomes, can transfer mRNA and proteins. A fragment of Y RNA was enriched in exosomes derived from cardiosphere-derived cells. These exosomes were delivered to macrophages and induced IL-10 production. Moreover, these Y RNA-loaded exosomes reduced infarct size and elicited a cardioprotective response following myocardial ischemia/reperfusion (I/R) [141]. Overall, a variety of DNA molecules encoding miRNA, siRNA, DNA and mRNA sequences can be loaded into exosomes, suggesting that MSC-exosomes represent a promising DNA 'drug' delivery system in regenerative medicine.

TABLE 1 Exosome-based gene delivery

Parental cells	Loading method	Gene biotype	Effects	Ref.
Dendritic cells	Electroporation	siRNA-GAPDH	Knockdown GAPDH in vitro and in vivo	[104]
Endothelial cells	Transfection	siRNA	Inhibit luciferase expression in target cells	[125]
Fibroblast L929 cells	Transfection	siRNA-TGF-beta1	Suppress TGF-beta1 expression and inhibit growth and metastases in mice	[126]
HEK293T cells	Transfection	siRNA-MOR (opioid receptor Mu)	Deliver siRNA to the brain and suppress MOR expression	[127]
MSCs	Transfection	miR-146b	Reduce glioma xenograft growth of rats	[128]
MSCs	Transfection	Anti-miR-9	Reverse the chemoresistance of glioblastoma multiforme cells	[129]
HEK293 cells	Transfection	let-7a	Deliver let-7a to EGFR-expressing breast cancer cells and tissues	[105]
Human adult liver stem cells	Transfection	miRNA	Inhibit the growth and survival of hepatocellular carcinoma in mice	[130]
HEK293 cells	Transfection	miR-124	Suppress target gene	[131]
Bone marrow MSCs	Electroporation	miR-124	Deliver miR-124 to the infarct site and protect against ischemic injury	[132]
Endothelial cells	Transfection	miR-195	Inhibit 5-HTT (5-hydroxytryptamine) expression	[133]
Adipose tissue MSCs	Transfection	miR-450a-5p	Promote adipogenesis through suppression of WISP2	[134]

Continued

TABLE 1 Exosome-based gene delivery—cont'd

Parental cells	Loading method	Gene biotype	Effects	Ref.
Human Wharton Jelly MSCs	Transfection	miR-30	Ameliorate acute renal IRI by inhibiting mitochondrial	[135]
HEK293T cells	Transfection	miR-29a/c	Control gastric cancer growth by blocking angiogenesis	[136]
Bone marrow MSCs	Transfection	miR-200b	Prevent EMT and alleviate colon fibrosis	[137]
HL-1	Transfection	DNA	Change gene expression	[138]
HEK293T cells, HUVEC cells, HRVT, MSCs	Electroporation	dsDNA	Deliver DNA	[139]
Human glioma cancer stem cells	Transfection	DNA	Extracellular vesicles cross the intact blood-brain barrier	[140]
Cardiosphere-derived cells	Transfection	Y RNA fragment	Cardioprotection via modulation of IL-10 expression and secretion	[141]
ES-2, SKOV3	Transfection	MMP1 mRNA	Induce apoptosis in mesothelial cells	[142]
HEK293T cells	Transfection	Cytosine deaminase uracil mRNA	Inhibit Schwannoma tumor growth	[143]

5 Clinical application of MSC-exosomes in regenerative medicine

Recently, exosomes have been approved for application in several clinical trials, and the process of exosome-based therapies in humans has been boosted. Dendritic cell (DC)-derived exosomes have attracted much interest as tumor vaccines due to their specific regulation of the immune system. It was revealed that DC-exosomes stimulated naïve CD4 + T cells and CD8 + T cells by presenting both MHC-I and MHC-II [146, 147]. Three phase I clinical trials confirmed the safety and feasibility of DC-exosomes in cancer therapies; however, the injected DC-exosomes exerted limited therapeutic effects in stimulating MHC-restricted T cell responses [148, 149]. Another Phase II clinical trial was conducted using exosomes derived from IFN-γ-matured DCs rather than immature DCs. It was demonstrated that activation of NK cells through NKp30 might be an effective immunomodulatory strategy for patients with stage IIIB/IV NSCLC [150]. Unfortunately, the end point to 50% nonprogressors by 4months post-chemotherapy was not reached. In addition, there is at least one company that has been focusing on the commercial use of DC-exosomes. Anosys, Inc. has produced an autologous DC-exosomes vaccine for cancer therapy (http://chromos.com). Other effective applications of exosomes from MSCs, endothelial progenitors, Tregs, and other types of cells include regenerative medicine and immunotherapy for nonmalignant disorders. Clinical trials of exosomes derived from MSCs are ongoing. Another two companies are developing commercial use of MSC-exosomes [151]. ReNeuron Group PLC is practical experience in neurological and ischemic fields (http://www.reneuron.com). Capricor, Inc. is investigating the therapeutic effects of exosomes in cardiac and muscle diseases (http://capricor.com).

6 Limitations and challenges

Currently, it has been demonstrated that exosomes derived from MSCs play a critical role in various physiological and pathological processes. It was initially suggested that MSCs are of great interest as candidates for cell-therapy in tissue repair and regeneration due to their differential potential and immunomodulation benefits [152–154]. However, accumulating evidence indicates that the paracrine action of MSCs tends to be primarily responsible for the therapeutic effects of MSCs in regenerative medicine [119, 155, 156]. Exosomes, one of the factors that MSCs secrete, mediate interactions between cells and mimic the biological functions of their parental cells. Due to their multiple and comprehensive characteristics, exosomes hold outstanding potential as an alternative for MSC-based cell therapy. For use in clinical trials, the first step is to prepare the source cells for exosome isolation. Second, expand the culture system to obtain the required number of exosomes. Then, the exosomes must be separated and purified via several methods. Finally, exosomes with therapeutic potential need to be administered to the patient intravenously or intranasally (Fig. 2) [104, 157, 158].

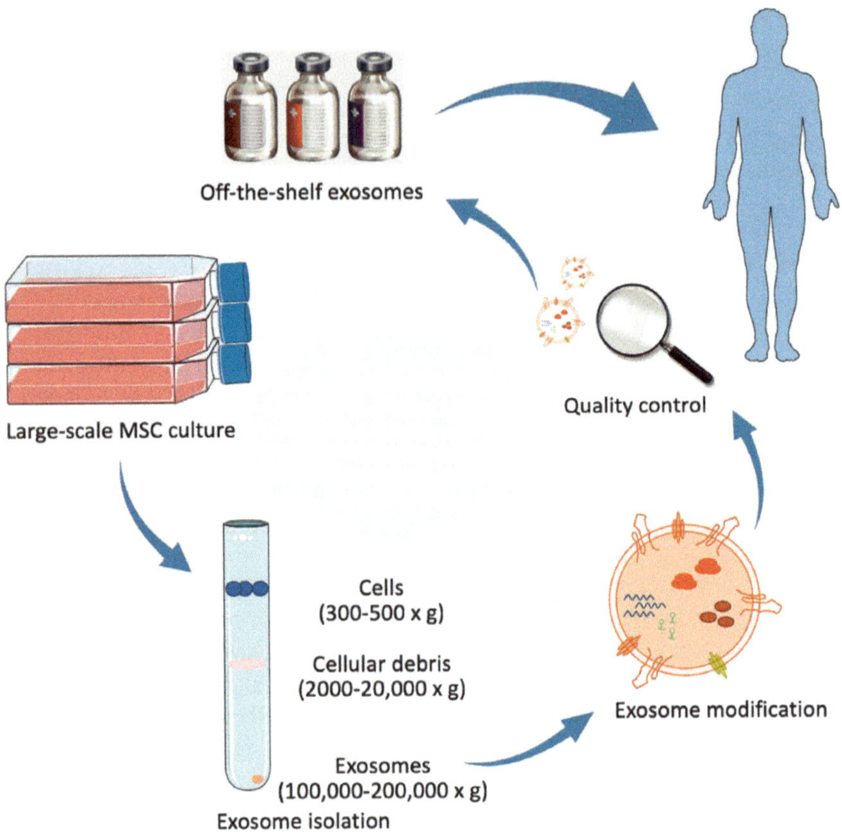

FIG. 2 Schematic of representative steps of clinical application of exosomes-based therapy.

As for application of MSC-exosomes in clinical trials, there are still several obstacles in production, modification, storage, transport, and quality control (Table 2).

For different types and stages of diseases, several questions need to be addressed. For instance, which type of stem cells should be the exosome source. It has been verified that the contents of exosomes, at least in part, mimic that in parental cells. Thus, distinct parental mesenchymal stromal cells may be selected depending on their tissue origin and the therapeutic goals. The cells can be isolated from the patient himself/herself or from an allogeneic donor, and the cells should be cultured in a common microenvironment or with special stimulation. Application of MSCs for clinical purposes takes advantage of their poor immunogenicity, which supports the possibility of obtaining MSCs or even exosomes from allogeneic donors [3]. Then, to implement accurate targeted delivery and/or to enhance therapeutic effects, exosomes may be modified with distinct peptides or receptors, and specific types of therapeutic cargos (genetic drugs and/or chemical compounds) or their combination may

TABLE 2 Challenges in clinical application of exosomes

Problems	Detailed questions
Cell source	The types of MSCs
	The culture and stimulation conditions
	Autologous/exogenous
Modification	Target molecules
	Bioactive contents
	Label
Quality control	cGMP (clinical good manufacturing practice)
	Homogenous
	A large scale
	Viability/efficiency
	Storage and transport

be loaded into exosomes. It is also necessary to select the loading strategy according to the contents to be loaded to ensure encapsulation efficiency. Furthermore, additional labeling markers are likely necessary to track the biodistribution of exosomes and their homing to target tissues and organs via in vivo imaging techniques. In addition, to obtain sufficient quantities of exosomes for large-scale application in clinical trials, the first approach is to expand the source MSC cells ex vivo. It was reported that TSAP6 enhanced the secretion of exosomes by cells with a p53 response to stress [159]. It is also been demonstrated that modulate modulating the expression of some genes or the pH of the culture environment may increase exosomes production by MSCs [160–162]. The second step is to select the isolation method. The current common methods for isolating and purifying exosomes are differential centrifugation, size exclusion chromatography, polymer precipitation, immunomagnetic isolation, and filtration.

Differential centrifugation is the most widely used method for exosome isolation [163]. It has been widely adopted as a reliable technique for isolating exosomes from biological fluids [164]. In this method, cells and apoptotic debris are removed via low-speed centrifugation, larger-sized microvesicles are removed at a higher speed, and finally exosomes are precipitated through high-speed ultracentrifugation. However, this time-consuming process is less efficient when the source biological liquids are plasma and serum [165]. Size-exclusion chromatography (SEC) is a separation method based on the size of the compound and depends on how efficiently a compound penetrates the pores of the stationary phase. Compared with differential centrifugation, this method achieves high yields and stable recovery of exosomes due to the complete removal of contaminating proteins. On the negative side, there is a possibility of contamination by the chromatography column, and a large number of fractions need to be collected and postanalyzed, representing a series of time-consuming steps [166]. The biofluid mixed with precipitation solution is incubated overnight at 4°C and then centrifuged at low speed to obtain the exosome pellet. This method is relatively easy to perform and does not require specialized equipment or an extensive length of time. However, it is expense due to the cost of commercial precipitation-based kits [167]. Immunomagnetic isolation isolates exosomes through binding of specific surface proteins or antigens and attachment to beads. Although the yield is low through this immunoaffinity approach, the isolated exosomes are much purer compared with those obtained through other methods. Taken together, each method has advantages and disadvantages, requires different sample preprocessing techniques, and produces exosome preparations of varying purity and quality. The user may choose a method for exosome isolation according to the intended downstream application. Further research is required to develop novel approaches for mass production of exosomes.

Exosomes for clinical application are required for clinical good manufacturing practice (cGMP) standards, including cell banking and storage, scalability, stability, lot and batch tracking, pathogen screening, and other physical, chemical, and biological methods for quality control [157, 168]. In addition to the production, and quality control of exosomes, other key points for future exploration should be proposed that the exosome treatment may correlates closely with the patient's condition and the safety considerations including toxicity and side effects. Hence, much work should be conducted to take full advantages of MSC-exosomes and support their potential for clinical and personalized treatment.

7 Conclusion and perspectives

In this chapter, we have introduced the biological characteristics and therapeutic functions of exosomes derived from MSCs and highlighted the potential of MSC-exosomes for clinical application in regenerative medicine. Since it has been well accepted that the biogenesis of exosomes is involved in intercellular communication, exosomes exert a variety of superior properties as therapeutics compared with cells, synthetic nanoparticles and single molecules. Because exosomes are released by most cell types, they can be engineered to be low or even nonimmunogenic, and they possess exceptional biocompatibility, biostability and safety. Due to their lack of a nucleus, exosomes avoid neoplastic transformation relative to whole cells. They remain viable after freeze/thaw cycles, long-distance transport and long-term storage and hold bioactive therapeutics inside a membrane, which protects the therapeutics from degradation and inactivation. In addition, they can be genetically modified with specific surface receptors to offer specific delivery to cells and organs. Hence, exosomes have attracted much attention as vehicles for drug delivery and gene therapy. For regenerative medicine, exosomes may hold promise for activating injured cells and tissues and promote restoration via their bioactive contents and functional molecules. Additionally, MSC-exosomes can be applied as biomarkers for identification of parental MSCs of high quality for transplantation using current regenerative medicine approaches.

Despite the increasing number of studies and encouraging results in vitro and in animal models, the approaches applied to utilize exosomes derived from MSCs for therapeutics need to be addressed and normalized. In particular, production scale-up to obtain sufficient quantities while maintaining cGMP standards of exosomes for use in clinical trials and treatments is a current barrier. It is especially crucial to further investigate the side effects of MSC-exosomes in animal studies and preclinical tests. Therefore, not only does their novel therapeutic potential need to be explored but also their safety and efficacy must be carefully assessed.

In summary, additional work should be conducted to determine the underlying mechanisms by which exosomes derived from mesenchymal stromal cells contribute to tissue repair and regeneration. Encouraging preclinical studies fuel the hope that MSC-exosomes will become promising candidates for regenerative medicine.

Competing interests

The authors declare that they have no competing interests.

References

[1] Bajaj P, Schweller RM, Khademhosseini A, West JL, Bashir R. 3D biofabrication strategies for tissue engineering and regenerative medicine. Annu Rev Biomed Eng 2014;16:247–76.

[2] Kassem M, Kristiansen M, Abdallah BM. Mesenchymal stem cells: cell biology and potential use in therapy. Basic Clin Pharmacol Toxicol 2004 Nov;95(5):209–14.

[3] Uccelli A, Moretta L, Pistoia V. Mesenchymal stem cells in health and disease. Nat Rev Immunol 2008;8(9):726–36.

[4] Trohatou O, Roubelakis MG. Mesenchymal stem/stromal cells in regenerative medicine: past, present, and future. Cell Reprogram 2017;19(4):217–24.

[5] Madrigal M, Rao KS, Riordan NH. A review of therapeutic effects of mesenchymal stem cell secretions and induction of secretory modification by different culture methods. J Transl Med 2014;12(1):260. Available from: http://translational-medicine.biomedcentral.com/articles/10.1186/s12967-014-0260-8. [cited 15 October 2018].

[6] Haynesworth SE, Baber MA, Caplan AI. Cytokine expression by human marrow-derived mesenchymal progenitor cells in vitro: effects of dexamethasone and IL-1α. J Cell Physiol 1996;166:585–92. https://doi.org/10.1002/(SICI)1097-4652(199603)166:3<585::AID-JCP13>3.0.CO;2-6.

[7] Basu J, Ludlow JW. Exosomes for repair, regeneration and rejuvenation. Expert Opin Biol Ther 2016;16(4):489–506.

[8] Seki A, Sakai Y, Komura T, Nasti A, Yoshida K, Higashimoto M, et al. Adipose tissue-derived stem cells as a regenerative therapy for a mouse steatohepatitis-induced cirrhosis model. Hepatology 2013;58(3):1133–42.
[9] Higashimoto M, Sakai Y, Takamura M, Usui S, Nasti A, Yoshida K, et al. Adipose tissue derived stromal stem cell therapy in murine ConA-derived hepatitis is dependent on myeloid-lineage and CD4+ T-cell suppression. Eur J Immunol 2013;43(11):2956–68.
[10] Katsuda T, Ochiya T. Potential application of mesenchymal stem cell-derived exosomes as a novel therapeutic drug. Drug Deliv Syst 2014;29(2):140–51.
[11] De Jong OG, Van Balkom BWM, Schiffelers RM, Bouten CVC, Verhaar MC. Extracellular vesicles: potential roles in regenerative medicine. Front Immunol 2014;5:608. Available from: http://journal.frontiersin.org/article/10.3389/fimmu.2014.00608/abstract. [cited 15 October 2018].
[12] Goradel NH, Jahangiri S, Negahdari B. Effects of mesenchymal stem cell-derived exosomes on angiogenesis in regenerative medicine. Curr Regen Med 2018;7(1):46–53.
[13] Cheng L, Zhang K, Wu S, Cui M, Xu T. Focus on mesenchymal stem cell-derived exosomes: opportunities and challenges in cell-free therapy. Stem Cells Int 2017;2017:1–10.
[14] Rani S, Ryan AE, Griffin MD, Ritter T. Mesenchymal stem cell-derived extracellular vesicles: toward cell-free therapeutic applications. Mol Ther 2015;23(5):812–23.
[15] Keshtkar S, Azarpira N, Ghahremani MH. Mesenchymal stem cell-derived extracellular vesicles: novel frontiers in regenerative medicine. Stem Cell Res Ther 2018;9(1):63. Available from: https://stemcellres.biomedcentral.com/articles/10.1186/s13287-018-0791-7. [cited 15 October 2018].
[16] Zhang H, Freitas D, Kim HS, Fabijanic K, Li Z, Chen H, et al. Identification of distinct nanoparticles and subsets of extracellular vesicles by asymmetric flow field-flow fractionation. Nat Cell Biol 2018;20(3):332–43.
[17] Pan B-T, Johnstone RM. Fate of the transferrin receptor during maturation of sheep reticulocytes in vitro: selective externalization of the receptor. Cell 1983;33(3):967–78.
[18] Stoorvogel W, Kleijmeer MJ, Geuze HJ, Raposo G. The biogenesis and functions of exosomes. Traffic 2002;3(5):321–30.
[19] Théry C, Zitvogel L, Amigorena S. Exosomes: composition, biogenesis and function. Nat Rev Immunol 2002;2:569.
[20] Valadi H, Ekström K, Bossios A, Sjöstrand M, Lee JJ, Lötvall JO. Exosome-mediated transfer of mRNAs and microRNAs is a novel mechanism of genetic exchange between cells. Nat Cell Biol 2007;9:654.
[21] Minciacchi VR, Freeman MR, Di Vizio D. Extracellular vesicles in cancer: exosomes, microvesicles and the emerging role of large oncosomes. Semin Cell Dev Biol 2015;40:41–51.
[22] Kalra H, Drummen GPC, Mathivanan S. Focus on extracellular vesicles: introducing the next small big thing. Tikkanen R, editor, Int J Mol Sci 2016;17(2):170.
[23] Tan SS, Yin Y, Lee T, Lai RC, Yeo RWY, Zhang B, et al. Therapeutic MSC exosomes are derived from lipid raft microdomains in the plasma membrane. J Extracell Vesicles 2013;2. https://doi.org/10.3402/jev.v2i0.22614.
[24] Liu S, Liu D, Chen C, Hamamura K, Moshaverinia A, Yang R, et al. MSC transplantation improves osteopenia via epigenetic regulation of notch signaling in lupus. Cell Metab 2015;22(4):606–18.
[25] Pashoutan Sarvar D, Shamsasenjan K, Akbarzadehlaleh P. Mesenchymal stem cell-derived exosomes: new opportunity in cell-free therapy. Adv Pharm Bull 2016;6(3):293–9.
[26] Kim H-S, Choi D-Y, Yun SJ, Choi S-M, Kang JW, Jung JW, et al. Proteomic analysis of microvesicles derived from human mesenchymal stem cells. J Proteome Res 2012;11(2):839–49.

[27] Vallabhaneni KC, Penfornis P, Dhule S, Guillonneau F, Adams KV, Mo YY, et al. Extracellular vesicles from bone marrow mesenchymal stem/stromal cells transport tumor regulatory microRNA, proteins, and metabolites. Oncotarget 2015;6(7):4953–67.

[28] Reis LA, Borges FT, Simões MJ, Borges AA, Sinigaglia-Coimbra R, Schor N. Bone marrow-derived mesenchymal stem cells repaired but did not prevent gentamicin-induced acute kidney injury through paracrine effects in rats. PLoS One 2012;7(9):e44092.

[29] Haraszti RA, Didiot M-C, Sapp E, Leszyk J, Shaffer SA, Rockwell HE, et al. High-resolution proteomic and lipidomic analysis of exosomes and microvesicles from different cell sources. J Extracell Vesicles 2016;5:32570. https://doi.org/10.3402/jev.v5.32570.

[30] Wang Y, Fu B, Sun X, Li D, Huang Q, Zhao W, et al. Differentially expressed microRNAs in bone marrow mesenchymal stem cell-derived microvesicles in young and older rats and their effect on tumor growth factor-β1-mediated epithelial-mesenchymal transition in HK2 cells. Stem Cell Res Ther 2015;6:185.

[31] Lei Q, Liu T, Gao F, Xie H, Sun L, Zhao A, et al. Microvesicles as potential biomarkers for the identification of senescence in human mesenchymal stem cells. Theranostics 2017;7(10):2673–89.

[32] Sakaida I, Terai S, Yamamoto N, Aoyama K, Ishikawa T, Nishina H, et al. Transplantation of bone marrow cells reduces CCl4-induced liver fibrosis in mice. Hepatology 2004;40(6):1304–11.

[33] Tögel F, Hu Z, Weiss K, Isaac J, Lange C, Westenfelder C. Administered mesenchymal stem cells protect against ischemic acute renal failure through differentiation-independent mechanisms. Am J Physiol-Ren Physiol 2005;289(1):F31–42.

[34] Kourembanas S. Exosomes: vehicles of intercellular signaling, biomarkers, and vectors of cell therapy. Annu Rev Physiol 2015;77(1):13–27.

[35] Yáñez-Mó M, Siljander PR-M, Andreu Z, Zavec AB, Borràs FE, Buzas EI, et al. Biological properties of extracellular vesicles and their physiological functions. J Extracell Vesicles 2015;4:27066. https://doi.org/10.3402/jev.v4.27066.

[36] Yoo J, Kim H-S, Hwang D-Y. Stem cells as promising therapeutic options for neurological disorders. J Cell Biochem 2012;114(4):743–53.

[37] Katsuda T, Tsuchiya R, Kosaka N, Yoshioka Y, Takagaki K, Oki K, et al. Human adipose tissue-derived mesenchymal stem cells secrete functional neprilysin-bound exosomes. Sci Rep 2013;3:1197.

[38] Jarmalavičiūtė A, Tunaitis V, Pivoraitė U, Venalis A, Pivoriūnas A. Exosomes from dental pulp stem cells rescue human dopaminergic neurons from 6-hydroxy-dopamine-induced apoptosis. Cytotherapy 2015;17(7):932–9.

[39] Xin H, Li Y, Cui Y, Yang JJ, Zhang ZG, Chopp M. Systemic administration of exosomes released from mesenchymal stromal cells promote functional recovery and neurovascular plasticity after stroke in rats. J Cereb Blood Flow Metab 2013;33(11):1711–5.

[40] Doeppner TR, Herz J, Görgens A, Schlechter J, Ludwig A-K, Radtke S, et al. Extracellular vesicles improve post-stroke neuroregeneration and prevent postischemic immunosuppression. Stem Cells Transl Med 2015;4(10):1131–43.

[41] Ophelders DR, Wolfs TG, Jellema RK, Zwanenburg A, Andriessen P, Delhaas T, et al. Mesenchymal stromal cell-derived extracellular vesicles protect the fetal brain after hypoxia-ischemia. Stem Cells Transl Med 2016;5(6):754–63.

[42] Drommelschmidt K, Serdar M, Bendix I, Herz J, Bertling F, Prager S, et al. Mesenchymal stem cell-derived extracellular vesicles ameliorate inflammation-induced preterm brain injury. Brain Behav Immun 2017;60:220–32.

[43] Xin H, Li Y, Liu Z, Wang X, Shang X, Cui Y, et al. MiR-133b promotes neural plasticity and functional recovery after treatment of stroke with multipotent mesenchymal stromal cells in rats via transfer of exosome-enriched extracellular particles. Stem Cells 2013;31(12):2737–46.

[44] Xin H, Katakowski M, Wang F, Qian J-Y, Shuang Liu X, Ali MM, et al. MiR-17-92 cluster in exosomes enhance neuroplasticity and functional recovery after stroke in rats. Stroke 2017;48(3):747–53.

[45] Mead B, Amaral J, Tomarev S. Mesenchymal stem cell-derived small extracellular vesicles promote neuroprotection in rodent models of glaucoma. Invest Ophthalmol Vis Sci 2018;59(2):702–14.

[46] Menon DK, Schwab K, Wright DW, Maas AI. Position statement: definition of traumatic brain injury. Arch Phys Med Rehabil 2010;91(11):1637–40.

[47] Zhang Y, Chopp M, Meng Y, Katakowski M, Xin H, Mahmood A, et al. Effect of exosomes derived from multipluripotent mesenchymal stromal cells on functional recovery and neurovascular plasticity in rats after traumatic brain injury. J Neurosurg 2015;122(4):856–67.

[48] Kim D, Nishida H, An SY, Shetty AK, Bartosh TJ, Prockop DJ. Chromatographically isolated CD63(+)CD81(+) extracellular vesicles from mesenchymal stromal cells rescue cognitive impairments after TBI. Proc Natl Acad Sci U S A 2016;113(1):170–5.

[49] Yang Y, Ye Y, Su X, He J, Bai W, He X. MSCs-derived exosomes and neuroinflammation, neurogenesis and therapy of traumatic brain injury. Front Cell Neurosci 2017;11:55.

[50] Hausenloy DJ, Yellon DM. Myocardial ischemia-reperfusion injury: a neglected therapeutic target. J Clin Invest 2013;123(1):92–100.

[51] Lai RC, Arslan F, Lee MM, Sze NSK, Choo A, Chen TS, et al. Exosome secreted by MSC reduces myocardial ischemia/reperfusion injury. Stem Cell Res 2010;4(3):214–22.

[52] Arslan F, Lai RC, Smeets MB, Akeroyd L, Choo A, Aguor ENE, et al. Mesenchymal stem cell-derived exosomes increase ATP levels, decrease oxidative stress and activate PI3K/Akt pathway to enhance myocardial viability and prevent adverse remodeling after myocardial ischemia/reperfusion injury. Stem Cell Res 2013;10(3):301–12.

[53] Bian S, Zhang L, Duan L, Wang X, Min Y, Yu H. Extracellular vesicles derived from human bone marrow mesenchymal stem cells promote angiogenesis in a rat myocardial infarction model. J Mol Med 2014;92(4):387–97.

[54] Teng X, Chen L, Chen W, Yang J, Yang Z, Shen Z. Mesenchymal stem cell-derived exosomes improve the microenvironment of infarcted myocardium contributing to angiogenesis and anti-inflammation. Cell Physiol Biochem 2015;37(6):2415–24.

[55] Yu B, Gong M, Wang Y, Millard RW, Pasha Z, Yang Y, et al. Cardiomyocyte protection by GATA-4 gene engineered mesenchymal stem cells is partially mediated by translocation of miR-221 in microvesicles. PLoS ONE 2013;8(8):e73304.

[56] Yu B, Kim HW, Gong M, Wang J, Millard RW, Wang Y, et al. Exosomes secreted from GATA-4 overexpressing mesenchymal stem cells serve as a reservoir of anti-apoptotic microRNAs for cardioprotection. Int J Cardiol 2015;182:349–60.

[57] Shao L, Zhang Y, Lan B, Wang J, Zhang Z, Zhang L, et al. MiRNA-sequence indicates that mesenchymal stem cells and exosomes have similar mechanism to enhance cardiac repair. Biomed Res Int 2017;2017:4150705.

[58] Wang K, Jiang Z, Webster KA, Chen J, Hu H, Zhou Y, et al. Enhanced cardioprotection by human endometrium mesenchymal stem cells driven by exosomal MicroRNA-21. Stem Cells Transl Med 2017;6(1):209–22.

[59] Romero-Bermejo FJ, Ruiz-Bailen M, Gil-Cebrian J, Huertos-Ranchal MJ. Sepsis-induced cardiomyopathy. Curr Cardiol Rev 2011;7(3):163–83.

[60] Wang X, Gu H, Qin D, Yang L, Huang W, Essandoh K, et al. Exosomal miR-223 contributes to mesenchymal stem cell-elicited cardioprotection in polymicrobial sepsis. Sci Rep 2015;5:13721.

[61] Zhang Z, Yang J, Yan W, Li Y, Shen Z, Asahara T. Pretreatment of cardiac stem cells with exosomes derived from mesenchymal stem cells enhances myocardial repair. J Am Heart Assoc 2016;5(1):e002856.

[62] Davidson SM, Yellon DM. Exosomes and cardioprotection—a critical analysis. Mol Aspects Med 2018;60:104–14.
[63] Li T, Yan Y, Wang B, Qian H, Zhang X, Shen L, et al. Exosomes derived from human umbilical cord mesenchymal stem cells alleviate liver fibrosis. Stem Cells Dev 2013;22(6):845–54.
[64] Tan CY, Lai RC, Wong W, Dan YY, Lim S-K, Ho HK. Mesenchymal stem cell-derived exosomes promote hepatic regeneration in drug-induced liver injury models. Stem Cell Res Ther 2014;5(3):76.
[65] Yan Y, Jiang W, Tan Y, Zou S, Zhang H, Mao F, et al. hucMSC exosome-derived GPX1 is required for the recovery of hepatic oxidant injury. Mol Ther 2017;25(2):465–79.
[66] Nong K, Wang W, Niu X, Hu B, Ma C, Bai Y, et al. Hepatoprotective effect of exosomes from human-induced pluripotent stem cell-derived mesenchymal stromal cells against hepatic ischemia-reperfusion injury in rats. Cytotherapy 2016;18(12):1548–59.
[67] Du Y, Li D, Han C, Wu H, Xu L, Zhang M, et al. Exosomes from human-induced pluripotent stem cell-derived mesenchymal stromal cells (hiPSC-MSCs) protect liver against hepatic ischemia/reperfusion injury via activating sphingosine kinase and sphingosine-1-phosphate signaling pathway. Cell Physiol Biochem 2017;43(2):611–25.
[68] Chen L, Xiang B, Wang X, Xiang C. Exosomes derived from human menstrual blood-derived stem cells alleviate fulminant hepatic failure. Stem Cell Res Ther 2017;8:9.
[69] Fujiyoshi M, Ozaki M. Molecular mechanisms of liver regeneration and protection for treatment of liver dysfunction and diseases. J Hepato-Biliary-Pancreat Sci 2010;18(1):13–22.
[70] Torres Crigna A, Daniele C, Gamez C, Medina Balbuena S, Pastene DO, Nardozi D, et al. Stem/stromal cells for treatment of kidney injuries with focus on preclinical models. Front Med 2018;5:179.
[71] van Koppen A, Joles JA, van Balkom BWM, Lim SK, de Kleijn D, Giles RH, et al. Human embryonic mesenchymal stem cell-derived conditioned medium rescues kidney function in rats with established chronic kidney disease. Dussaule J-C, editor PLoS ONE 2012;7(6):e38746.
[72] Zhou Y, Xu H, Xu W, Wang B, Wu H, Tao Y, et al. Exosomes released by human umbilical cord mesenchymal stem cells protect against cisplatin-induced renal oxidative stress and apoptosis in vivo and in vitro. Stem Cell Res Ther 2013;4(2):34.
[73] Zou X, Zhang G, Cheng Z, Yin D, Du T, Ju G, et al. Microvesicles derived from human Wharton's jelly mesenchymal stromal cells ameliorate renal ischemia-reperfusion injury in rats by suppressing CX3CL1. Stem Cell Res Ther 2014;5(2):40.
[74] Li L, Huang L, Sung S-SJ, Vergis AL, Rosin DL, Rose CE, et al. The chemokine receptors CCR2 and CX3CR1 mediate monocyte/macrophage trafficking in kidney ischemia–reperfusion injury. Kidney Int 2008;74(12):1526–37.
[75] Shen B, Liu J, Zhang F, Wang Y, Qin Y, Zhou Z, et al. CCR2 positive exosome released by mesenchymal stem cells suppresses macrophage functions and alleviates ischemia/reperfusion-induced renal injury. Stem Cells Int 2016;2016:1–9.
[76] Eirin A, Zhu X-Y, Puranik AS, Tang H, McGurren KA, van Wijnen AJ, et al. Mesenchymal stem cell-derived extracellular vesicles attenuate kidney inflammation. Kidney Int 2017;92(1):114–24.
[77] Collino F, Bruno S, Incarnato D, Dettori D, Neri F, Provero P, et al. AKI recovery induced by mesenchymal stromal cell-derived extracellular vesicles carrying microRNAs. J Am Soc Nephrol 2015;26(10):2349–60.
[78] Nassar W, El-Ansary M, Sabry D, Mostafa MA, Fayad T, Kotb E, et al. Umbilical cord mesenchymal stem cells derived extracellular vesicles can safely ameliorate the progression of chronic kidney diseases. Biomater Res 2016;20:21.

[79] Kowada T, Kikuta J, Kubo A, Ishii M, Maeda H, Mizukami S, et al. In vivo fluorescence imaging of bone-resorbing osteoclasts. J Am Chem Soc 2011;133(44):17772–6.

[80] Xie Y, Gao Y, Zhang L, Chen Y, Ge W, Tang P. Involvement of serum-derived exosomes of elderly patients with bone loss in failure of bone remodeling via alteration of exosomal bone-related proteins. Aging Cell 2018;17(3):e12758.

[81] Manolagas SC. Steroids and osteoporosis: the quest for mechanisms. J Clin Invest 2013;123(5):1919–21.

[82] Lu Z, Chen Y, Dunstan C, Roohani-Esfahani S, Zreiqat H. Priming adipose stem cells with tumor necrosis factor-alpha preconditioning potentiates their exosome efficacy for bone regeneration. Tissue Eng Part A 2017;23(21–22):1212–20.

[83] Zhang J, Liu X, Li H, Chen C, Hu B, Niu X, et al. Exosomes/tricalcium phosphate combination scaffolds can enhance bone regeneration by activating the PI3K/Akt signaling pathway. Stem Cell Res Ther 2016;7(1):136.

[84] Zhang S, Chu W, Lai R, Hui J, Lee E, Lim S, et al. Human mesenchymal stem cell-derived exosomes promote orderly cartilage regeneration in an immunocompetent rat osteochondral defect model. Cytotherapy 2016;18(6):S13.

[85] Furuta T, Miyaki S, Ishitobi H, Ogura T, Kato Y, Kamei N, et al. Mesenchymal stem cell-derived exosomes promote fracture healing in a mouse model. Stem Cells Transl Med 2016;5(12):1620–30.

[86] Li H, Yang F, Wang Z, Fu Q, Liang A. MicroRNA-21 promotes osteogenic differentiation by targeting small mothers against decapentaplegic 7. Mol Med Rep 2015;12(1):1561–7.

[87] Meng Y-B, Li X, Li Z-Y, Zhao J, Yuan X-B, Ren Y, et al. microRNA-21 promotes osteogenic differentiation of mesenchymal stem cells by the PI3K/β-catenin pathway. J Orthop Res 2015;33(7):957–64.

[88] Sun Y, Xu L, Huang S, Hou Y, Liu Y, Chan K-M, et al. mir-21 overexpressing mesenchymal stem cells accelerate fracture healing in a rat closed femur fracture model. Biomed Res Int 2015;2015:412327.

[89] Zhang S, Chu WC, Lai RC, Lim SK, Hui JHP, Toh WS. Exosomes derived from human embryonic mesenchymal stem cells promote osteochondral regeneration. Osteoarthr Cartil 2016;24(12):2135–40.

[90] Qi X, Zhang J, Yuan H, Xu Z, Li Q, Niu X, et al. Exosomes secreted by human-induced pluripotent stem cell-derived mesenchymal stem cells repair critical-sized bone defects through enhanced angiogenesis and osteogenesis in osteoporotic rats. Int J Biol Sci 2016;12(7):836–49.

[91] Nakamura Y, Miyaki S, Ishitobi H, Matsuyama S, Nakasa T, Kamei N, et al. Mesenchymal-stem-cell-derived exosomes accelerate skeletal muscle regeneration. FEBS Lett 2015;589(11):1257–65.

[92] Wang X, Zhang X, Ren X-P, Chen J, Liu H, Yang J, et al. MicroRNA-494 targeting both pro-apoptotic and antiapoptotic proteins protects against ischemia/reperfusion-induced cardiac injury. Circulation 2010;122(13):1308–18.

[93] Ferreira E, Porter RM. Harnessing extracellular vesicles to direct endochondral repair of large bone defects. Bone Joint Res 2018;7(4):263–73.

[94] Lee DE, Ayoub N, Agrawal DK. Mesenchymal stem cells and cutaneous wound healing: novel methods to increase cell delivery and therapeutic efficacy. Stem Cell Res Ther 2016;7(1):37.

[95] Shabbir A, Cox A, Rodriguez-Menocal L, Salgado M, Van Badiavas E. Mesenchymal stem cell exosomes induce proliferation and migration of normal and chronic wound fibroblasts, and enhance angiogenesis in vitro. Stem Cells Dev 2015;24(14):1635–47.

[96] Zhang J, Guan J, Niu X, Hu G, Guo S, Li Q, et al. Exosomes released from human induced pluripotent stem cells-derived MSCs facilitate cutaneous wound healing by promoting collagen synthesis and angiogenesis. J Transl Med 2015;13:49.

[97] Fang S, Xu C, Zhang Y, Xue C, Yang C, Bi H, et al. Umbilical cord-derived mesenchymal stem cell-derived exosomal MicroRNAs suppress myofibroblast differentiation by inhibiting the transforming growth factor-β/SMAD2 pathway during wound healing. Stem Cells Transl Med 2016;5(10):1425–39.

[98] Liang X, Zhang L, Wang S, Han Q, Zhao RC. Exosomes secreted by mesenchymal stem cells promote endothelial cell angiogenesis by transferring miR-125a. J Cell Sci 2016;129(11):2182.

[99] Zhang B, Wu X, Zhang X, Sun Y, Yan Y, Shi H, et al. Human umbilical cord mesenchymal stem cell exosomes enhance angiogenesis through the Wnt4/β-catenin pathway. Stem Cells Transl Med 2015;4(5):513–22.

[100] Pu C-M, Liu C-W, Liang C-J, Yen Y-H, Chen S-H, Jiang-Shieh Y-F, et al. Adipose-derived stem cells protect skin flaps against ischemia/reperfusion injury via IL-6 expression. J Invest Dermatol 2017;137(6):1353–62.

[101] Rhim T, Lee DY, Lee M. Drug delivery systems for the treatment of ischemic stroke. Pharm Res 2013;30(10):2429–44.

[102] Thompson BJ, Ronaldson PT. Chapter six—Drug delivery to the ischemic brain. In: Davis TP, editor. Advances in pharmacology. Academic Press; 2014. p. 165–202. [Internet]. Available from: http://www.sciencedirect.com/science/article/pii/S1054358914000143.

[103] Tian T, Zhang H-X, He C-P, Fan S, Zhu Y-L, Qi C, et al. Surface functionalized exosomes as targeted drug delivery vehicles for cerebral ischemia therapy. Biomaterials 2018;150:137–49.

[104] Alvarez-Erviti L, Seow Y, Yin H, Betts C, Lakhal S, Wood MJA. Delivery of siRNA to the mouse brain by systemic injection of targeted exosomes. Nat Biotechnol 2011;29:341.

[105] Ohno S, Takanashi M, Sudo K, Ueda S, Ishikawa A, Matsuyama N, et al. Systemically injected exosomes targeted to EGFR deliver antitumor microRNA to breast cancer cells. Mol Ther 2013;21(1):185–91.

[106] Fuhrmann G, Serio A, Mazo M, Nair R, Stevens MM. Active loading into extracellular vesicles significantly improves the cellular uptake and photodynamic effect of porphyrins. In: 3rd Symposium on Innovative Polymers Control Delivery SIPCD 2014;205.2015. p. 35–44.

[107] Luan X, Sansanaphongpricha K, Myers I, Chen H, Yuan H, Sun D. Engineering exosomes as refined biological nanoplatforms for drug delivery. Acta Pharmacol Sin 2017;38(6):754–63.

[108] Sun D, Zhuang X, Xiang X, Liu Y, Zhang S, Liu C, et al. A novel nanoparticle drug delivery system: the anti-inflammatory activity of curcumin is enhanced when encapsulated in exosomes. Mol Ther 2010;18(9):1606–14.

[109] Haney MJ, Klyachko NL, Zhao Y, Gupta R, Plotnikova EG, He Z, et al. Exosomes as drug delivery vehicles for Parkinson's disease therapy. J Control Release 2015;207:18–30.

[110] Didiot M-C, Hall LM, Coles AH, Haraszti RA, Godinho BM, Chase K, et al. Exosome-mediated delivery of hydrophobically modified siRNA for huntingtin mRNA silencing. Mol Ther 2016;24(10):1836–47.

[111] Wahlgren J, De L, Karlson T, Brisslert M, Vaziri Sani F, Telemo E, Sunnerhagen P, et al. Plasma exosomes can deliver exogenous short interfering RNA to monocytes and lymphocytes. Nucleic Acids Res 2012;40(17):e130.

[112] Shtam TA, Kovalev RA, Varfolomeeva EY, Makarov EM, Kil YV, Filatov MV. Exosomes are natural carriers of exogenous siRNA to human cells in vitro. Cell Commun Signal 2013;11:88.

[113] Kooijmans SAA, Stremersch S, Braeckmans K, de Smedt SC, Hendrix A, Wood MJA, et al. Electroporation-induced siRNA precipitation obscures the efficiency of siRNA loading into extracellular vesicles. J Control Release 2013;172(1):229–38.
[114] Momen-Heravi F, Bala S, Bukong T, Szabo G. Exosome-mediated delivery of functionally active miRNA-155 inhibitor to macrophages. Nanomedicine Nanotechnol Biol Med 2014;10(7):1517–27.
[115] Tian Y, Li S, Song J, Ji T, Zhu M, Anderson GJ, et al. A doxorubicin delivery platform using engineered natural membrane vesicle exosomes for targeted tumor therapy. Biomaterials 2014;35(7):2383–90.
[116] Johnsen KB, Gudbergsson JM, Skov MN, Christiansen G, Gurevich L, Moos T, et al. Evaluation of electroporation-induced adverse effects on adipose-derived stem cell exosomes. Cytotechnology 2016;68(5):2125–38.
[117] Kim MS, Haney MJ, Zhao Y, Mahajan V, Deygen I, Klyachko NL, et al. Development of exosome-encapsulated paclitaxel to overcome MDR in cancer cells. Nanomedicine Nanotechnol Biol Med 2016;12(3):655–64.
[118] Sato YT, Umezaki K, Sawada S, Mukai S, Sasaki Y, Harada N, et al. Engineering hybrid exosomes by membrane fusion with liposomes. Sci Rep 2016;6:21933.
[119] Jamur MC, Oliver C. Permeabilization of cell membranes. In: Oliver C, Jamur MC, editors. Immunocytochemical methods and protocols. Totowa, NJ: Humana Press; 2010. p. 63–6. [Internet]. Available from https://doi.org/10.1007/978-1-59745-324-0_9.
[120] Batrakova EV, Kim MS. Using exosomes, naturally-equipped nanocarriers, for drug delivery. J Control Release 2015;219:396–405.
[121] Pascucci L, Coccè V, Bonomi A, Ami D, Ceccarelli P, Ciusani E, et al. Paclitaxel is incorporated by mesenchymal stromal cells and released in exosomes that inhibit in vitro tumor growth: a new approach for drug delivery. J Control Release 2014;192:262–70.
[122] Lv L-H, Wan Y-L, Lin Y, Zhang W, Yang M, Li G-L, et al. Anticancer drugs cause release of exosomes with heat shock proteins from human hepatocellular carcinoma cells that elicit effective natural killer cell antitumor responses in vitro. J Biol Chem 2012;287(19):15874–85.
[123] Jang SC, Kim OY, Yoon CM, Choi D-S, Roh T-Y, Park J, et al. Bioinspired exosome-mimetic nanovesicles for targeted delivery of chemotherapeutics to malignant tumors. ACS Nano 2013;7(9):7698–710.
[124] Lee J, Kim J, Jeong M, Lee H, Goh U, Kim H, et al. Liposome-based engineering of cells to package hydrophobic compounds in membrane vesicles for tumor penetration. Nano Lett 2015;15(5):2938–44.
[125] Banizs AB, Huang T, Dryden K, Berr SS, Stone JR, Nakamoto RK, et al. In vitro evaluation of endothelial exosomes as carriers for small interfering ribonucleic acid delivery. Int J Nanomedicine 2014;9:4223–30.
[126] Zhang Y, Li L, Yu J, Zhu D, Zhang Y, Li X, et al. Microvesicle-mediated delivery of transforming growth factor β1 siRNA for the suppression of tumor growth in mice. Biomaterials 2014;35(14):4390–400.
[127] Liu Y, Li D, Liu Z, Zhou Y, Chu D, Li X, et al. Targeted exosome-mediated delivery of opioid receptor Mu siRNA for the treatment of morphine relapse. Sci Rep 2015;5:17543.
[128] Katakowski M, Buller B, Zheng X, Lu Y, Rogers T, Osobamiro O, et al. Exosomes from marrow stromal cells expressing miR-146b inhibit glioma growth. Cancer Lett 2013;335(1):201–4.
[129] Munoz JL, Bliss SA, Greco SJ, Ramkissoon SH, Ligon KL, Rameshwar P. Delivery of functional anti-miR-9 by mesenchymal stem cell-derived exosomes to glioblastoma multiforme cells conferred chemosensitivity. Mol Ther Nucleic Acids 2013;2(10):e126.

[130] Fonsato V, Collino F, Herrera MB, Cavallari C, Deregibus MC, Cisterna B, et al. Human liver stem cell-derived microvesicles inhibit hepatoma growth in SCID mice by delivering antitumor microRNAs. Stem Cells 2012;30(9):1985–98.

[131] Lee S-T, Im W, Ban J-J, Lee M, Jung K-H, Lee SK, et al. Exosome-based delivery of miR-124 in a Huntington's disease model. J Mov Disord 2017;10(1):45–52.

[132] Yang J, Zhang X, Chen X, Wang L, Yang G. Exosome mediated delivery of miR-124 promotes neurogenesis after ischemia. Mol Ther Nucleic Acids 2017;7:278–87.

[133] Gu J, Zhang H, Ji B, Jiang H, Zhao T, Jiang R, et al. Vesicle miR-195 derived from endothelial cells inhibits expression of serotonin transporter in vessel smooth muscle cells. Sci Rep 2017;7:43546.

[134] Zhang Y, Yu M, Dai M, Chen C, Tang Q, Jing W, et al. miR-450a-5p within rat adipose tissue exosome-like vesicles promotes adipogenic differentiation by targeting WISP2. J Cell Sci 2017;130(6):1158.

[135] Gu D, Zou X, Ju G, Zhang G, Bao E, Zhu Y. Mesenchymal stromal cells derived extracellular vesicles ameliorate acute renal ischemia reperfusion injury by inhibition of mitochondrial fission through miR-30. Stem Cells Int 2016;2016:2093940.

[136] Zhang H, Bai M, Deng T, Liu R, Wang X, Qu Y, et al. Cell-derived microvesicles mediate the delivery of miR-29a/c to suppress angiogenesis in gastric carcinoma. Cancer Lett 2016;375(2):331–9.

[137] Yang J, Zhou C-Z, Zhu R, Fan H, Liu X-X, Duan X-Y, et al. miR-200b-containing microvesicles attenuate experimental colitis associated intestinal fibrosis by inhibiting epithelial-mesenchymal transition. J Gastroenterol Hepatol 2017;32(12):1966–74.

[138] Waldenström A, Gennebäck N, Hellman U, Ronquist G. Cardiomyocyte microvesicles contain DNA/RNA and convey biological messages to target cells. PLoS ONE 2012;7(4):e34653.

[139] Lamichhane TN, Raiker RS, Jay SM. Exogenous DNA loading into extracellular vesicles via electroporation is size-dependent and enables limited gene delivery. Mol Pharm 2015;12(10):3650–7.

[140] García-Romero N, Carrión-Navarro J, Esteban-Rubio S, Lázaro-Ibáñez E, Peris-Celda M, Alonso MM, et al. DNA sequences within glioma-derived extracellular vesicles can cross the intact blood-brain barrier and be detected in peripheral blood of patients. Oncotarget 2016;8(1):1416–28.

[141] Cambier L, de Couto G, Ibrahim A, Echavez AK, Valle J, Liu W, et al. Y RNA fragment in extracellular vesicles confers cardioprotection via modulation of IL-10 expression and secretion. EMBO Mol Med 2017;9(3):337–52.

[142] Yokoi A, Yoshioka Y, Yamamoto Y, Ishikawa M, Ikeda S, Kato T, et al. Malignant extracellular vesicles carrying MMP1 mRNA facilitate peritoneal dissemination in ovarian cancer. Nat Commun 2017;8:14470.

[143] Mizrak A, Bolukbasi MF, Ozdener GB, Brenner GJ, Madlener S, Erkan EP, et al. Genetically engineered microvesicles carrying suicide mRNA/protein inhibit schwannoma tumor growth. Mol Ther 2013;21(1):101–8.

[144] Sugars KL, Rubinsztein DC. Transcriptional abnormalities in Huntington disease. Trends Genet 2003;19(5):233–8.

[145] Cai J, Han Y, Ren H, Chen C, He D, Zhou L, et al. Extracellular vesicle-mediated transfer of donor genomic DNA to recipient cells is a novel mechanism for genetic influence between cells. J Mol Cell Biol 2013;5(4):227–38.

[146] Théry C, Duban L, Segura E, Véron P, Lantz O, Amigorena S. Indirect activation of naïve CD4+ T cells by dendritic cell-derived exosomes. Nat Immunol 2002;3:1156.

[147] Hsu D-H, Paz P, Villaflor G, Rivas A, Mehta-Damani A, Angevin E, et al. Exosomes as a tumor vaccine: enhancing potency through direct loading of antigenic peptides. J Immunother 2003;26(5):440–50. [Internet]. Available from: https://journals.lww.com/immunotherapy-journal/Fulltext/2003/09000/Exosomes_as_a_Tumor_Vaccine__Enhancing_Potency.7.aspx.
[148] Morse MA, Garst J, Osada T, Khan S, Hobeika A, Clay TM, et al. A phase I study of dexosome immunotherapy in patients with advanced non-small cell lung cancer. J Transl Med 2005;3(1):9.
[149] Dai S, Wei D, Wu Z, Zhou X, Wei X, Huang H, et al. Phase I clinical trial of autologous ascites-derived exosomes combined with GM-CSF for colorectal cancer. Mol Ther 2008;16(4):782–90.
[150] Besse B, Charrier M, Lapierre V, Dansin E, Lantz O, Planchard D, et al. Dendritic cell-derived exosomes as maintenance immunotherapy after first line chemotherapy in NSCLC. Oncoimmunology 2015;5(4):e1071008.
[151] Nooshabadi VT, Mardpour S, Yousefi-Ahmadipour A, Allahverdi A, Izadpanah M, Daneshimehr F, et al. The extracellular vesicles-derived from mesenchymal stromal cells: a new therapeutic option in regenerative medicine. J Cell Biochem 2018;119(10):8048–73.
[152] Park JS, Suryaprakash S, Lao Y-H, Leong KW. Engineering mesenchymal stem cells for regenerative medicine and drug delivery. Methods 2015;84:3–16.
[153] Schäfer R, Spohn G, Baer PC. Mesenchymal stem/stromal cells in regenerative medicine: can preconditioning strategies improve therapeutic efficacy? Transfus Med Hemother 2016;43(4):256–67.
[154] Shyam H, Singh SK, Kant R, Saxena SK. Mesenchymal stem cells in regenerative medicine: a new paradigm for degenerative bone diseases. Regen Med 2017;12(2):111–4.
[155] Kusuma GD, Carthew J, Lim R, Frith JE. Effect of the microenvironment on mesenchymal stem cell paracrine signaling: opportunities to engineer the therapeutic effect. Stem Cells Dev 2017;26(9):617–31.
[156] Park WS, Ahn SY, Sung SI, Ahn J-Y, Chang YS. Strategies to enhance paracrine potency of transplanted mesenchymal stem cells in intractable neonatal disorders. Pediatr Res 2017;83:214.
[157] Katsuda T, Kosaka N, Takeshita F, Ochiya T. The therapeutic potential of mesenchymal stem cell-derived extracellular vesicles. Proteomics 2013;13(10–11):1637–53.
[158] Zhuang X, Xiang X, Grizzle W, Sun D, Zhang S, Axtell RC, et al. Treatment of brain inflammatory diseases by delivering exosome encapsulated anti-inflammatory drugs from the nasal region to the brain. Mol Ther 2011;19(10):1769–79.
[159] Amzallag N, Passer BJ, Allanic D, Segura E, Théry C, Goud B, et al. TSAP6 facilitates the secretion of translationally controlled tumor protein/histamine-releasing factor via a nonclassical pathway. J Biol Chem 2004;279(44):46104–12.
[160] Alonso R, Rodríguez MC, Pindado J, Merino E, Mérida I, Izquierdo M. Diacylglycerol kinase α regulates the secretion of lethal exosomes bearing Fas ligand during activation-induced cell death of T lymphocytes. J Biol Chem 2005;280(31):28439–50.
[161] Islam A, Shen X, Hiroi T, Moss J, Vaughan M, Levine SJ. The brefeldin A-inhibited guanine nucleotide-exchange protein, BIG2, regulates the constitutive release of TNFR1 exosome-like vesicles. J Biol Chem 2007;282(13):9591–9.
[162] Parolini I, Federici C, Raggi C, Lugini L, Palleschi S, De Milito A, et al. Microenvironmental pH is a key factor for exosome traffic in tumor cells. J Biol Chem 2009;284(49):34211–22.
[163] Greening DW, Xu R, Ji H, Tauro BJ, Simpson RJ. A protocol for exosome isolation and characterization: evaluation of ultracentrifugation, density-gradient separation, and immunoaffinity capture methods. In: Posch A, editor. Proteomic profiling: methods and protocols. New York, NY: Springer New York; 2015. p. 179–209. [Internet]. Available from: https://doi.org/10.1007/978-1-4939-2550-6_15.

[164] Helwa I, Cai J, Drewry MD, Zimmerman A, Dinkins MB, Khaled ML, et al. A comparative study of serum exosome isolation using differential ultracentrifugation and three commercial reagents. PLoS ONE 2017;12(1):e0170628.

[165] Sunkara V, Woo H-K, Cho Y-K. Emerging techniques in the isolation and characterization of extracellular vesicles and their roles in cancer diagnostics and prognostics. Analyst 2016;141(2):371–81.

[166] Muller L, Hong C-S, Stolz DB, Watkins SC, Whiteside TL. Isolation of biologically-active exosomes from human plasma. J Immunol Methods 2014;411:55–65.

[167] Soares Martins T, Catita J, Martins Rosa I, A B Da Cruz E Silva O, Henriques AG. Exosome isolation from distinct biofluids using precipitation and column-based approaches. PLoS ONE 2018;13(6):e0198820.

[168] Aiastui A. Should cell culture platforms move towards EV therapy requirements? Front Immunol 2015;6:8.

Chapter 20

The potential of exosomes as theragnostics in various clinical situations

Ju-Seop Kang
Department of Pharmacology & Clinical Pharmacology Lab, College of Medicine, Hanyang University, Seoul, South Korea

1 Exosomes as biomarkers of diseases and therapeutic or vaccine candidates for infections

Exosomes are small intracellular membrane-based vesicles that are naturally released by eukaryotic cells into the circulation. These so-called extracellular vesicles (EV) have important functions in cell-to-cell communication, and bodily fluids display different proteins and other cellular contents such as mRNA and microRNA in healthy subjects and patients with various diseases, which can be measured as potential diagnostic markers (see [1–3]). Tumor-derived exosomes are abundant in miRNAs that may serve as tumor markers (see [4–6]). For example, the RNA contents in serum EVs of patients with glioblastoma multiforme markedly differed from those in healthy subjects (see [7, 8]), which showed its diagnostic potential as a biomarker.

The levels of EGFRvIII mRNA are increased in circulating exosomes from patients with glioblastoma multiforme, and thus it can be used as a diagnostic biomarker for this disease (see [9]). This method may serve the role of a "liquid-biopsy," thus avoiding the need to remove tissue samples from the brain for detection of EGFRvIII protein. EGFR localized to exosome membranes also was detected as a possible diagnostic biomarker for lung cancer (see [10]). Proteoglycan glypican-1 (GP1)-positive exosomes have been detected in the serum of patients with pancreatic cancer with absolute specificity and sensitivity, distinguishing them from healthy subjects and patients with benign pancreatic disease (see [11]). Levels of GP1-positive exosomes correlated with tumor effect and survival of pre- and post-surgical patients, suggesting a valuable prognostic biomarker for pancreatic cancer. In mice, GP1-positive exosomes were reliable for detecting pancreatic intraperitoneal lesions despite negative

Exosomes. https://doi.org/10.1016/B978-0-12-816053-4.00020-1

magnetic resonance imaging findings. Plasma exosomes derived from prostate acinar cells were investigated as new biomarkers for diagnosis of prostate cancer (see [12]). Proteomics profiling of exosomes showed potential biomarkers of the disease (see [13]). Patients who experienced recurrence of hepatocellular carcinoma after liver transplantation showed increased levels of a miRNA biomarker in serum exosomes (see [14]).

Exosomes have also been evaluated as biomarkers for various non-cancer diseases of multiple organs including the central nervous system (CNS) (see [15]), liver (see [16]), kidney (see [17]), lung (see [18]), and arteries (see [19]). In the CNS, tau-mediated neuropathology is the result of extracellular accumulation of abnormally processed tau protein (see [20]). In the M1C cell neuroblastoma tauopathy model, tau protein appeared to be spread via exosomes, and AT270 phosphorylated-tau was increased in the cerebrospinal fluid (CSF), a biomarker for early stages of Alzheimer's disease (AD). The elevated levels of AT270-phosphorylated tau in the CSF seen in mild to moderate cases of sporadic AD resulted from selective abundance of phosphorylated tau in the exosome portion of the CSF relative to total CSF tau. The increase in CSF phosphorylated tau level with attack of AD was associated with an increase in the exosome-associated portion in the CSF. In another study, levels of autolysosomal proteins (cathepsin D, lysosomal-associated membrane protein 1, and ubiquitinylated proteins) in neuron-derived serum exosomes distinguished patients with pre-clinical AD from matched controls and patients with frontotemporal dementia (see [21]). Altered microRNA profiles in CSF/blood exosomes associated with neurodegenerative disorders are possible new biomarkers in the early diagnosis of AD and PD (see [22]), and the fact that exosomes can deliver siRNA offers a therapeutic potential in AD (see [23, 24]). Proteomics profiling of serum exosomes identified proteins that were abundant in patients with PD compared with healthy subjects (see [25]).

With regard to lung disease, exosomes isolated from bronchoalveolar lavage fluids of patients with asthma compared with healthy subjects showed different miRNA profiles (see [18]). Exosomes are released from the key cells implicated in asthma such as mast cells, eosinophils, dendritic cells (DCs), T-cells, and bronchial epithelial cells. These in turn can provoke the activation or repression of other asthma-related cells and stimulate allergic responses (see [26, 27]). The DC-derived exosomes (Dex, dexosome) have costimulatory molecules on their surfaces that can stimulate allergen-specific Th2 cells (see [28, 29]). The eosinophil-derived exosomes also have important roles in the modulation of asthma, and their numbers are increased in asthmatic patients (see [30, 31]). The exosomal miRNA content in patients with severe asthma was significantly different compared with healthy subjects (see [32]). The dysregulated miRNAs were associated with pathways related to airway integrity as well as being correlated with certain clinical features such as eosinophil count or FEV1 (see [33]). In a different study, the exosomal miRNA profile in patients with severe asthma was related to the TGF-ß and ErbB signaling pathway and focal adhesion (see [34]).

Exosomes may also serve as vaccines for allergic diseases (see [35]). Exosomes isolated from the bronchoalveolar lavage fluid of mice after respiratory exposure to the olive pollen allergen induced tolerance and protection against allergic sensitization in mice (see [31]).

Increased miR-192 levels in serum exosomes predicted the development of heart failure after acute myocardial infarction (see [36]). Finally, urinary exosomes have been used as starting material for diagnostic biomarkers for renal, urogenital, and systemic diseases (see [17, 37]).

Despite the number of studies showing the relevance of EV biomarkers with diverse diseases, the results of individual studies have shown inconsistent trends. Methodological differences in EV purification may explain this contradiction (see [38]). For a given application, it is mandatory to inspect the method in terms of its sensitivity and specificity including quality control measures under well-defined settings.

Exosomes are preferential candidates for use in vaccines for infectious diseases such as toxoplasmosis, diphtheria, tuberculosis, and atypical severe acute respiratory syndrome (SARS). It has been reported that immunization using DCs with *Toxoplasma gondii* antigens (T-Ag) in healthy mice induced protection against a virulent strain of *T. gondii* after oral application, but it was difficult to obtain a sufficient amount of DCs pertinent to vaccination (see [39–41]). Murine bone marrow-derived DCs pulsed in vitro with intact diphtheria toxin (DT)-released exosomes after injection into mice showed induction of IgG2b and IgG2a responses specific for DT (see [42]). Infection with *Mycobacterium tuberculosis* excites macrophages to stimulate the release of exosomes, and it should be noted that exosomes containing *M. tuberculosis* peptide-MHC-II complexes can induce antimicrobial T-cell responses (see [43, 44]). Exosomes as vaccination materials have also been studied in SARS-related coronavirus (CoV), an infection that causes a fatal atypical pulmonary disease. Kuate et al. [35] found that exosomes with the SARS-CoV spike S protein produced neutralizing antibody titers, which was further reinforced by priming with the SARS-S exosome vaccine and then boosting with the presently applied adenoviral vector vaccine (see [35]).

Stimulating a potent and general cytotoxic T lymphocyte (CTL) immune reaction has therapeutic potential for various diseases, including viral infections. For example, inducing anti-Ebola virus (EboV)-specific CTL immunity could have benefits in both therapeutic and preventive settings (see [45]). In fact, stimulation of virus-specific CTLs has been recognized in survivors of acute EboV infections (see [46]), and virus-specific CTL immunity plays a crucial role in protection in several nonhuman primates, including macaques (see [47]). Furthermore, transfusion of $CD8^+$ T lymphocytes from mice infected with mouse-adapted EboV to naïve recipient mice defended them against EboV infection (see [48]). Consistently, a powerful CTL-related immunity response could also have pertinent therapeutic effects with influenza viruses A (Flu) (see [49]) and hepatitis C (HCV) virus infections (see [50]). Anticoli et al. [45] suggested an exosome-based vaccine platform to design exosomes in vivo with the E7 protein of human papilloma virus (HPV). This method involves intramuscular injection of a DNA vector encoding HPV-E7 fused at the C-terminus of an exosome-anchoring Nef mutant protein (Nef^{mut}). Human immunodeficiency virus type-1 (HIV-1) Nef^{mut} is a 27-kDa protein (see [51]) connecting with raft microdomains at cellular membranes (see [52]). Nef^{mut} lacks several anti-cellular effects generally caused by wild-type Nef, including CD4 down-regulation, increase of HIV-1 infectivity, PAK-2 stimulation, and MHC Class I down-regulation, and is found in exosomes at very high levels (see [53, 54]). In this alignment, the ≈11-kDa E7 protein produced both potent and effective antigen-specific CTL immunity. To establish the general application of this technology, immunogenicity studies were performed with an array of viral products of various origins and sizes including EboV, West Nile Virus NS3 and HCV NS3. All antigens were stable upon fusion with Nef^{mut}, and were transferred into exosomes at levels compared to Nef^{mut}. When injected into mice, DNA vectors expressing the various fusion products produced a clearly detectable antigen-specific CD8+ T cell response with sufficient cytotoxicity to kill peptide-loaded and/or antigen-expressing syngeneic cells (see [45]).

DCs are the most competent cells at presenting antigens, and are the only antigen-presenting cell able to stimulate naïve T cells, creating the adaptive immune reaction (see [55]). Indeed, we can define cancer immunosurveillance as a stage of stepwise results leading to the effective killing of cancer cells by T cells: specifically, DC capturing and processing of tumor neoantigens is the first phase, a process that depends on molecular signals such as pro-inflammatory cytokines, co-stimulatory ligands, dying tumor cells-derived molecules, and gut microbiome products (see [56]). Accordingly, potent DC-based cancer vaccinations have been researched for some time; some positive results using these technologies have emerged, such as Sipuleucel-T immunotherapy for castration-resistant prostate cancer (see [57]). However, the diverse application of DC-based cancer vaccines shows some main limitations (see [58, 59]). Fig. 1 describes DC-based immunotherapeutic strategies. Dexosome (Dex)-based cancer vaccines have recently emerged as an alternative that may overcome some of these obstacles. First, the Dex molecular component is simple to analyze, thus enabling the rigid definition of validation parameters (see [60]). Second, Dex components are more plentiful in peptide-MHC class II complexes, allowing for higher yields (see [58, 60]). Third, Dex compared with DC can tolerate longer-term frozen storage, for up to 6 months (see [58]). In addition to these merits, the immunosuppressive tumor microenvironment often inhibits antigen presentation and T cell stimulation by DCs, but this should not affect Dex (see [61, 62]). Finally, Dex are not associated with most of the risks related to the administration of viable cells, such as generation of immune dysfunction or microvascular occlusions (see [63]). Tumor peptide-pulsed Dex produced in vivo CTL priming, tumor growth repression, and tumor remission. Indeed, single intradermal injections stimulated significant tumor growth repression after a week, and 40–60% of the animals were tumor-free after 60 days (see [62]). Furthermore,

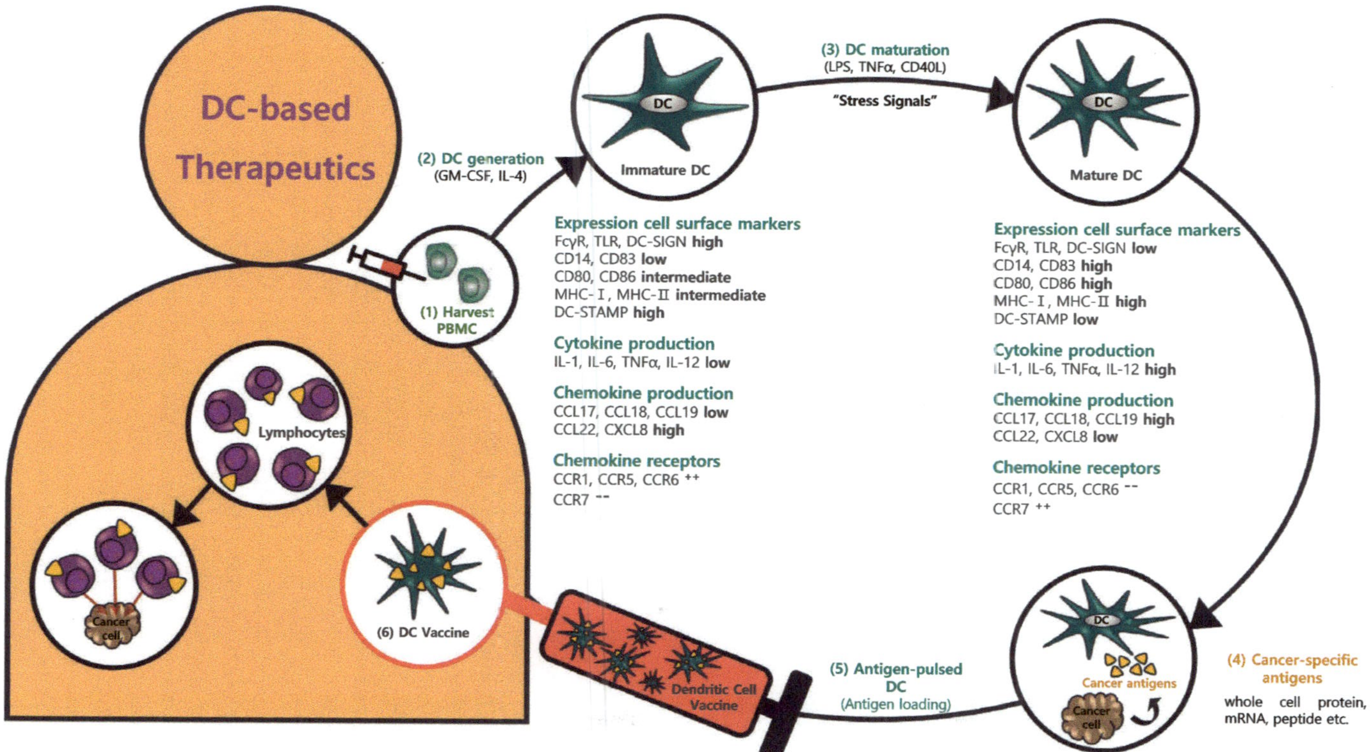

FIG. 1 DCs-based immunotherapeutic strategies: (1) to harvest peripheral blood mononuclear cells, (2) to generate immature DCs with cytokine stimulation, (3) to mature DCs by sensing the presence of a potential pathogen ("stress signal") via detection of PAMPs (exogenous signal) or infection-induced alteration in self-markers (endogenous signal), (4) cancer cells brought into contact with dendritic cells are consumed, thereby imprinting the dendritic cells with the cancer marker (5) to transfer activated antigen-presenting DCs back to the patients, (6) to stimulate robust anti-tumor immune effector cells such as T cells and NK cells.

these cell-free immunotherapeutic vaccines were more potent than directly administered viable DC vaccines, with which only 20% of the mice were tumor-free after 60 days. These differences may illustrate the exosomes' resistance to the immunomodulatory effects of the tumor microenvironment, which can block the ability of DCs to present antigens (see [62]). In the past decade, several successful clinical trials were performed assessing the feasibility, safety and efficacy of Dex-based cancer vaccines in patients with non-small cell lung cancer (NSCLC) (see [64]) and metastatic melanoma (see [65]), and in general the results were promising. In both trials, the patients received four doses of vaccine that consisted of autologous Dex loaded with several different MHC class II peptides. Vaccine production was shown to be practical, and the therapy was well tolerated with only minor grade 1–2 adverse events (see [64, 65]). A more recent phase II clinical trial evaluated the use of IFN-γ-Dex, Dex derived from IFN-γ-stimulated mature DC, as maintenance immunotherapy after the use of first-line chemotherapy in patients with advanced NSCLC (see [66]). This study showed the feasibility of production and safety of IFN-γ-Dex, with only one of 26 patients developing a grade 3 hepatotoxicity. This trial did not show any objective tumor response among clinical outcomes, according to the Response Evaluation Criteria in Solid Tumors. However, it did show that the patients with the longest progression-free survival had a notable improvement in NK cell function after Dex treatment, showing that Dex can stimulate the NK cell arm of antitumor immunity in patients with advanced NSCLC (see [66]).

Tumor-derived exosomes also function as an antigen delivery system, capable of blocking tumor development in a CD4+ and CD8+ T cell-dependent pattern (see [67]). Because of this, cell-free vaccines based on the use of tumor-derived exosomes are another possibility for clinical application. However, the isolation of tumor-derived exosomes is inconvenient and has a low preparation efficiency, with a low yield from in vitro culture of the patients' tumor cells (see [68]). However, malignant effusions from patients with melanoma are rich in exosomes, which can transmit tumor antigens to DCs, which in turn stimulate tumor-specific CTL capable of an effective in vitro antitumor response (see [69]). A phase I clinical trial examined the effects of exosomes harvested from the ascites of patients with advanced CRC as immunotherapy, and showed that a combination of tumor exosomes with GM-CSF permitted a more effective induction of systemic anti-tumor immunity and CTL responses compared with tumor exosomes alone. The patients treated with only tumor exosomes showed no therapeutic response, while one patient with stable disease and one patient with a minor clinical response were observed in the group treated with ascites-derived exosomes combined with GM-CSF (see [68]). Despite the attempts engaged thus far, Dex-based immunotherapy as a novel cancer control remains a highly encouraging possibility. Dex are proficient mediators of immune responses and the technical simplicity of managing their immunostimulatory characteristics (via the donor DC) along with their advantages over whole cell-based applications, confirms their therapeutic promise (see [60]).

Of particular interest to this section, exosomes can serve as both promoters of tumor growth and invasion by establishment of an immunosuppressive microenvironment and as agents for cancer immunosurveillance by inducing antigen presentation and stimulating destruction of tumor cells by CD4+ (see [70]) and CD8+ (see [69]) T cells and by components of the innate immune system, such as NK cells (see [71]) (Fig. 1).

2 Extracellular vesicles (EVs) as a drug delivery system

The most important characteristic for a successful nanocarrier is satisfactory in vivo behavior. The development of EVs as delivery systems requires comprehension of their in vivo kinetics after administration. However, understanding of extracellular behavior, pathways of cell uptake, and subcellular paths of EVs remains obscure. EVs released by various types of cells can be found and are relatively stable in the blood circulation and biological fluids. This suggests that EVs are more slowly cleared and remain longer than synthetic nanocarriers in biological systems. However, different in vivo pharmacokinetic studies have shown that when EVs are injected into the circulation, they are rapidly cleared. EVs derived from B16 melanoma cells and splenocytes underwent rapid clearance and showed a very short half-life, approximately 2min, after intravenous administration in mice (see [72, 73]). Increased levels of fetuin-A in urinary exosomes correlated with acute urinary injury (see [74]).

Exosomes and microvesicles participate in a large variety of body processes. They are carriers of concentrated genetic and proteomic information, and thus are believed to play important roles in cell-to-cell communication. Secreted vesicles can carry their messages in different ways. Firstly, they may stimulate recipient target cells via ligands expressed on their surface. For example, it has been verified that antigen-presenting exosomes derived from DC induce the T cell-mediated immune response in vivo (see [75]). In addition, ligand-receptor signaling via exosomes can also play a role in other regulatory processes, such as angiogenesis (see [76]), hemostasis (see [77]), cancer progression (see [78]), and metastasis (see [79]). Secondly, secreted EVs may transfer surface receptors from one cell to another by fusion with the plasma membrane of target cells (see [80, 81]). With this mechanism, HIV increase susceptibility to infection by transferring CD4 receptors from infected cells to intact cells (see [82]). EVs appear to have multiple obvious advantages, such as high delivery capacity, innate targeting properties, and low immunogenic potentiality, which position them as efficient biological delivery systems for therapeutics ranging from small molecules to macromolecular nucleic acids and proteins (see [83]) (Table 1).

Despite the development of various methodologies for EV-based delivery, a major hurdle is the lack of standardized, efficient, and reasonable approaches for isolation of EVs. Isolation methods need to be validated when considering the reproducibility, yield, purity, and functional properties of EVs for its general application. Also, there is no validated standard procedure for storage time,

TABLE 1 Pros and cons of extracellular vesicles for therapeutic delivery (see [1, 24, 68, 83–87])

Pros (advantages)	Cons (disadvantages)
- Nanoscale vesicles for biocompatibility and stability in body fluids. - Naturally derived low immunogenicity. - Stealth capacity against immune system. - Inherent target properties with reduced off-target effects. - Ability of guiding therapeutic cargo across biological barriers, especially BBB. - Capacity to be loaded with specific small molecules such as miRNA and drugs. - Unique composition allows direct membrane fusion with target cell, for efficient cell uptake. - Safe in clinical trials. - Low inherent toxicity.	- Secretion and uptake mechanism, composition, and biological functions are not yet understood. - Impact on the target cell is unknown. - Isolation techniques with high efficiency and robust yield are lacking. - Scalable production difficult-there are no optimal purification methods: large-scale production is expensive and challenging. - Efficient loading methods without damaging EV integrity are lacking. - In vivo data have been less studied, and in vivo tracking requires further studies. - Clinical studies on therapeutic delivery are lacking. - Vesicles with heterogeneous constituents can be immunogenic.

which needs to be carefully evaluated for EV-based products (see [88]). Fig. 2 shows the overall scheme of different modalities for using exosome-based formulations (see [89]).

Also, the loading efficiency of therapeutics into EVs remains inadequate. The relatively tight and ordered lipid bilayer impedes efficient loading of drugs into EVs without reducing membrane integrity. Such damage might alter the immune-oriented characteristics of EVs and make them visible to the mononuclear phagocyte system (see [89]). Therefore, ideal loading methods should not only possess high loading efficiency but also preserve the structural integrity of EVs and the functional integrity of therapeutics (see [83]).

A growing area of interest is the application of diverse nanotechnology-based DDS such as liposomes, polymeric nanoparticles, dendrimers, and magnetic nanoparticles (see [90]). These delivery systems are being used to deliver various types of cargo including chemotherapeutics, anti-inflammatory drugs and miRNAs (see [91, 92]). The nano-sized diameters of these DDS facilitate delivery through the blood and lymphatic systems with effective drug-loading capacity (see [93]). Moreover, our understanding has progressed regarding the

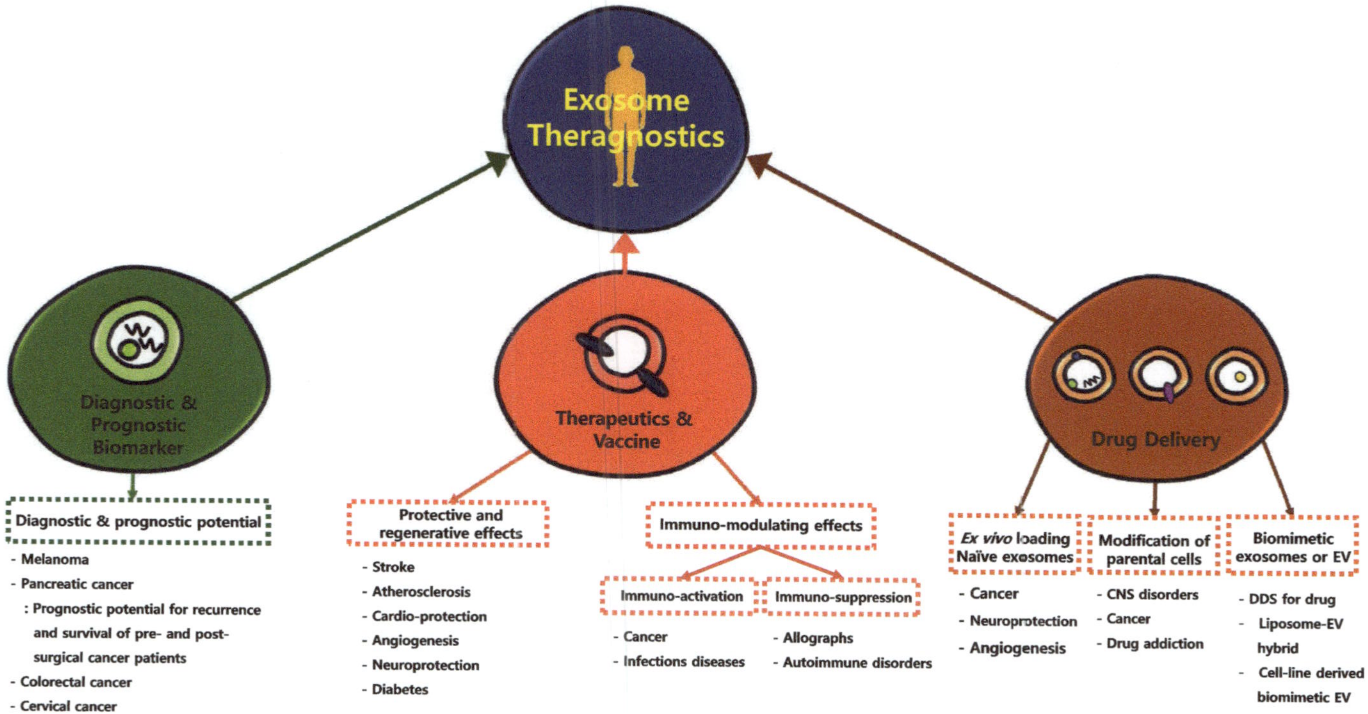

FIG. 2 Schematic representation of various strategies of exosome-based diagnostic and therapeutic approaches in translational and clinical medicine. EV, extracellular vesicle.

in vivo pharmacokinetic behaviors of EVs such as circulatory half-life, tissue distribution in vivo, cellular uptake, and intracellular fates. This understanding is crucial to the clarification of the biological functions of exosomes and practical application of exosome-based therapeutics. To define the pharmacokinetics of exosomes clearly, the first step includes the evaluation of tissue distribution in vivo, in other words the so-called biodistribution of exosomes. Several labeling methods with small lipophilic fluorescence dyes have been introduced and used for that purpose for in vivo tracking. Although the reliability of in vivo analysis would be decreased by the release of free dye from exosomes, this strategy is a useful approach to assess the localization of exosomes to tissues (see [94, 95]). PKH67, a lipophilic fluorescent dye, was used to label highly metastatic B16F10 murine melanoma cell-derived exosomes that accumulated in the lung, bone marrow, spleen, and liver, enhanced endothelial permeability in the lung, and facilitated tumor metastasis to the lung (see [96]). In addition to fluorescent dyes, lipophilic near-infrared dyes such as 1,1′-dioctadecyl-3,3,3′,3′-tetramethylindocarbocyanine perchlorate (DiD) and 1,1′-dioctadecyltetramethylindotricarbocyanine iodide (DiR) have been widely used for the imaging of exogenously administered exosomes. DiD-labeled MSC exosomes were well distributed to the spleen and liver after intravenous administration in normal mice (see [94]). On the other hand, in a mouse model of acute kidney injury, these exosomes accumulated in the kidney, as well as the spleen and liver, after intravenous injection. This finding may explain how the intravenous administration of MSC-derived exosomes stimulates the recovery from cisplatin-induced acute kidney injury in SCID mice (see [97]). In the study of in vivo behavior of DiR-labeled exosomes of various cell types such as B16F10 murine melanoma cells, C2C12 murine myoblast cells, bone marrow-derived DCs, and HEK293T human embryonic kidney cells, exosomes were mainly distributed in the liver, spleen, lung, and gastrointestinal tract after intravenous administration. Among these exosomes, B16F10 murine melanoma cell-derived exosomes mainly accumulated in the lung compared with the exosomes derived from the other two types of murine cells. The highest accumulation in the spleen and liver was observed with DC exosomes and C2C12 exosomes, respectively. HEK293T exosomes mainly accumulated in the liver after intravenous injection and the liver, pancreas, or gastrointestinal tract after intraperitoneal or subcutaneous injection, respectively (see [98]). Because of its higher sensitivity and stability, the labeling of exosomes with radiotracer is a more appropriate method for quantitative assessment of the pharmacokinetics and tissue distribution of exosomes compared with labeling with fluorescence dyes or chemiluminescent proteins. For example, ^{111}In-labeled PC3 exosomes rapidly disappeared from blood circulation and were primarily distributed in the liver (12% injection dose [ID]/g at 24h) after intravenous injection (see [99]). Various types of cells recognize and take up exosomes. Therefore, identifying these cells is important for further exploration of the biology of exosomes and for the development of exosome-based therapeutics. The mouse DC-derived exosomes were picked up by macrophages in the spleen and liver (see [90]), and exosomes derived from MDA-MB-231 breast cancer cells were taken up by macrophages in the lung and brain after intravenous administration, respectively. Exosomes derived from C2C12 cells, NIH3T3 cells, MAEC cells, and RAW264.7 cells were mainly picked up by macrophages in the liver after intravenous administration (see [100]). These results suggest that macrophages are the main cells that actively take up exogenous exosomes (see [101]). It has been predicted that exosomes are taken up by cells through the recognition of surface molecules on the membranes of the exosomes. Several studies investigated the molecules that may contribute to the in vivo pharmacokinetics of exosomes. An in vitro study proved that carbohydrate moieties on the membranes of exosomes contributed to the cellular uptake of exosomes (see [102]). Exosomes derived from tumor cells that metastasized to the lung (MDA-MB-231 and 4175) or to the liver (BxPC-3 and HPAF-II) mainly gathered in the lung and liver, respectively. A proteomic analysis of exosomes showed high expression of integrins $\alpha6\beta4$ and $\alpha V\beta5$, respectively. Exosomes collected from integrin $\beta4$-knocked down 4175 cells showed reduced accumulation in the lung. These results demonstrate that integrins play a key role in the pharmacokinetics and tissue distribution of exosomes in vivo (see [101]). Exosomes derived from genetically modified immature DCs expressing the iRGD peptide showed a selective distribution to αv integrin-positive tumor tissues (see [103]).

In addition to delivery of small RNA molecules, which depend on intracellular delivery to perform intrinsic functions, EVs have also been applied to deliver chemotherapeutic agents with the aim to increase their efficacy and reduce adverse effects. One study involved the use of exosomes to deliver curcumin to treat an inflammatory disease (see [104]). Exosomes are applied to form a complex with curcumin to enhance curcumin anticancer activity (see [105]). The intravenous injection of integrin-targeted, dendritic cell-derived EVs with the chemotherapeutic doxorubicin led to significant repression of tumor growth compared to free doxorubicin in a mouse model of breast tumor. Moreover, doxorubicin when loaded into EVs was shown to cause less cardiac damage, which is otherwise its most important dose-limiting adverse effect (see [103]). The advantage of exosomal doxorubicin versus liposomal doxorubicin involves the natural orientation of exosomal membrane proteins and their ability to interact with the receptors in the target cell plasma membrane (see [106, 107]). Furthermore, exosome-encapsulated paclitaxel compared with free taxol was shown to be more effective in controlling the growth of Lewis lung carcinoma metastases, and it holds significant potential for the delivery of various chemotherapeutics to treat drug-resistant cancers (see [108]). Repeated intraperitoneal injections of cisplatin-loaded EVs improved long-term survival of ovarian cancer-bearing mice as compared to free cisplatin, and intravenous injection of doxorubicin-loaded EVs delayed growth of established subcutaneous hepatic cancer (see [109]). Importantly, these exosome-encapsulated drugs did not adversely affect liver or kidney function, which is frequently observed after administration of free drugs.

Regrettably, the movement of large proteins through the blood-brain barrier (BBB) is severely limited. In fact, 98% of all potent drugs that may be better options for various CNS diseases are not applied in the clinic because they cannot cross the BBB (see [110]). Various nano-sized drug formulations have been developed to overcome this hurdle (see [111, 112]). Parkinson's disease (PD) is known to be associated with brain inflammation, microglia activation and secretory neurotoxic activities, including reactive oxygen species (ROS) (see [113, 114]). Samples of brain tissue from patients with PD have shown reduced levels of redox enzymes, catalase, and superoxide dismutase, and other antioxidants (see [115–117]), which indicates an impaired defense against oxidative stress and neurodegeneration in these patients. Among these molecules, catalase is one of the most effective natural antioxidants: it scavenges one million free radicals through a catalytic reaction. Therefore, successful delivery of catalase into the brain may be an important and possible approach to PD therapy (see [84]). EVs have been proposed as therapeutic delivery vehicles for the treatment of PD. Exosomes are readily taken up by neuronal cells in vitro, and a considerable number of exosomes was detected in PD mouse brain after intranasal administration. Catalase-loaded exosomes (ExoCAT) were shown to suppress microglial activation and protect neurons against ROS more efficiently compared to free catalase in in vitro and in vivo models of PD (see [84]). Therefore, ExoCAT is a more adaptable strategy for treating inflammatory and degenerative disorders such as PD (see [84, 118]). Although these reports are preliminary, the results have shown that exosomes are promising candidate drug delivery systems for the treatment of a variety of diseases.

Exosomes derived from human adipose tissue-derived mesenchymal stem cells (MSCs) are considered to have therapeutic value for treating AD (see [119]). Exosomes can be used in vivo as a vehicle to carry active neprilysin (NEP), the most important enzyme for β-amyloid (Aβ) peptide degradation in the brain. MSC-derived exosomes also decrease intracellular and extracellular Aβ levels in the neuroblastoma cell line N2A in vitro. Therefore, human adipose tissue-derived MSC-originated exosomes are proposed as a potential therapy against AD from the point of view of their Aβ-degrading capacity. Recent studies have demonstrated that multipotent mesenchymal stromal cells (MSCs) hold great promise for neurovascular remodeling and neurological function recovery following a stroke. It has been demonstrated that MSC-derived exosomes have a neuroprotective effect against stroke due in part to changing the miRNA profile of exosomes during and after stroke (see [120]).

When we consider the problems associated with many current nanoparticulate delivery systems, exosomes as a mimic of "natural delivery systems" are a potential alternative for delivery of the biological molecules described above. Because of their small size and host-derived cellular product, these vesicles can avoid phagocytosis or degradation by macrophages and also circulate for long periods of time in the body. One of the interesting advantages of these delivery vehicles includes their ability to cross the BBB and arrive in the CNS (see [121]). Insufficient knowledge of the nature of exosomes and their role in the pathophysiology of overall health and disease makes it complicated and difficult to predict long-term safety and therapeutic effects. In vivo trafficking of exosomes, their biological fate and their impact on targeted organs need to be understood with regard to therapeutic cargo loading and assembly for drug delivery (see [122]). Currently, there is no distinct optimal purification technique for isolation of exosomes with high purity (see [123]). The current isolation methods yield low quantities of exosomes, and their large-scale production for clinical studies and post drug approval is expensive and complicated (see [38]). It is highly likely that future clinical use will demand hybrid designs of exosomes (see [124]), and when combined with therapeutic cargo they may show undesirable effects. Even though extensive exosome biology is already known, exosomes comprise heterogeneous constituents and may show immunogenicity (stimulatory or suppressive) effects based on the nature of the donor cells. Exosomes provide immense promise and are a new therapeutic area for delivery of various synthetic and biological molecules in cellular therapy. Exosomes as drug delivery systems provide a major advantage as there is no undesired aggregation or homing of exosomes in the liver and/or first-pass effect before arriving in target sites. Well-characterized exosomes with long-term safety that deliver nucleic acids and therapeutic molecules between cells and through difficult-to-cross membranes such as BBB would have major practical significance. However, before these drug delivery systems become a therapeutic reality, components and processes including immune reactions need to be clarified (see [122]). The emerging evidence that tumor cell-derived exosomes have unique properties may be used to develop an exosome-based drug delivery system that is better than synthetic drug carriers. However, some limitations and hurdles must be overcome before exosome-based drug delivery systems can be used in the clinic. Important issues still need to be autologous or can non-immunogenic exosome factories. Therefore, a strategy needs to be developed for manufacturing vesicles for therapeutic application, with establishment of the producer cell type, physical methods to produce vesicular nanoparticles, enhancement of EV yield and scale-up, measurement of potency of EV-based products, and EV-inspired bioengineered artificial vesicles. Recently, a group of researchers developed a method to produce exosome-mimetic vesicles, which can overcome natural exosome limitations as like low drug-loading efficiency and low exosome production yields (see [125]). These chemotherapeutic-loaded nanovesicles, which are 100–200nm in diameter, were generated by breaking down cells by serial extrusion through filters with diminishing pore size (see [126]). It was further suggested that these nano-vesicles with exosome-mimetic properties can be used as a platform for RNAi transfer to the cell cytoplasm (see [127]). However, the high level of cholesterol, ganglioside, and sphingomyelin in exosomal membranes leads to a more rigid bilayer structure than that of their parent cells (see [128]), which suggests that their fusion with lipid-based particles requires rough conditions (see [129]), such as aggressive freeze-thaw processes (see [130]). To avoid the need for such conditions, Yang et al. [131] designed a virus-mimetic fusogenic exosome platform to deliver membrane proteins to target cell membranes, involving integrated vascular stomatitis virus G protein, a viral fusogen (see [131]). Interestingly, these methods allow easy exosome modification by fusing exosomes derived from modified cells with liposomes inserted with peptides, antibodies or polyethylene glycol.

Although there are currently many challenges in the treatment of cancer and other refractory diseases, exosomes, including exosome-mimetic nanovesicles, are considered valid diagnostic biomarkers and potential therapeutic tools. Moreover, along with chemical, physical, cellular and genetic engineering techniques, many existing exosome modification strategies are promising for application in various clinical situations (Fig. 2).

References

[1] Clayton A, Harris CL, Court J, et al. Antigen-presenting cell exosomes are protected from complement-mediated lysis by expression of CD55 and CD59. Eur J Immunol 2003;33(2):522–31.

[2] Pant S, Hilton H, Burczynski ME. The multifaceted exosome: Biogenesis, role in normal and aberrant cellular function, and frontiers for pharmacological and biomarker opportunities. Biochem Pharmacol 2012;83(11):1484–94.

[3] Revenfeld ALS, Bæk R, Nielsen MH, et al. Diagnostic and prognostic potential of extracellular vesicles in peripheral blood. Clin Ther 2014;36(6):830–46.

[4] Kumar D, Gupta D, Shankar S, et al. Biomolecular characterization of exosomes released from cancer stem cells: possible implications for biomarker and treatment of cancer. Oncotarget 2015;6(5):3280.

[5] Mishra PJ. Non-coding RNAs as clinical biomarkers for cancer diagnosis and prognosis. Expert Rev Mol Diagn 2014;14(8):917–9.

[6] Schwarzenbach H. The clinical relevance of circulating, exosomal miRNAs as biomarkers for cancer. Expert Rev Mol Diagn 2015;15(9):1159–69.

[7] Noerholm M, Balaj L, Limperg T, et al. RNA expression patterns in serum microvesicles from patients with glioblastoma multiforme and controls. BMC Cancer 2012;12(1):22.

[8] Taylor DD, Gercel-Taylor C. MicroRNA signatures of tumor-derived exosomes as diagnostic biomarkers of ovarian cancer. Gynecol Oncol 2008;110(1):13–21.

[9] Skog J, Würdinger T, Van Rijn S, et al. Glioblastoma microvesicles transport RNA and proteins that promote tumour growth and provide diagnostic biomarkers. Nat Cell Biol 2008;10(12):1470.

[10] Yamashita T, Kamada H, Kanasaki S, et al. Epidermal growth factor receptor localized to exosome membranes as a possible biomarker for lung cancer diagnosis. Pharmazie 2013;68(12):969–73.

[11] Melo SA, Luecke LB, Kahlert C, et al. Glypican-1 identifies cancer exosomes and detects early pancreatic cancer. Nature 2015;523(7559):177.

[12] Tavoosidana G, Ronquist G, Darmanis S, et al. Multiple recognition assay reveals prostasomes as promising plasma biomarkers for prostate cancer. Proc Natl Acad Sci U S A 2011;108(21):8809–14.

[13] Duijvesz D, Burnum-Johnson KE, Gritsenko MA, et al. Proteomic profiling of exosomes leads to the identification of novel biomarkers for prostate cancer. PLoS One 2013;8(12):e82589.

[14] Sugimachi K, Matsumura T, Hirata H, et al. Identification of a bona fide microRNA biomarker in serum exosomes that predicts hepatocellular carcinoma recurrence after liver transplantation. Br J Cancer 2015;112(3):532.

[15] Kawikova I, Askenase PW. Diagnostic and therapeutic potentials of exosomes in CNS diseases. Brain Res 2015;1617:63–71.

[16] Masyuk AI, Masyuk TV, LaRusso NF. Exosomes in the pathogenesis, diagnostics and therapeutics of liver diseases. J Hepatol 2013;59(3):621–5.

[17] Spanu S, van Roeyen CR, Denecke B, et al. Urinary exosomes: a novel means to non-invasively assess changes in renal gene and protein expression. PLoS ONE 2014;9(10):e109631.

[18] Levänen B, Bhakta NR, Paredes PT, et al. Altered microRNA profiles in bronchoalveolar lavage fluid exosomes in asthmatic patients. J Allergy Clin Immunol 2013;131(3):894–903. e898.

[19] Hoefer IE, Steffens S, Ala-Korpela M, et al. Novel methodologies for biomarker discovery in atherosclerosis. Eur Heart J 2015;36(39):2635–42.

[20] Saman S, Kim W, Raya M, et al. Exosome-associated tau is secreted in tauopathy models and is selectively phosphorylated in cerebrospinal fluid in early Alzheimer disease. J Biol Chem 2012;287(6):3842–9.

[21] Goetzl EJ, Boxer A, Schwartz JB, et al. Altered lysosomal proteins in neural-derived plasma exosomes in preclinical Alzheimer disease. Neurology 2015;85(1):40–7.

[22] Gui Y, Liu H, Zhang L, et al. Altered microRNA profiles in cerebrospinal fluid exosome in Parkinson disease and Alzheimer disease. Oncotarget 2015;6(35):37043–53.

[23] Cheng L, Doecke JD, Sharples RA, et al. Prognostic serum miRNA biomarkers associated with Alzheimer's disease shows concordance with neuropsychological and neuroimaging assessment. Mol Psychiatry 2014;20:1188.

[24] van den Boorn JG, Schlee M, Coch C, et al. SiRNA delivery with exosome nanoparticles. Nat Biotechnol 2011;29(4):325.

[25] Tomlinson PR, Zheng Y, Fischer R, et al. Identification of distinct circulating exosomes in Parkinson's disease. Ann Clin Transl Neurol 2015;2(4):353–61.

[26] Fujita Y, Yoshioka Y, Ito S, et al. Intercellular communication by extracellular vesicles and their microRNAs in asthma. Clin Ther 2014;36(6):873–81.

[27] van den Berge M, Tasena H. Role of microRNAs and exosomes in asthma. Curr Opin Pulm Med 2019;25(1):87–93.

[28] Admyre C, Grunewald J, Thyberg J, et al. Exosomes with major histocompatibility complex class II and co-stimulatory molecules are present in human BAL fluid. Eur Respir J 2003;22(4):578–83.

[29] Admyre C, Telemo E, Almqvist N, et al. Exosomes–nanovesicles with possible roles in allergic inflammation. Allergy 2008;63(4):404–8.

[30] Mazzeo C, Cañas JA, Zafra MP, et al. Exosome secretion by eosinophils: a possible role in asthma pathogenesis. J Allergy Clin Immunol 2015;135(6):1603–13.

[31] Prado N, Marazuela EG, Segura E, et al. Exosomes from bronchoalveolar fluid of tolerized mice prevent allergic reaction. J Immunol 2008;181(2):1519–25.

[32] Mortaz E, Alipoor SD, Varahram M, et al. Exosomes in severe asthma: update in their roles and potential in therapy. Biomed Res Int 2018;2018:2862187. 1–10.

[33] Francisco-Garcia A, Martinez-Nunez RT, Rupani H, et al. LSC abstract-altered small RNA cargo in severe asthma exosomes. Eur Respir J 2016;48:101.

[34] Suzuki M, Konno S, Makita H, et al. Altered circulating exosomal RNA profiles detected by next-generation sequencing in patients with severe asthma. Eur Respir Soc 2016;48(S60): PA3410.

[35] Kuate S, Cinatl J, Doerr HW, et al. Exosomal vaccines containing the S protein of the SARS coronavirus induce high levels of neutralizing antibodies. Virology 2007;362(1):26–37.
[36] Matsumoto S, Sakata Y, Nakatani D, et al. Circulating p53-responsive microRNAs are predictive indicators of heart failure after acute myocardial infarction. Circ Res 2013;113(3):322–6.
[37] Huebner AR, Somparn P, Benjachat T, et al. Exosomes in urine biomarker discovery urine proteomics in kidney disease biomarker discovery. Springer; 2015;43–58.
[38] Taylor DD, Shah S. Methods of isolating extracellular vesicles impact down-stream analyses of their cargoes. Methods 2015;87:3–10.
[39] Aline F, Bout D, Amigorena S, et al. Toxoplasma gondii antigen-pulsed-dendritic cell-derived exosomes induce a protective immune response against T. gondii infection. Infect Immun 2004;72(7):4127–37.
[40] Beauvillain C, Juste MO, Dion S, et al. Exosomes are an effective vaccine against congenital toxoplasmosis in mice. Vaccine 2009;27(11):1750–7.
[41] Beauvillain C, Ruiz S, Guiton R, et al. A vaccine based on exosomes secreted by a dendritic cell line confers protection against T. gondii infection in syngeneic and allogeneic mice. Microbes Infect 2007;9(14–15):1614–22.
[42] Colino J, Snapper CM. Exosomes from bone marrow dendritic cells pulsed with diphtheria toxoid preferentially induce type 1 antigen-specific IgG responses in naive recipients in the absence of free antigen. J Immunol 2006;177(6):3757–62.
[43] Ramachandra L, Qu Y, Wang Y, et al. Mycobacterium tuberculosis synergizes with ATP to induce release of microvesicles and exosomes containing major histocompatibility complex class II molecules capable of antigen presentation. Infect Immun 2010;78(12):5116–25.
[44] Singh PP, LeMaire C, Tan JC, et al. Exosomes released from *M. tuberculosis* infected cells can suppress IFN-γ mediated activation of naïve macrophages. PLoS ONE 2011;6(4):e18564.
[45] Anticoli S, Manfredi F, Chiozzini C, et al. An exosome-based vaccine platform imparts cytotoxic T lymphocyte immunity against viral antigens. Biotechnol J 2018;13(4):1700443.
[46] McElroy AK, Akondy RS, Davis CW, et al. Human Ebola virus infection results in substantial immune activation. Proc Natl Acad Sci U S A 2015;112(15):4719–24.
[47] Sullivan NJ, Hensley L, Asiedu C, et al. CD8+ cellular immunity mediates rAd5 vaccine protection against Ebola virus infection of nonhuman primates. Nat Med 2011;17(9):1128.
[48] Bradfute SB, Warfield KL, Bavari S. Functional CD8+ T cell responses in lethal Ebola virus infection. J Immunol 2008;180(6):4058–66.
[49] Fontana JM, Christos PJ, Michelini Z, et al. Mucosal immunization with integrase-defective lentiviral vectors protects against influenza virus challenge in mice. PLoS ONE 2014;9(5):e97270.
[50] Liang TJ. Current progress in development of hepatitis C virus vaccines. Nat Med 2013;19(7):869.
[51] Pereira EA, daSilva LL. HIV-1 Nef: taking control of protein trafficking. Traffic 2016;17(9):976–96.
[52] Lenassi M, Cagney G, Liao M, et al. HIV Nef is secreted in exosomes and triggers apoptosis in bystander CD4+ T cells. Traffic 2010;11(1):110–22.
[53] D'Aloja P, Santarcangelo AC, Arold S, et al. Genetic and functional analysis of the human immunodeficiency virus (HIV) type 1-inhibiting F12-HIVnef allele. J Gen Virol 2001;82(11):2735–45.
[54] Lattanzi L, Federico M. A strategy of antigen incorporation into exosomes: comparing cross-presentation levels of antigens delivered by engineered exosomes and by lentiviral virus-like particles. Vaccine 2012;30(50):7229–37.

[55] Théry C, Amigorena S. The cell biology of antigen presentation in dendritic cells. Curr Opin Immunol 2001;13(1):45–51.
[56] Chen DS, Mellman I. Oncology meets immunology: the cancer-immunity cycle. Immunity 2013;39(1):1–10.
[57] Kantoff PW, Higano CS, Shore ND, et al. Sipuleucel-T immunotherapy for castration-resistant prostate cancer. N Engl J Med 2010;363(5):411–22.
[58] Andre F, Escudier B, Angevin E, et al. Exosomes for cancer immunotherapy. Ann Oncol 2004;15(Suppl. 4):iv141–4.
[59] Pitt JM, Charrier M, Viaud S, et al. Dendritic cell–derived exosomes as immunotherapies in the fight against cancer. J Immunol 2014;193(3):1006–11.
[60] Pitt JM, André F, Amigorena S, et al. Dendritic cell–derived exosomes for cancer therapy. J Clin Invest 2016;126(4):1224–32.
[61] Gabrilovich DI, Ciernik IF, Carbone DP. Dendritic cells in antitumor immune responses: I. Defective antigen presentation in tumor-bearing hosts. Cell Immunol 1996;170(1):101–10.
[62] Zitvogel L, Regnault A, Lozier A, et al. Eradication of established murine tumors using a novel cell-free vaccine: dendritic cell derived exosomes. Nat Med 1998;4(5):594.
[63] Zhang B, Yin Y, Lai RC, et al. Immunotherapeutic potential of extracellular vesicles. Front Immunol 2014;5:518.
[64] Morse MA, Garst J, Osada T, et al. A phase I study of dexosome immunotherapy in patients with advanced non-small cell lung cancer. J Transl Med 2005;3(1):9.
[65] Escudier B, Dorval T, Chaput N, et al. Vaccination of metastatic melanoma patients with autologous dendritic cell (DC) derived-exosomes: results of thefirst phase I clinical trial. J Transl Med 2005;3(1):10.
[66] Besse B, Charrier M, Lapierre V, et al. Dendritic cell-derived exosomes as maintenance immunotherapy after first line chemotherapy in NSCLC. Oncoimmunology 2016;5(4):e1071008.
[67] Wolfers J, Lozier A, Raposo G, et al. Tumor-derived exosomes are a source of shared tumor rejection antigens for CTL cross-priming. Nat Med 2001;7(3):297.
[68] Dai S, Wei D, Wu Z, et al. Phase I clinical trial of autologous ascites-derived exosomes combined with GM-CSF for colorectal cancer. Mol Ther 2008;16(4):782–90.
[69] Andre F, Schartz NE, Movassagh M, et al. Malignant effusions and immunogenic tumour-derived exosomes. Lancet 2002;360(9329):295–305.
[70] Raposo G, Nijman HW, Stoorvogel W, et al. B lymphocytes secrete antigen-presenting vesicles. J Exp Med 1996;183(3):1161–72.
[71] Viaud S, Terme M, Flament C, et al. Dendritic cell-derived exosomes promote natural killer cell activation and proliferation: a role for NKG2D ligands and IL-15Rα. PLoS ONE 2009;4(3):e4942.
[72] Saunderson SC, Dunn AC, Crocker PR, et al. CD169 mediates the capture of exosomes in spleen and lymph node. Blood 2014;123(2):208–16.
[73] Takahashi Y, Nishikawa M, Shinotsuka H, et al. Visualization and in vivo tracking of the exosomes of murine melanoma B16-BL6 cells in mice after intravenous injection. J Biotechnol 2013;165(2):77–84.
[74] Zhou H, Pisitkun T, Aponte A, et al. Exosomal Fetuin-A identified by proteomics: a novel urinary biomarker for detecting acute kidney injury. Kidney Int 2006;70(10):1847–57.
[75] Théry C, Duban L, Segura E, et al. Indirect activation of naïve CD4+ T cells by dendritic cell-derived exosomes. Nat Immunol 2002;3(12):1156.
[76] Martinez MC, Andriantsitohaina R. Microparticles in angiogenesis: therapeutic potential. Circ Res 2011;109(1):110–9.

[77] Aharon A, Brenner B. Microparticles, thrombosis and cancer. Best Pract Res Clin Haematol 2009;22(1):61–9.
[78] Cho JA, Park H, Lim EH, et al. Exosomes from ovarian cancer cells induce adipose tissue-derived mesenchymal stem cells to acquire the physical and functional characteristics of tumor-supporting myofibroblasts. Gynecol Oncol 2011;123(2):379–86.
[79] Xu J, Lamouille S, Derynck R. TGF-β-induced epithelial to mesenchymal transition. Cell Res 2009;19(2):156.
[80] Al-Nedawi K, Meehan B, Kerbel RS, et al. Endothelial expression of autocrine VEGF upon the uptake of tumor-derived microvesicles containing oncogenic EGFR. Proc Natl Acad Sci U S A 2009;106(10):3794–9.
[81] Baj-Krzyworzeka M, Majka M, Pratico D, et al. Platelet-derived microparticles stimulate proliferation, survival, adhesion, and chemotaxis of hematopoietic cells. Exp Hematol 2002;30(5):450–9.
[82] Ratajczak J, Wysoczynski M, Hayek F, et al. Membrane-derived microvesicles: important and underappreciated mediators of cell-to-cell communication. Leukemia 2006;20(9):1487.
[83] Lu M, Xing H, Yang Z, et al. Recent advances on extracellular vesicles in therapeutic delivery: challenges, solutions, and opportunities. Eur J Pharm Biopharm 2017;119:381–95.
[84] Haney MJ, Klyachko NL, Zhao Y, et al. Exosomes as drug delivery vehicles for Parkinson's disease therapy. J Control Release 2015;207:18–30.
[85] Klumperman J, Raposo G. The complex ultrastructure of the endolysosomal system. Cold Spring Harb Perspect Biol 2014;6(10):a016857.
[86] Rufino-Ramos D, Albuquerque PR, Carmona V, et al. Extracellular vesicles: novel promising delivery systems for therapy of brain diseases. J Control Release 2017;262:247–58.
[87] Viaud S, Ploix S, Lapierre V, et al. Updated technology to produce highly immunogenic dendritic cell-derived exosomes of clinical grade: a critical role of interferon-γ. J Immunother 2011;34(1):65–75.
[88] Lener T, Gimona M, Aigner L, et al. Applying extracellular vesicles based therapeutics in clinical trials–an ISEV position paper. J Extracell Vesicles 2015;4(1):30087.
[89] Batrakova EV, Kim MS. Development and regulation of exosome-based therapy products. Wiley Interdiscip Rev Nanomed Nanobiotechnol 2016;8(5):744–57.
[90] Kim OY, Lee JW, Gho YS. Extracellular vesicle mimetics: novel alternatives to extracellular vesicle-based theranostics, drug delivery, and vaccines. Semin Cell Dev Biol 2017;67:74–82.
[91] De Jong WH, Borm PJ. Drug delivery and nanoparticles: applications and hazards. Int J Nanomedicine 2008;3(2):133.
[92] Wilczewska AZ, Niemirowicz K, Markiewicz KH, et al. Nanoparticles as drug delivery systems. Pharmacol Rep 2012;64(5):1020–37.
[93] Sercombe L, Veerati T, Moheimani F, et al. Advances and challenges of liposome assisted drug delivery. Front Pharmacol 2015;6:286.
[94] Grange C, Tapparo M, Bruno S, et al. Biodistribution of mesenchymal stem cell-derived extracellular vesicles in a model of acute kidney injury monitored by optical imaging. Int J Mol Med 2014;33(5):1055–63.
[95] Kotmakçı M, Çetintaş VB. Extracellular vesicles as natural nanosized delivery systems for small-molecule drugs and genetic material: steps towards the future nanomedicines. J Pharm Pharm Sci 2015;18(3):396–413.
[96] Peinado H, Ale?kovi M, Lavotshkin S, et al. Melanoma exosomes educate bone marrow progenitor cells toward a pro-metastatic phenotype through MET. Nat Med 2012;18(6):883–91.
[97] Bruno S, Grange C, Collino F, et al. Microvesicles derived from mesenchymal stem cells enhance survival in a lethal model of acute kidney injury. PLoS One 2012;7(3):e33115.

[98] Wiklander OP, Nordin JZ, O'Loughlin A, et al. Extracellular vesicle in vivo biodistribution is determined by cell source, route of administration and targeting. J Extracell Vesicles 2015;4(1):26316.

[99] Smyth TJ. Exploration of the drug delivery potential of tumor-derived exosomes. (Doctor of Philosophy). University of Colorado at Denver, Anschutz Medical Campus; 2014. Pharmaceutical Sciences Program.

[100] Charoenviriyakul C, Takahashi Y, Morishita M, et al. Cell type-specific and common characteristics of exosomes derived from mouse cell lines: yield, physicochemical properties, and pharmacokinetics. Eur J Pharm Sci 2017;96:316–22.

[101] Morishita M, Takahashi Y, Nishikawa M, et al. Pharmacokinetics of exosomes—an important factor for elucidating the biological roles of exosomes and for the development of exosome-based therapeutics. J Pharm Sci 2017;106(9):2265–9.

[102] Hao S, Bai O, Li F, et al. Mature dendritic cells pulsed with exosomes stimulate efficient cytotoxic T-lymphocyte responses and antitumour immunity. Immunology 2007;120(1):90–102.

[103] Tian Y, Li S, Song J, et al. A doxorubicin delivery platform using engineered natural membrane vesicle exosomes for targeted tumor therapy. Biomaterials 2014;35(7):2383–90.

[104] Sun D, Zhuang X, Xiang X, et al. A novel nanoparticle drug delivery system: the anti-inflammatory activity of curcumin is enhanced when encapsulated in exosomes. Mol Ther 2010;18(9):1606–14.

[105] Dhillon N, Aggarwal BB, Newman RA, et al. Phase II trial of curcumin in patients with advanced pancreatic cancer. Clin Cancer Res 2008;14(14):4491–9.

[106] Adlakha YK, Saini N. Brain microRNAs and insights into biological functions and therapeutic potential of brain enriched miRNA-128. Mol Cancer 2014;13(1):33.

[107] Xitong D, Xiaorong Z. Targeted therapeutic delivery using engineered exosomes and its applications in cardiovascular diseases. Gene 2016;575(2):377–84.

[108] Kim MS, Haney MJ, Zhao Y, et al. Development of exosome-encapsulated paclitaxel to overcome MDR in cancer cells. Nanomedicine 2016;12(3):655–64.

[109] Tang K, Zhang Y, Zhang H, et al. Delivery of chemotherapeutic drugs in tumour cell-derived microparticles. Nat Commun 2012;3:1282.

[110] Pardridge WM. Drug transport across the blood-brain barrier. J Cereb Blood Flow Metab 2012;32(11):1959–72.

[111] Kabanov AV, Batrakova EV. Polymer nanomaterials. In: Neuroimmune Pharmacology. Springer; 2008. p. 691–707.

[112] Silva GA. Nanotechnology applications and approaches for neuroregeneration and drug delivery to the central nervous system. Ann N Y Acad Sci 2010;1199(1):221–30.

[113] Ebadi M, Srinivasan SK, Baxi MD. Oxidative stress and antioxidant therapy in Parkinson's disease. Prog Neurobiol 1996;48(1):1–19.

[114] McGeer P, Itagaki S, Boyes B, et al. Reactive microglia are positive for HLA-DR in the substantia nigra of Parkinson's and Alzheimer's disease brains. Neurology 1988;38(8):1285.

[115] Abraham S, Soundararajan C, Vivekanandhan S, et al. Erythrocyte antioxidant enzymes in Parkinson's disease. Indian J Med Res 2005;121(2):111–5.

[116] Ambani LM, Van Woert MH, Murphy S. Brain peroxidase and catalase in Parkinson disease. Arch Neurol 1975;32(2):114–8.

[117] Riederer P, Sofic E, Rausch WD, et al. Transition metals, ferritin, glutathione, and ascorbic acid in parkinsonian brains. J Neurochem 1989;52(2):515–20.

[118] Kim SH, Bianco N, Menon R, et al. Exosomes derived from genetically modified DC expressing FasL are anti-inflammatory and immunosuppressive. Mol Ther 2006;13(2):289–300.

[119] Katsuda T, Oki K, Ochiya T. Potential application of extracellular vesicles of human adipose tissue-derived mesenchymal stem cells in Alzheimer's disease therapeutics. In: Stem cell renewal and cell-cell communication. Springer; 2014. p. 171–81.
[120] Li Y, Cheng Q, Hu G, et al. Extracellular vesicles in mesenchymal stromal cells: a novel therapeutic strategy for stroke. Exp Ther Med 2018;15(5):4067–79.
[121] Tran T-H, Mattheolabakis G, Aldawsari H, et al. Exosomes as nanocarriers for immunotherapy of cancer and inflammatory diseases. Clin Immunol 2015;160(1):46–58.
[122] Ha D, Yang N, Nadithe V. Exosomes as therapeutic drug carriers and delivery vehicles across biological membranes: current perspectives and future challenges. Acta Pharm Sin B 2016;6(4):287–96.
[123] Petersen KE, Manangon E, Hood JL, et al. A review of exosome separation techniques and characterization of B16-F10 mouse melanoma exosomes with AF4-UV-MALS-DLS-TEM. Anal Bioanal Chem 2014;406(30):7855–66.
[124] Rana S, Bajaj A, Mout R, et al. Monolayer coated gold nanoparticles for delivery applications. Adv Drug Deliv Rev 2012;64(2):200–16.
[125] He C, Zheng S, Luo Y, et al. Exosome theranostics: biology and translational medicine. Theranostics 2018;8(1):237–55.
[126] Jang SC, Kim OY, Yoon CM, et al. Bioinspired exosome-mimetic nanovesicles for targeted delivery of chemotherapeutics to malignant tumors. ACS Nano 2013;7(9):7698–710.
[127] Lunavat TR, Jang SC, Nilsson L, et al. RNAi delivery by exosome-mimetic nanovesicles—implications for targeting c-Myc in cancer. Biomaterials 2016;102:231–8.
[128] Parolini I, Federici C, Raggi C, et al. Microenvironmental pH is a key factor for exosome traffic in tumor cells. J Biol Chem 2009;284(49):34211–22.
[129] Armstrong JP, Holme MN, Stevens MM. Re-engineering extracellular vesicles as smart nanoscale therapeutics. ACS Nano 2017;11(1):69–83.
[130] Sato YT, Umezaki K, Sawada S, et al. Engineering hybrid exosomes by membrane fusion with liposomes. Sci Rep 2016;6:21933.
[131] Yang Y, Hong Y, Nam GH, et al. Virus-mimetic fusogenic exosomes for direct delivery of integral membrane proteins to target cell membranes. Adv Mater 2017;29(13):1605604.

Chapter 21

Exosomes, gemmules, pangenesis and Darwin

Denis Noble
Department of Physiology, Anatomy & Genetics, University of Oxford, Oxford, United Kingdom

1 Introduction

This chapter takes the form of a historical note. I will first explain why the recent discovery of exosomes and their importance is strongly compatible with some of the ideas of Charles Darwin as expressed in his books and correspondence. I will then explain why those ideas were deliberately removed from standard evolutionary biology as it developed after Charles Darwin's death in 1882, and through to the formulation of the neo-Darwinist Modern Synthesis during the first half of the twentieth century. Finally, I will comment on current trends in Evolutionary Biology that can be viewed as returning to a modern form of Darwin's more nuanced multi-mechanism position. I will then briefly review modern work showing that the mechanisms for transgenerational transmission via sperm exist and are functional.

2 The Darwinist view

If Charles Darwin were alive today, what would he think of the discovery of exosomes and their importance? What would he say about this book? To answer these questions we need to understand Darwin's character as a scientist and thinker, and how he reacted to his own doubts about his theories.

2.1 The origin of species

Darwin was a cautious and slow thinker. When he published his ground-breaking work *The Origin of Species* in 1859 [1], the book was finished in a hurry, partly because Alfred Russel Wallace was hot on his heels with important information supporting the same ideas from his work in the Malay Archipelago. The ideas that Darwin developed into the theory of evolution by natural selection had previously slowly matured through three decades since his famous voyage on *The Beagle* in 1831–1836. For all that period of time he allowed his ideas

Exosomes. https://doi.org/10.1016/B978-0-12-816053-4.00021-3

to mature before publishing them, and even then he was not sure he had fully arrived at his goal.

After finishing *The Origin of Species* Darwin wrote to the geologist Charles Lyell:

> *I suppose that I am a very slow thinker, for you would be surprised at the number of years it took me to see clearly what some of the problems were, which had to be solved—such as the necessity of the principle of divergence of character—the extinction of intermediate varieties on a continuous area with graduated conditions—the double problem of sterile first crosses & sterile hybrids, &c &c—.*

One consequence of the sudden rush to publish is that Darwin did not fully acknowledge his predecessors until the third edition of his book. His predecessors even included his grandfather, Erasmus Darwin. So Charles Darwin knew that he was by no means the first to propose a theory of the transformation of species, as evolution was called in his time. When he did get around to acknowledging his predecessors, in the third (1861) edition of his book, he referred to no fewer than 30 of them. He was generous in his selection. Even Aristotle was listed. In the fourth edition (1866) the list had grown to 38. Prominent in this list was Jean-Baptiste Lamarck, of whom Darwin wrote:

"this justly celebrated naturalist....who upholds the doctrine that all species, including man, are descended from other species."

He had good reason to praise Lamarck. Half a century earlier in his *Philosophie Zoologique* [2] published in 1809, exactly 50 years before Darwin's *Origin*, Lamarck had laid out the reasons for transformationism, which he had to defend against severe critics amongst his scientific colleagues, just as Darwin would have to do. Darwin was greatly helped in this task by Thomas Henry Huxley, 'Darwin's bulldog', who acted as the public face of Darwinism, while Darwin himself worked quietly away at his country home. Lamarck had to fight his battles in the intellectually challenging Parisian culture largely alone against powerful opponents like Georges Cuvier, who wrote an obituary oration that systematically trashed Lamarck's reputation.[a] That oration was read at Lamarck's pauper burial and it was to reverberate like a death knell across time. It was written from a highly biased perspective. Cuvier proposed a form of creationism in which new species were separately created following global catastrophes, and he was strongly opposed to Lamarck's more radical ideas of evolution as a gradual transformation from one species to another.

A further example of Darwin's caution is found in a letter that he sent to the explorer and natural historian Moritz Wagner in 1876:

> *In my opinion, the greatest error which I have committed, has not been allowing sufficient weight to the direct action of the environment, i.e. food, climate, etc., independently of natural selection.*

---

a. http://www.victorianweb.org/science/science_texts/cuvier/cuvier_on_lamarck.htm.

It is significant that this was written later than the preface to the fourth edition of *The Origin of Species*. His doubts seem to have increased with time, not the reverse. Note that he refers to the *direct* action of the environment, *independently* of natural selection. This is extraordinary, given the strong bias in favor of attributing everything to natural selection in the neo-Darwinist synthesis. This is one, but by no means the only, reason that it would be wrong to view Darwin as a neo-Darwinist.

It is often thought that, apart from both favoring gradual transformation of species, Lamarck and Darwin were poles apart on the question of the processes involved. There are two key defining aspects. First, Lamarck is usually represented as favoring a 'Ladder of Life', a single continuum of increasing complexity, while Darwin is famous for his sketch of the 'Tree of Life', a branching process of differentiation radiating from a common origin (Fig. 1). Second, Lamarck is famous (or infamous) for espousing the inheritance of acquired characteristics, while Darwin is famous for his theory of natural selection, a process which, in the neo-Darwinist view, no longer requires the inheritance of acquired characteristics.

2.2 Trees of life

The historical facts do not support this common misunderstanding of the difference between the Lamarckian and Darwinian positions.

First, consider who developed the Ladders and Trees of Life. It is true that, when Lamarck wrote his great work, *Zoologie Philosophique*, published in 1809, he described his idea of a Ladder of Life, moving upwards to increasing complexity through a tendency that he called "le Pouvoir de la Vie". Translated as "the Force of Life" this has been misinterpreted to mean that Lamarck was a vitalist favoring the idea of a "vital force". This is far from the truth. In his introduction to the Flammarion (1994) reprint, the French historian of genetics, André Pichot, wrote [3]:

> *Lamarck's claim that ... there is a radical difference between living beings and inanimate objects might lead people to think that he was a vitalist. But he is not. On the contrary, his biology is a mechanistic reply to the physiological vitalism of Bichat, which was then the dominant theory (my translation of Pichot's French).*

But even more importantly, toward the end of the book, in a not very widely known addendum, Lamarck completely changed his mind. His research on worms led him inexorably to the conclusion that what he called 'internal worms' (e.g. tapeworms) and external worms (e.g. earthworms) could not be part of a single ladder of life. Once he had realized that, many other facts fell into place. The outcome is shown in Fig. 2: he constructed an absolutely clear Tree of Life, to the extent that was possible from scientific knowledge in 1809. This discovery is 28 years before Darwin's 1837 notebook. Darwin cannot have known of this Table in Lamarck's book.

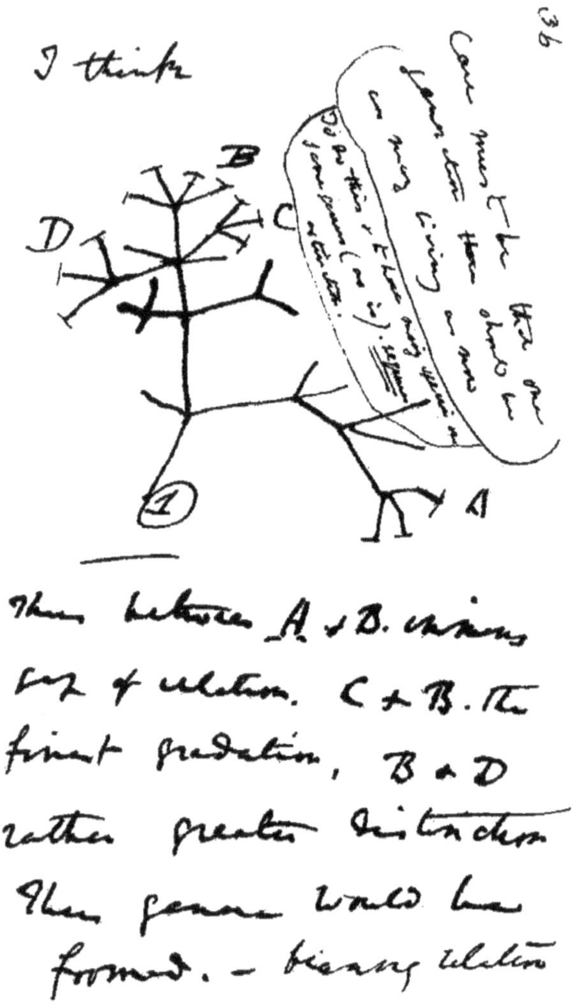

FIG. 1 Darwin's first sketch of the evolutionary Tree of Life. *From his "B" notebook begun in 1837.*

Not only did Lamarck change his theory while writing his 1809 book, he explicitly repeated the branching Tree of Life theory in two later books published in 1815 [4] and 1820 [5]. The 1809 addendum was not therefore just a token passing whim.

Reviewing this aspect of Lamarck's research Stephen J Gould wrote:

> *How can we view his [Lamarck's] slow acknowledgement of logical error, and his willingness to construct an entirely new and contrary explanation, as anything other than a heroic act, worthy of our greatest admiration and identifying Lamarck as one of the finest intellects in the history of biology? [6]*

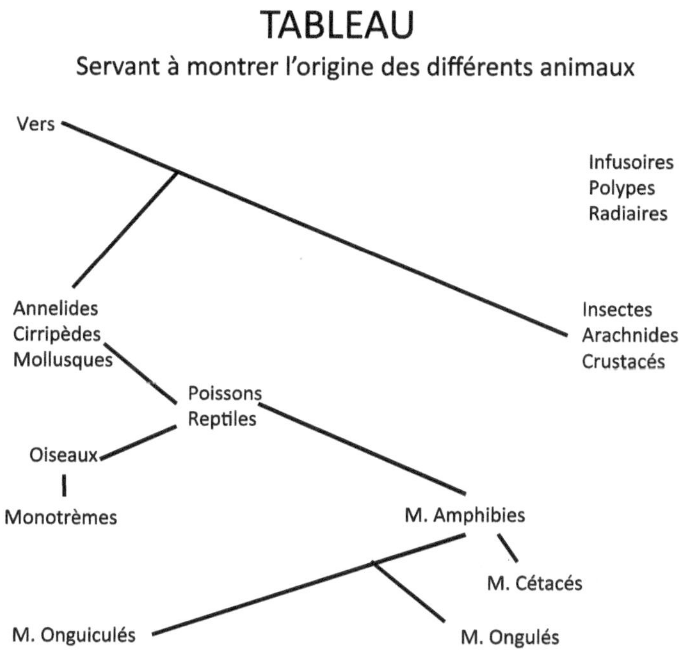

FIG. 2 Lamarck's Tree of Life. *Redrawn from the addendum to Philosophie Zoologique, 1809.*

On the Tree of Life theory Lamarck unambiguously predated Darwin. Now let's turn to the other widely-assumed difference.

2.3 The inheritance of acquired characteristics

Why did Darwin write to Moritz Wagner (see above) that he (Darwin) had not given "sufficient weight to the direct action of the environment, i.e. food, climate, etc., independently of natural selection." What influence could the environment exert "independently of natural selection" if it was not to be a form of the inheritance of acquired characteristics? And so, indeed, we find the evidence in *The Origin of Species* itself. In his introduction to Harvard's 1964 republication of *The Origin of Species*, Ernst Mayr wrote:

> *Curiously few evolutionists have noted that, in addition to natural selection, Darwin admits use and disuse as an important evolutionary mechanism. In this he is perfectly clear. For instance,... on page 137 he says that the reduced size of the eyes in moles and other burrowing mammals is 'probably due to gradual reduction from disuse, but aided perhaps by natural selection'. In the case of cave animals, when speaking of the loss of eyes he says, 'I attribute their loss wholly to disuse' (p. 137). On page 455 he begins unequivocally, 'At whatever period of life disuse or selection reduces an organ...' The importance he gives to use or disuse is indicated by the frequency with which he invokes this agent of evolution in the*

Origin. I find references on pages 11, 43, 134, 135, 136, 137, 447, 454, 455, 472, 479, and 480. [7]

As with Lamarck's Tree of Life in his 1809 book, we can ask whether this was just a passing phase in Darwin's thinking. Many evolutionary biologists have thought that it must have been. This idea is also completely incorrect. Darwin did much more than refer to use and disuse as a mechanism in his 1859 book. In fact he developed it into a theory for how it might happen.

The problem that Darwin saw is that it is not obvious how such a process could happen in multicellular organisms possessing a separate germ line. If changes in the soma occur as a consequence of adaptation to the environment, there is no reason why this should change the germ line cells unless the adaptation is somehow transmitted to them. How for example could an adaptation in a sense organ like the eye, forming part of the soma, have an effect on sperm or egg cells far away in the reproductive organs containing the germ cells? Something would have to transmit the information to the germ cells so that it could be inherited by the next generation. To deal with this problem Darwin invented the idea of gemmules, little particles that he supposed to travel (presumably through the blood stream) to carry the relevant influences. The gemmule theory of what is usually called pangenesis is found in Darwin's 1868 book *The Variation of Animals and Plants under Domestication* [8].

Readers of this book on exosomes do not have to go far to identify what might correspond to Darwin's gemmules.

I think therefore that we may answer the question at the beginning of this section of my chapter: Darwin would surely be thrilled to see his idea mirrored by modern discoveries in this way. Of course, discovering exosomes does not itself prove that they are used in the way Darwin envisaged.

3 The neo-Darwinist (modern synthesis) view

Darwin's work was done with no knowledge of genetics. Mendel's work on the genetics of hybridization in peas was not rediscovered until the late nineteenth century, when the first experiments to test the inheritance of acquired characteristics were also done.

3.1 Early experiments to test inheritance of acquired characteristics

August Weismann was the first to do this in 1890 [9]. He decided to do so not by exposing animals, and particularly their embryos, to different environments, but rather by treating them surgically. This fact is crucial.

The experiments consisted in amputating the tails of mice and then observing whether this had any effect on the progeny. The answer was a clear 'no'. Since this work forms a foundation stone of Neo-Darwinism it is important to ask whether it really answers the relevant question. Even on the older versions of Lamarckism, as expressed by Lamarck himself, this is a curiously inappropriate way of testing the idea. The idea is that Lamarckian inheritance may occur in a *functional* interaction between the organisms *and their environment*, through use and disuse of the organism's structures and functions, not whether the *non-functional results of surgery* can be inherited.

Furthermore, Darwin must have known already that such inheritance did not occur from the work of animal breeders. Tail amputation in dogs for aesthetic reasons does not result in stunted tails in the offspring, no matter how many generations are bred from the animals. To put the question in a more modern form, it is whether the germ-line is or is not isolated from environmental influences. The relevant way to do a tail-cutting experiment or any other experiment to answer that question would be to change the environment in a way that makes tail-lessness a functional advantage. Quite apart from the obvious question why a surgical change should be inherited, even a standard Lamarckian would notice that the environment, apart from the surgery, is not different. Furthermore, even if there were environments that would favor tail-lessness the experiment would not test for that.

The work of Conrad Waddington in the 1950s showed the more successful way forward for such experiments, which is to investigate environmental effects on embryos. His classic 1957 book *The Strategy of the Genes* [10] contains his experimental work showing that acquired characteristics can be assimilated into the genome after just a dozen or so generations of exposing embryos to environmental change.

Nevertheless, the tail-cutting experiment convinced Weismann and others that Lamarckism is impossible. The Weismann Barrier, i.e. the isolation of the germ cells from the soma, then became a cornerstone of the development of Neo-Darwinism into the Modern Synthesis. To quote Weismann directly:

> *In my opinion this [the hereditary substance] can only be the substance of the germ cells; and this substance transfers its hereditary tendencies from generation to generation, at first unchanged, and always uninfluenced in any corresponding manner, by that which happens during the life of the individual which bears it. If these views... be correct, all our ideas upon the transformation of species by means of exercise (use and disuse), as proposed by Lamarck, and accepted in some cases by Darwin, entirely collapses.*
>
> (1883 Lecture *On Inheritance*)

Weismann was therefore fully aware of Darwin's acceptance of the inheritance of acquired characteristics as part of evolutionary theory.

It is curious, to say the least, that such a distinguished and widely praised scientist should have put up such an inappropriate Straw Man to knock down, and even more curious that so many other eminent scientists should have accepted it as the basis of a major evolutionary theory. No major publishable science today could be based on such flimsy and inappropriate evidence. Why did Weismann and his successors do that?

In Weismann's case, part of the answer lies in the little-known fact that he fully acknowledged that his experiments disproved only the inheritance of surgical mutilation. He performed these experiments because he was aware of 'Lamarckian' claims that such mutilations could be inherited. The Straw Man had already been set up by over-imaginative Lamarckians, who had claimed, for example, that repeated circumcision in the generations of populations that practice this particular surgery, could lead to babies born without foreskins. He well knew the limitations of his experiments and that their crudeness was responding to a similarly crude and incorrect alternative.

The answer to this puzzle is that Weismann was *already* convinced of the correctness of his other assumption, the randomness of variations. In the same 1883 lecture (before his tail-cutting experiments) he rejected the inheritance of acquired characteristics and proposed alternative explanations for the use and disuse examples Darwin gave in the *Origin*. But showing that there are alternative explanations for Darwin's examples does not prove that they are the correct explanations. So far as I am aware, the tail-cutting experiments Weismann performed are the only direct experimental tests.

3.2 Randomness of mutations

Weismann is also credited with the idea, which he developed in his *Essays upon Heredity* [11] as early as 1889, that changes in the germ line cells were largely random, which, like the Weismann Barrier, also became a kind of dogma. He was therefore responsible for the two main assumptions of the Modern Synthesis and it is not surprising that he is often judged to be the most important evolutionary biology thinker forming the link between Darwin and the formulation of the Modern Synthesis in the 1930s and 1940s. Ernst Mayr, author of the magisterial 1982 book *The Growth of Biological Thought* [12], described him as "one of the great biologists of all time."

3.3 Is the Weismann barrier now "embodied by the Central dogma of molecular biology"?

We now know that the Weismann Barrier is not absolute. It can be bypassed in many ways. This is how maternal and paternal effects are transmitted across generations [13] and it is now an important question to what extent exosomes may mediate such transmission. Evolutionary biologists have therefore known for many years that a theory based on absolute isolation of the germline cannot be correct.

So, why was neo-Darwinism not abandoned when these effects became known during the later part of the twentieth century? After all, the main reason for distinguishing neo-Darwinism from Darwin's own theories was the exclusion of the inheritance of acquired characteristics. This was the very essence of the distinction, and it was also supported by Alfred Russel Wallace who, like Weismann, took a hard line on this issue.

The issue was never debated in Darwin's lifetime since Darwin died in 1882, just a year before Weismann's seminal 1883 lecture.

This realization that the cellular barrier is permeable led to a subtle shift in the Barrier concept. Instead of defending the supposed absolute nature of the Weismann Barrier, some neo-Darwinist theorists proposed that it had become embodied in the Central Dogma of Molecular Biology, which is the fact that sequences of nucleotides can specify sequences of amino acids, but change in sequences of amino acids cannot be used to determine changes in nucleotide sequences.

The idea that the Weismann Barrier is now "embodied by the central dogma" is widespread. It even appears on the Simple Wikipedia entry on the central dogma: "The dogma is a modern version of the Weismann barrier".[b]

The statement on the main Wikipedia page is more circumspect: "This [the Weismann Barrier] does not predict the central dogma, but does anticipate its gene-centric view of life, albeit in non-molecular terms." The grossly misleading statement on Simple Wikipedia is repeated on a website designed for schoolchildren.[c]

This confusion is unfortunate. The difference between germ line cells and the genome is fundamental. The cell contains much more than its genome. In most of life on earth, it is the complete organism. Moreover, it can be shown that the information content of the rest of the cell matches that of the genome [14]. So how did the idea that the barrier could be embodied in the dogma come about?

3.4 Shifting definition of a gene

This development is explained by a shift in the definition of a gene. When Johannsen first introduced the word in 1909, it was defined as an inheritable phenotype characteristic [15]. This was also essentially Mendel's concept. Their concept of a gene would therefore have included anything that went through the germline *cells*. Johannsen made this clear when he explained that the gene could be anything (*ein etwas*) in the organism that was responsible for inheritance of the characteristic. Had he known of them, RNAs, and epigenetic marks on DNAs and histones, would have been included, as would cellular structures that make the replication and inheritance possible. This is clearly not the modern molecular biological definition of a gene, which is restricted to a DNA sequence forming a template for a protein [16, 17].

This major shift in definition was not known at the time the Modern Synthesis was formulated. DNA was not even known to be the genetic material, so it is not surprising that the shift in definition did not matter to those who formulated the Modern Synthesis. The great advances in, for example, the mathematical

b. https://simple.wikipedia.org/wiki/Central_dogma_of_molecular_biology.

c. https://wiki.kidzsearch.com/wiki/Central_dogma_of_molecular_biology.

theories of population genetics [18] worked perfectly well with the gene being defined as a phenotype characteristic. In fact, for most applications of genetics to the social sciences, such as economics and sociology, retaining the phenotype definition is important and even necessary.

But, to return to the definition of a gene, the difference between phenotype and genotype definitions matters enormously to versions of neo-Darwinism, such as selfish gene theory, based on distinguishing the replicator (regarded as DNA) from the vehicle (the phenotype) [19].

There are two fatal difficulties in the *selfish gene* version of neo-Darwinism. The first is that, from a physiological viewpoint, it doesn't lead to a testable prediction. The problem is that the central definition of selfish gene theory is not independent of the only experimental test of the theory, which is whether genes, defined as DNA sequences, are in fact selfish, for example, whether their frequency in the gene pool increases [20]. The second difficulty is that DNA can't be regarded as a replicator separate from the cell [16]. The cell, and specifically its living physiological functionality, is what enables DNA to be replicated faithfully.

This difficulty leads to the next error in the development toward the Central Dogma, since the fact that the cell, not its DNA, is the real replicator is fundamental. This is the core issue so I will now explain how the development toward the Central Dogma got this part of the story wrong.

When the Central Dogma was first formulated, Watson and Crick acknowledged their indebtedness to Schrödinger's famous book *What is Life?* [21]. In that book, Schrödinger made two predictions, one of which was spectacularly successful, the other is necessarily incorrect [22] (pp. 176–181).

The correct prediction was that the genetic material would be found to be what he called an aperiodic crystal. If one allows a polymer to be regarded as a kind of crystal, this is a good description of DNA.

The incorrect prediction was that the molecule would behave like a determinate crystal. This idea leads directly to strong interpretations of the Central Dogma that attribute faithful replication and determinate qualities to DNA alone.

If you compare the DNA in a daughter cell with that of its parent cell, you will indeed find a copying error rate of less than one base pair in a complete genome. The error rate is around only 1 in 10^{10} base pairs, which is remarkably accurate. But now suppose that we could compare the DNA sequence *immediately after being copied*. We would find an error rate of around 1 in 10^4, which in a genome of 3 billion base pairs would mean millions of errors. No eukaryotic cell could survive such an error rate. Schrodinger was therefore wrong about the molecule itself behaving like a determinate crystal. Crystal structure growth is indeed determinate, but that is the wrong metaphor for DNA. That is not how it grows and replicates. On its own, it would be left with millions of errors.

It replicates accurately only in a complete cell containing all the objective functionality that enable cells to be alive. Cells achieve this accurate outcome in

DNA copying through a very complex three-stage process in which the millions of errors are detected and corrected. The famed 'immortality' of the genome is actually a function of the complete cell, not of the genome alone.

Moreover the fact that the error-correcting process is under control by the cell, not by the genome itself, leads to the genome being much more than a "ready-only" database. Varying the efficiency of error correction, and allowing insertions and deletions, enables organisms to perform what James Shapiro calls Natural Genetic Engineering [23–25].

4 The new trends view

There is now considerable debate about whether neo-Darwinism needs fine-tuning, extending, or perhaps abandoned altogether. Before I try to answer that question I want to acknowledge the fact that the neo-Darwinist modern Synthesis was very useful. Whole fields of mathematical biology, such as population genetics, would probably not have flourished in the twentieth century without the modern synthesis as a framework.

But, I also think that we have reached a watershed in relation to the issue of the utility of the neo-Darwinist modern synthesis. As I have argued in detail elsewhere [22], there are too many experimental breaks with the original theory as formulated by Weismann and Wallace. The time has come to see that evolutionary biology would progress faster if we used a different framework to develop a more inclusive theory, as illustrated in Figs. 3 and 4. Fig. 3 shows the Extended Evolutionary Synthesis, EES (Pigliucci and Mueller, [26]).

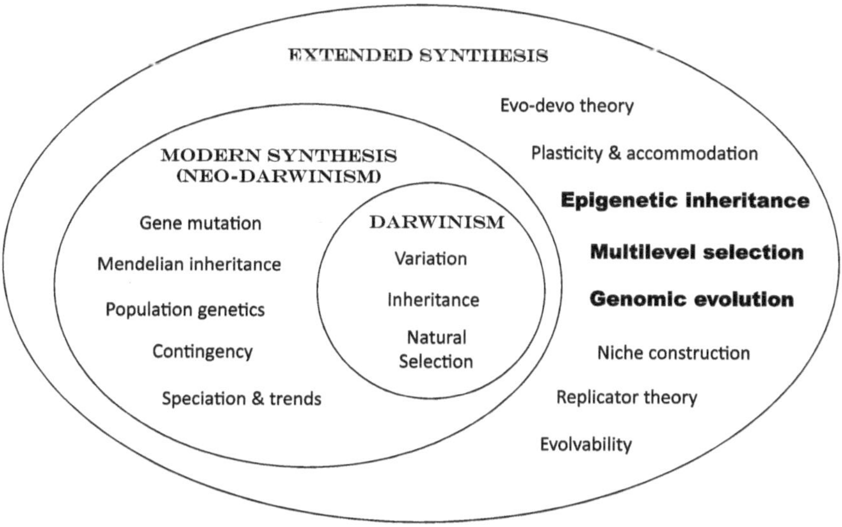

FIG. 3 The extended evolutionary synthesis. Each extension includes all that went before. *Based on Pigliucci and Mueller [26].*

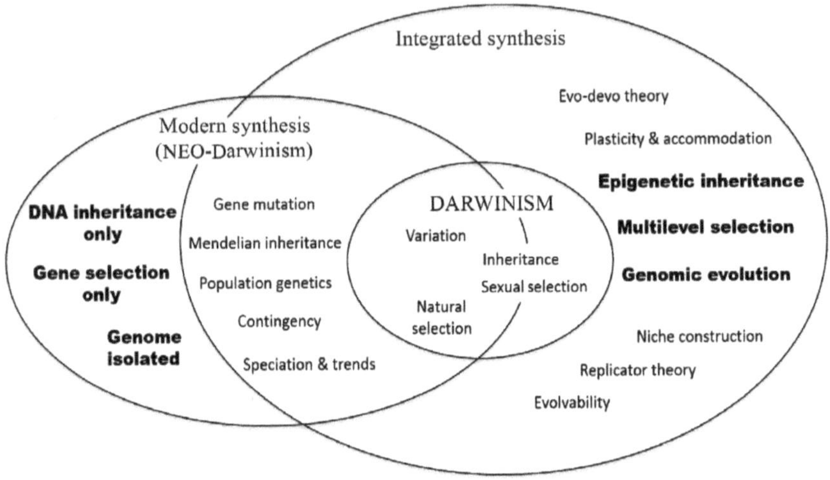

FIG. 4 The integrated evolutionary synthesis. Darwinism no longer completely contained within neo-Darwinism, but is contained within the integrated synthesis. *From Noble 2017 [27].*

Fig. 4 shows the version of this diagram that I think better represents the conclusions of this chapter. There are several important differences. First, it represents the fact that Darwin's view of inheritance included the inheritance of acquired characteristics, which was excluded by neo-Darwinism. Darwin's concept of inheritance is therefore shown as being partly outside the neo-Darwinist modern synthesis. So also is his ideas on sexual selection [28]. Second, it represents the features of the extended synthesis (highlighted in bold in both Figs. 3 and 4) that lie outside the range of neo-Darwinism as defined by Weismann and Wallace. The features of that theory that were excluded are shown as corresponding bold-face items. The highlighted items on the far left correspond with the highlighted items at the far right. Sexual selection is also included as overlapping Darwinism and Integrated Synthesis (see Ref. [29]).

As Fig. 4 makes clear, the integrated synthesis flows more naturally as extensions of Darwin's ideas than as extensions of neo-Darwinism. I believe that a key field of experimental facts that will enable us to more accurately decide on a more inclusive theory is that of exosome research, i.e. the focus of this book. The field is at the very frontier of questions concerning the fundamentals of modern biology, which is why this book will command widespread attention. I therefore finish this chapter with a brief review of modern work on exosomes and transmission of DNA and microRNA by sperm cells.

5 Modern work on trans-generational role of exosomes and microRNAs

The reason for the great interest in this possible role for exosomes is that any transmission of exosome RNAs, DNAs, protein transcription factors, etc. via the germline would cross the Weismann Barrier [30]. Furthermore, the transmission of a *pattern* of such molecules will represent, at least in part, the state of genome regulation by the cells from which the exosomes come. Smith and Spadafora reviewed this field in 2005 and concluded:

> *It is now well established that spermatozoa can play a role in transgenesis in virtually all species. ...the underlying mechanism of SMGT [Sperm Mediated Gene Transfer] is best viewed essentially as a retrotransposition-mediated process. Inasmuch as sperm cells are vectors of exogenous genetic information, there is little doubt that they have the potential to cause both genetic and phenotypic modification in individuals of a variety of species and are worthy of further methodological investigation for optimal use in biotechnology. [31]*

A more recent review by Spadafora reinforces this conclusion:

> *Epigenetics is increasingly regarded as a potential contributing factor to evolution. Building on apparently unrelated results, here I propose that RNA-containing nanovesicles, predominantly small regulatory RNAs, are released from somatic tissues in the bloodstream, cross the Weismann barrier, reach the epididymis, and are eventually taken up by spermatozoa; henceforth the information is delivered to oocytes at fertilization. [32]*

Lavitrano et al. [33] have demonstrated that SMGT can be used to generate transgenic pigs, leading to the possible development of multigene transgenic animals suitable for organ transplantation to humans. Similar methods had previously been demonstrated in mice [34]. In humans, the expression profiles of RNAs in exosomes from seminal fluid can be used as a marker for mechanisms of male infertility [35, 36].

That the microRNAs transmitted by exosomes can be functional when transferred between different species has been shown by Valadi et al. [37], who found that "After transfer of mouse exosomal RNA to human mast cells, new mouse proteins were found in the recipient cells, indicating that transferred exosomal mRNA can be translated after entering another cell."

This is a rapidly developing field. We have probably not heard the last from Darwin's gemmules idea. Nor have the echoes of Lamarck died away.

Acknowledgements

Parts of this chapter draw on some of my recent publications, which are included in the references.

References

[1] Darwin C. On the origin of species by means of natural selection, or the preservation of favoured races in the struggle for life. London: John Murray; 1859.

[2] Lamarck J-B. Philosophie zoologique, original edition of 1809 with introduction by Andre Pichot. Paris: Flammarion; 1994.

[3] Pichot A. Introduction. In: Philosophie zoologique. Paris: Flammarion; 1994.
[4] Lamarck, J.-B., Histoire naturelle des animaux sans vertebres. 1815–1822, Paris: Verdiere.
[5] Lamarck J-B. Système analytique des connaissances positives de l'homme. Paris: Flammarion; 1820.
[6] Gould SJ. A tree grows in Paris: Lamarck's division of worms and revision of nature. In: The lying stones of Marrakech. Harmony Books; 2000.
[7] Mayr E. Introduction. In: On the origin of species. Cambridge, MA: Harvard; 1964. p. xxv–xxvi.
[8] Darwin C. The variation of animals and plants under domestication. London: John Murray; 1868.
[9] Weismann A. The germ-plasm: a theory of heredity. New York: Charles Scribner's Sons; 1893.
[10] Waddington CH. The strategy of the genes. London: Allen and Unwin; 1957.
[11] Weismann A. Essays upon heredity and kindred biological problems. Oxford University Press; 1889.
[12] Mayr E. The growth of biological thought. Cambridge, Mass: Harvard; 1982.
[13] Hanson M, Skinner M. Developmental origins of epigenetic transgenerational inheritance. Environ Epigenet 2016; https://doi.org/10.1093/eep/dvw002.
[14] Noble D. Differential and integral views of genetics in computational systems biology. Interface Focus 2011;1:7–15.
[15] Johannsen W. Elemente der Exakten Erblichkeitslehre. Jena: Gustav Fischer; 1909.
[16] Noble D. Central dogma or central debate? Physiology 2018;33:246–9. https://doi.org/10.1152/physiol.00017.2018.
[17] Kohl P, et al. Systems biology: an approach. Clin Pharmacol Ther 2010;88:25–33.
[18] Fisher RA. The genetical theory of natural selection. Oxford: Oxford University Press; 1930.
[19] Dawkins R. The selfish gene. vol. 2006. Oxford: OUP; 1976.
[20] Noble D. Neo-Darwinism, the modern synthesis, and selfish genes: are they of use in physiology? J Physiol 2011;589:1007–15.
[21] Schrödinger E. What is life? Cambridge: Cambridge University Press; 1944.
[22] Noble D. Dance to the tune of life. In: Biological relativity. Cambridge: Cambridge University Press; 2016.
[23] Shapiro JA. Revisiting the central dogma in the 21st century. Ann N Y Acad Sci 2009;1178:6–28.
[24] Shapiro JA. Evolution: a view from the 21st century. Upper Saddle River, NJ: Pearson Education Inc.; 2011.
[25] Shapiro JA. Biological action in read-write genome evolution. Interface Focus 2017;7. https://doi.org/10.1098/rsfs.2016.0115.
[26] Pigliucci M, Müller GB. Evolution—the extended synthesis. Cambridge, MA: MIT Press; 2010.
[27] Noble D. Evolution viewed from physics, physiology and medicine. Interface Focus 2017;7. https://doi.org/10.1098/rsfs.2016.0159.
[28] Darwin C. The descent of man, and selection in relation to sex. London: John Murray; 1871.
[29] Noble R, Noble D. Harnessing stochasticity. How organisms make choices. Chaos 2018;28. https://doi.org/10.1063/1.5039668.
[30] Smythies J, Edelstein L, Ramachandran V. Molecular mechanisms for the inheritance of acquired characteristics—exosomes, microRNA shuttling, fear and stress: Lamarck resurrected? Front Genet 2014;5:133.
[31] Smith K, Spadafora C. Sperm-mediated gene transfer: applications and implications. Bioessays 2005;27:551–62.
[32] Spadafora C. The "evolutionary field" hypothesis. Non-Mendelian transgenerational inheritance mediates diversification and evolution. Prog Biophys Mol Biol 2018;134:27–37.
[33] Lavitrano M, et al. Sperm-mediated gene transfer. Reprod Fertil Dev 2006;18:19–23.

[34] Lavitrano M, et al. Sperm cells as vectors for introducing foreign DNA into eggs: genetic transformation of mice. Cell 1989;57:717–23.

[35] Abu-Halima M, et al. Altered micro-ribonucleic acid expression profiles of extracellular microvesicles in the seminal plasma of patients with oligoasthenozoospermia. Fertil Steril 2016;106:1061–9.

[36] Barcelo M, et al. Exosomal microRNAs in seminal plasma are markers of the origin of azoospermia and can predict the presence of sperm in testicular tissue. Hum Reprod 2018;33:1087–98.

[37] Valadi H, et al. Exosome-mediated transfer of mRNAs and microRNA is a novel mechanism of genetic exchange between cells. Nat Cell Biol 2007;9:654–9.

Index

Note: Page numbers followed by *f* indicate figures and *t* indicate tables.

F

G

H

I

K

L

M

S

T

U

V

W

Y

Z

Kolibri 360 GmbH
Ferdinand-Jühlke-Straße 7
99095 Erfurt, Deutschland
produktsicherheit@kolibri360.de